ɔ® Practices, Certification, and ·editation Handbook

LEED® Practices, Certification, and Accreditation Handbook

Sam Kubba Ph.D., LEED AP

ELSEVIER

AMSTERDAM • BOSTON • HEIDELBERG • LONDON
NEW YORK • OXFORD • PARIS • SAN DIEGO
SAN FRANCISCO • SINGAPORE • SYDNEY • TOKYO
Butterworth–Heinemann is an imprint of Elsevier

Butterworth–Heinemann is an imprint of Elsevier
30 Corporate Drive, Suite 400, Burlington, MA 01803, USA
Linacre House, Jordan Hill, Oxford OX2 8DP, UK

Library of Congress Cataloging-in-Publication Data
Application submitted

British Library Cataloguing-in-Publication Data
A catalogue record for this book is available from the British Library

ISBN: 978-1-85617-691-0

For information on all Butterworth–Heinemann publications visit our Web site at *www.elsevierdirect.com*

Typeset by: diacriTech, India

Printed in the United States of America
09 10 11 12 13 10 9 8 7 6 5 4 3 2 1

To my mother and father,
Who bestowed on me the gift of life…
And to my wife and four children,
Whose love and affection inspired me on…

Contents

Acknowledgments

A book of this scope would not have been possible without the active and passive support of many friends and colleagues who have contributed greatly to my thinking and insights during the writing of this book and who were in many ways instrumental in the crystallization and formulation of my thoughts on the subjects and issues discussed within. To them I am heavily indebted, as I am to the innumerable people and organizations that have contributed ideas, comments, photographs, illustrations, and other items that have helped make this book a reality instead of a pipe dream.

I must also unequivocally mention that without the unfailing fervor, encouragement, and wisdom of Mr. Roger Woodson, president of Lone Wolf Enterprises Ltd., and Mr. Kenneth McCombs, Senior Acquisitions Editor, Elsevier Science and Technology Books, this book would still be on the drawing board. It is always a great pleasure working with them. I must likewise acknowledge the wonderful work of Ms. Maria Alonso, Assistant Editor, Elsevier Science & Technology Books, and thank her for her unwavering commitment and support. I also wish to thank Ms. Wendy Lochner, a highly valued and dedicated member of the Lone Wolf team, for copyediting the first drafts, and Robert Swanson for indexing the book. I would especially like to salute and express my deepest appreciation to Sarah Binns, who saw the book through production, diacriTech, who laid out the manuscript, and Alisa Marie Andreola, for the excellent cover design.

I am particularly indebted to the U.S. Green Building Council (USGBC) and its staff for their assistance, continuous updates, and support on the new LEED™ 2009 version 3 Rating System. Likewise, my special thanks are extended to Mr. Paul Fair, LEED™ AP of Contracting Services, for reviewing many of the chapters and for his expert comments and suggestions.

I would really be amiss if I failed to acknowledge my wife Ibtesam, for her loving companionship and continuous support and for helping me prepare some of the line illustrations. And last but not least, I wish to record my gratitude to all those who came to my rescue during the final stretch of this work – the many nameless colleagues – architects, engineers, and contractors who kept me motivated with their ardent enthusiasm, support, and technical expertise. To these wonderful professionals, I can only say, "Thank you." I relied upon them in so many ways, and while no words can reflect the depth of my gratitude to all of the above for their assistance and advice, in the final analysis, I alone must bear responsibility for any mistakes, omissions or errors that may have found their way into the text.

Introduction – The Green Movement: Past, Present, and Future

What is the Green Movement?

Our perception of the green movement has evolved considerably since its early formative days. So much so that Jerry Yudelson, author of the *Green Building Revolution*, says, "The green building revolution is sweeping across not only the United States but most of the world. It's a revolution inspired by an awakened understanding of how buildings use resources, affect people, and harm the environment." But to understand the modern green movement, we need to trace its origins back to the beginning. It is not always easy to define exactly when a movement may have started; some associate its beginning with Rachel Carson's (1907–1964) book *Silent Spring* and the legislative fervor of the 1970s or with Henry David Thoreau and his book *The Maine Woods*, where he called for an awakening to the conservation of and respect for nature and the federal preservation of virgin forests. Others believe sustainability and environmentalism are actually rooted in the intellectual thought of the 1830s and 1840s.

The 20th century witnessed the beginning of federal government action to preserve lands, and much of the credit is due to Teddy Roosevelt and John Muir for popularizing conservation. Roosevelt's visit to Yosemite in 1903 gained national publicity, and by 1916 the National Park Service had been established. Aldo Leopold (1887–1948) is one of many innovative philosophers whose theories have significantly influenced the North American green and environmental movements. Unfortunately, the World Wars and the Great Depression pushed environmental concerns out of the mainstream and into the background of public consciousness. After World War II, environmental efforts continued to be focused on land conservation. The Sierra Club, however, continued to grow and became pivotal in establishing many parks throughout the nation during these years.

Following the 1930s, new building technologies began to fundamentally impact the urban landscape. The introduction of air conditioning, structural steel, vertical transport, and reflective glass suddenly made it possible to build multistory enclosed glass-and-steel-framed structures with controlled heating and cooling. The postwar economic boom was partly motivated by international architects such as Mies van der Rohe, whose International Style "glass box" provided a further catalyst in accelerating the pace of this phenomenon.

But in the 1970s, a small group of enlightened architects, and environmentalists, started to question the wisdom of this method of building. Their efforts, however, were not taken seriously until 1973, when the imposition of an oil embargo by OPEC catapulted the nascent "environmental movement" into the attention and imagination of the general public. As the American public lamely watched the upward spike of gasoline prices and, for the first time, long lines of cars at gas stations, many began to reexamine the wisdom of our reliance on fossil fuels for transportation and buildings.

Thus, for all practical purposes most scholars would probably agree that the green-building movement was greatly impacted by the energy crisis of the 1970s and the creative approaches to saving energy that followed, including the use of active and passive solar design and tighter building envelopes. The 1970s saw numerous legislative steps to clean up the environment, such as the National Environmental Policy Act, the Clean Air Act, the banning of DDT, the Water Pollution Control Act, the Endangered Species Act, and the founding of Earth Day. Added to this, the disasters at Love Canal in 1978 and Three Mile Island in 1979 horrified an oblivious public with the visible consequences of toxic waste, pollution, and contamination.

Also as a direct response to this crisis, the American Institute of Architects formed an energy task force, later followed by an AIA Committee on Energy. According to committee member Dan Williams, two groups were formed – one looked mainly at passive systems, such as reflective roofing materials and environmentally appropriate siting of buildings, to achieve energy savings, while the other looked primarily into solutions that involved the use of new technologies.

Even as the immediate impact of the energy crisis began to ebb, pioneering efforts in energy conservation for buildings continued to advance. In England, Norman Foster's Willis Faber and Dumas Headquarters (1977) incorporated a grass roof, a daylighted atrium, and mirrored windows.

In California eight energy-efficient state office buildings were commissioned, including the Gregory Bateson Building (1978) in Sacramento (Figure I.1), which employed photovoltaics, underfloor rock-store cooling systems, and a computerized solar-tracking shading system. In 1977, a new cabinet department, the Department of Energy, was created to address issues related to energy usage and conservation, and in the same year the Solar Energy Research Institute (later to be renamed the National Renewable Energy Laboratory) was founded to look at new energy technologies, such as photovoltaics.

In the United States the 1980s and early 1990s saw significant efforts by sustainability proponents such as Robert Berkebile, and Sandra Mendler. Internationally, designers such as Germany's Thomas Herzog, Malaysia's Kenneth Yeang, and England's Norman Foster and Richard Rogers were experimenting with prefabricated energy-efficient wall systems, water-reclamation systems, and modular construction units that were designed to minimize construction waste. At this time Scandinavian governments set minimum standards for access to daylight and operable windows in workspaces. In 1987 the UN World Commission on Environment and Development, under Norwegian prime minister Gro Harlem Bruntland, was perhaps the first to define the term "sustainable development" as development that "meets the needs of the present without compromising the ability of future generations to meet their own needs."

Figure I.1 Photo of the Gregory Bateson Building in Sacramento, California (1978), which was a model energy-efficient state office building designed by Sim Van der Ryn and which employed several energy-saving tools including an automated solar-tracking shading system with exterior ElectroShades®.
Source: MechoShade Systems, Inc.

In 1989 the AIA Energy Committee was transformed into a more broadly scaled AIA Committee on the Environment (COTE), and the following year, the AIA (through COTE) and the AIA Scientific Advisory Committee on the Environment obtained funding from the Environmental Protection Agency (EPA) to embark on the development of a building-products guide based on life-cycle analysis. This would be the first such assessment of its kind to be undertaken in the U.S. Individual product evaluations were ultimately compiled in the *AIA Environmental Resource Guide*, which was published in 1992.

In June 1992 Susan Maxman, president of the AIA, attended a UN Conference on Environment and Development in Rio de Janeiro. The conference was a spectacular success and drew delegations from 172 governments and 2400 representatives of non-governmental organizations. This momentous event witnessed the passage of the Rio Declaration on Environment and Development, a blueprint for achieving global sustainability. Following on the heels of the Rio de Janeiro summit, the AIA president-elect chose sustainability as her theme for the June 1993 UIA/AIA World Congress of Architects. An estimated 6000 architects from around the world attended the Chicago event. Another milestone and turning point in the history of the green-building movement that is worthy of mentioning was the Architecture at the Crossroads convention.

With the election of Bill Clinton to the presidency in November 1992, a number of proponents of sustainability began to circulate the idea to "green" the White House itself. On Earth Day, April 21, 1993, President Clinton announced to the nation his plans to make the White House "a model for efficiency and waste reduction." The "greening of the White House" also included the 600,000-square-foot. Old Executive Office Building across from the White House and embarked with an energy audit by the Department of Energy (DOE), an environmental audit led by the Environmental Protection Agency (EPA), and a series of design charrettes in which nearly 100 design professionals, engineers, environmentalists, and government officials participated. They were required to formulate energy-conservation strategies using off-the-shelf technologies. Within three years, the results of these energy-conservation strategies produced numerous improvements to the nearly 200-year-old residence, such as an estimated $300,000 in annual energy and water savings, landscaping expenses, and solid-waste costs while at the same time reducing atmospheric emissions from the White House by an estimated 845 tons of carbon per year.

The flurry of federal greening projects was not the sole force propelling the sustainability movement in the 1990s. On September 14, 1998, President Clinton issued the first of three executive orders; it called upon the federal government to improve its use of recycled and "environmentally preferred" products (including building products). Another executive order in June 1999 was issued to encourage government agencies to improve energy management and reduce emissions in federal buildings through improved design, construction, and operation. A final report (under Chairman Ray C. Anderson) was issued recommending 140 specific actions to improve the nation's environment, many of which were related to building sustainability. The third executive order was issued in April 2000; it charged federal agencies to integrate environmental accountability into their daily decision making and long-term planning.

The overtly successful greening of the White House encouraged the participants to green other properties in the massive federal portfolio, including the Pentagon, the Presidio, and the U.S. Department of Energy Headquarters as well as three national parks: Grand Canyon, Yellowstone, and Alaska's Denali. In 1996 the U.S. Department of Energy signed a memorandum of understanding with AIA/COTE to work together on research and development, seeking to establish a program that contained a series of roadmaps for the construction of sustainable buildings in the 21st century. And on Earth Day 1998, the then-chair of AIA/COTE, Gail Lindsey, announced the first "Top 10 Green Projects," designed to increase the public's awareness of successful sustainable design, a program that has continued to the present day.

Individual federal departments were also making significant headway. For example, the Navy undertook eight pilot projects, including the Naval Facilities Engineering Command (NAVFAC) headquarters at the Washington Navy Yard. This large 156,000-square-foot structure, which was built some 150 years earlier as a gunnery assembly plant, went through a major transformation that resulted in a reduced energy consumption of 35 percent, thereby saving $58,000 a year. The General Services Administration undertook similar pilot projects such as the 1995 federal courthouse in Denver, and the Environmental Protection Agency (EPA) revamped its facilities in Research Triangle Park, N.C.

In 1997, the Navy initiated development of an online resource, the *Whole Building Design Guide* (WBDG), with the object of incorporating sustainability requirements into mainstream specifications and guidelines. Seven other federal agencies have joined this project, which is now managed by the National Institute of Building Sciences (NIBS).

From the above, it is evident that the green movement has been influenced by many people from all walks of life. These visionaries, both past and present, have recognized the need for serious changes in how we treat our environment. The championing of green issues by forward-thinking politicians, celebrities, and visionaries duch as Al Gore, Ralph Nader, Robert Redford, and Leonardo DiCaprio, as well as the consequences of the 9/11 tragedy, gas shortages, visible climate change, the wars in Iraq and Afghanistan, and scientific consensus, have all validated environmental concerns during the early years of this century. Former Vice President Al Gore's release in May 2006 of his acclaimed documentary film *An Inconvenient Truth* projected the climate crisis into the popular consciousness and raised public awareness of global warming, water contamination by toxic chemicals, and the exhaustion of our resources, and skyrocketing gas prices, among other consequences. To many people's surprise, the documentary was an instant hit and one of the many factors that helped change American attitudes toward the present climate predicament.

Lindsay McDuff, a green movement activist, says: "When politicians create or formulate policies, the business industries are consequently affected. With the rise in green policy, business executives from every arena are jumping on the green-movement bandwagon, basically out of the growing market demand. Being green has become a selling advantage in the business world, and eager companies are starting to jump at the chance to get ahead." The green movement has become an international conglomeration, seeking solutions to environmental concerns around the world.

Another environmental activist is Robert Redford, who for over four decades has served as a trustee for the Natural Resources Defense Council and was the founder of the Institute for Resource Management. Redford also created the Sundance Institute, which recently developed and produced a television channel, "The Green," whose primary function is to program environmentally themed shows and documentaries. Like Redford, DiCaprio serves on the board of the Natural Resources Defense Council and is a member of Global Green USA. DiCaprio is also producing, co-writing, and narrating a documentary regarding environmental problems and possible solutions. Robert Redford and Leonardo DiCaprio are but two of the many celebrities who are able of taking advantage of their status to advance environmental issues and awareness.

In many ways the green-building movement can be deemed to be a reaction to crises, and it came into being as a result, fostered by efforts to make buildings more efficient and revamp the way they use energy, water, and materials. It is also about enhancing our communities and minimizing the impact of buildings on the environment through better site location, design, construction, operation, maintenance, and removal.

Exactly What is Green Building?

The phrase "green building" is a relatively new term in our vocabulary. Its core message is to improve current design and construction practices and standards so that the buildings we build today will last longer, be more efficient, cost less to operate, and contribute to healthier living and working environments for occupants, thereby helping to increase productivity. Green building is also about increasing the efficiency with which buildings and their sites utilize energy, water, and materials; protect natural resources; and improve the built environment so that ecosystems, people, and communities can thrive and prosper.

It is evident that the green-building phenomenon has over the last decade or more not only significantly impacted the U.S. construction market but also the global marketplace as well. However, its total impact on the building-construction industry and its suppliers has yet to be fully quantified and may not be fully grasped for some time. It is important to remember that we are witnessing a movement that has become truly prolific as state and local governments increasingly scramble to pass legislation or adopt regulations that will promote green building-construction practices.

Green building represents a model shift in the way we understand, design and construction today. Studies have consistently shown that buildings in the United States consume roughly one-third of all primary energy produced and close to two-thirds of electricity produced. Studies have also shown that roughly 30 percent of all new and renovated buildings in the U.S. have poor indoor environmental quality due to several factors, including noxious emissions, pathogens, and outgassing of harmful substances present in building materials. Efforts to address these environmental impacts and other issues are ongoing, and one of the solutions devised was the inclusion of sustainability practices in construction project objectives. The added sense of urgency caused by continually rising energy costs has inspired many in the construction industry to finally turn to the green movement for solutions.

However, the concept of incorporating sustainable practices into conventional design and construction procedures requires a redefinition and reevaluation of the existing roles of project participants to be able to contribute effectively to sustainable project objectives. One of the primary characteristics of successful sustainable design is the implementation of a multidisciplinary and integrated team approach, particularly during the early design phases. An integrated team approach, early involvement, and greater participation of the various project members and stakeholders, help ensure an end product that is more efficient and healthier for both owner and occupants.

Meanwhile, a parallel effort was also taking shape – the founding of the U.S. Green Building Council (USGBC). Outside the U.S., the Building Research Establishment was developing its own building-assessment method, known as BREEAM. The most ambitious international effort of the period was the Green Building Challenge (October 1998), held in Vancouver, B.C. It was a well-attended affair with representatives from 14 nations: Austria, Canada, Denmark, Finland, France, Germany, Japan, the Netherlands,

Norway, Poland, Sweden, Switzerland, the U.K., and the U.S. Subsequent conferences took place in Maastricht, the Netherlands; Oslo, Norway; and other venues. The goal of these conferences was to create an international assessment tool that takes into account regional and national environmental, economic, and social-equity conditions.

Even in the United States the green building-movement encouraged the creation of several green-building rating systems. At the forefront is the Leadership in Energy and Environmental Design (LEED™) Green Building Rating System, developed by the United States Green Building Council (USGBC), which is the most widely accepted standard. Many building owners as well as the design and construction industry have gradually incorporated the LEED™ rating system into mainstream practice, as evidenced by the dramatic growth of LEED™ projects over the past few years.

As the LEED™ rating system continues to make inroads into the mainstream design and construction industry, contractors are starting to realize that they can make important contributions towards a project's green objectives by first understanding LEED™ and their own role in achieving LEED™ credits, followed by early involvement and more substantial participation during the project's major phases. There is also a pressing need to define the role of contractors and their early involvement in the green-building process.

The increase in the number of LEED™-certified buildings reflects a growing demand for green buildings in the design and construction industry. During the early stages of the green-building movement, many builders were reluctant to embrace it. However, this reluctance has diminished dramatically over the last decade. Today, an increasing number of projects are seeking LEED™ certification from the U.S. Green Building Council (USGBC).

The LEED™ NC rating system has recently been introduced in countries such as India, China, and Canada. Among the Indian Green Building Council's stated objectives, for example, is to achieve 1 square foot of green building for every Indian resident by 2012. According to Prem C. Jain, the council's chairman, India already has an estimated 240 million square feet of green buildings in place.

The green-building goal as set by the Indian Green Building Council (IGBC) envisages 1 billion square feet of green-building footprint to be registered for certification by 2012, 1000 green buildings to be registered by 2010, a major share of the $40 billion market for green-building materials by 2012, and training of 5000 IGBC-accredited green-building professionals by 2010.

However, some organizations such as the North American Coalition on Green Building, which consists of 32 associations and is a strong supporter of green-building concepts, is not a champion of the USGBC's LEED™ program as currently developed and implemented. The coalition believes that the present LEED™ rating system was created in the absence of a nationally recognized consensus process that allows for meaningful participation by all interested parties.

The coalition has voiced its concerns with the LEED™ program to regulators as well as policy makers at the federal, state, and local levels of government and has started to explore alternatives. It is working with GBI on the possibility of endorsing the Green

Globes program as a preferable alternative to LEED™. Nevertheless, LEED™ continues to move forward and has taken important steps to address many of these concerns in the latest 2009 modifications (LEED™ V3).

An NPR news report by Chris Arnold in July 2006 highlighted the impact the green movement is having on the construction industry. One example given was a project in Boston where "in addition to nice-looking fixtures and appliances and wood floors, the bathroom faucets going in the project will save water, and the AC and heat systems were designed to save energy and boost indoor air quality." The 140-unit luxury condo development reportedly consumes 30 percent less electricity than a conventional building.

Research on green building has been minimal and presently constitutes an estimated 0.2 percent of all federally funded research, which comes to about $193 million per year. This is the approximate equivalent of a mere 0.02 percent of the estimated $1 trillion value of buildings constructed in the U.S. annually, even though the building-construction industry represents over 10 percent of the U.S. GDP.

Thus, federal funding for relevant research is relatively insignificant compared to funding for other research topics, even after taking into account the major impact of buildings and the built environment on our quality of life, natural environment, and economy. The federal government and other relevant funding sources should be encouraged to provide appropriate support to these research programs and readily achievable strategies. Unless we move forward by significantly increasing and improving green-building practices, we are likely to experience a dramatic increase in the negative impacts of the built environment on human and environmental health in the years ahead.

Funding green-building research is certainly considered a wise investment. In 2001 the National Research Council released a report assessing the benefits of DOE's energy efficiency and other research over the past few decades, and found that not only do the benefits of building energy-efficiency research significantly outweigh the investment but that energy savings in the building sector are one of the most significant benefits of all DOE research programs undertaken. Despite the many benefits of energy research and development, which included green buildings, federal energy-research funding has decreased significantly over the last few decades; accounting for inflation, it now is less than half of what it was 25 years ago. On the other hand, federal support for medical and military research has increased by an estimated 400 and 260 percent, respectively.

A substantial increase in green-building research is vital if design and construction firms are to meet the many challenges pivotal to the nation's economy, health, and wellbeing. These challenges include critical issues such as global warming, water shortages, indoor environmental-quality problems, and destruction of our ecosystem. Funding for green-building research should be increased to reflect the severity of the problems that buildings create. The current 0.2 percent of federal research dollars is totally inadequate to address these critical issues in a timely manner.

It has been consistently shown that buildings contribute substantially to environmental problems in the U.S. For example, building operations account for some

38 percent of U.S. carbon-dioxide emissions, 71 percent of electricity use, and, according to the Environmental Information Administration (2008), nearly 40 percent of total energy use; the latter number increases to an estimated 48 percent if the energy required to make building materials and construct buildings is included. Buildings consume 13.6 percent of the country's potable water, and EPA estimates of waste from demolition, construction, and remodeling amount to 136 million tons of landfill additions annually. Construction and remodeling of buildings account for 3 billion tons, or roughly 40 percent, of raw-material use globally each year. They also negatively impact human health; up to 30 percent of new and remodeled buildings may experience acute indoor-air-quality problems in which indoor air pollutants are at concentrations typically between two and five times greater than those of outdoor air.

One example of the significant impact of green-building research by DOE's National Renewable Energy Lab (NREL) and other organizations is to help drive down the costs of photovoltaic modules dramatically – from $30 per peak watt in 1975 to under $5 in the late 1980s and 1990s. However, these costs will need to go lower still for solar electricity to be fully competitive when access to grid power is available.

To cite another important example, the impact of carbon emissions on global warming has recently become an issue of national concern, resulting in part in the AIA, ASHRAE, USGBC, Construction Specifications Institute (CSI), and the U.S. Conference of Mayors' collective adoption of the 2030 Challenge, a series of goals whose main objective is that all new construction will have net zero carbon emissions by the year 2030 and that an equivalent amount of existing square footage will be renovated to use half of its previous energy use.

A March 2007 UN report clearly reaffirms buildings' role in global warming; according to Achim Steiner, UN Under-Secretary General and UNEP Executive Director: "Energy efficiency, along with cleaner and renewable forms of energy generation, is one of the pillars upon which a decarbonized world will stand or fall. The savings that can be made right now are potentially huge and the costs to implement them relatively low if sufficient numbers of governments, industries, businesses, and consumers act."

He goes on to say: "This report focuses on the building sector. By some conservative estimates, the building sector worldwide could deliver emission reductions of 1.8 billion tonnes [1 tonne = 1000 kilograms = 2025 pounds] of CO_2. A more aggressive energy-efficiency policy might deliver over two billion tonnes or close to three times the amount scheduled to be reduced under the Kyoto Protocol." However, it will not be possible to meet the 2030 Challenge without a fundamental change in our present approach and improved knowledge of building energy issues.

A growing demand for building projects that use environmentally friendly and energy-efficient materials has motivated a green movement in the construction industry. It is estimated that over $10 billion of green buildings are in the process of construction in the United States.

Today's construction industry is facing unprecedented and growing pressures, originating from a global economic crisis, lack of available resources, rising material costs, and an increase in natural disasters, among other factors. Together these trends have motivated the industry to increasingly adopt sustainable design and construction methods in an effort to construct more efficient buildings designed to conserve energy and water, improve building operations, and enhance the health and wellbeing of the global population.

One report analyzed the global activity of green building from the perspective of early market adopters and construction-industry professionals in 45 countries; it confirms that green building is achieving an elevated adoption rate in most regions of the world and that the global marketplace is gradually but consistently undergoing a broad transformation to green. The green transformation can be attributed to a series of global pressures and trends, ranging from natural disasters to the dramatic increase and impact of green consumers.

As the cost of resources such as oil skyrocket, the construction industry is being forced to reevaluate its position and explore alternatives to traditional methods of design, energy use, and resource-heavy activities.

The seaside city of San Diego, for example, appears to be growing greener by the month as it taps into the $12 billion sustainable building industry while earning recognition and high marks for environmentally responsible development.

Green construction market share will continue to rise, particularly since state and local governments are using a variety of incentive-based techniques to encourage green-building practices. These efforts have encountered various obstacles and challenges along the way, including the cost of these new incentive programs, lack of available resources, and implementation difficulties. In order to help communities circumvent these obstacles, the American Institute of Architects (AIA) commissioned a report, *Local Leaders in Sustainability – Green Incentives*, which defines various incentive programs, examines the main barriers to success, and highlights best-practice examples from around the country.

Size and Impact of the U.S. Built Environment

According to the Department of Commerce (2008) the construction market accounts for 13.4 percent of the $13.2 trillion U.S. GDP, of which the value of green-building construction is projected to increase to $60 billion by 2010, according to McGraw-Hill's "Key Trends in the European and U.S. Construction Marketplace" *SmartMarket Report*. Furthermore, it is estimated that 82 percent of corporate America will be greening at least 16 percent of real-estate portfolios in 2009; of these corporations, 18 percent are expected to be greening more than 60 percent of their real-estate portfolios.

There are a number of sectors in the green-building industry that are expected to grow in the coming years, including office, education, healthcare, government, hospitality, and industrial. Of these, the three largest segments for nonresidential green-building construction – office, education, and healthcare—in 2008 accounted for more than 80 percent of total nonresidential green construction (*Source*: FMI 2008 U.S. Construction Overview).

Among the factors that are driving green building and expediting its growth are the unprecedented level of government interest and initiatives, the increase in residential awareness and demand for green construction, and the dramatic improvements in and availability of sustainable materials in the market.

Owners of LEED™-registered and -certified projects represent a diverse cross-section of the industry. For example, roofing companies have turned their focus to technologies that will allow their customers to harness the energy a rooftop solution can provide. But these companies realize that to meet the demands and challenges that green technology brings requires a serious commitment financially and in terms of manpower, technology, training, and equipment.

Likewise, new green technologies are having a tremendous positive economic impact on the plumbing industry and are spurring economic growth for plumbing contractors around the country. We have recently witnessed an increased demand for new technologies in the plumbing industry, which has proven to be a positive economic stimulus to its members. Plumbing contractors are in the forefront of advocating for and installing energy-efficient water systems, and promoting water conservation and energy efficiency through the installation and use of green technologies.

Designers and property owners, both public and private, are playing a critical role in pursuing sustainable design and construction practices. However, since both the source (designer) and the end user (owner) are readily adopting sustainable design practices, it becomes essential for the contractor/builder to become an active team member if green-building projects are to be successfully implemented. An experienced builder can provide important input on aspects such as, material selection and specification, system performance, minimizing construction waste, ways to improve indoor air quality, and other factors. The contractor can also assist in streamlining construction and value-engineering methods, to achieve the project goals.

Studies on the costs and benefits of green buildings found that energy and water savings alone outweigh the initial cost premium in most green buildings and that green buildings may cost, on average, less than 2 percent more to build than conventional buildings. This stands in contrast to public perception: a 2007 survey by the World Business Council for Sustainable Development found that business leaders believe green buildings to be on average 17 percent more expensive than conventionally designed buildings.

One of the key findings in a landmark international study, *Greening Buildings and Communities: Costs and Benefits,* concludes: "Most green buildings cost 0 to 4 percent more than conventional buildings, with the largest concentration of reported "green premiums" between 0 and 1 percent. Green premiums increase with the level of greenness, but most LEED™ buildings, up through gold level, can be built for the same cost as conventional buildings." This contrasts sharply with the common belief that green buildings tend to be much more expensive than conventional buildings.

"The deep downturn in real estate has not reduced the rapid growth in demand for and construction of green buildings," says Greg Kats, the study's lead author and a managing director of Good Energies, one of the study's main supporters. "This suggests a flight to quality as buyers express a market preference for buildings that are more energy-efficient, more comfortable, and healthier." A more detailed discussion of the green movement and green building is provided in Chapter 1.

1 Defining "Green" and "Sustainability"

1.1 Green Design – Concepts and Definitions

It is exciting to see the many changes that have taken place in the construction industry and the architectural/engineering professions over the last decade in the promotion of environmentally responsible buildings. Sustainable buildings use valuable resources such as energy, water, materials, and land more efficiently than buildings that are simply built to code. These green buildings are also kinder to the environment, and provide indoor spaces that are generally more healthy, comfortable, and productive. As the environmental impact of buildings becomes more apparent, we are witnessing a growing movement of sustainable building leading the way to reduce this impact at the source.

Architects, designers, builders, and building owners are increasingly becoming interested and involved in green building. National and local programs encouraging green building are flourishing throughout the nation as well as globally. Thousands of projects have been built over the last decade, providing tangible evidence of what green building can accomplish in terms of improved comfort levels, aesthetics, and energy and resource efficiency.

The main benefits of building green include:

- Reducing energy consumption
- Protection of ecosystems
- Improved occupant health

A major contributing factor to the growth of building efficiency is the demand from occupants and tenants who have to live and work inside these structures. There is a growing body of evidence linking more efficient buildings with improved working conditions, leading to increased productivity, reduced turnover and absenteeism, and other benefits. The reason for these results is that in many cases building operating expenses represent less than 10 percent of an organization's cost structure, whereas the cost of personnel comprises the remaining 90 percent. This is a strong indicator that even small improvements in worker comfort can result in substantial dividends in performance and productivity.

But although sustainable or green building is a strategy for creating healthier and more resource-efficient models for construction, renovation, operation, maintenance, and demolition in which energy remains a pivotal component, green design needs to consider other environmental impacts. Recent research and experience have clearly

DOI: 10.1016/B978-1-85617-691-0.00001-1

shown that when buildings are designed and operated with their life-cycle impacts in mind, they usually provide substantially greater environmental, economic, and social benefits. This forces us to conclude that green-building success requires the implementation of an integrated approach to building design.

This whole-building approach is pivotal to a project's success and means that all aspects of a project, from site selection to structure to floor finish, are carefully considered. And focusing on a single component of a building can profoundly impact the project negatively with unforeseen and unintended environmental, social, and/or economic consequences. For example, designing a building envelope that is not energy-efficient can have a significantly adverse impact on indoor environmental quality. Likewise, exposure to materials such as asbestos, lead, and formaldehydes, which can have high volatile-organic-compound (VOC) emissions in a building, can precipitate significant health problems due to poor indoor-air quality, creating what is known as "sick building syndrome." An interdisciplinary team is thus considered a prerequisite to building green.

The related concepts of sustainable development and sustainability have today become integral to green building. But it is critical to consider any and all green options during the early design programming and planning phases. Furthermore, incorporating green strategies and materials during the early design phase is a great way to increase a project's potential market value. Continuing research has shown that sustainable developments can help lower operating costs over the life of the building by increasing productivity and utilizing less energy and water; green developments can also provide tenants and occupants with a healthier and more productive working environment. They can also significantly reduce environmental impacts by, for example, lessening storm-water runoff and the heat-island effect. Practitioners of green building often seek to achieve not only ecological but aesthetic harmony between a structure and its surrounding natural and built environment. The outward appearance and style of green buildings are not always immediately distinguishable from their less sustainable counterparts that are built to code.

History shows us that buildings have an enormous impact on the environment – both during the construction phase and throughout their operation. Building operation typically requires expending substantial amounts of energy, water, and materials while leaving behind large amounts of waste. Rob Watson, author of the *Green Building Impact Report,* issued in November 2008, states: "The construction and operation of buildings require more energy than any other human activity. The International Energy Agency (IEA) estimated in 2006 that buildings used 40 percent of primary energy consumed globally, accounting for roughly a quarter of the world's greenhouse-gas emissions. Commercial buildings comprise one-third of this total. Urbanization trends in developing countries are accelerating the growth of this sector relative to residential buildings, according to the World Business Council on Sustainable Development (WBCSD)."

As mentioned earlier, how these buildings are built can strongly impact the ecosystems around us. Moreover, the buildings themselves create new indoor environments that present new environmental problems and challenges. Indeed, sustainable/

green-building strategies and practices offer a unique opportunity to create environmentally sound and resource-efficient buildings. An integrated approach to design can achieve this by inviting architects, engineers, land planners, building owners and operators, and members of the construction industry to work as a team to design the project. This is discussed more fully in Chapter 5.

The federal government is the nation's largest single landlord and has become one of the leaders in building green. In this respect, the General Services Administration recently announced that it was applying stringent green-building standards to its $12 billion construction portfolio of courthouses, post offices, border stations, and other buildings.

This role is echoed by Richard Fedrizzi, CEO and Founding Chairman of the U.S. Green Building Council, who comments: "The federal government has been at the forefront of the sustainable building movement since its inception, providing resources, pioneering best practices, and engaging multiple federal agencies in the mission of transforming the built environment." And on January 24 and 25, 2006, the first-ever conference took place; over 150 federal facility managers and decision makers attended. In addition, 21 government agencies participated in formulating and witnessed the signing of the Federal Leadership in High Performance and Sustainable Buildings Memorandum of Understanding (MOU). With this Memorandum of Understanding, signatory agencies committed to federal leadership in the design, construction, and operation of high-performance and sustainable buildings. The MOU represents a significant achievement by the federal government through its cumulative efforts to define common strategies and guiding principles of green building. The signatory agencies will also coordinate their efforts with others in the private and public sectors to achieve these goals.

Green building essentially consists of several strategies relating to land use, design, construction, and operation that in aggregate help minimize or mitigate the overall impact on the environment. Green buildings therefore are intended to improve the efficiency with which buildings use available natural resources such as energy, water, and materials, while at the same time minimizing a building's adverse impact on human health and the environment. This is typically achieved through better design, siting, construction, operation, maintenance, and removal.

The manner in which green buildings can reduce the overall impact of the built environment on human health and the natural environment is by the following strategies:

- The efficient use of energy, water, and other resources
- Protecting the health of a building's occupants
- Improving employee productivity
- Reducing waste, pollution, and environmental degradation

Leah B. Garris, senior associate editor at *Buildings* magazine, says that "Myth and misinformation surround the topic of sustainability, clouding its definition and purpose, and blurring the lines between green fact and fiction." But once you set aside the myths and misinformation relating to sustainability and green design, a number of pertinent strategies will become apparent that will help you achieve your green building objectives. Alan Scott, principal, Green Building Services, Portland, OR, says that "You can have a green building that doesn't really 'look' any different than any other building." A level of sustainability can easily be achieved by designing a green building that looks "normal." According to Ralph DiNola, principal, Green Building Services: "People don't really talk about the value of aesthetics in terms of the longevity of a building. A beautiful building will be preserved by a culture for a greater length of time than an ugly building." And, to be sustainable, it is important for a building to have a long-term, useful life. Aesthetics is a pivotal ingredient for longevity, and longevity is pivotal for sustainability.

In reality, sustainability is about conscious choices and not about spending more on superfluous options in hopes of earning an increased return on investment. It is about working with nature – not against it. Furthermore, it is not about buildings that are supposedly environmentally responsible but that ultimately sacrifice tenant/occupant comfort. This does not imply that purchasing green products or recycling assets at the end of their useful lives is not sustainable, because it is. It is good for both the environment and for the health of a building's occupants.

Before making a final decision to go green, it is important to take the time to research what will work best for your project and offer the best possible return on investment. Instead of blindly buying into the myths that can defeat all your efforts to be sustainable, it is prudent to first thoroughly research the numerous options available to adequately evaluate them before you decide. When there is little or no time set aside for sorting through the various sustainable technology options, a decision is often made at the last minute that turns out to be both costly and inappropriate.

1.1.1 Sustainability Begins with Climate

Perhaps the main reason that green strategies are considered green is because they typically work in harmony with the surrounding climatic and geographic conditions and not against them. This means that you need to fully understand the environment in which you are designing and building in order to utilize them to your advantage. Successful architects and designers fully understand that, in order to succeed in sustainable design, they must be familiar with year-round weather conditions such as temperature, humidity, rainfall, site topography, prevailing winds, and indigenous plants. All impact sustainability in one way or another and depend to some extent on where you are. But in order to measure the degree of success of a sustainable design, the building's performance should be compared to a baseline condition. That baseline condition relates to where your building or project is located and the microclimate and environmental conditions of that specific location.

To attain sustainability, one cannot overemphasize the need to clearly identify and reduce a building's need for resources that are scarce or unavailable locally (e.g., water and energy) and increase the use of readily available resources, such as the sun, rain, wind, etc. A proficient understanding of the local climate is important because it means that you comprehend what is available and at your disposal: e.g., the sun for heating and

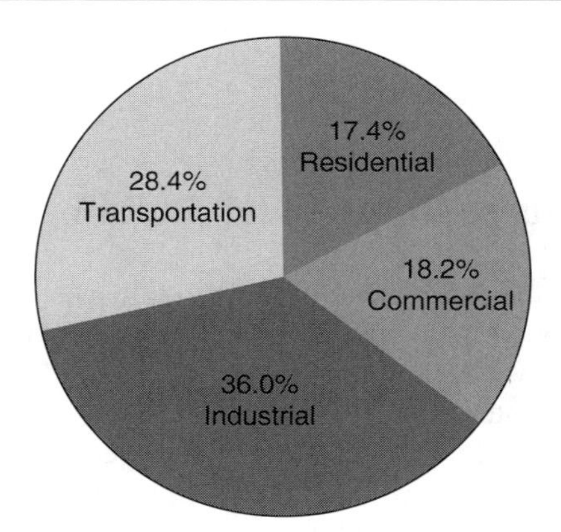

Figure 1.1 Pie chart showing greenhouse-gas emissions by sector.
Source: Greener World Media, Inc.

lighting, the wind for ventilation, and rain for irrigation and other water requirements (Figure 1.1).

When discussing sustainability, five areas immediately come to mind (this list is modeled after the Washington, D.C.-based U.S. Green Building Council's LEED™ rating system):

- Sustainable sites
- Water conservation
- Energy efficiency
- Materials and resource conservation
- Indoor environmental quality

Two of the above five areas are almost entirely independent of climate (indoor environment quality and materials and resource conservation). Site sustainability does depend on climate but more specifically on the specifications and microelements that are particular to a given site. As for water conservation and energy efficiency, both rely more heavily on the climate. Each region or location may have a different climate – from hot and arid to freezing and windy. Thus, in order to avoid using inappropriate techniques that will invariably increase costs, it is important from the outset to understand the resources that are readily available and those that are not in the project area.

In promoting LEED™, the USGBC emphasizes the simplicity of the program. But the uniqueness of the LEED™ certification system is that it typically mandates performance over process. In fact, over the past decade or so the USGBC, through the application of its widely circulated scoring system and other efforts, has dramatically impacted the way many contractors and their subcontractors operate.

To get the optimum indoor-air quality, builders should use adhesives, paints, and sealants that don't release harmful chemicals into the air. Sustainability is facilitated by the use of high-performance insulation and installation of high-efficiency water, gas, and electrical systems. Passive cooling and "daylighting" are systems that can have a great impact on energy savings.

1.2 Recent Upsurge in the Green Building Movement

The continuing growing demand for building projects that use environmentally friendly and energy-efficient materials has spurred a green movement in the construction industry. It is estimated that over $10 billion in "green buildings" are presently under construction in the United States.

It is unfortunate, however, that the quantity and quality of data on business and the environment remain wanting, to say the least. In this respect, government agencies, nonprofit groups, academic institutions, and corporations have remained lethargic and produced relatively little to quantify, let alone assess, simple measures of business environmental impact. The federal government in particular has been wanting in its funding of research into green building. Figure 1.2 gives an indication of the amount of funding allocated to green building as compared with other functions. According to Mara Baum, "other functions," within which sustainability research is allocated, make up 1.57 percent of total allocations. They "include education, training, employment, social services; income security; and commerce. Green-building data was compiled from agencies and the Office of Management and Budget (OMB); baseline

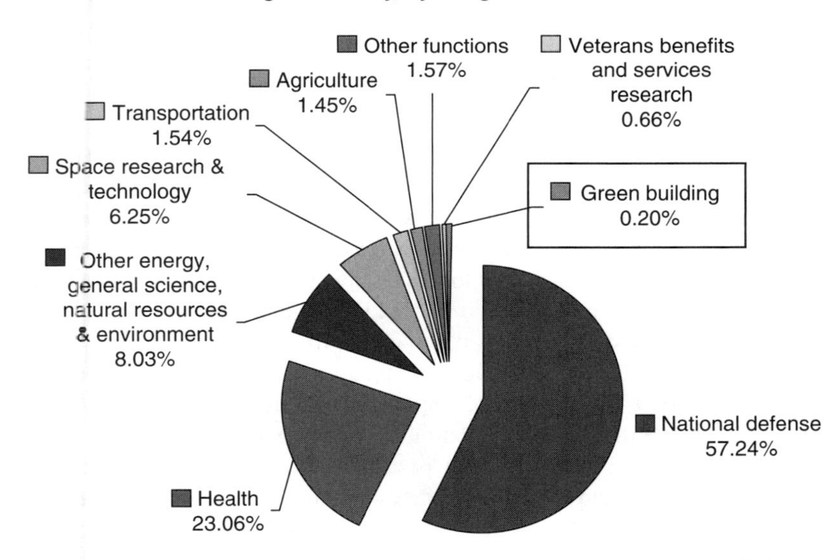

Figure 1.2 Pie chart showing federal research-and-development budget appropriations according to budget function and the minimal amount allocated to green building.
Source: Mara Baum, USGBC.

federal R&D budget data comes from the National Science Foundation (NSF). The 0.2 percent funding toward green building does not include money from the Department of Defense." This is particularly alarming since, according to the American Institute of Architects (AIA), buildings are the leading source of greenhouse-gas emissions in the United States, and a recent online survey conducted by Harris Interactive shows that only 4 percent of U.S. adults were aware of this fact.

With this in mind, Frank Hackett, an energy-conservation sales consultant for Mayer Electric Supply Co., Inc., says that one of the most basic things that a business can do to improve its efficiency is to update or retrofit its lighting system. One example is to modify and update existing lighting fixtures to use the more energy-efficient T-5 or T-8 fluorescent lamps as opposed to the T-12 models that are often used. Replacement of the magnetic ballasts in the lighting can also increase the system's energy efficiency. According to Department of Energy estimates, more than a modest percentage of a business's typical power bill is made up of lighting costs; reductions can have a significant economic impact.

With many state and local governments and federal agencies becoming increasingly eager to hop on the "green" bandwagon, it is likely that updating and retrofitting existing lighting systems will eventually become mandatory, because government regulations will leave businesses with no alternative. But other energy-saving options should be looked into as well, such as automatic controls that take advantage of natural light and automatically switch off the lights when no one is around. Sometimes tax deductions for these procedures are available, giving them an added advantage.

Among the more popular methods people are looking at to save money and energy over the long term is the application of so-called "green building" techniques to new homes or retrofitting existing homes. Green-building materials can be anything from straw bale to low-flow toilets and blue-jean insulation made from recycled waste.

The U.S. Department of Energy (DOE) recently issued a ruling that states must now certify that their building codes meet the requirements in ASHRAE/IESNA's 2004 energy-efficiency standard. The DOE has determined that Standard 90.1-2004 will save more energy than the previously referenced Standard 90.1-1999.

On October 3, 2008, President Bush signed into law H.R. 1424, which extended the Energy-Efficient Commercial Building Tax Deduction as part of the Emergency Economic Stabilization Act of 2008. This tax deduction is not a tax credit – i.e., the amount will not be directly subtracted from the tax owed, but an amount will be subtracted from the gross taxable income. Tax deductions offer benefits to the taxpayer; thus this newly created program can be used as an incentive to assist in choosing energy-efficient building systems.

In October 2005 California passed its Revised Title 24 Code, and since then most major cities throughout the United States have initiated some form of energy-efficiency standards for new construction and existing buildings. Additionally, a new law that took effect on January 1, 2009, states that owners of all nonresidential properties in California are mandated to make available to tenants, lenders, and potential buyers the energy consumption of their buildings as part of the state's participation in the federal ENERGY STAR program. This information will then be given to the Environmental

Protection Agency's ENERGY STAR portfolio manager, who will benchmark the data under its standards. Once this data is compiled, commencing in 2010, building owners will be required to disclose the data and ratings.

In April 2005 Washington State began requiring that all state-funded construction projects of more than 5000 square feet, including school district buildings, are to be built green. And in May 2006, Seattle approved a plan offering incentives to encourage site-appropriate packages of greening possibilities that would include green roofs, interior green walls, exterior vertical landscaping, air filtration, and storm-water runoff management. Seattle also became the first municipality in the United States to adopt the U.S. Green Building Council's Leadership in Energy and Environmental Design (LEED™) Silver rating for its own major construction projects.

New York City's Local Law 86 (sometimes called "the LEED™ Law") took effect in January 2007. It basically requires that many of New York City's new municipal buildings, as well as additions and renovations to its existing municipal buildings, achieve certain standards of sustainability that would meet various LEED™ criteria.

Washington, D.C.'s Green Building Act of 2006 went into effect in March 2007; it requires compliance with the LEED™ rating system for both private and public projects within the nation's capital. Washington, D.C., has become the first major U.S. city to require LEED™ compliance for private projects. These new green-building standards became mandatory in the District in 2009 for privately owned, nonresidential construction projects with 50,000 square feet or more; public projects will also have to comply with these new standards.

In Maryland, the Baltimore City Planning Commission voted in April 2007 to mandate developers to incorporate green-building standards into their projects by 2010. Boston has taken it a step further, amending its zoning code to require green building for all public and private development projects in excess of 50,000 square feet.

Pennsylvania, which presently ranks second only to California in the number of LEED™-certified buildings with 83, currently has four state funds, including a $20 million Sustainable Energy Fund, that provide grants and loans for energy-efficient and renewable-energy projects. Philadelphia has recently enacted a Green Roofs Tax Credit for costs incurred by installing a roof that supports living vegetation. Philadelphia has also proposed a sustainable zoning ordinance that mandates incorporating green roofs for buildings that occupy a minimum of 90,000 square feet.

Many other states – including Arizona, Arkansas, Colorado, Connecticut, Florida, Michigan, and Nevada – have followed suit, in addition to nearly 60 cities and counties nationwide. Soon, green or sustainable building may cease to be an option but rather a requirement. According to Stacey Richardson, a product specialist with the Tremco Roofing & Building Maintenance division: "It is the way of the future, and industry developments in new green technology will provide building owners increasing access to energy-saving, environmentally friendly systems and materials. Everything from bio-based adhesives and sealants, low-VOC or recycled-content building products, to the far-reaching capabilities of nanotechnology – the movement of building 'renewable' and 'energy-efficient' will only continue to strengthen".

Many colleges and universities, including Pennsylvania State, the University of Florida, the University of South Carolina, and the University of California at Merced, have also adopted steps to go green. Harvard University, for example, has already built 12 green buildings.

While the United States is the uncontested global leader in green-building construction, other countries around the world have also started to make substantial investments in sustainability. European Union (EU) leaders, for example, were able to agree on a new sustainable-development strategy that could potentially be pivotal in defining how the EU economy evolves in the coming decades. In the Gothenburg European Council meeting of June 2001, sustainable development was defined as a means of meeting the needs of the present generation without compromising those of the future. There are several green-building assessment systems used around the world, such as Building Research Establishment's Environmental Assessment Method (BREEAM), Comprehensive Assessment System for Building Environmental Efficiency (CASBEE), Green Globes™ US, and Green Building (GB) Tool (Mago and Syal, 2007).

1.3 Incentives for Building Green

1.3.1 Incentives and Tax Deductions

More than 65 local governments have already made a commitment to LEED™ standards in building construction, with some reducing the entitlement process by up to a year in addition to providing various tax credits. Annual energy costs have become a major office-building expense, but they can be reduced by as much as 30 percent, and these reductions will gradually increase with the development of new technologies.

The many benefits of developing and owning a green building are undeniable. Beyond the health and environmental benefits of living and working in a green building, many local and state governments, utility companies, and other entities nationwide offer rebates, tax breaks, and other incentives to encourage the incorporation of eco-friendly elements in building projects. In fact, the majority of large cities in the United States today provide financial incentives for building green. Some of these are code- or tax-based, while others are cash paid for energy and water conservation.

Through the Energy Policy Act of 2005, the U.S. government now provides an Energy-Efficient Commercial Building Tax Deduction and other possible tax breaks, which serve as an important incentive for commercial developers to construct energy-efficient buildings as well as efficiency upgrades to homes.

The U.S. Environmental Protection Agency (EPA) website provides links to various funding sources for green building that are available to homeowners, industry, government organizations, and nonprofits in the form of grants, tax credits, loans, and other sources, both nationally and at the state and local levels.

The Database of State Incentives for Renewables and Efficiency (DSIRE), which is a nonprofit project funded by the U.S. Department of Energy through the North Carolina Solar Center and the Interstate Renewable Energy Council, also has information regarding local, state, federal, and utility incentives available for switching to renewable or efficient energy use. The U.S. government's ENERGY STAR® site can also light the path for consumers, homebuilders, and others to obtain federal tax credits for using energy-efficient products.

Property owners now generally agree that one of the more tangible benefits of attaining LEED™ certification for a building is the ability to use that achievement as a marketing tool. It has become apparent that certifiably green buildings are more likely to attract quality tenants and get higher rents. Likewise, designers and contractors with LEED™-certified buildings in their portfolios often find that they have a distinct competitive edge.

1.3.2 Green Building Programs

- Throughout the United States, numerous cities are now promoting the use of various external green building programs. The City of Seattle is one such example, promoting a number of green building programs including:
  - Built Green: This is a nonprofit residential green-building program developed by the Master Builders Association of King and Snohomish Counties in partnership with Seattle, King County, and a number of local environmental groups. BUILT GREEN™ for residential developments, multifamily, and single family new construction and remodeling.
- Energy Star Homes: A program for new homes that was created by the U.S. EPA and Department of Energy.
- LEED™ for Homes: A newly proposed residential rating system from the U.S. Green Building Council that is currently in the pilot stage but is due to be operational shortly.

Seattle has also started implementing a Sustainable Building Policy that requires all new city-funded projects and renovations with over 5000 square feet of occupied space to attain a LEED™ Silver rating. This policy affects all city departments that are involved with construction, such as the Department of Planning and Development (DPD), which monitors implementation of the policy. Seattle's green-building program is now called CITY Green Building and resides within DPD.

1.3.3 Defining Sustainable Communities

The definition of sustainable communities remains somewhat elusive and continues to evolve over time as our knowledge about sustainability increases. Some community planners are now trying to articulate a vision of how their community will grow in ways that will sustain its citizens' core values, which include:

- The community
- Environmental stewardship and responsibility
- Economic prosperity, opportunity, and security
- Social equity

Several cities in the United States are adopting a comprehensive plan that includes goals and policies intended to help guide development toward a more sustainable future.

This emerging group of sustainable communities, sometimes called green urbanism, seeks to apply leading-edge tools, models, strategies, and technologies to assist cities in meeting sustainability goals and policies. By applying an integrated, whole-systems approach to the planning of communities or neighborhoods, a city can achieve a still greater level of environmental protection.

Other compelling incentives for building owners and property developers to invest in green buildings and especially the LEED™ certification program include the financial benefit of operating a more efficient and less expensive facility. Adhering to LEED™ guidelines will help ensure that the facilities are designed, constructed, and operated more effectively, mainly because LEED™ focuses the design- and construction team on operating life-cycle costs, not initial construction costs.

Another compelling incentive involves the tax benefits that many states are now offering for green building and LEED™ compliance. An example is the State of New York, where former Governor Pataki established the New York State Green Building Tax Credit Program to facilitate the funding of concepts and ideas that will encourage green building. Through this program, the Hearst Building project in New York City is eligible for close to $5,000,000 in tax credits if required green-building standards are achieved.

It is surprising that until recently no single organization has taken a strong initiative to bring green construction to the American home market. Previously, various residential green-building programs were in place and were sponsored for the most part by local homebuilder associations (HBAs), nonprofit organizations, and/or municipalities. Realizing that it was only a matter of time before the USGBC launched a LEED™-like program in this market, the National Association of Homebuilders (NAHB) and the NAHB Research Center (NAHB RC) took preemptive action and produced the Model Green Home Building Guidelines in addition to various utility programs.

While the NAHB program provides many of the answers for the residential building market, LEED™ was for many years virtually the only game in town with respect to commercial construction. However, an alternative from Canada has been introduced that hopes to provide the U.S. commercial construction industry with a simpler, less expensive approach to assessing and rating a building's environmental performance.

Green Globes is a web-based auditing tool that was developed by a Toronto-based environmental consultant, Energy and Environment Canada. Green Globes' strength is purported to be based on its reportedly rapid and economical method for assessing and rating the environmental performance of new and existing buildings. The Green Building Initiative (GBI) recently purchased the rights to market the program in the United States, and GBI budgeted over $800,000 as a first step to create national awareness of Green Globes as a viable alternative to the LEED™ program throughout the construction and development community and capture a viable percentage of its market share.

California and New York building codes are among the national leaders in sustainable development. California Governor Arnold Schwarzenegger's solar initiatives were instrumental in kick-starting the solar industry in his state, which just happen to be the nation's largest market. On December 14, 2004, Schwarzenegger signed Executive Order #S-20-04, requiring the design, construction, and operation of all new and renovated state-owned facilities to be LEED™ Silver-certified. The state is pursuing LEED™ Silver certification for new construction projects and LEED™ certification for existing buildings and facilities. California was followed by New Mexico, which passed a major green-building tax credit in 2007, and Oregon, which passed a 35 percent tax credit for employing solar-energy systems.

Green-building tenant attraction and retention continues to grow stronger, as major tenants increasingly favor healthier air quality over luxury amenities in premium properties, making a green building a sound investment and excellent long-term value.

The main economic benefits of using energy more efficiently include:

- Saves operating costs on utility bills over the life of the building
- Reduces the unit cost of manufactured goods and services
- Increases resale and lease value of real estate

The main economic benefits of utilizing the environment more efficiently, reducing environmental impact, include:

- Reduces water usage
- Reduces materials waste and disposal costs
- Reduces chemical use and disposal costs
- Encourages recycling and reuse of materials
- Encourages development of local markets for locally produced materials, saving on transportation costs

The main economic benefits of fostering social equity, improving indoor environment, and producing healthier places to work include:

- Increases productivity
- Reduces absenteeism
- Increases morale and corporate loyalty
- Reduces employee turnover

Green buildings continue to enjoy an enviable reputation for excellence when it comes to a healthy work environment, and their developers continue to enjoy the well-deserved public perception of goodwill toward employees and the community in addition to the potential financial benefits that green buildings have to offer.

1.4 Emerging Directions: Where Do We Go From Here?

The construction industry is currently witnessing rapidly changing conditions. The world of building design has been increasingly going green so that today, it has become an integral part of the mainstream and our global culture. Architects, designers, engineers, contractors, and manufacturers and federal, state, and local governments have all become willing participants in this emerging green phenomenon. According to a 2008 Green Building Market Barometer on-line survey of commercial real-estate executives conducted by New York City-based Turner Construction, even the credit collapse of 2008 did not affect developers' desire to go green. Green building has not only become global, but the trend towards sustainable design has shown that it is indeed now much more than a nominal trend. By 2010, the value of green-building construction is projected to reach $60 billion (*Source*: McGraw-Hill Construction Key Trends in the European and U.S. Construction Marketplace *SmartMarket Report*).

In addition to the U.S., LEED™ buildings can be found the world over in countries as diverse as Australia, Britain, Canada, China, India, Japan, Mexico, and Spain, to name but a few examples where the movement is well underway. Likewise, green building continues to impact and transform the building market, while revolutionizing our perception of how we design, inhabit, and operate our buildings. There are several factors that are accelerating this push towards building green. These include an increased demand for green construction, particularly in the residential sector; increasing levels of government initiatives; and improvements in the quality and availability of sustainable materials.

A report, Greening Buildings and Communities: Costs and Benefits, is purported to be the largest international study of its kind; it is based on extensive financial and technical analysis of 150 green buildings in 33 U.S. states and 10 countries built from 1998 to 2008. It provides the most detailed findings to date on the costs and financial benefits of building green. The conclusions of this report are outlined in Chapter 10.

While green construction is booming, achieving sustainability on a large scale remains elusive, as green buildings have to date only been able to capture a very small percentage of the construction industry's current market share. LEED™-registered projects today represent roughly 5 percent of the total square footage in U.S. new construction. The ultimate target, according to *Environmental Design + Construction* magazine, is 25 percent of the entire market. Figure 1.3 shows some of the largest buildings certified by LEED™ to date.

Owner/Project	Location	Gross Square Feet
Johnson Diversey/Global Headquarters	Sturtevant, WI	2,316,996
State of Illinois/McCormick Place West Expansion	Chicago, IL	2,226,000
State of California/Capitol Area East End Complex	Sacramento, CA	1,728,702
Silverstein Properties/7 World Trade Center	New York, NY	1,682,000
Nitze-Stagen & Co./Starbucks Center	Seattle, WA	1,650,000
Goldman Sachs/Goldman Sachs Tower	Jersey City, NJ	1,556,915
General Motors/Lansing Assembly Plant	Lansing, MI	1,500,000
General Dynamics/Roosevelt C4 Facility	Scottsdale, AZ	1,500,000
Union Investment/111 South Wacker Drive	Chicago, IL	1,400,000
LaSalle Street Capital/Abn Amro Plaza	Chicago, IL	1,375,058

Figure 1.3 Table showing LEED™'s largest certified buildings in the U.S.
Source: U.S. Green Building Council/Greener World Media, Inc.

Although the U.S Green Building Council (USGBC), a nonprofit membership organization, was founded in 1993, it was only in the last several years that it has become a potent driving force in the green-building construction movement. It was able to do this largely through the development of its commercial-building rating system known as the Leadership in Energy and Design (LEED™). Earning LEED™ certification commences in the early planning stage, where the interested parties make the decision to pursue certification. Once this is accomplished, the next step is to register the project, which requires payment of an initial fee. Upon completion and commissioning of the project, with all the numbers and supporting documentation in, the project is submitted for evaluation and certification.

Today, USGBC has emerged as the leader in fostering and furthering green-building efforts throughout the world. The LEED™ Green Building Rating System is increasingly becoming the national standard for green building in the United States; it is also recognized worldwide as an invaluable tool for the design and construction of high-performance, sustainable buildings.

In looking to the future, the USGBC has formulated a strategic plan for the period 2009 to 2013 in which it outlines the key strategic issues that face the green-building community:

- Shift in emphasis from individual buildings toward the built environment and broader aspects of sustainability, including a more focused approach to social equity
- Need for strategies to reduce contribution of the built environment to climate change
- Rapidly increasing activity of government in the green building arena
- Lack of capacity in the building trades to meet the demand for green building
- Lack of data on green-building performance
- Lack of education about how to manage, operate, and inhabit green buildings
- Increasing interest in and need for green-building expertise internationally

One of the indicators reflecting international interest in the USGBC and LEED™ is evidenced each year by the large attendance and number of different countries represented at Greenbuild, the USGBC's International Conference and Expo. Nearly 30,000 people attended the Greenbuild 2008 conference in Boston (*Source*: USGBC Green Building Facts), compared to only 4200 who attended the same event in Austin, TX, in 2002. The 2008 Greenbuild attendees included representatives from all 50 states, 85 countries, and 6 continents. This venue has thus emerged as an important forum for international leaders in green building, one that will facilitate the exchange of ideas and information on the subject.

Many project teams from different countries continue to successfully apply LEED™ as developed in the United States. However, the USGBC also recognizes that certain criteria, processes, or technologies may not be appropriate for all countries. To address this reality, the USGBC has agreed to sanction other countries to license LEED™ and to allow them to adapt the rating system to their specific needs while maintaining LEED™'s high standards. To date, many countries have expressed an interest in being LEED™-licensed, and several countries including Canada and India have already done so. In India LEED™ has been adopted by the Confederation of Indian Industry.

The USGBC also recognizes that successful strategies for encouraging and practicing green building will by necessity differ from one country to another, depending on local traditions and practices.

Internationally, the USGBC works through the World Green Building Council (WorldGBC), which was formed in 1999 by David Gottfried and officially launched at Greenbuild 2002 in Austin, TX. One of WorldGBC's goals is to help other countries to establish their own councils and find means to work effectively with local industry and policy makers.

The WorldGBC currently consists of a federation of 10 national green-building councils including the United States and Canada and is devoted to transforming the global property industry to sustainability. Part of its mission is to support and promote individual green building councils, and it serves as a forum for knowledge transfer among councils. Its mission also includes encouraging the development of market-based environmental rating systems and recognizing global green-building leadership. As one of the founding members of the WorldGBC, the USGBC is firmly committed to this mission.

The international importance of green building continues to be highlighted at the annual Greenbuild Conferences and Expos. Among the many international delegations was a recent high-level group from China that included the Vice Minister of the Ministry of Construction. Over the past decade, China's economy has been expanding and growing at an alarming rate and is on track to becoming the largest economy in the world by 2020. With such growth, however, comes a mixed bag of severe environmental problems, including a looming energy crisis. China has made impressive inroads to addressing these threats and reversing these environmental trends. To do so, China announced a new energy-efficiency strategy, of which green building is a primary component. Representatives of the Chinese Ministry of Construction and of the USGBC also signed a Memorandum of Understanding identifying points of dual interest for collaboration in the promotion of environmentally responsible buildings in both China and the U.S. The combination of China's booming construction industry and its critical need for green buildings makes it a very exciting international opportunity for the USGBC.

These global developments reflect a few of the many ways the U.S. Green Building Council and its members are contributing to the international green-building movement. And they represent but a small percentage of the opportunities that will appear in the future. But despite the enormous progress that has been achieved, there is still a great amount of work to be done to transform the built environment, both at home and aboard.

Great architecture has always been a delicate balance of form and function. Yet high-rise buildings are being constructed globally with increasing ferocity, often without any particular concern for due diligence, the environment, or aesthetics. At the same time, cities monitor the construction in awe, hoping that they will create distinctive skylines that will foster civic pride and attract greater international recognition.

Following the attacks on the World Trade Center in New York and in Mumbai, India, global concern regarding the risk of terrorism has dramatically increased. This has created a greater demand for buildings that protect both the structure and the inhabitants. In many countries, architects and building owners are now demanding that their facilities be designed to have greater blast resistance and to better withstand the effects of violent storms such as tornadoes and hurricanes – for example, by the use of blast-resistant windows with protective glazing. The windows of federal buildings are now required to be designed to provide protection against potential threats. Also from the perspective of governments and the public alike, there are increasing demands for structures that are sustainable and meet environmental requirements. Taken together, these emerging trends require greater industry expertise, innovation, and solutions capable of addressing a wide range of issues.

Since the launching of LEED™ in 2000, the number of certified and registered projects has witnessed dramatic growth – to the tune of nearly 700 percent a year. By the end of 2007, the square footage of U.S. office and commercial space registered or certified under LEED™ increased by more than 500 percent from two years earlier, reaching 2.3 billion. Certified projects have been completed and verified through the USGBC's process, while registered projects are those that are still in the process of design or construction. The astonishing increase in LEED™ project registration for new construction – roughly 40 percent in 2008 – is very significant, as it is a clear indicator of things to come.

This is echoed by Bob Schroeder, Industry Director (Americas) for Dow Corning's construction business, who says: "Today, sustainable design has been recognized by the industry and the public as a critical factor in achieving high-quality architecture and benefiting the building owners, the companies that occupy these structures, and the wider community." Architects also need to work more on bringing the outdoors inside with designs that increase natural daylight, giving a sense of spaciousness and harmony with nature.

1.4.1 Liability

Ward Hubbell, President, Green Building Initiative, suggests that liability issues have become one of our most pressing problems. This is largely because some buildings that were designed to be green fail to live up to expectations. And often in the building industry, where there are failed expectations, lawsuits tend to follow. Hubbell says that "problems arise because building owners, designers, and builders often have very different ideas of what constitutes a successful green building, and they fail to explicitly communicate their thoughts at the outset of the project. As one might expect, these issues are compounded when the parties are relatively new to the green-building process." Hubbell further maintains that the two main areas for dissatisfaction are a building's failure to achieve a promised level of green-building certification and its operational performance. Liability issues are discussed in greater detail in Chapter 10.

It suffices to say here that this is precisely why an integrated design process is needed, one that from the outset involves all the disciplines in a project working together, including the architect, engineers, contractor, building owner, suppliers, and others. A major benefit is that synergies are often achieved when the various parties contribute their advice and expertise from the earliest phases of the design, allowing them to perform their own functions with enhanced knowledge of the project in its entirety. The need for an integrated process continues from the building's design through construction, commissioning, and subsequent use. It further allows all of the relevant disciplines to agree on the scope of the project, its goals in terms of sustainability, cost, time, and accountability if the specified goals are not realized.

1.4.2 Spectacular Landmarks

Another emerging trend is that of building national landmarks. Typical examples are the Sydney Opera House (Figure 1.4) and the Burj Dubai (Figure 1.5). The latter is expected to be completed by December 2009 and will reportedly be the tallest building in the world. Sometimes the driving factor is the desire for increased tourism, but more often cities are looking for symbols to foster local pride.

Figure 1.4 Night photo of the Sydney Opera House which is considered to be one of the most recognizable buildings in the world and has become the city's landmark with its iconic 14 triangular vaulted shells resembling white sails. In addition to representing Sydney, the opera house has also become a symbol for the country of Australia throughout the world. The original plan to build the opera house was won in competition in 1957 by the late Danish architect John Utzon, whose vision and design were too advanced for the architectural and engineering capabilities at the time. It wasn't until 1973 that the Opera House was finally opened by Queen Elizabeth II. *Source*: beingaustralian.wordpress.com.

Figure 1.5 Illustration of the Burj Dubai, the tallest building in the world. When finished in December 2009, the Symbol of Dubai skyscraper will have more than 160 habitable floors, 56 elevators, apartments, shops, swimming pools, spas, corporate suites, Italian fashion designer Giorgio Armani's first hotel, and an observation platform on the 124th floor. The estimated height of the completed tower, which is designed by Skidmore, Owings & Merrill and developed by Emaar Properties, is estimated to be 2,275 feet and will cost about $800,000,000.
Source: Skidmore, Owings & Merrill.

2 Basic LEED™ Concepts

2.1 Overview: Establishing Measurable Green Criteria

The creation of the World Commission on Environment and Development (WCED) by the United Nations in December 1983 was intended to address growing concerns "about the accelerating deterioration of the human environment and natural resources and the consequences of that deterioration for economic and social development." In establishing the commission, the UN General Assembly recognized that environmental problems were global in nature and determined that it was in the common interest of all nations to establish policies for sustainable development (Report of the World Commission on Environment and Development: Our Common Future; http://www.un-documents.net/wced-ocf.htm). The WCED was soon followed by the Brundtland Commission in 1987, which produced the Brundtland Report in August of the same year. The findings of this report are particularly troubling; it states among other things:

> *The 'greenhouse effect,' one such threat to life-support systems, springs directly from increased resource use. The burning of fossil fuels and the cutting and burning of forests release carbon dioxide (CO_2). The accumulation in the atmosphere of CO_2 and certain other gases traps solar radiation near the Earth's surface, causing global warming. This could cause sea-level rises over the next 45 years large enough to inundate many low-lying coastal cities and river deltas. It could also drastically upset national and international agricultural production and trade systems.*

Another threat arises from the depletion of the atmospheric ozone layer by gases released during the production of foam and the use of refrigerants and aerosols. A substantial loss of such ozone could have catastrophic effects on human and livestock health and on some life forms at the base of the marine food chain. The 1986 discovery of a hole in the ozone layer above the Antarctic suggests the possibility of a more rapid depletion than previously suspected.

The report goes to say: "A variety of air pollutants are killing trees and lakes and damaging buildings and cultural treasures, close to and sometimes thousands of miles from points of emission. The acidification of the environment threatens large areas of Europe and North America. Central Europe is currently receiving more than one gram of sulfur on every square meter of ground each year. The loss of forests could bring in its wake disastrous erosion, siltation, floods, and local climatic change. Air-pollution damage is also becoming evident in some newly industrialized countries."

Until very recently, we witnessed a substantial building boom that was frequently supported by inferior design and construction strategies as well as highly inefficient

DOI: 10.1016/B978-1-85617-691-0.00002-3

HVAC systems, making buildings the largest contributors to global warming. Several federal and private organizations have tried (and continue to do so) to address these problems, and due partly to these efforts we are now seeing a surge of interest in green concepts and sustainability to the extent that "green" has now entered the mainstream of the construction industry. Many project owners are aware of the numerous benefits of incorporating green strategies and are increasingly aspiring to achieve LEED™ certification for their buildings. The green-building rating systems serve two principal functions: to promote high-performance buildings and to help create demand for sustainable construction. Green buildings are economically viable, ecologically benign, and sustainable in operation over the long term.

With the above in mind, the Partnership for Achieving Construction Excellence and the Pentagon Renovation and Construction Program Office recently put out a *Field Guide for Sustainable Construction*, which has been developed to assist and educate field workers, supervisors, and managers in making decisions that help the project team meet sustainable project goals. The main points of this guide are summarized below:

1. Procurement: Specific procurement strategies to ensure sustainable construction requirements are addressed.
2. Site/environment: Methods to reduce the environmental impact of construction on the project site and surrounding environment are identified.
3. Material selection: Environmentally friendly building materials as well as harmful and toxic materials that should be avoided are identified.
4. Waste prevention: Methods to reduce and eliminate waste on construction projects are identified.
5. Recycling: Materials to recycle at each phase of construction and methods to support the on-site recycling effort are identified.
6. Energy: Methods to ensure and improve the building's energy performance, reduce energy consumed during construction, and identify opportunities to use renewable energy sources are specified.
7. Building and material reuse: Reusable materials and methods to facilitate the future reuse of a facility, systems, equipment, products, and materials are identified.
8. Construction technologies: Technologies that can be used during construction to improve efficiency and reduce waste (especially paper) are identified.
9. Health and safety: Methods to improve the quality of life for construction workers are identified.
10. Indoor environmental quality: Methods to ensure that indoor environmental-quality measures during construction are managed and executed properly are specified.

The Department of Energy (DOE) also has an Environmental Protection Program, the goals and objectives of which are "to implement sound stewardship practices that are protective of the air, water, land, and other natural and cultural resources impacted by DOE operations and by which DOE cost-effectively meets or exceeds compliance with applicable environmental, public-health, and resource-protection laws, regulations, and DOE requirements. This objective must be accomplished by implementing Environmental Management Systems (EMSs) at DOE sites. An EMS is a continuing cycle of planning, implementing, evaluating, and improving processes and actions undertaken to achieve environmental goals." Some of these goals and objectives are outlined below:

1. Goal: Protect the environment and enhance mission accomplishment through waste prevention.
   Objective: Reduce environmental hazards, protect environmental resources, minimize life-cycle cost and liability of DOE programs, and maximize operational capability by eliminating or minimizing the generation of wastes that would otherwise require storage, treatment, disposal, and long-term monitoring and surveillance (i.e., future environmental legacies).
2. Goal: Protect the environment and enhance mission accomplishment through reduction of environmental releases.
   Objective: Reduce environmental hazards, protect environmental resources, minimize life-cycle cost and liability of DOE programs, and maximize operational capability by eliminating or minimizing the use of toxic chemicals and associated releases of pollutants to the environment that would otherwise require control, treatment, monitoring, and reporting.
3. Goal: Protect the environment and enhance mission accomplishment through environmentally preferable purchasing.
   Objective: Reduce environmental hazards, conserve environmental resources, minimize life-cycle cost and liability of DOE programs, and maximize operational capability through the procurement of recycled-content, bio-based-content, and other environmentally preferable products, thereby minimizing the economic and environmental impacts of managing toxic byproducts and hazardous wastes generated in the conduct of site activities.
4. Goal: Protect the environment and enhance mission accomplishment through incorporation of environmental stewardship in program planning and operational design.
   Objective: Reduce environmental hazards, conserve environmental and energy resources, minimize life-cycle cost and liability of DOE programs, and maximize operational capability by incorporating sustainable environmental stewardship in the commissioning of site operations and facilities.
5. Goal: Protect the environment and enhance mission accomplishment through post-consumer material recycling.
   Objective: Protect environmental resources, minimize life-cycle cost of DOE programs, and maximize operational capability by diverting materials suitable for reuse and recycling from landfills, thereby minimizing the economic and environmental impacts of waste disposal and long-term monitoring and surveillance.

In this respect, it is important for readers to have a clear understanding of LEED™ certification and why it will help property owners remain competitive in an increasingly green market. Certification gives independent verification that a building has met accepted guidelines in these areas, as outlined in the LEED™ Green Building Rating System. LEED™ certification of a project provides recognition of its quality and environmental stewardship. The LEED™ green-building rating system is widely accepted by public and private owners, further fueling the demand for green-building certification.

As the LEED™ rating system makes inroads into the mainstream design and construction industry, contractors and property developers are realizing that they can contribute towards a project's success in achieving green objectives by, first, understanding the LEED™ process and their specific role in achieving LEED™ credits, and, then, through early involvement and participation throughout the various project phases by incorporating a team approach in an integrated design process. It also goes without saying that measureable benchmarks are required in order to accomplish verification and confirm a building's acceptable performance. It is interesting to note that ASHRAE puts the responsibility squarely on the shoulders of the owner to define design intent requirements. But to be in a position to correctly evaluate a building or project, certain information must be made available regarding the criteria on which the project's design and execution were based. So unless the project's plans and specifications are prepared in a manner that has measurable results, it would not be possible to make a meaningful evaluation of a project to see if it has met the original design intent. Furthermore, in terms of sustainability, before measureable green criteria can be established, it is necessary to first agree on a finite definition of green construction and to articulate exactly what is required.

The National Association of Home Builders (NAHB) has put forward a set of green home-building guidelines that "should be viewed as a dynamic document that will change and evolve as new information becomes available, improvements are made to existing techniques and technologies, and new research tools are developed." The NAHB says that its Model Green Home Building Guidelines were written to help move environmentally-friendly home building concepts further into the mainstream marketplace, and they are one of two rating systems that make up NAHBGreen, the National Green Building Program.

The NAHB point system, for example, consists of three different levels of green building, Bronze, Silver, and Gold, which are available to builders wishing to use these guidelines to rate their projects. "At all levels, there are a minimum number of points required for each of the seven guiding principles to assure that all aspects of green building are addressed and that there is a balanced, whole-systems approach. After reaching the thresholds, an additional 100 points must be achieved by implementing any of the remaining line items." Figure 2.1 outlines the points required for the three different levels thresholds of green building.

	Bronze	Silver	Gold
Lot Design, Preparation, and Development	8	10	12
Resource Efficiency	44	60	77
Energy Efficiency	37	62	100
Water Efficiency	6	13	19
Indoor Environmental Quality	32	54	72
Operation, Maintenance, and Homeowner Education	7	7	9
Global Impact	3	5	6
Additional Points from Sections of Your Choice	100	100	100

Figure 2.1 The NAHB point system, which is available to builders wishing to use these guidelines to rate their projects.

Although the appearance of a green building may be similar to other building forms, the conceptual approach to the design is different in that it revolves around a concern for the environment by extending the life span of natural resources and providing human comfort and wellbeing, security, productivity, and energy efficiency. In turn, this approach will result in reduced operating costs including energy and water as well as other intangible benefits. According to the Indian Green Building Council (IGBC), which administers the LEED™ India rating system, there are a number of salient attributes of a green building, as listed below:

- Minimal disturbance to landscapes and site condition
- Use of recycled and environmentally friendly building materials
- Use of nontoxic and recycled/recyclable materials
- Efficient use of water and water recycling
- Use of energy-efficient and eco-friendly equipment
- Use of renewable energy
- Indoor-air quality for human safety and comfort
- Effective controls and building-management systems

The *Whole Building Design Guide* (WBDG) also sets out certain objectives and principles of sustainable design. The objectives are:

1. Avoid resource depletion of energy, water, and raw material.
2. Prevent environmental degradation caused by facilities and infrastructure throughout their life cycles.
3. Create built environments that are livable, comfortable, safe, and productive.

The principles are:

1. Optimize site potential.
2. Optimize energy use.
3. Protect and conserve water.
4. Use environmentally preferred products.
5. Enhance indoor environmental quality (IEQ).
6. Optimize operations and maintenance procedures.

James Woods, executive director of the Building Diagnostics Research Institute, notes: "Building performance is a set of facts and not just promises. If the promises are achieved and verified through measurement, beneficial consequences will result and risks will be managed. However, if the promises are not achieved, adverse consequences are likely to lead to increased risks to the occupants and tenants, building owners, designers and contractors; and to the larger interests of national security and climate change."

Another sustainable design expert, Alan Bilka, with ICC Technical Services, quite rightly points out: "Over time, more and more 'green' materials and methods will appear in the codes and/or have an effect on current code text. But the implications of green and sustainable building are so wide and far-reaching that their effects will most certainly not be limited to one single code or standard. On the contrary, they will affect virtually all codes and will spill beyond the codes. Some green-building concepts may become hotly contested political issues in the future, possibly requiring the creation of new legislation and/or entirely new government agencies."

2.2 USGBC LEED™ Green Building Rating System

2.2.1 General

The most appropriate way to contribute to the success of a LEED™ project is to become familiar with the requirements and opportunities offered by the new program. Comprehensive documentation can be found on the USGBC website (www.LEEDbuilding.org), from accreditation requirements to careers and e-newsletters. To be successful in earning LEED™ certification, the process must begin in the initial planning stage, where the stakeholders involved make the decision to pursue certification. The official first step must be to register the project and pay an initial fee. Upon completion of the project, including all the numbers and preparation of all supporting documentation, the project is submitted for evaluation and certification. Once this has been determined, the project is listed on the LEED™ project list. The summary sheet showing the tally of credits earned becomes available for most certified projects.

2.2.2 Process Overview

The LEED™ 2009 Green Building Rating System consists of a set of performance standards used in the certification of commercial, institutional, and other building types in both the public and private sectors with the intention of promoting healthy, durable, and environmentally sound practices. A LEED™ certification provides independent, third-party verification that a building project has achieved the highest green-building and performance measures according to the level of certification achieved. Setting up an integrated project team to include the major stakeholders of the project, such as the developer/owner, architect, engineer, landscape architect, contractor, and asset and property management staff, will help jump-start the process. This implementation of an integrated, systems-oriented approach to green project design, development, and operations can yield significant synergies while enhancing the overall performance of a building. During the initial project team meetings, the project's goals will be articulated and the LEED™ certification level sought will be determined.

2.2.3 LEED™ 2009 Minimum Program Requirements (MPRs)

MPRs describe the eligibility for each system and are intended to "evolve over time in tandem with the LEED™ rating systems." Though there are eight requirements that are standardized for all systems, the thresholds and levels apply differently for each system and are defined in the rating systems noted above. To clarify the minimum program requirements, one of the categories will be used as an example, New Construction and Major Renovations:

1. Total compliance with all applicable federal, state, and local building-related environmental laws and regulations is mandated where the project is located.

2. Project must consist of a complete, permanent building or space on already existing land. LEED™ projects are required to include new, ground-up design and construction or major renovation of at least one complete building.
3. The project must be executed within a reasonable site boundary. The LEED™ project boundary must include all contiguous land that is associated with and supports normal building operations for the project building. The project boundary must normally only include land that is owned by the party that owns the project. Projects located on a campus must contain project boundaries so that if all the campus buildings become certified, then 100 percent of the gross campus land area would be included within a LEED™ boundary. Any given parcel of real property may only be attributed to a single project building. E. Tampering with a LEED™ project boundary is completely prohibited.
4. Project must comply with minimum floor-area requirements by incorporating a minimum of 1000 square feet (93 square meters) of gross floor area.
5. Projects must comply with minimum full-time-equivalent occupancy rates (FTE).
6. Project owners must consent to sharing whole-building energy and water-usage data with USGBC and/or GBCI for a period of at least five years.
7. The gross floor area of the LEED™ project must conform to a minimum building-area to site-area ratio – building must not be less than 2 percent of the gross land area within the project boundary.
8. Registration and certification activity must comply with reasonable timetables and rating-system sunset dates, which basically means that if a LEED™ 2009 project is inactive for four years, the GBCI reserves the right to cancel the registration.

2.2.4 How LEED™ Works

The LEED™ Green Building Rating System™ inspires and instigates global adoption of sustainable green-building practices through the adoption and execution of universally understood and accepted tools and performance criteria. The LEED™ system is based on awarding points relative to performance including sustainable site development, energy and atmosphere, water efficiency, indoor environmental quality, materials and resources, regional priority, and innovation in design. Designers can select the points that are most appropriate to their projects to achieve a LEED™ rating. A total of 100 + 10 points is possible. Depending on the points, Platinum, Gold, Silver, or Certified ratings are awarded. Moreover, LEED™ 2009 alignment provides a continuous improvement structure that will enable USGBC to develop LEED™ in a predictable way.

When the U.S. Green Building Council (USGBC) first introduced the Leadership in Energy and Environmental Design (LEED™) green-building rating system, Version 1.0, in December 1998, it was a pioneering effort. Today, LEED™ has become the predominant means for certifying green buildings and has recently released a new version, LEED™ 2009, formerly known as LEED™ V3, which is the first major LEED™ overhaul since Version 2.2 came out in 2005.

LEED™ 2009 has been significantly transformed by many changes, both major and minor, to the rating system. The major LEED™ systems will be affected, and the overlap between them will increase substantially. Some of these changes include an entirely new weighting system for LEED™ 2009, which refers to the process of redistributing the available points in LEED™ in a manner that a credit's point value so that it more accurately reflects the potential to either mitigate negative or promote positive environmental impacts of a building. Thus in LEED™ 2009, credits that most directly address the most significant impacts are given the greatest weight, subject to the system-design parameters described above. This has resulted in a significant change in allocation of points compared with earlier LEED™ rating systems. Generally speaking, the modifications reflect a greater relative emphasis on the reduction of energy consumption and greenhouse-gas emissions associated with building systems, transportation, the embodied energy of water and materials, and, where applicable, solid waste (e.g., for the operation and maintenance of existing buildings).

Other changes are an increased opportunity for innovation credits and a new opportunity for achieving bonus points for regional priority credits. A less obvious revision in LEED™ 2009 is the reduction of possible exemplary performance credits from a maximum of four to a maximum of three. The intention here was to return to the original intent of the credit, which is to encourage projects to pursue innovation in green building.

2.2.5 The LEED™ Points System

LEED™ is a continually evolving simple point-based structure that has set the green-building standard and made it the most widely accepted green program in the United States. The various LEED™ categories differ in their scoring systems based on a set of required "prerequisites" and a variety of "credits" in the seven major categories listed below. In LEED™ v2.2 for new construction and major renovations for commercial buildings there were 69 possible points, and buildings were able to qualify for four levels of certification.

With LEED™ 2009, it has become much less complicated to figure out how many points a building receives and where that places it in the continuum of green-building achievement. The new USGBG LEED™ Green Building certification levels for all systems are more consistent and are shown below (with previous rating system shown in brackets):

- Certified: 40–49 (26–32) points
- Silver: 50–59 (33–38) points
- Gold: 60–79 (39–51) points
- Platinum: 80+ (52–69) points

The number of points available per LEED™ system has been increased so that all LEED™ systems have 100 base points as well as 10 possible innovation and regional bonus points, bringing the possible total points achievable for each category to 110. Figures 2.2, 2.3, 2.4, and 2.5 depict buildings that have received various levels of LEED™ certification.

As can be seen from the above, the previous maximum achievement in LEED™ NC was 69 points, which in LEED™ 2009 has increased to 100. Unfortunately, it remains unclear sometimes how the added 31 points are distributed. Aurora Sharrard,

Figure 2.2 The West Building of the Phoenix Convention Center in Phoenix, Arizona, was the first phase of a $600 million expansion to be completed and has been awarded a LEED™ Silver certification by the U.S. Green Building Council (USGBC). The West Building, which opened in June 2006, was designed by Leo A Daly and HOK and achieved the LEED™ certification for incorporating energy use, lighting, water, and material use as well as a variety of other sustainable strategies.
Source: AV Concepts.

research manager at Green Building Alliance (GBA), says: "The determination of which credits achieve more than 1 point (and how many points they achieve) is actually the most complex part of LEED™ 2009. LEED™ has always implicitly weighted buildings' impacts by offering more credits in certain sections. However, in an effort to drive greater (and more focused) reduction of building impact, the USGBC is now applying explicit weightings to all LEED™ credits. The existing weighting scheme was developed by the National Institute of Standards and Technology (NIST). The USGBC hopes to have its own weighting system for future LEED™ revisions, but currently LEED™ credits are proposed to be weighted based on the following categories, which are in order of weighted importance":

- Greenhouse-gas emissions
- Indoor environmental quality fossil-fuel depletion
- Particulates
- Water use
- Human health (cancer-related)
- Ecotoxicity
- Land use

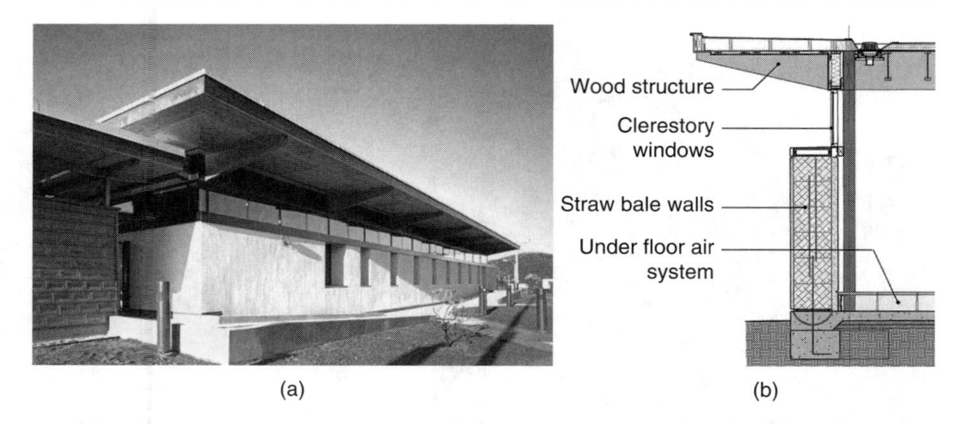

Figure 2.3 (a) The Santa Clarita Transit Maintenance is one of the first LEED™ Gold-certified straw-bale buildings in the world. The resource- and energy-efficient transit facility was designed by HOK and exceeds California Energy Efficiency Standards by more than 40 percent, securing a new standard for straw-bale in high-performance building design. (b) Diagram shows a section taken through the exterior wall of the transit facility. The designers reportedly opted for a solar photovoltaic canopy to shade buses and provide nearly half of the building's annual energy needs. An electronic monitoring system is in place to track thermal comfort, energy efficiency, and moisture levels.
Source: HOK Architects.

- Eutrophication
- Smog formation
- Human health (non-cancer-related)
- Acidification
- Ozone depletion

As will be seen, the new weighting preferences in the LEED™ 2009 system put much greater emphasis on energy, which is appropriate and addresses some of the criticisms levied against the earlier version. There has also been an increase in the Innovation and Design (ID) credits from 4 points to 5. An additional point can be achieved for having a LEED™ Accredited Professional (LEED™ AP) on the project team (bringing the total ID points to 6). The introduction of the new category of Regional Priority adds another potential 4 bonus points (bringing the total points possible to 110).

Mike Opitz, Vice President of LEED™ Implementation for the USGBC, highlights many of the changes incorporated (by category) in the new LEED™ V3. These include the following.

Sustainable site categories include:

- SSp1. Construction Activity Pollution Prevention (NC, CS, Schools): Compliance verification now required through photodocumentation, inspection logs, or reports.

(a)

(b)

Figure 2.4 (a) Birdseye view of Donald Bren School of Environmental Science and Management at UCSB, which is a LEED™ Gold Pilot project. It utilized silt fencing, straw bale catch basins, and scheduled grading activities in accordance with the project's erosion control plan. (b) Bren Hall, a facility for faculty and students on the third-floor terrace.
Source: Bren School website: http://www.esm.ucsb.edu/.

(a)

(b)

Figure 2.5 Interior of BP America's new Government Affairs Office in Washington, D.C., designed by Fox Architects. The 22,000-square-foot building achieved a LEED™ Platinum level.
Source: Fox Architects.

- SSc4.1. Alternative Transportation, Public Transportation Access (NC, CS, Schools): Exemplary performance option re "doubling transit ridership" has been formally incorporated into the reference guide.
- SSc4.3. Alternative Transportation, Low-Emitting and Fuel-Efficient Vehicles (NC, CS, Schools): Defines preferred parking through discounts for spaces as at least 20 percent below the normal price. For NC projects, it defines a compliance path option for a low-emitting vehicle-sharing program.
- SSc5.1. Maximize Open Space (NC, CS, Schools): In the implementation section, there is a clause that states that if a 10-acre site contains 5 acres of greenfield and 5 acres of

previously developed land, site disturbance must be limited in the greenfield area, and native and adapted vegetation must be protected or restored for at least 50 percent (excluding the building footprint) of the previously developed site area.

- SSc7.1. Heat-Island Effect, Non-Roof (NC, CS, Schools): Allows areas shaded by solar panels or other renewable-energy sources (e.g., over parking) to be added to qualifying area calculation.

Water-efficiency categories include:

- WEp1 and WEc2. Water-Use Reduction (NC, CS, Schools): The 20 pecent requirement that was formerly a point is now a prerequisite. Points begin at a 30 percent reduction and rise up to 45 percent for exemplary performance. There is some helpful information about how the EPA WaterSense program requirements relate to the LEED™ requirements.

Energy and atmosphere categories include:

- EAp1 (EAc3). Fundamental (Enhanced) Commissioning of Building Energy Systems (NC, CS, and Schools): Provides much greater detail of the step-by-step process for moving through owner's project requirements, basis of design, and commissioning coordination as well as about who is and is not allowed to act as the commissioning authority.
- EAp2 (EAc1). Minimum (Optimized) Energy Performance (NC, CS, and Schools): Calculations are now based on ASHRAE 90.1-2007 instead of the less stringent 2004 edition, and the required benchmarks have been lowered as well. The v3 prerequisite calls for a 10 percent reduction over the 2007 standard, whereas in v2.2 a 14 percent reduction over the 2004 standard is required. Additional ASHRAE Advanced Energy Design Guide compliance paths have been added for retail, small warehouse buildings, and schools.
- EAc2. On-Site Renewable Energy (NC, CS, and Schools): This credit now offers regular points generating as little as 1 percent and as much as 13 percent of the power needed by the building; exemplary performance is set at 15 percent. There is also a clause that clearly forbids including energy that is generated but not used or energy that is to be sold back to the grid (presumably via net-metering). Furthermore, the owner cannot sell Renewable Energy Credits (RECs) for the power they are claiming as part of the calculations. There is also a documentation requirement in regard to "any incentives that were provided to support the installation of on-site renewable energy systems."
- EAc5.2. Measurement and Verification, Tenant Submetering (CS only): In addition to v2 requirements, the project team must "provide a process for corrective action if the results of the M&V plan indicate that energy savings are not being achieved."

This edition of the reference guide refers to MasterFormat 2004 in its definitions of what is and is not included in the qualified materials for MRc3-7. Instead of the 1997 divisions 2-10 definition, it now cites 2004 divisions 3-10, 31.6, 32.1, 32.3, and 32.9. Division 12 (furniture and furnishings) is still optional.

Materials and resources categories include:

- MRp1. Storage and Collection of Recyclables (NC, CS, and Schools): Clearly indicates that "it may be possible to create a central collection area that is outside of the building footprint or project site boundary" if you can "document how the recyclable materials will be transported to the separate collection area."
- MRc1.1. Building Reuse – Maintain Existing Walls, Floors, and Roof (NC, CS, and Schools): Because additional points are assigned to these credits, the thresholds for achievement have

changed to 55 to 95 percent for NC, 25 to 75 percent for CS, and 75 to 95 percent for schools. Only CS has an exemplary performance option at 95 percent.

- MRc7. Certified Wood: Goes into much greater detail about FSC chain-of-custody reporting requirements and how to report them for LEED™ purposes.

Indoor environmental-quality categories include:

- IEQc4.3. Low-Emitting Materials, Flooring Systems (NC, CS, and Schools): This was formerly the "carpet systems" standard, now expanded to included hard surface flooring. The standard carpets remain the same. FloorScore is used for other surfaces, and any stains or sealants on wood or other materials must comply with IEQc4.1 requirements. An alternative option is to use products meeting CHiPS requirements.
- IEQc5. Indoor Chemical and Pollutant Source Control (NC, CS, and Schools): Metal grates by main entries must now be 10 feet long instead of 6 feet. There is a new requirement to "provide containment for appropriate disposal of hazardous liquid wastes in places where water and chemical concentrate mixing occurs." There are also additional requirements for separating "battery banks used to provide temporary back-up power."
- IEQc7.2. Thermal Comfort, Verification (NC and Schools): There is now a clause that states that earning this credit "is contingent on achieving IEQc7.1, Thermal Comfort, Design."
- IEQc8.1. Daylight and Views, Daylight (NC, CS, and Schools): This credit has abandoned the 2 percent daylight-factor calculation in favor of one that uses a more simple multiplication of glazing visible transmittance and window-to-floor area ratio. There are also now maximum footcandle levels (500 fc) if a simulation is being used. There is a clause that frees you from the maximum levels if you "incorporate view-preserving automated shades for glare control." Finally, there is a fourth "combination" option that allows you to use any of the above to document individual compliant spaces.

2.2.6 Building Certification Model

For LEED™ v3, the project-certification process has moved to the Green Building Certification Institute (GBCI), which is an independent nonprofit organization established in 2007 with the support of USGBC. The revised LEED™ v3 process is an improved ISO-compliant certification process that is designed to grow with the green-building movement. USGBC, however, states that it will continue to administer the development and ongoing improvement of the LEED™ rating system and will remain the primary source for LEED™ and green-building education.

The new LEED™ v3 building-certification infrastructure, based on ISO standards and administered by the Green Building Certification Institute (GBCI), now has 10 new certification bodies assigned to the certification process. These new LEED™ certification bodies include:

- ABS Quality Evaluations, Inc. (http://www.abs-qe.com)
- BSI Management Systems America, Inc. (http://www.bsigroup.com)
- Bureau Veritas North America, Inc. (http://www.us.bureauveritas.com)
- DNV Certification (http://www.dnv.us/certification/managementsystems/index.asp)
- Intertek (http://www.intertek-sc.com)
- KEMA-Registered Quality, Inc. (http://www.kema.com)
- Lloyd's Register Quality Assurance, Inc. (http://www.lrqausa.com)
- NSF-International Strategic Registrations (http://www.nsf.org)

- SRI Quality System Registrar, Inc. (http://www.sri-i.com)
- Underwriters Laboratories-DQS, Inc. (http://www.ul.com/mss)

2.2.7 *What's New?*

LEED™ Online v3 has become a greatly improved system by providing enhanced functionality. For example, the USGBC notes some of the new project-management improvement tools incorporated into version 3, such as:

- Project organization: The ability to sort, view, and group LEED™ projects according to a number of project traits, such as location, design or management firm, etc.
- Team-member administration: Increased functionality and flexibility in making credit assignments, adding team roles, and assigning them to team members. For example, credits are now assigned by team-member name rather than by project role.
- Status indicators and timeline: Clearer explanation of the review and certification process and highlighting of steps as they are completed in specific projects. The system now displays specific dates related to each phase and step, including target dates that each review is to be returned to the customer.
- LEED™ support for certification review and submittals.
- LEED™ Online v3 offers many other enabling features to support the LEED™ certification review process, as well as enhancements to the functionality of submittal documentation and certification forms:
    - End-to-end process support: The new system will guide project teams through the certification process from initial project registration through the various review phases. Furthermore, it will provide assistance to beginners during the registration phase to help them determine the type of LEED™ rating system that is best suited for their project.
    - Improved midstream communication: A mid-review clarification page allows a LEED™ reviewer to contact the project team through the system when minor clarifications are required to complete the review.
    - Data linkages: LEED™ Online v3 automatically fills out fields in all appropriate forms after user inputs data the first time, which saves time and helps ensure project-wide consistency. Override options are available when required.
    - Automatic data checks: New system alerts users when incomplete or required data is missing, allowing user to correct errors before application submission, thus avoiding delays.
    - Progressive, context-based disclosure of relevant content: Upon selection of an option, the new system will simplify process of completing forms by only showing data fields that are relevant to the customer's situation and hiding all extraneous content.

2.3 LEED™ Variants and Other Systems Used Worldwide

Rating and certification systems are required to facilitate the definition of green buildings in the market. They tell us how environmentally sound a building is, clarify the extent to which green components have been incorporated, and identify the sustainable principles and practices that have been employed. There are many different green-building rating systems in use around the world, and each has its pros and cons, depending on the type of certification targeted for a specific building. While many people may agree with the definition of a "green" building as being in the eye of the beholder, rating or certifying a green building helps remove some of that subjectivity. Moreover, rating a green building makes a property more marketable by informing tenants and the public about its environmental benefits and also discloses the additional innovation and effort that the owner has invested to achieve a high-performance building.

Energy and resource conservation are the general logic behind the design and construction of a green-building project. A holistic approach to design means strategic integration of mechanical, electrical, and materials systems, which often creates substantial efficiencies whose complexity is not always apparent. Rating a green building identifies those differences objectively and quantifies their contribution to energy and resource efficiency. Rating buildings can reduce implied risks, because rating systems usually require independent third-party testing of the various elements, which means there is less risk that these systems will not perform as predicted. Likewise, when a building is formally rated or certified, the risk of the project being marketed under the perception that it is green when in fact it is not is much less.

Several examples of major rating systems used in the United States are outlined below:

- LEED™: This rating system is a product of the U.S. Green Building Council and is the most widely applied rating system in the United States for commercial buildings. The LEED™ framework consists of several rating categories, applicable to different points in a building's life cycle, as discussed in other parts of this chapter. Numerous municipalities and government departments, including the General Services Administration (GSA) as well as an increasing number of private investors and owners, have instituted policies requiring LEED™ certification for new construction projects.
- Green Globes™: This rating system (www.thegbi.org/greenglobes) is discussed in the next section. It is basically an interactive, web-based commercial green-building assessment protocol offered by the Green Building Initiative (GBI). It offers immediate feedback on the building's strengths and weaknesses and automatically generates links to engineering, design, and product sources. The system evaluates buildings in seven areas.
- ENERGY STAR®: This is a joint program of the Environmental Protection Agency (EPA) and the U.S. Department of Energy (www.energystar.gov). The program is designed for existing buildings and consists of an Energy Performance Rating System that is free and an online tool that focuses on energy performance. The impacts of other factors such as materials, indoor-air quality, or recycling are not taken into consideration. The system essentially compares the energy performance of a particular building to that of a national stock of similar buildings. Data entered into the ENERGY STAR Portfolio Manager tool will model energy consumption based on a building's size, occupancy, climate, and space type. A minimum of one year of utility-information input is required, after which the property is assigned a rating from 1 to 100. Buildings that acquire a score of 75 or more can apply and receive the ENERGY STAR label.

Many countries have developed their own standards of energy efficiency for buildings. Only one or two of these systems are currently available in the U.S., but they may still prove influential in the emerging green-building industry. Examples include BREEAM in the UK, Green Star in Australia, and BOMA Go Green Plus in Canada.

The Green Building Council of Australia (GBCA) has developed a green-building standard known as Green Star. The Green Star environmental rating system is accepted as the Australian industry standard for green buildings. In three states it has been mandated as a minimum for government office accommodation. The Green Star environmental rating tools for buildings benchmark the potential of buildings based on nine environmental-impact categories. Other standards used include EER (Energy Efficiency Rating) and NABERS (National Australian Built Environment Rating System), which is a government initiative to measure and compare the environmental performance of Australian buildings.

Green Globes and LEED™ are the two main rating systems used in Canada. Established in December 2002, the Canada Green Building Council acquired an exclusive license in 2003 from the USGBC to adapt the LEED™ rating system to Canadian circumstances. The Canadian LEED™ for Homes rating system was released on March 3, 2009. In 1982 Canada implemented the R-2000 standard to promote construction that exceeds the building code to increase energy efficiency and promote sustainability. An optional feature of the R-2000 home program is the EnerGuide rating service, which is available across Canada and which allows homebuilders and homebuyers to measure and rate the performance of their homes and confirm that those specifications have been met.

Regional initiatives based on R-2000 include Energy Star for New Homes, Built Green, Novoclimat, GreenHome, Power Smart for New Homes, and GreenHouse. In March 2006, Canada's first green-building point of service, Light House Sustainable Building Centre, opened in Vancouver, BC, funded by Canadian government departments and businesses to help implement green-building practices and to recognize the economic value of green building as a new regional economy.

In China, there are two sets of national building energy standards, one for public buildings and another for residential buildings. Although China has developed mandatory building energy standards, they are narrow in their scope and lack a strong regulatory framework. Moreover, Ministry of Construction (MoC) enforcement remains problematic, and in 2005 the central government put in place a building-inspection program to monitor the implementation of building energy efficiency. Under this program design institutions, developers, and construction companies will lose their licenses or certificates if they do not comply with the regulations.

The MoC recently introduced the Evaluation Standard for Green Building (GB/T 50378-2006), which is similar in structure and rating process to the USGBC's LEED™ (which itself is also being used). The building energy-consumption data will be collected by MoC and used to assess building performance; a three-star Green Building certificate will be awarded to qualified buildings. The Green Olympic Building Assessment System (GOBAS) is another green-building rating system, developed from Japan's Comprehensive Assessment System for Building Environment Efficiency (CASBEE). High-performance building projects are being supported both by the government and business. WBGC (an SBCI member) has assisted the Ministry of Construction in China to establish the China Green Building Council.

The French government formed six working groups to find ways to redefine France's environment policy. The proposed recommendations were then put to public consultation, leading to a set of recommendations released at the end of October 2007. This process was named "Le Grenelle de l'Environnement," and recommendations were put to the French parliament in early 2008. The six working groups addressed climate change, biodiversity and natural resources, health and the environment, production and consumption, democracy and governance, and competitiveness and employment. These developments are intended to match European and international regulations and frameworks.

In January 2009 the first German standard for the new certificates for sustainable buildings was developed by the DGNB (Deutsche Gesellschaft für nachhaltiges Bauen e.V., or German Society for Sustainable Construction) and the BMVBS (Bundesministeriums für Verkehr, Bau, und Stadtentwicklung, or Federal Ministry of Transport, Building and Urban Affairs).

There are a number of German organizations that employ green-building techniques:

- The Solarsiedlung (Solar Village) in Freiburg, Germany, which features energy-plus houses
- The Vauban development, also in Freiburg
- Houses designed by Baufritz, incorporating passive-solar design, heavily insulated walls, triple-glaze doors and windows, nontoxic paints and finishes, summer shading, heat recovery ventilation, and gray-water treatment systems
- The new Reichstag building in Berlin, which produces its own energy

The Energy and Resource Institute of India plays a key role in developing green-building awareness and strategies in that country. A rating system called GRIHA was adopted by the government of India as the National Green Building Rating System for the country, and measures are being taken to spread awareness. GRIHA aims at ensuring that all types of buildings become green buildings. One of the strengths of GRIHA is that it puts great emphasis on local and traditional construction knowledge and even rates non-air-conditioned buildings as green.

Another organization that is playing an active role in promoting sustainability in the Indian construction sector is the Confederation of Indian Industry (CII). The CII is the central pillar of the Indian Green Building Council or IGBC. The IGBC is licensed by the LEED™ Green Building Standard from the U.S. Green Building Council and is currently responsible for certifying LEED™ new construction and LEED™ core and shell buildings in India. All other projects are certified through the U.S. Green Building Council. There are many energy-efficient buildings in India, situated in a variety of climatic zones (Figure 2.6).

In February 2007 the Indian Bureau of Energy Efficiency (BEE) launched the Energy Conservation Building Code (ECBC), which set energy-efficiency standards for design and construction of any building with a minimum conditioned area of 1000 square meters and a connected demand of power of 500 KW or 600 KVA. On February 25, 2009, the BEE launched a five-star rating scheme for office buildings operated only during the day in three climatic zones, composite, hot and dry, and warm and humid.

Figure 2.6 The Sohrabji Godrej Green Business Centre in Hyderabad, India, in 2001 encouraged the development of green building in the country. This was the first Platinum-rated green building under the LEED™ rating system, outside the U.S., boasting energy savings of 63 percent.
Source: Confederation of Indian Industry.

Israel has recently implemented a voluntary new standard, "Buildings with Reduced Environmental Impact," SI-5281; this standard is based on a point rating system (55 = certified and 75 = excellence), and, together with complementary standards 5282-1, 5282-2 for energy analysis, and 1738 for sustainable products, provides a system for evaluating environmental sustainability of buildings. United States Green Building Council LEED™ rating system has also been implemented on several building projects in Israel, and there is a strong industry drive to introduce an Israeli version of LEED™ in the near future.

In 2001 a joint industrial/government/academic project was created with the support of the Housing Bureau and the Ministry of Land, Infrastructure, Transport, and Tourism (MLIT). This led to the creation of the Japan GreenBuild Council (JaGBC)/Japan Sustainable Building Consortium (JSBC), which in turn created the CASBEE system. CASBEE certification is currently available for new construction, existing buildings, renovations, heat islands, urban development, urban areas plus buildings, and homes.

CASBEE is composed of four assessment tools that correspond to the building life cycle. The collective name for these four tools and expanded tools for specific purposes is the CASBEE Family; they are designed to serve at each stage of the design process. Each tool is intended for a separate purpose and target user and is designed to accommodate a wide range of uses (offices, schools, apartments, etc.) in the evaluated buildings. Likewise, the process of obtaining CASBEE certification differs from

LEED™. The LEED™ certification process starts at the beginning of the design process, with review and comments taking place throughout the design and construction of a project. Though CASBEE's latest version for new construction ranking uses pre-design tools, certification consists mainly of site visits upon the building's completion.

The main organization in Malaysia promoting green practices and building techniques is the Standards and Industrial Research Institute of Malaysia (SIRIM).

The Mexico Green Building Council (MexicoGBC) is the principal organization promoting sustainable-building technology, policy, and best practices. It is an independent nonprofit, non-governmental organization that works from within the construction industry in order to promote a broad-based transition towards sustainability.

The New Zealand Green Building Council was founded July 2005. In 2006 and 2007 several major milestones were achieved, including membership in the World GBC and the launch of the Green Star NZ (office design tool).

The Green Building Council of South Africa was launched in 2007 and has since developed Green Star SA rating tools, based on the Green Building Council of Australia tools, to provide the property industry with an objective measurement tool for green and sustainable buildings and to recognize and reward environmental leadership. Each Green Star SA rating tool reflects a different market sector (e.g., office, retail, multi-unit residential, etc.). Green Star SA Office was the first tool developed and was released in final form (version 1) at the Green Building Council of South Africa Convention and Exhibition '08 on November 2008.

South Africa is in the process of incorporating an energy standard, SANS 204, which aims to provide energy-saving practices as a basic standard in the South African context. Green Building Media, which was launched 2007, has also played an instrumental role in green building in South Africa.

The Association for Environment Conscious Building (AECB) has promoted sustainable building in the UK since 1989. Now under the Energy Performance of Building Directive (EPBD), Europe has required energy certification since January 4, 2009. A mandatory certificate called the Building Energy Rating system (BER) and a certification Energy Performance Certificate (EPC) are needed by all buildings that measure more than 1000 square meters (approximately 10,765 square feet) in all the European nations.

In March 2009 the UK Green Building Council (UK-GBC) also called for the introduction of a Code for Sustainable Buildings to cover all nondomestic buildings, both new and existing. The Code for Sustainable Buildings is owned by the government but developed, managed, and implemented by industry and covers refurbishment as well as new construction. The Welsh Assembly Government planning policy sets a national standard for sustainability for most new buildings proposed in Wales beginning September 1, 2009.

BREEAM (BRE Environmental Assessment Method) is a leading and widely used environmental-assessment method for buildings, setting the standard for best practices in sustainable design and establishing a measure used to describe a building's environmental performance. There have been some dramatic changes to BREEAM 2008, mainly in response to an evolving construction industry and the public agenda, which came into force on August 1, 2008. Some of the major changes include:

- A new two-stage assessment process, design stage and post-construction stage
- Introduction of mandatory credits
- A new rating level, BREEAM Outstanding
- Modification of environmental weightings
- Benchmarks set for CO_2 emissions to align with the new EPC (Environmental Performance Certificate)
- Changes to certain specific credits
- Updated Green Guide Ratings, which will be available online
- Shell-only assessments
- New schemes: BREEAM Healthcare and BREEAM Further Education

The United States has in place several sustainable-design organizations and programs. The most prevalent is the U.S. Green Building Council (USGBC), which is a nonprofit trade organization that promotes sustainability in building design, construction, and operation. The USGBC is best known for the development of the LEED™ rating system and Greenbuild, a green-building conference that promotes the industry.

By September 2008 the USGBC had more than 17,000 member organizations from all sectors of the building industry; it works to promote buildings that are environmentally responsible, profitable, and healthy places to live and work. As discussed here and in Chapter 4, the USGBC, through its Green Building Certification Institute, offers industry professionals the chance to develop expertise in the field of green building and to receive accreditation as green-building professionals.

The National Association of Home Builders, which is a trade association representing home-builders, remodelers, and suppliers to the industry, has formed a voluntary residential green-building program called NAHBGreen (www.nahbgreen.org). This program incorporates an online scoring tool, national certification, industry education, and training for local verifiers. The online scoring tool is free to both builders and homeowners.

The Green Building Initiative is a nonprofit network of building industry leaders working to mainstream building approaches that are environmentally progressive but also practical and affordable for builders to implement. The GBI has introduced a web-based rating tool called Green Globes™, which is discussed in the following section.

The United States Environmental Protection Agency's Energy Star program rates commercial buildings for energy efficiency and provides Energy Star qualifications for new homes that meet its standards for energy-efficient building design.

2.4 The Challenge of Green Globes™

Although there are a number of green-building rating systems in the United States, the two national systems most widely used for commercial structures are LEED™ (Leadership in Energy and Environmental Design) and Green Globes™ (Go Green Plus). The LEED™ Green Building Rating System® is focused on assessing *new* construction high-performance, sustainable buildings. Go Green targets *existing* building owners who want to have a more environmentally friendly building. In this section we will compare Green Globes™ with the LEED™ Rating System. While Green

Globes™ currently has a minute share of certified buildings in the United States (about 55 buildings), that pales compared to LEED™ certified buildings, it is moving energetically forward to increase its market share within the United States. In March 2009 the GBI and the American Institute of Architects (AIA) signed a memorandum of understanding, pledging to work together to promote the design and construction of energy-efficient and environmentally responsible buildings. Likewise, a new memorandum of understanding was signed between ASHRAE and the GBI to work together to accelerate the adoption of sustainability principles in the built environment. Green Globes® rating system is also on track to become the first American national standard for commercial green buildings.

2.4.1 Green Globes™ Emerges to Challenge LEED™

The birth of the Green Globes™ system lies in the Building Research Establishment's Environmental Assessment Method (BREEAM), which was brought to Canada in 1996 in cooperation with ECD Energy and Environment and was initially developed as a rating and assessment system to monitor and assess green buildings in Canada.

The Green Globes™ environmental-assessment and rating system represents more than a decade of research and refinement by a wide range of prominent international organizations and experts. Among the persons that were instrumental to the formation of this project are Jiri Skopek, John Doggart, and Roger Baldwin, who were the principal authors of the BREEAM Canada document. The Canadian Standards Association, that same year, published BREEAM Canada for Existing Buildings. This was followed in 1999 by the formation of the BREEAM Green Leaf eco-rating program, and in 2000 BREEAM Green Leaf took another giant step forward in its development by becoming an online assessment and rating tool under the name Green Globes for Existing Buildings. In the same year, BREEAM Green Leaf for the Design of New Buildings was developed for the Canadian Department of National Defense and Public Works and Government Services Canada. The product underwent a further iteration in 2002 by a panel of experts including representatives from Arizona State University, the Athena Institute, BOMA, and several federal departments in Canada.

During 2002 Green Globes for Existing Buildings went online in the United Kingdom as the Global Environmental Method (GEM), and efforts began to incorporate BREEAM Green Leaf for the Design of New Buildings into the online Green Globes for New Buildings. In Canada, Go Green (for Existing Buildings) is owned and operated by the Canada Building Owners and Manufacturers Association (BOMA). All other Green Globes™ products in Canada are owned and operated by ECD Energy and Environment Canada. BOMA Canada adopted the system in 2004 under the name Go Green Comprehensive (now known as Go Green Plus). The Canadian federal government also later announced plans to adopt Go Green Plus for its entire real-estate portfolio.

Also in 2004 the Green Building Initiative (GBI) purchased the rights to promote and develop Green Globes™ in the United States. The GBI committed itself to refining the system to ensure that it reflects changing opinions and ongoing advances in research and technology and, in so doing, to involve multiple stakeholders in an open and transparent process. To that end, GBI in 2005 became the first green-building organization to be accredited as a standards developer by the American National Standards Institute (ANSI) and started the process to establish Green Globes™ as an official ANSI standard.

Green Globes™ is an environmental-assessment, education, and rating system that is promoted in the United States under the auspices of the Green Building Initiative (GBI), a Portland, Oregon – based nonprofit organization. Canada's federal government has been using the Green Globes™ suite of tools for several years under the Green Globes™ name, and it has been the basis for the Building Owners and Manufacturer's Association of Canada's Go Green Plus program. Adopted by BOMA Canada in 2004, Go Green Plus was chosen by Canada's Department of Public Works and Government Services, which has an estimated 300 buildings in its existing portfolio. The Green Globes™ system has also been used by the Continental Association for Building Automation (CABA) to power a building-intelligence tool called Building Intelligence Quotient (BiQ).

2.4.2 Green Globes™, an Alternative to LEED™

Green Globes™ (Canada) is a product of the Building Research Establishment Environmental Assessment Method (BREEAM), which was developed in the UK. One of BREEAM's creators, ECD Consultants, Ltd., used it as the basis for a Canadian assessment method called BREEAM Green Leaf. The original intention of creating BREEAM Green Leaf was to allow building owners and managers to self-assess the performance of their existing buildings. This was followed by Green Globes™ as a Web-based application of Green Leaf.

The Green Globes™ system is a Web-based interactive green-building performance software tool that came from Canada and competes with the better known, and perhaps more cumbersome and more expensive LEED™ system from the U.S. Green Building Council (a nonprofit organization based in Washington, DC). Green Globes™ was introduced to the U.S. market as an alternative to the U.S. Green Building Council's LEED™® Rating System. The Green Building Initiative (GBI) was established to promote the use of the National Association of Homebuilders' (NAHB) Model Green Home Building Guidelines and has expanded into the nonresidential building market by licensing Green Globes™ for use in the U.S. GBI is supported by a number of industry groups, including the Wood Promotion Network, that object to some provisions in LEED™ and that, as trade associations, are not allowed to join the U.S. Green Building Council.

Green Globes™ was released in Canada in January 2002 and consists of a series of questionnaires, customized by project phase and the role of the user in the design team (e.g., architect, mechanical engineer, or landscape architect). A total of eight design phases are supported. A separate Green Globes™ model for assessing the performance of existing buildings has not been licensed into the U.S. yet. The questionnaires produce design guidance appropriate to each team member and project phase. Green Globes™ users can order a Green Globes™ third-party assessment, upon purchasing a subscription, at any time during or after the completion of the questionnaire. After an online self-assessment is completed and payment is made, a GBI representative contacts the project manager or owner to schedule the assessment and provide the assessor's name and contact information. Completion of the preassessment checklist helps prepare for the assessment process; this can be downloaded from the Green Globes™ Customer Training area.

Formal rating and certification programs are necessary to provide a mechanism to ensure that new-construction project teams or facilities-management staff are fully aware of the environmental impact of design and/or operating decisions. It also offers a visible means to quantify/measure their performance, and allows recognition for their achievements and hard work at the end of the process (Figure 2.7). While investment in environmental/sustainability ratings and certification is currently a choice for the leading proponents of ecological responsibility, emerging standards such as GBI Proposed ANSI Standard 01-200 XP and legislative recognition of green building construction and operation practices may signify and facilitate broader adoption in the future.

Green Globes™ is designed to be cost-effective in that, through its value-added online system and a comprehensive yet streamlined in-person third-party review process, significant savings on consulting fees are made that were normally associated with green certification. There is an annual per-building license fee for use of the online tool as well as a third-party assessment fee. Rates are based on a number of factors including size of project (hectares/acres), number of integrated developments, and location (environmentally sensitive areas). Users can register/subscribe for both the annual license for the online Green Globes™ tools and choose to purchase a third-party assessment (required for certification). Third-party assessor travel expenses are billed separately. Below are some of the costs currently involved with Green Globes™:

- Green Globes Existing Building Rating/Certification Package: $5,270* per building
- Green Globes New Construction Rating/Certification Package: $7,270* per building
- Green Globes CIEB Existing Building one-year subscription $1,000
- Green Globes New Construction one-project subscription $500
- Green Globes New Construction three-project subscription $1,500
- Green Globes New Construction ten-project subscription $2,500
- Green Globes CIEB assessment/rating $3,500*
- Green Globes NC Stage I assessment $2,000
- Green Globes NC Stage II assessment/rating $4,000*
- Green Globes NC Stage I and II assessment/rating $6,000*

Note: A Green Globes™ subscription is required for third-party assessment/certification.
*Pricing for buildings over 250,000 square feet in size or departing significantly from standard commercial-building complexity will be custom-quoted prior to assessment services being performed. Travel for the GBI assessor to/from building location is billed at invoice actual expenses plus 20 percent after the assessment or a flat fee of $1,000 upfront.

As a LEED™ alternative, the Green Globes™ building rating system provides an assessment tool for characterizing a building's energy efficiency and environmental

Figure 2.7 Newell Rubbermaid Corporate Headquarters in Atlanta, GA, measuring 365,000 square feet, which achieved a two Globe rating using the Green Globes New Construction module.
Source: Green Building Initiative.

performance. The system also provides guidance for green-building design, operation, and management. When compared to LEED™, some feel that the appeal of Green Globes™ may be enhanced by the flexibility and affordability the system may provide while simultaneously providing market recognition of a building's environmental attributes through recognized third-party verification. And from a practical and marketing perspective, it should not be necessary to pursue LEED™ certification in order to demonstrate to tenants, customers, clients, and building visitors that a building's owners and management are taking steps to be more environmentally responsible.

According to a 2007 study by the University of Minnesota that compared LEED™ (pre-v3) with Green Globes™, the systems were very similar. For example, "nearly 80 percent of the categories available for points in Green Globes™ are also addressed in LEED™ v2.2 and that over 85 percent of the categories specified in LEED™ v2.2 are addressed in Green Globes™." The study further indicated that LEED™ was characterized as being more rigorous, rigid, and quantitative, whereas Green Globes™, while also rigorous, nevertheless maintained greater flexibility and primarily focused on energy efficiency as a goal. Green Globes™ was also found to be easier to work with, less costly, and less time-consuming than LEED™. The same study concluded that there was only moderate dissimilarity between the two rating standards but that LEED™ has a slightly greater emphasis on materials choices and Green Globes™ has a greater emphasis on energy saving. (Green Globes™ more heavily weights energy systems, up to 36 percent of the total points needed, whereas LEED™ initially limited the energy category to about 25 percent of the total in the rating system.) This has now been appropriately addressed in the LEED™ v3 version.

Of the many buildings that have been evaluated with both systems, in all but two instances the systems generated comparable ratings. The other two buildings were only marginally different. It should be noted here that LEED™ 2009 has addressed many of these issues. In the final analysis, it appears that the primary differences between the two approaches boil down to ease of use and cost.

The University of Minnesota study also concluded: "From a process perspective, Green Globes™ simpler methodology, employing a user-friendly interactive guide for assessing and integrating green-design principles for buildings, continues to be a point of differentiation to LEED™'s more complex system. While LEED™ has introduced an online-based system, it remains more extensive and requires expert knowledge in various areas. Green Globes™ web-based self-assessment tool can be completed by any team member with general knowledge of the building's parameters."

The Green Building Initiative (GBI) currently oversees Green Globes™ in the United States. GBI has also become an accredited standards developer under the American National Standards Institute (ANSI) and is in the process of establishing Green Globes™ as an official ANSI standard.

The ANSI process has always been a consensus-based process, involving a balanced committee of varying interests including users, producers, interested parties, and NGOs who basically conduct a thorough technical review through an ANSI-approved, open and transparent process. The standard continues to be monitored by this committee and will continue to follow ANSI approved rules and procedures for updating the standard.

Surprisingly, neither LEED™ nor Green Globes™ (or Energy Star for that matter) provides continuous, longitudinal monitoring of energy efficiency or building performance. This indicates that building measurements and ratings are concluded on a one-off basis that must then be re-verified later on. This is a significant shortcoming in terms of the practicality of greening existing real estate, since buildings are dynamic and rarely perform in an identical manner week after week.

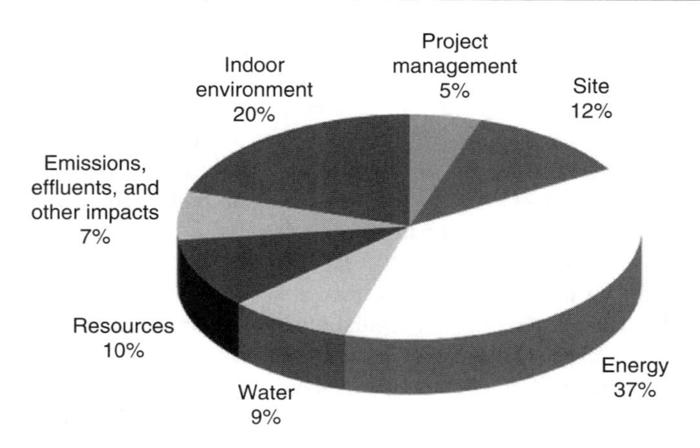

Figure 2.8 Pie chart showing the distribution of points in the Green Globes™ system. *Source*: Building Green LLC.

Green Globes™ generates numerical assessment scores at two of the eight project phases; these are the schematic-design phase and the construction-documents phase. These scores can either be used as self-assessments internally or verified by third-party certifiers. Projects that have had their scores independently verified can use the Green Globes™ logo and brand to promote their environmental performance. The Green Globes™ questionnaire corresponds to a checklist with a total of 1000 points listed in seven categories as opposed to LEED™'s 100 points in seven categories (Figure 2.8).

Green Globes™, however, differs from LEED™ by not holding projects accountable for strategies that are not applicable, which is why the actual number of points available varies by project. For example, points are available for designing exterior lighting to avoid glare and sky glow, but for a project with no exterior lighting a user can select "N/A," which removes those points from the total number available so as not to penalize the project. A rating of one or more Green Globes is applied to projects based on the percentage of applicable points they have achieved. In Canada the ratings range from one to five Green Globes, while in the U.S. the lowest rating was eliminated and the rest adjusted so that the highest rating is four Globes. Ward Hubbell, executive director of GBI, says that the objective was to have something that people are accustomed to, a four-stage system, which is roughly comparable to the four levels of LEED™.

Green Globes™ is reportedly broader than LEED™ in terms of technical content, including points for topics such as optimized use of space, acoustical comfort, and an integrated design process. It is difficult to compare the levels of achievement required to claim points in the two systems, because they are organized differently and also because the precise requirements within Green Globes™ are not publicly available. One of the main attractions of Green Globes™ for industry groups supporting GBI in the U.S. is that it recognizes all the mainstream forest-certification systems, while LEED™ previously referenced only the Forest Stewardship Council's program. Green

Globes™ also awards points for the use of life-cycle assessment methods in product selection, although it doesn't specify how those methods should be applied.

Hubbell also claims that Green Globes™ is on a par with LEED™ with respect to overall achievement levels, and notes, "We did carry out a harmonization exercise with LEED™ – not credit-by-credit; we compared objectives." The actual development of the Green Globes™ system in Canada, as well as its subsequent adaptation for the U.S., has involved many iterations and participation by a wide range of organizations and individuals. Changes originally made to adapt Green Globes™ for the U.S. market do not appear to be substantive – e.g., converting units of measurement, referencing U.S. rather than Canadian standards and regulations, and incorporation of U.S. programs such as the EPA's Target Finder.

Supporters of Green Globes™ tried to block the introduction of LEED™ into Canada but lost a close vote in a committee of the Royal Architectural Institute of Canada that led the creation of the Canada Green Building Council (CaGBC) in 2003. Not surprisingly, Alex Zimmerman, president of CaGBC, has some criticisms of Green Globes™ and notes that in Canada Jiri Skopek, president of ECD Energy and Environment Canada, has been the primary developer of Green Globes™ and in the past was its sole certifier. Zimmerman also says, "While there are more certifiers now, it is not clear who they are, how they were chosen, or who they are answerable to." GBI responded to this criticism in the U.S. by training a network of independent certifiers to verify Green Globes™ ratings; they have access to the report generated by the Green Globes™ website as well as other relevant information such as the project drawings, results of an energy simulation, specifications, and commissioning plan.

Green Globes™ has been advanced as a green certification system in areas with a strong timber-industry lobbying presence because it opposes favoring FSC over SFI forest certification. The consensus is that legislation to encourage green building in states like Virginia and Arkansas is likely to include Green Globes™ in addition to LEED™. Furthermore, a number of federal agencies such as the Department of the Interior are also reportedly considering an endorsement of Green Globes™. It is possible that the presence of Green Globes™ on the American scene has had a beneficial impact on LEED™, perhaps prompting it to improve its rating system and release LEED™ 2009 v3. It is also important to recognize that Green Globes™ can attract a significant following with people who for various reasons are alienated by LEED™ certification's costs and complexity. This must be good for the green-building industry and the environment.

2.4.3 Comprehensive Environmental Assessment and Rating Protocol

The Green Globes™ assessment protocol covers six different areas, with each area having an assigned number of points that are utilized to quantify overall building performance. These are shown in Figure 2.9.

The scoring for the six Green Globes™ categories is based on approximately 150 questions that are completed via the online questionnaire within the Green Globes Tool. There are pop-up "tool tips" embedded within the questionnaire to address frequently

Assessment Category	Points	Description
Energy	350	Performance, Efficiency, Management, CO_2, transportation
Indoor Environment	185	Air Quality, Lighting, Noise
Emissions and Effluents	175	Boilers, Water Effluents, Hazmat
Resources	110	Waste Reduction, Recycling
Environmental Management	100	EMS Documentation, Purchasing, Environmental Awareness
Water	80	Performance, Conservation, Management
TOTAL	**1000**	

Figure 2.9 Table depicting the six different assessment categories of Green Globes™. The table shows a clear emphasis on energy, which takes up more than a third of the total points.

asked questions and add clarifications regarding the input data requirements that will appear during the survey. The time normally required to input data and complete the survey is roughly two to three hours per building; this doesn't include time required to research and gather required information for the survey.

In order to earn a formal Green Globes™ certification, a building must be evaluated by an independent third party that is recognized, trained, and affiliated with the Green Building Initiative. Both new construction and existing buildings can be formally rated or certified within the Green Globes™ system. Buildings that achieve 35 percent or more of the 1000 points possible in the rating system are eligible candidates for a rating of one or more Green Globes. A summary of rating levels and how they relate to environmental achievement can be seen in Figure 2.10.

GBI states that the process for obtaining formal Green Globes™ Certification is fairly straightforward and essentially consists of implementing the following steps:

- Step 1: Purchase a subscription to either Green Globes™ NC or CIEB.
- Step 2: Log in to Green Globes™ at the GBI website with your username and password.
- Step 3: Select the tool you have purchased (NC or CIEB) to go to Green Globes™.
- Step 4: Add a building and enter basic building information.
- Step 5: Use step-through navigation and building dashboard to complete the survey.
- Step 6: Print your report to see your building projected rating and get feedback using automatic reports.
- Step 7: Order a third-party assessment and Green Globes™ rating/certification (if automated report indicates a predicted rating of at least 35 percent of 1000 points).
- Step 8: Schedule and complete a third-party building assessment. Third-party assessment for Green Globes NC occurs in two comprehensive stages: The first stage includes a review of the construction documents developed through the design and delivery process. The second stage includes a walkthrough of the building postconstruction.
- Step 9: Receive the Green Globes™ Rating and Certification.

Green Globes™ is marketed as an economical, practical, and convenient means for obtaining comprehensive environmental and sustainability certification for new or

85–100%	Reserved for select buildings that serve as national or world leaders in reducing environmental impacts and efficiency of buildings.
70–84%	Demonstrates leadership in energy and environmentally efficient buildings and a commitment to continual improvement.
55–69%	Demonstrates excellent progress in reducing environmental impacts by applying best practices in energy and environmental efficiency.
35–54%	Demonstrates movement beyond awareness and a commitment to good energy and environmental efficiency practices.

Figure 2.10 Green Globes™ Rating Levels in the United States.
Source: Green Building Initiative.

existing commercial buildings. It provides a complete, integrated system that has been developed to enable design teams and property managers to focus their resources on the processes of actual environmental improvement of facilities and operations rather than on costly, cumbersome, and lengthy certification and rating processes.

Other advantages of Green Globes™ relative to other rating/certification systems include:

- Projects can be evaluated and informally self-assessed for a low registration fee.
- Consultants are not required for the certification process, thereby reducing costs.
- Submission requirements are generally less complicated than other rating/certification systems.
- Online web tools provide a convenient, proven, and effective way to complete the assessment process.
- Upfront commitment to a lengthy and costly rating/certification process is not required.
- The entire process is fairly rapid, with minimal waiting for final rating/certification.
- The estimated rating number of Green Globes™ that will be achieved is largely known in advance of the decision to pursue certification because the self-assessed score is always available to users.

On February 7, 2008, New Jersey joined Arkansas, Connecticut, Hawaii, Maryland, Minnesota, North Carolina, Pennsylvania, South Carolina, Kentucky, Illinois, and Wisconsin as states that have formally recognized the Green Building Initiative's (GBI) Green Globes™ environmental assessment and rating system in legislation.

3 LEED™ Documentation and Technical Requirements

3.1 General Overview

The Leadership in Energy and Environmental Design (LEED™) Green Building Rating System reflects the U.S. Green Building Council's continuing effort to provide a global standard for what constitutes a "green building." It is also intended to be used as a design guideline and third-party certification tool to improve contractor performance and occupant wellbeing as well as environmental performance and economic returns. The LEED™ Green Building Rating System is essentially a voluntary, consensus-based national standard that aims at developing high-performance, sustainable buildings. Many of the LEED™ programs can be downloaded at no cost from the USGBC website, www.usgbc.org.

As previously mentioned, for a project to be certified by the U.S. Green Building Council (USGBC) under the LEED™ system, it must be registered with the USGBC. According to the USGBC, the entire LEED™ documentation and certification process has been greatly streamlined and has also migrated online (i.e., become paperless) in an effort to simplify the process and reduce the time and cost of achieving LEED™ certification. Project teams are no longer required to submit binders of documentation that often take weeks to prepare but now have the option of submitting 100 percent of their documentation online. LEED™ Online is a user-friendly interface that enables project team members to upload credit templates, track Credit Interpretation Requests (CIRs), manage key project details, contact customer service, and communicate with reviewers throughout the design and construction reviews. With LEED™ Online, all LEED™ information, resources, and support are accessible in a centralized location.

Streamlining the certification process is but one of a series of recently implemented LEED™ process modifications that will simplify the documentation and certification process, making it more user-friendly and easier for project teams to manage without negatively impacting the technical rigor and quality of the LEED™ process. The LEED™ credit requirements remain essentially unchanged. Project teams are still required to verify their achievements through third-party validation and ensure that the building is built as designed.

Recent process refinements are the resultant harvest of continuous market surveys, discussions with various organizations and individuals who apply LEED™ practices, and the recently formed technical partnership between USGBC and Adobe.

DOI: 10.1016/B978-1-85617-691-0.00003-5

The changes to the accreditation process announced by USGBC toward the end of 2008 going forward include the replacement of the LEED™ NC designation with the LEED™ Building Design and Construction (BD+C) designation and the LEED™ CI designation with LEED™ for Interior Design and Construction (ID+C). This change was made in order to maintain alignment with the latest LEED™ 2009 Rating System. Details of the changes to the LEED™ AP program can be found at the Green Building Certification Institute (GBCI) website, www.gbci.org. As of 2009, LEED™ eligibility for certification is available for New Construction, Existing Buildings: Operations and Maintenance, Commercial Interiors, Core and Shell Development, Homes, Schools, Healthcare (pilot stage), Retail (pilot stage), and Neighborhood Development (pilot stage). These categories include subcategories that cover specific building types.

3.2 Credit Categories

There are several ways to denote a building's "greenness." In the U.S. and now in many countries around the world the most widely recognized method for certifying green buildings is through an independent third party, the Green Building Council's program, Leadership in Energy and Environmental Design (LEED™).

As stated earlier, 2009 heralded the implementation of a much-anticipated and urgently needed modification of the United States Green Building Council's LEED™ green-building rating system. The USGBC's latest revamped rating system includes a series of major technical advancements focused on improving energy efficiency, reducing carbon emissions, and addressing additional environmental and human health concerns. This overhaul is the outcome of eight years of user feedback including 7000 comments and Credit Interpretation Rulings from USGBC members and stakeholders as well as years of committee meetings. Brendan Owens, vice president for LEED™ technical development, sums it up nicely: "The conclusion of the balloting process marks the culmination of tireless work done by representatives from all corners of the building industry."

The latest modifications could not have come sooner, as numerous criticisms were being raised about the rating system's inadequacy. For example, Charles Crawford, a San Diego architect and professor at the New School of Architecture and Design, while agreeing to the LEED™ program's importance, believes that the program has significant shortcomings. He says: "The points program they administer is heavily tilted toward rewarding greener technology over basic 'passive' principles. So, for example, you get more points for an energy-efficient air conditioner than you do for a naturally ventilated AC-free building." Another example cited by Crawford is LEED™'s apparent lack of consideration for the vast amounts of fuel expended in overseas transportation. He goes on to say that "A green product manufactured by a multinational overseas and shipped to the U.S. is often worth more points than a product manufactured only a few miles away."

The revised rating system addresses many of these criticisms. The new system aligns the prerequisite/credit structure of each category into a common denominator so that

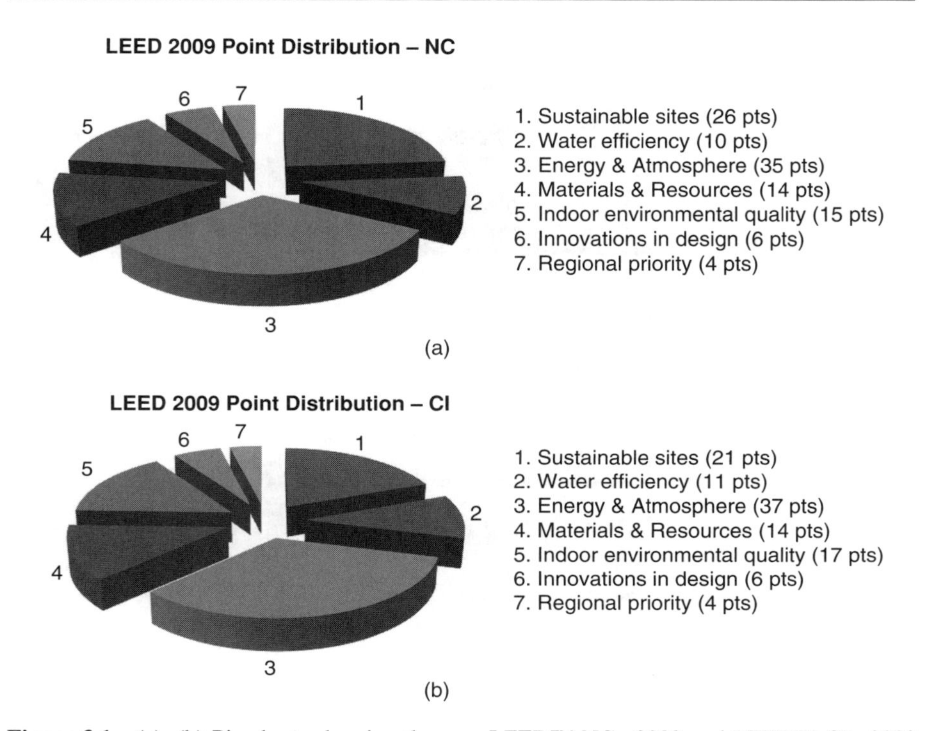

Figure 3.1 (a), (b) Pie charts showing the new LEED™ NC v2009 and LEED™ CI v2009 point-distribution system, which incorporates a number of major technical advancements focused on improving energy efficiency, reducing carbon emissions, and addressing additional environmental and human health concerns.

the same set of credits is typically offered under each rating system (Figure 3.1). Other significant changes to LEED™ 2009 are the incorporation of Regional Priority Credits and revision of the credit weightings. These are discussed further in Chapter 4. The intent of the USGBC is for LEED™ to transition into what the Council terms a "predictable development cycle" to help drive continuous improvements and allow the market to effectively participate in LEED™'s growth and development.

Under the latest LEED™® certification program green-building design focuses on seven basic categories, as shown in Figure 3.2. The total point distribution for the seven categories is 100 points plus 6 possible points for Innovations in Design and 4 possible points for Regional Priority.

3.2.1 Sustainable Sites (SS) Credit Category

The planning and design of green buildings requires a conscious decision from the start to build in an environmentally friendly manner. The intent of the Sustainable Sites category is to encourage good stewardship of the land and the minimizing of any adverse project impacts on surrounding areas during and after construction.

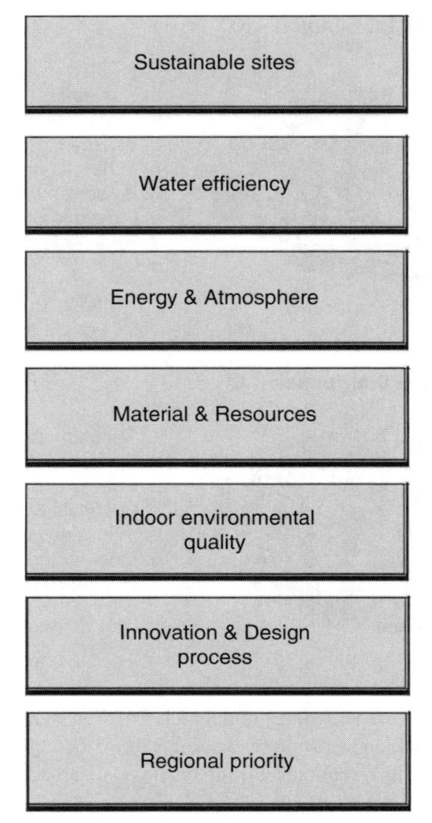

Figure 3.2 The seven LEED™ 2009 credit categories: Sustainable sites, Water efficiency, Energy and Atmosphere, Materials and Resources, Indoor environmental quality, Innovations in design, and Regional priority.

An excellent example of a contemporary building designed to achieve a gold rating in the LEED™ certification program is the San Elijo Lagoon Nature Center in San Diego County, California (Figure 3.3). This nature center reportedly features state-of-the-art, museum-quality exhibits interpreting the natural and cultural resources of the San Elijo Lagoon as well as sustainable-design features of the building. The two-story 5525-square-foot building is constructed from recycled and environmentally friendly building materials including block, steel, rebar, and decorative glass. More than 90 percent of the debris from demolition and construction was recycled, and the building's insulation is made from recycled denim. To ensure good indoor-air quality, the structure was built using few or no volatile organic compounds, materials, paints, and sealers. The center also boasts a green roof that filters pollutants from storm water. High-efficiency lighting is incorporated into the design, and more than half of the building's energy

Figure 3.3 The San Elijo Lagoon Nature Center building in San Diego County, California, incorporates a number of green-building elements including radiant floor heating, green planted roof, recycled cotton insulation, certified renewable lumber, on-site renewable energy in the form of photovoltaics designed to provide 52 percent of energy requirements, natural daylighting and ventilation, storm-water filtering, native vegetation, and recycled water for both irrigation and toilet facilities.

is provided by solar panels. Native, drought-tolerant plants are used for landscaping around the center to reduce water usage, and the building makes use of recycled water for irrigation and restrooms. The nature center also provides bike racks in its parking lot.

To achieve maximum credits under the LEED™ rating system, a building owner should whenever possible consider choosing an appropriate site in terms of urban and brownfield redevelopment. It is important that, upon identifying a site, the building design proceeds to utilize the site conditions to maximum advantage and to minimize site disturbances to the surrounding ecosystem. Careful consideration should be given to the positioning of buildings on the site and their relationship to the site's topography, landscape, and proximity to amenities, as these are key factors that influence a building's impact on the environment and the community. The Sustainable Sites category also encourages the use of alternative transportation options, reducing site disturbance, and storm-water management. Credit is also given to projects that achieve a reduced heat-island effect, reduced light pollution, and incorporation of on-site renewable energy.

A summary of best practices for the Sustainable Sites catergory is provided in the following lists.

For project selection/siting:

- Identify and develop on appropriate sites to minimize the environmental impact of construction.
- Select an urban-infill or a brownfield site to build on as opposed to an undisturbed greenfield, farmland, or wetland. Building footprints are thus minimized and land used efficiently.
- Buildings should preferably be located in dense urban areas to take advantage of existing infrastructure.
- Disruption of existing habitats should be avoided; ample open space is to be provided.
- Protect and retain existing landscaping and natural features. Plants should be selected that have low water and pesticide needs and that generate minimum plant trimmings. The use of compost and mulches should be encouraged to save water and time.
- Recycled-content paving materials and furnishings should be used whenever possible.

For transportation:

- Encourage alternatives to driving to reduce the adverse environmental impact of automobile use; look for a site that is well suited to take advantage of nearby public transportation.
- Provide preferred parking for alternative-fuel vehicles.
- Reduce parking capacity and encourage carpooling.
- Include adequate secure bicycle storage, showers, and changing rooms.

To reduce the heat-island effect:

- Reduce heat islands (heating of a site from heat captured by dark-colored surfaces), as they can disturb local microclimates and increase overall summer cooling loads, leading to increased levels of greenhouse gas and air pollution.
- Use cool roofing strategies such as light-colored or living "green" roofs.
- Parking garages should be located underground when possible.
- Utilize light-colored pavement and shade trees whenever possible.

To reduce light pollution:

- Design lighting to prevent excessive emissions to the night sky, which would negatively affect the comfort of neighbors and the habits of migratory birds. This can be achieved by using low-intensity, shielded fixtures with proper cutoffs in addition to ensuring that lights are turned off or dimmed during nonbusiness hours.

3.2.2 Water Efficiency (WE) Credit Category

Water efficiency can best be defined as achieving a desired result or level of service or the accomplishment of a function, task, or process with the least necessary water. Water efficiency reflects the relationship between the amount of water needed for a particular purpose or function and the amount of water actually used or delivered.

There is a distinct difference between water conservation and water efficiency, although the two terms are frequently used interchangeably. Water conservation can be considered to be water-management practices that improve the use of water resources to benefit people or the environment. It basically constitutes a beneficial reduction in water use, loss, or waste. Water efficiency differs from water conservation in that it focuses on reducing waste. It also means that the emphasis is on reducing waste, not restricting use. Consumers have a major role to play in water efficiency by making small behavioral changes to reduce wastage and choosing more water-efficient products. Water efficiency can be enhanced by various means such as fixing leaking taps, taking showers rather than baths, installing displacements devices inside toilet cisterns, and only running dishwashers and washing machines with full loads.

The intent of the Water Efficiency category is to encourage the thoughtful use of water. LEED™ credits are given to building and landscape designs that reduce the use of potable water for irrigation through the use of a gray-water system that recovers rain water or other nonpotable water. Credits are also given when there is a total reduction in potable-water use in the building through various water-conservation strategies.

Because potable water is becoming a limited resource and is pivotal to sustaining life, economic development, and the environment, water efficiency is an essential element of green-building practices. Water-efficiency and storm-water management practices can be developed and implemented through a collaborative process in many areas and often complement site-related strategies to improve multiple building systems.

Other methods for reducing waste water by increasing water efficiency (and saving on utility bills) are through the implementation of water-efficient plumbing strategies such as ultra-low-flush or dual-flush toilets, low-flow automatic faucets, low-flow shower heads, and low-flow or water-free urinals (Figure 3.4).

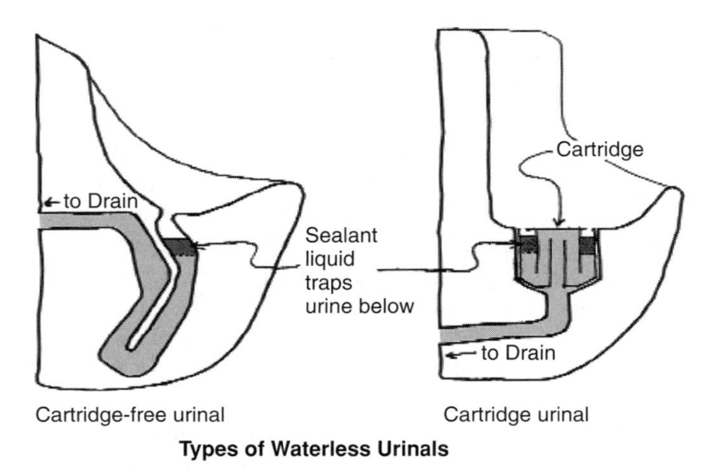

Figure 3.4 Two types of waterless urinals currently on the market. Waterless urinals do not use water and have no flush valves, which can save considerably on water and maintenance. However, local code requirements should be checked prior to installation, as some jurisdictions do not allow waterless urinals.

A summary of best practices for the Water Efficiency catergory is provided in the following list:

- Employ storm-water management techniques to prevent/minimize pollution, sedimentation, and flooding of receiving waters.
- Employ recirculating systems for centralized hot-water distribution and install point-of-use hot-water heating systems for more distant locations.
- Capture and reuse storm water for nonpotable applications such as landscape irrigation and toilet flushing.
- Employ a water budget approach that schedules irrigation for landscaping when possible.
- Encourage use of bioretention systems such as rain gardens and bioswales into landscaping strategies to store and treat storm water.
- Landscaped areas and buildings should be metered separately. Micro-irrigation (which excludes sprinklers and high-pressure sprayers) should be used to supply water in nonturf areas.
- Use irrigation controllers and self-closing nozzles on hoses.
- Eliminate or minimize the need for irrigation through use of native and adaptive species. Where irrigation is unavoidable, techniques should be used that have proven to be more efficient and use less water, such as drip irrigation systems, low-volume/low-angle sprinklers, and nighttime watering.
- Install a vegetated green roof to filter storm water and reduce runoff.
- Incorporate water-efficient plumbing technologies and strategies into the design: for instance, composting toilets and waterless urinals use no water; ultra-high-efficiency plumbing fixtures such as low-flow lavatory faucets with automatic controls are very useful.

3.2.3 Energy and Atmosphere (EA) Credit Category

The first prerequisite in the Energy and Atmosphere (EA) category is that the building undergoes fundamental commissioning to ensure that its systems are operating in the manner they were designed to operate. Other prerequisites for this category include attaining a minimum energy performance (in compliance with ASHRAE 90.1-2004) and eliminating the use of CFC-based (chlorinated fluorocarbon) refrigerants in new building HVACR equipment. LEED™ credits are given for projects that further optimize energy performance, employ renewable energy generated on site, and purchase green power from a Green-e utility program or other acceptable supply source such as Green-e Tradable Renewable Certificates. Additional credits are given for more enhanced commissioning and long-term continuous measurement and verification of building performance.

Green buildings employ high-performance systems and strategies to achieve increased energy efficiency. There are numerous aspects of a building design that can influence its energy performance, and many of the most significant utility cost savings can be realized through implementing energy-efficiency techniques.

ASHRAE 90.1-2004 is the baseline that registered projects are generally required to meet to satisfy the prerequisite requirements. Where local codes are more stringent, these must be followed (and you should include an explanation with the submittal for LEED™ certification). Of particular relevance is the California Title 24 standard, which is more stringent than the new Standard 90.1-2007 and which is accepted without further evaluation. Attaining and surpassing the energy-efficiency levels of the California Title 24 standard is not difficult, yet most projects strive only to meet the standard rather than surpassing it. With regard to this standard, the Energy Commission agreed to a California Building Standards Commission request that all parts of the California Building Standards Code (Title 24), including the Building Energy Efficiency Standards, be effective on August 1, 2009, so that the entire code update will have the same effective date. The 2008 Energy Efficiency Standards have incorporated new measures to reduce energy use and greenhouse-gas emissions.

Implementation of passive-design strategies can have a dramatic effect on building energy performance. Such measures would typically include building shape and orientation, passive-solar design, and the appropriate use of natural lighting. Studies have consistently shown that providing natural lighting has a positive impact on productivity and wellbeing.

ENERGY STAR® Portfolio Manager is an online tool for assessing a building's energy and water consumption. It can be used for a single building or for an entire portfolio. It is a useful benchmarking tool that can help building owners identify areas for efficiency improvements, track the performance of these improvements over time, and compare their building's performance with national averages. It is discussed further in Chapter 4.

A summary of best practices for the Energy and Atmosphere catergory is provided in the following list:

- Correct sizing of HVAC equipment is imperative for the building demand. Effective strategies such as demand-control ventilation, variable-speed pumping, heat recovery, and economizer cycles should whenever possible be incorporated.
- Eliminate all use of CFC and HCFC refrigerants in order to reduce project's potential adverse impact on ozone depletion and global warming.
- Employ high-efficiency ENERGY STAR® – rated appliances, computers, and equipment whenever possible.
- Employ an appropriately sized and energy-efficient heating/cooling system in conjunction with a thermally efficient building shell.
- Minimize the electric loads from lighting, equipment, and appliances. Thus, for example, great savings can be achieved by switching to energy-efficient lightbulbs, which use up to 75 percent less energy and which reportedly have a life cycle 10 times longer than incandescent bulbs.
- Coordinate energy-efficient lighting with daylighting strategies, coupled with installation of high-efficiency lighting systems with advanced lighting controls. Motion and occupancy sensors tied to dimmable lighting controls should be included. Task lighting also helps reduce general overhead light levels.
- Maximize the use of light colors for roofing and wall finish materials. Wall and ceiling insulation should have high R-values. Minimal glass to be used on east and west exposures (because of the low angle of the sun) to reduce heat gain.
- Ensure that proper building-systems commissioning is conducted to verify that the project's energy-related systems are designed, installed, calibrated, and operating properly and as intended.

- Check that the building envelope and systems are designed to maximize energy performance by addressing insulation, glazing ratios, and glass efficiency.
- Consider use of alternative energy sources including on-site power-generation technologies such as solar thermal, photovoltaic panels, and wind turbines. Renewable energy sources are symbolic of emerging technologies for the future. The use of green power should be encouraged by purchasing electricity generated from such renewable sources as wind and solar.
- Employ measurement and verification systems to monitor energy used by the various energy-consuming systems within the building and provide useful data about the building-systems performance while identifying potential maintenance requirements for performance optimization.
- Computer modeling has great potential and is an invaluable tool in optimizing design of electrical and mechanical systems and the building shell.

3.2.4 Materials and Resources (MR) Credit Category

The Materials and Resources category requires that a designated area be provided for the collection and storage of recyclables on the site, preferably during design and operations. The category also gives credit and encourages the use of recycled materials, rapidly renewable materials (that are harvested within a 10-year or shorter cycle), locally manufactured environmentally responsible materials, and certified wood. Credit is also given to projects that reuse parts of the existing building on the site and to projects that implement a construction waste-management plan, thus reducing waste going to landfills.

The CI Reference Guide quite rightly states that "Building materials choices are important in sustainable design because of the extensive network of extraction, processing, and transportation steps required to process them." Indeed, the energy and resources required extend a project's impact far beyond the building itself. This is why careful selection and disposal of materials can immensely benefit a building's environmental impact and promote wellness (Figure 3.5). Toward this end, Penny Bonda and Katie Sosnowchik state in their book *Sustainable Commercial Interiors* that "First and foremost, designers must remember that sustainable materials must have certain attributes: they must be healthy, durable, appropriate, and easily maintained with minimal environmental impacts throughout their life cycle."

A summary of best practices for the Materials and Resources catergory is provided in the following list:

- Locate in an existing building and reuse as much of the building materials as possible.
- Provide easily accessible collection and storage points for recyclable materials.
- Apply construction waste-management operations in building out a tenant space to reduce debris by recycling these materials. Property-management firms such as Cassidy & Pinkard Colliers, for example, require contractors to recycle a minimum of 50 percent of the waste generated on site.
- In building out new office spaces, preference should be given to materials that are harvested and manufactured locally or regionally.
- Preference should be given to materials that contain a high percentage of recycled content.

Figure 3.5 An interior space in Jupiter, Florida, illustrating the impact that a careful selection of furnishings and finishes can have on a project. Diligent choices were made, such as cork for flooring and walls, low-VOC paints, natural lighting, reuse of furnishings, and green cleaning and maintenance of the interior.
Source: Denise Robinette, www.healthylivinginteriors.com.

- Priority should be given to rapidly renewable materials and materials that are reclaimed, salvaged, or refurbished.
- Whenever possible, use wood that has been certified by the Forest Stewardship Council (FSC) or other acceptable organization.

When selecting sustainable building materials and products for a project, start by evaluating some of the important inherent characteristics that could adversely impact the environment. These include zero or low offgassing of harmful air emissions and zero or low toxicity in addition to reused and recycled content, high recyclability, durability, longevity, and local production. Such products also promote resource conservation and efficiency. The use of recycled-content products also helps develop markets for recycled materials that are being diverted from landfills. It is particularly important to achieve LEED™ credits. Thus, the ability to maintain 40 or 60 percent of a building's interior nonstructural components, for example, can earn LEED™ MR Credits.

The use of dimensional planning and other material-efficiency strategies help reduce the amount of building materials needed and cut construction costs. For example, designing rooms on four-foot modules minimizes waste by conforming to standard-sized wallboard and plywood sheets. A further example is the reuse and recycling of construction and demolition materials (say, as a base course for the building foundation); this reduces the materials designated for landfills and cuts costs. Plans should be prepared for managing materials through deconstruction, demolition, and construction.

3.2.5 Indoor Environmental Quality (IEQ) Credit Category

The Indoor Environmental Quality category seeks to ensure that green buildings have, among other elements, optimal lighting, thermal comfort, and healthy indoor-air quality for their occupants. Prerequisites in this category are that the building meets a minimum IAQ performance standard (thus contributing to the comfort and wellbeing of the occupants) and includes environmental tobacco-smoke control. Credits are given for implementing outdoor-air delivery monitoring; installation of CO_2 sensors; increased ventilation effectiveness; indoor-air quality management during construction and prior to occupancy; the use of low-emitting materials, adhesives, paints, and finishes; and allowing occupants to control the systems in their personal workspaces.

Indoor environmental quality is a pivotal component of green buildings. Numerous studies have confirmed the effect of the indoor environment on the health and productivity of building occupants, as outlined in Chapter 7 of this book. Ventilation, thermal comfort, air quality, and access to daylight and views are all cardinal factors that play a critical role in determining indoor environmental quality.

Recent research leaves no doubt that buildings with good overall environmental quality can reduce the rate of respiratory disease, allergy, asthma, and sick-building symptoms in addition to enhancing worker performance. The potential financial benefits of improving indoor environments exceed costs by a factor of between 8 and 14 (Fisk and Rosenfeld, 1998).

Many building materials and cleaning/maintenance products emit toxic gases, including volatile organic compounds (VOCs) and formaldehyde. These gases can have an adverse impact on occupants' health and productivity.

Minimize the potential of indoor microbial contamination by selecting materials resistant to microbial growth and by providing effective drainage from the roof and surrounding landscape. In addition adequate ventilation for all bathrooms, proper drainage of air-conditioning coils, and installation of other building systems to control humidity are required.

A summary of best practices for the Indoor Environmental Quality catergory is provided in the following list:

- Efficient ventilation systems and a high-efficiency, in-duct filtration system help to prevent the development of indoor-air-quality problems and contribute to the comfort and wellbeing of building occupants. Heating and cooling systems that maintain adequate ventilation and proper filtration can have a significant and positive impact on indoor-air quality.
- Facilitate environmental tobacco-smoke control (ETS) by prohibiting smoking within buildings or near building entrances. Outdoor smoking areas should be designated at least 25 feet from openings serving occupied spaces and air intakes.
- Install carbon-dioxide and airflow sensors in order to provide occupants with adequate fresh air when required.
- Use of zero or low-emitting construction materials and interior finish products (i.e., that contain minimal or no VOCs) will improve indoor-air quality. Such materials include adhesives, sealants, paints, carpet and flooring, furniture, composite-wood products, and insulation.
- On-site stored or installed absorptive materials should be protected during construction from moisture damage and particulates through the use of air filters.
- If air handlers are to be used during construction, filtration media with a minimum efficiency reporting value (MERV) of 8 are to be installed at each return grille, as determined by ASHRAE 52.2-1999.
- All filtration media are to be replaced immediately prior to occupancy. Conduct when possible, a minimum two-week flush out with new filtration media with 100 percent outside air after construction ends and prior to occupancy of the affected space. Where building occupants may be exposed to potentially hazardous particulates, biological contaminants and chemical pollutants that adversely impact air and water quality, new air-filtration media with a MERV of 13 or better should be provided.
- Install permanent entryway systems such as grilles or grates to prevent occupant-borne contaminants from entering the building.
- Incorporate design strategies that maximize daylight and views for building occupants.
- Occupants' thermal comfort can be maintained by incorporating adjustable features such as thermostats or operable windows.
- HVAC systems and building envelope should be designed to meet requirements of ASHRAE 55-2004, Thermal Comfort Conditions for Human Occupancy.
- During construction, meet or exceed the recommended design approaches of the Sheet Metal and Air Conditioning Contractors' National Association (SMACNA) IAQ Guidelines for Occupied Buildings Under Construction, 1995, Chapter 3.
- All adhesives used are to meet or exceed the requirements of the South Coast Air Quality Management District (SCAQMD) Rule #1168. Aerosol adhesives should meet the requirements of Green Seal Standard GC-36.

Figure 3.6 Kitchen by EcoFriendly™ Cabinetry using composite woods that have manufacturer guarantees to emit 0 percent VOCs. Avoid use of solvent-based stains or coatings, which are known for emitting formaldehyde.
Source: Executive Kitchens, Inc.

- Paints and coatings used on interior surfaces should comply with VOC limits. Architectural paints and primers for walls and ceilings should comply with Green Seal Standard GS-11; anticorrosive and antirust paints for interior ferrous surfaces should comply with Green Seal Standard GC-03; clear wood finishes, coatings, stains, sealers, and shellacs should comply with SCAWMD Rule 1113.
- All carpet systems must meet or exceed the requirements of the Carpet and Rug Institute's (CRI) Green Label Indoor Air Quality Test Program; all carpet cushion installed should meet requirements of the CRI Green Label Plus Program;
- Composite wood and agrifiber products must be void of added urea-formaldehyde resins (Figure 3.6).

3.2.6 Innovation and Design Process Credit Category

The Innovation and Design Process category provides an opportunity to receive additional points for performance that exceeds LEED™ requirements. A credit may also be achieved by having a LEED™ AP as a principal participant of the project team.

As outlined in the NC and CI reference guides: "Sustainable-design strategies and measures are constantly evolving and improving. New technologies are continually introduced to the marketplace, and up-to-date scientific research influences building-design strategies. The purpose of this LEED™ category is to recognize projects for innovative building features and sustainable building knowledge." It also helps foster creative thinking and research into new areas that have yet to be explored.

Green-building measures cannot attain their goals unless they work as intended. Building commissioning includes testing and adjusting the mechanical, electrical, and plumbing systems to ensure that all equipment meets the design criteria set out by the owner. It also includes instructing the staff on the operation and maintenance of equipment. Over time, building performance can be assured through measurement, adjustment, and upgrading. Proper maintenance is necessary to ensure that a building continues to perform as designed and commissioned.

3.3 Project Documentation, Submittals, and Certifications

The intent of LEED™ certification is to provide independent, third-party verification that a building project meets the highest green-building and performance measures. All LEED™ – certified projects receive a LEED™ plaque, which is a nationally recognized symbol stating that a building is environmentally friendly, responsible, profitable, and a healthy place to live and work – which is why LEED™ certification is sought after by many building owners.

Before LEED™ Online was available, the registration of projects was via a paper certification process. However, applications and documentation for LEED™ project certification can now be submitted completely online in an easy-to-use interface featuring Adobe LiveCycle® technology (see https://LEED™ online.usgbc.org/). The new facility of LEED™ Online, as noted earlier in this chapter, means that project team members are now able to upload credit templates, track Credit Interpretation Requests (CIRs), manage key project details, contact customer service, and communicate with reviewers throughout the design and construction reviews.

The LEED™ certification process consists of five basic stages: registering the project, integrating LEED™ requirements, obtaining technical support, documenting the project for certification, and the certification notification.

3.3.1 Step 1: Register the Project

Registering a LEED™ project is the first step toward earning LEED™ certification and should be completed as early as possible, preferably during schematic design. Registration can be completed by submitting an online registration form via the USGBC website, which also provides clear instructions on the information required. Registration essentially consists of inputting project contact information and other relevant information such as project name, location, square footage, site area, and building type; a brief narrative description; and a preliminary LEED™ credit scorecard. Images can be uploaded for inclusion with the project posting on the USGCB website list of all LEED™-registered projects. Architects and owners can choose to withhold registration data from public view if they have any confidentiality concerns. Upon registering, project teams receive information, tools, and communications that will help guide them through the certification process.

An important advantage of early registration – i.e., during early phases of project design – is that it ensures maximum potential for achieving the targeted certification. Another advantage is that it establishes a point of contact with USGBC and provides access to essential information, software tools, and communications. Furthermore, registration allows access to a database of existing Credit Interpretation Requests and Rulings as well as to the four sections of online Credit Templates:

1. Template status
2. Template management
3. Required documents
4. Documentation status

The LEED™ registration fee depends on member status. It is currently a flat fee of $450 for USGBC members and $600 for nonmembers and is paid at the time of registration.

Fees for LEED™ certification vary and are based on the rating system that the project is being considered under (e.g., NC, CI, EB, CS, etc.), project type, project size, member status, and level of review required; review can include construction-phase drawings, design-phase drawings, or both. This fee is paid when the project team submits documentation for review. Certification fees are waived for projects that receive Platinum LEED™ certification; they will receive a rebate for all fees paid at the completion of the certification process. The table on the next page, provided by the USGBC (Figure 3.7), outlines current rates.

3.3.2 Step 2: Integrate LEED™ Requirements

To earn LEED™ certification, a project must satisfy all LEED™ prerequisites and earn a minimum number of points outlined in the LEED™ Rating System under which it is registered. Architects use the LEED™ scorecard early on to reveal the project's potential and to explore possibilities for integrated solutions. When a proposed design solution can be used to achieve or contribute to more than one credit, that usually suggests that it is an effective solution. Integrated design typically mandates that appropriate time and effort are exerted to carry out the quantifiable performance and cost analysis during the earliest design phase in orderto save time, money, and effort in later phases. It should be noted that the LEED™ 2009 rating systems differs in several aspects (such as credit weightings and the introduction of regional priority) from earlier LEED™ rating-system versions. This will be clarified throughout the book.

3.3.3 Step 3: Obtain Technical Support

Obtaining technical support during the design and certification process for LEED™ registered projects is facilitated by the allocation of two free requests for a Credit Interpretation Ruling (CIR) on any technical or administrative questions that may arise during the design phase and that may require clarification. The CIRs provide a database

	Less than 50,000 Square Feet	50,000–500,000 Square Feet	More than 500,000 Square Feet	Appeals (if applicable)
LEED™ for New Construction, Commercial Interiors, Schools, and Core and Shell Full Certification	**Fixed Rate**	**Based on Square Footage**	**Fixed Rate**	**Per Credit**
Design Review				
Members	$1,250.00	$0.025/sf	$12,500.00	$500.00
Nonmembers	$1,500.00	$0.030/sf	$15,000.00	$500.00
Expedited fee*	$5,000.00 regardless of square footage			$500.00
Construction Review				
Members	$500.00	$0.010/sf	$5,000.00	$500.00
Nonmembers	$750.00	$0.015/sf	$7,500.00	$500.00
Expedited fee*	$5,000.00 regardless of square footage			$500.00
Combined Design and Construction Review				
Members	$1,750.00	$0.035/sf	$17,500.00	$500.00
Nonmembers	$2,250.00	$0.045/sf	$22,500.00	$500.00
Expedited fee*	$10,000.00 regardless of square footage			$500.00
LEED™ for Existing Buildings	**Fixed Rate**	**Based on Square Footage**	**Fixed Rate**	**Per Credit**
Initial Certification Review				
Members	$1,250.00	$0.025/sf	$12,500.00	$500.00
Nonmembers	$1,500.00	$0.030/sf	$15,000.00	$500.00
Expedited fee*	$10,000.00 regardless of square footage			$500.00
Recertification Review**				
Members	$625.00	$0.0125/sf	$6,250.00	$500.00
Nonmembers	$750.00	$0.015/sf	$7,500.00	$500.00
Expedited fee*	$10,000.00 regardless of square footage			$500.00
LEED™ for Core and Shell: Precertification	**Fixed rate for All Projects**			**Per Credit**
Members	$2,500.00			$500.00
Nonmembers	$3,500.00			$500.00
Expedited fee*	$5,000.00			$500.00

* In addition to regular review fee.

** The Existing Building Recertification Review fee is due upon submission of an application for recertification review. Before submitting, customer service (LEEDinfo@usgbc.org) should be contacted to receive a promotion code.

Figure 3.7 Table outlining current LEED™ certification rates.
Source: USGBC.

of requests or rulings that are available on the USGBC website to assist CIR customers in understanding LEED™ credits and how they may apply to their specific projects and how technologies and strategies may be successfully applied to earn points. This can greatly assist project teams during the certification process, especially when it is unclear whether or not a particular strategy applies to a given credit; a CIR can be submitted and the resultant ruling will determine the suitability of the approach. It should be emphasized that CIR rulings will not guarantee or award any credits – they merely provide specific information and guidance regarding applicability. The fee for each CIR submittal is $220.

The USGBC strongly advises that, before submitting a CIR request, project teams should:

1. Review the intent of the prerequisite credit in question to ensure that the project satisfies that intent.
2. Check the appropriate LEED™ Reference Guide to see if it contains the required answers.
3. Review the large online CIR database for previous CIRs logged by other projects on relevant credits to see if any listed CIRs address the raised issues.
4. If a satisfactory answer cannot be found, consider contacting LEED™ customer service to confirm whether it warrants a new CIR.

If a new CIR is warranted, it should be submitted via the LEED™ Online form. When submitting a new CIR, there are several points to remember:

1. Do not include the credit name of your contact information, as the database automatically tracks this information.
2. Confidential information should not be included, as the submitted text will be posted on the USGBC website.
3. The CIR should not be formatted as a letter. Only the inquiry and essential background information should be included.
4. Each CIR should refer to only one LEED™ credit or prerequisite (unless there is technical justification to do otherwise). The CIR should preferably be presented in the context of the credit intent.
5. The CIR inquiry should only include essential project strategy and relevant background and/or supporting information.
6. Your CIR submission text should not exceed 600 words or 4000 characters including spaces.
7. Submissions of plans, drawings, cut sheets, or other attachments are not permitted.
8. Note that CIRs can be viewed by all USGBC members, nonmembers with registered projects, and workshop attendees.
9. CIRs can only be requested by LEED™ Registered Project Team Members.

It is advisable to proofread the CIR text for clarity, readability, and spelling before submission. Credit language or achievement thresholds cannot be modified through the CIR process. In cases where the Technical Advisory Group cannot satisfactorily address the inquiry at hand, the USGBC states that "the Council reserves the right to circulate the interpretation request to the LEED™ Steering Committee and/or relevant LEED™ Committees as required." Should the project team that submitted the CIR disagree with the credit-interpretation ruling, the submitter has recourse to appeal. The CIR and ruling must be submitted with the LEED™ application to ensure an effective credit review.

Furthermore, once registered, project team members are given access to the electronic LEED™ "letter template" file that is now in place and is designed to help streamline the certification process. These templates provide the primary means for documenting key LEED™ credit-certification data and represent the core of a LEED™ certification submittal.

3.3.4 Step 4: Documenting the Project to be Certified

The USGBC has transformed this process into a paperless one using online submissions. This has made it possible for teams to upload and access documents on a Web location designated for their project.

The USGBC has established a two-phase application process allowing LEED™ certification documentation to be executed in two stages: a preliminary submission, which may be submitted for comment, after which it is revised and then resubmitted; and a final submission. Preliminary certification submissions should include two copies of the letter template file on CD-ROM and two three-ring binders containing the project's LEED™ scorecard, project narrative, letter templates, illustrative drawings and photographs, and any other relevant backup documentation required for all targeted credits, tabbed by credit. The USGBC also allows complete submittal in one phase (design and construction submittals together).

LEED™ Letter Templates consist of preformatted submittal sheets that are required for documentation of each LEED™ prerequisite and credit in the USGBC certification process. The Letter Templates outline the specific project data needed to demonstrate achievement of the LEED™ performance requirements and include calculation formulas where applicable. Sample Letter Templates are available for download from the USGBC website for review purposes only in a nonexecutable format. Participants in projects registered with the USGBC may access fully executable Templates through LEED™ Online. In addition USGBC template examples and LEED™ for New Construction and Major Renovation 2009 Project Scorecard can be found in the Exhibits section of this book.

The following is pertinent information required to be submitted online with the application for LEED™ certification so that the application process can proceed:

1. The LEED™ rating system under which you are submitting (NC, CI, EB, CS, etc.)
2. Project contact information, type, size, number of occupants, and anticipated date of construction completion (forms on USGBC website)
3. Project narrative including descriptions of at least three project highlights
4. LEED™ Project Checklist, which should include project prerequisites, credits, and total anticipated score (samples of LEED™ Project Checklist Templates in Exhibits)
5. Copies of latest LEED™ Letter Templates and supporting documentation (samples of LEED™ Letter Templates in Exhibits)
6. Complete list of all CIRs used, including dates of applied rulings
7. Drawings, photos, and diagrams (8.5 × 11 or 11 × 17 inches) that illustrate and explain the project, including:
    - Site plan
    - Typical floor plan
    - Typical building section

- Typical or primary elevation
- Photo or rendering of project as required

8. Payment of the appropriate certification fees, payable in different stages (i.e., design and construction phase); the application will not be reviewed by the USGBC until payment is fully processed.

Upon completing the design phase and after the USGBC receives the preliminary project submittal, it performs a technical review of the documentation and issues a Preliminary LEED™ Review. This review essentially assesses the initial status of the targeted credits and includes relevant comments. Credits are not actually awarded after the design phase. However, the USGBC will mark each credit and place it into one of four categories: likely to be earned, denied, audited, or pending. The pending category includes technical advice to project teams regarding clarifications that will be required in the final submission. The audited category will indicate additional documentation that will be required. Project teams are then required to respond by submitting a final, clarified, and updated compilation of the project's documentation.

3.3.5 Step 5: Receipt of Certification

Following the second submission, the USGBC issues a Final LEED™ Review report that outlines the final status of all targeted credits, along with the project's level of certification. Project teams have the right to appeal if they disagree with the final assessment.

The new LEED™ process innovations will facilitate the documentation and certification process and render it more user-friendly without diminishing the technical rigor and quality that the community has come to expect. The LEED™ credit requirements themselves have not changed, and project teams are still required to verify their achievements through third-party validation and documentation to ensure that the building is built as it was designed.

Other recently introduced LEED™ process refinements, such as the two-part documentation submittals, reflect the manner in which project teams work. It gives design teams an interim opportunity to modify design documents prior to commencing construction, and ensure that the project is on track for its certification goals. Likewise, with the revised instruments of service procedure, documentation requirements have been aligned with existing instruments of service to reduce additional project documentation. The USGBC has also introduced a "building in a feedback loop," which provides improved customer service throughout the LEED™ process in addition to USGBC's continued implementation of procedures to make customer feedback and interaction an integral part of that process.

A project team has the option to appeal a ruling (within 25 days after Final LEED™ Review) if in their opinion sufficient grounds exist to overturn a credit denial in the Final LEED™ Review. The cost of an appeal is $500 per credit. Appeal submittals are all done via LEED™ Online.

Because a different review team will undertake the appeal, the following is required to be submitted with the appeal:

- LEED™ registration information, including project contact, project type, project size, number of occupants, date of construction completion, etc.
- An overall project narrative including at least three project highlights
- The LEED™ Project Checklist Scorecard indicating project prerequisites and credits and the total score for the project
- Drawing and photos illustrating the project, including:
  - **a.** Site plan
  - **b.** Typical floor plan
  - **c.** Typical building section
  - **d.** Typical or primary elevation
  - **e.** Photo or rendering of project
  - **f.** Complete list of all CIRs used
- Original, resubmittal and appeal submittal documentation for only those credits that are being appealed, including narratives.

3.4 Greening Your Specifications

Construction documents typically consist of working drawings and specifications and are essential to convey the building design concept to the contractor. They provide the contractor with the necessary information to bid and build a project. The more accurately a concept is conveyed, the more likely it is to be realized. It is important, therefore, that the building specifications be an integral part of the written documents and that they go hand in hand with the drawings; they describe the materials to be used as well as the methods of installation. They also prescribe the quality and standards of construction required to be achieved on the project.

Thus, in order to facilitate communication of the building design concept, the construction industry has standardized the format for construction documents. The working drawings describe the location, size, and quantity of materials, whereas the specifications (the written documents that accompany the working drawings) describe the quality of construction. For example, if a working drawing shows a plaster wall, the specifications would include the description of the plaster mix, lath and paper backing, and finishing techniques. To do this more effectively, several standard formats were developed. However, the most widely used today is the standard organizational format for specifications developed by the Construction Specifications Institute (CSI), which is now used by manufacturers, architects, engineers, interior designers, contractors, and building officials throughout the United States for construction specifications in building contracts. The obvious purpose of this format is to assist the user in locating specific types of information.

Likewise, when specifying green-building materials, it is best to follow the Construction Specifications Institute's (CSI) *MasterFormat*™, as most specifications are organized according to it. Moreover, green specifications can be used to benchmark the efficacy of other environmental specifications. Likewise, environmental goals for a specific project can easily be implemented into CSI *MasterFormat*™. There is also a wealth of information on greening your specifications on the Internet. The EPA format in particular is designed to help supplement project specifications.

Furthermore, the EPA team will reportedly assist in the development and modification of project specifications to meet LEED™ credit requirements. But in order to achieve this, a clear understanding is necessary of how the specifications can best be used as a proactive mechanism to assist in procuring materials that are environmentally friendly and in collecting required LEED™ information from subcontractors and suppliers.

BuildingGreen.com has extensive information on sustainable design based on the CSI *MasterFormat*™ (2004) hierarchy. The specifications are general guidelines as to product selection and installation and may not be appropriate for a specific project, which is why before using the BuildingGreen.com Guideline Specifications the reader should read the disclaimer. Further information regarding *GreenSpec*® can be found on the BuildingGreen.com website, http://www.buildinggreen.com/menus/divisions.cfm.

While *GreenSpec*® is organized according to the Construction Specifications Institute (CSI) *MasterFormat*™ (2004) structure, the seventh edition of *GreenSpec*® introduced an entirely new approach to guideline specifications. The previous guideline specifications for a range of sections throughout the various divisions have been replaced with a much more comprehensive set of guideline specifications for four sections in Division 1 only. These sections are organized in the new *MasterFormat*™ 2004 structure.

These Division 1 sections are adapted from drafts of specifications that were developed by green-building consultants 7group and BuildingGreen for the U.S. Environmental Protection Agency's Research Triangle Park Campus. They contain a significant amount of new material and product-specific guidance in addition to updated language on other topics. It is intended that these guideline specifications be modified as needed for new developments, retrofits, and maintenance. They are organized into four basic Division 01 sections:

- 01 74 19 Construction Waste Management
- 01 81 09 Testing for Indoor Air Quality
- 01 81 13 Sustainable Design Requirements
- 01 91 00 General Commissioning Requirements

These four sections cumulatively provide an overview of sustainable-design requirements that may be applied to a wide variety of projects. When these sections are applied in actual project specifications, specific requirements must be inserted throughout the construction documents to ensure compliance with the sustainable-design intent.

In the revised *GreenSpec*® format the product listings in each division are divided into separate sections, following, as much as possible, the *MasterFormat*™ system. Each section starts with a summary of environmental considerations relating to products in that section. This is followed by an alphabetical listing of the products (by company name), with full contact information and brief descriptions. In some cases, the *MasterFormat*™ numbering had to be modified slightly to better fit the green products covered.

Below is the latest *GreenSpec*® Guideline Specifications based on the CSI classification menu:

- 01: General Requirements
- 02: Existing Conditions
- 03: Concrete
- 04: Masonry
- 05: Metals
- 06: Wood, Plastics, and Composites
- 07: Thermal and Moisture Protection
- 08: Openings
- 09: Finishes
- 10: Specialties
- 11: Equipment
- 12: Furnishings
- 13: Special Construction
- 14: Conveying Equipment
- 22: Plumbing
- 23: Heating, Ventilating, and Air Conditioning (HVAC)
- 26: Electrical
- 27: Communications
- 28: Electronic Safety and Security
- 31: Earthwork
- 32: Exterior Improvements
- 33: Utilities
- 34: Transportation
- 35: Waterway and Marine Construction
- 42: Process Heating, Cooling, and Drying Equipment
- 44: Pollution Control Equipment
- 48: Electric Power Generation

You can get more detailed information by simply clicking on one of the above divisions on the BuildingGreen website. For example, by clicking Division 22: Plumbing, you would get additional information as show below:

22 01 40: Operation and Maintenance of Plumbing Fixtures (4 products)
22 07 00: Plumbing Insulation (1 product)
22 11 16: Domestic Water Piping (4 articles, 1 product)
22 13 00: Sanitary Sewerage, Facilities (1 article, 10 products)
22 13 16: Sanitary Waste and Vent Piping (2 articles, 8 products)
22 14 00: Facility Storm Drainage (1 product)
22 16 00: Graywater Systems (3 articles, 9 products)
22 32 00: Domestic Water Filtration Equipment (6 products)
22 34 01: Fuel-Fired Residential Water Heaters (12 articles, 27 products)
22 34 02: Fuel-Fired Commercial Water Heaters (1 article, 34 products)
22 35 00: Domestic Water Heat Exchangers (5 articles, 10 products)

22 41 13: Residential Toilets (12 articles, 49 products)
22 41 14: Composting Toilet Systems (7 articles, 8 products)
22 41 15: Residential Urinals (17 articles, 1 product)
22 41 16: Residential Lavatories and Sinks (2 articles, 1 product)
22 41 39: Residential Faucets, Supplies, and Trim (12 articles, 15 products)
22 41 42: Residential Showerheads (12 products)
22 42 00: Commercial Plumbing Fixtures (8 articles)
22 42 13: Commercial Toilets (25 articles, 22 products)
22 42 14: Commercial Composting Toilet Systems (7 articles, 4 products)
22 42 15: Commercial Urinals (19 articles, 11 products)
22 42 39: Commercial Faucets, Supplies, and Trim (13 articles, 21 products)
22 42 42: Commercial Showerheads (1 article, 11 products)
22 42 43: Flushometers (9 articles, 5 products)
22 47 16: Pressure Water Coolers (1 article, 2 products)
22 51 00: Swimming Pool Plumbing Systems (5 products)

For designers seeking LEED™ certification or who wish to track their project's performance against LEED™, the specifications include details on how LEED™ requirements relate to the expressed requirements. Many LEED™ credits may not be addressed directly in the Guideline Specifications primarily because attaining those credits is determined by choices made in site selection or design and are not affected by product choices or other activities governed by these sections. The responsibility lies with the designer to ensure that any such credits have been satisfactorily addressed in the design and construction process.

Earlier versions of LEED™ required the submittal of extensive documentation from contractors and subcontractors to verify compliance with credit requirements. However, with the shift to online submissions, documentation requirements have been dramatically reduced.

For projects pursuing LEED™ certification, the contractor should be provided with a LEED™ Submittal Form for each LEED™ credit that the contractor is to provide documentation. The contractor would then complete the form and attach any additional documentation to it. Project managers sometimes link receipt of the completed forms to payment requests from the contractor at appropriate points in the construction process. In addition, there may be submittals required for LEED™ or for the client that are not typically within the scope of the specifications document.

Below is a sample guidance document that is based on the Whole Building Design Guide – Federal Green Construction Guide for Specifiers. It consists of sample specification language intended to be inserted into project specifications on this subject as appropriate for "greening" your specifications. Certain provisions, where indicated, are required for U.S. federal agency projects. Sample specification language is numbered to clearly distinguish it from advisory or discussion material. Each sample is preceded by identification of the typical location in a specification section where it would appear using the *SectionFormat*™ of the Construction Specifications Institute; the six-digit section number cited is per CSI *Masterformat*™ 2004 (The previous CSI

Masterformat™ 1995 used a five-digit section number.) For a more complete set, visit the Whole Building Design Guide website at: http://fedgreenspecs.wbdg.org.

Section 03 40 00: Precast Concrete

Specifier Note

Resource Management: Plant fabrication handles raw materials and byproducts at a single location that typically allows greater efficiency and better pollution prevention than job-site fabrication.

Aggregates for use in concrete include normal sand and gravel, crushed stone, expanded clay, expanded shale, expanded slate, pelletized or extruded fly ash, expanded slag, perlite, vermiculite, expanded polystyrene beads, processed clay, diatomite, pumice, scoria, or tuff.

Architectural items (e.g., planters, lintels, bollards) fabricated from lightweight and recycled content aggregates are available. The quantity and type of recycled materials vary from manufacturer to manufacturer and include: cellulose, fiberglass, polystyrene, and rubber.

Autoclaved aerated concrete (AAC) is a type of lightweight precast concrete prevalent in Europe, Asia, and the Middle East and recently available through manufacturing facilities in the United States. It is made with portland cement, silica sand or fly ash, lime, water, and aluminum powder or paste. The aluminum reacts with the products of hydration to release millions of tiny hydrogen-gas bubbles that expand the mix to approximately five times the normal volume. When set, the AAC is cut into blocks or slabs and steam-cured in an autoclave.

Toxicity/IEQ: Refer to Section 03 30 00: Cast-In-Place Concrete. Precast concrete generally requires less portland cement per volume of concrete for similar performance due to better quality control.

Performance: Performance is more predictable in precast operations since more exact dimensions, placement of reinforcing, and surface finishing can be obtained. Precast concrete can be fabricated with continuous insulation. AAC is significantly lighter (about 1/5 the weight of traditional concrete) than normal concrete and can be formed into blocks or panels. Lighter-weight concretes generally have greater fire and thermal resistance but less strength than traditional normal-weight concrete. A full range of lightweight concretes is available, and strength and weight are determined by the aggregates used.

Part 1: General

1.1 Summary

A. This section includes:
   1. Autoclaved Aerated Concrete (AAC).

B. Related sections:
   1. 03 30 00: Cast-In-Place Concrete.

1.2 Submittals

A. Product data. Unless otherwise indicated, submit the following for each type of product provided under work of this section:

Specifier Note

Green building rating systems often include credit for materials of recycled content. USGBC-LEED™ v3, for example, includes credit for materials with recycled content, calculated on the basis of pre-consumer and post-consumer percentage content and it includes credit for use of salvaged/recovered materials (Attention should be give to the latest LEED™ requirements and recommendations).

Green Globes US also provides points for reused building materials and components and for building materials with recycled content.

1. Recycled Content:
  a. Indicate recycled content; indicate percentage of pre-consumer and post-consumer recycled content per unit of product.
  b. Indicate relative dollar value of recycled content product to total dollar value of product included in project.
  c. If recycled content product is part of an assembly, indicate the percentage of recycled content product in the assembly by weight.
  d. If recycled content product is part of an assembly, indicate relative dollar value of recycled content product to total dollar value of assembly.

Specifier Note

Specifying local materials may help minimize transportation impacts; however, it may not have a significant impact on reducing the overall embodied energy of a building material because of efficiencies of scale in some modes of transportation.

Green-building rating systems frequently include credit for local materials. Transportation impacts include: fossil-fuel consumption, air pollution, and labor.

USGBC LEED™ includes credits for materials extracted/harvested and manufactured within a 500-mile radius from the project site. Green Globes™ US also provides points for materials that are locally manufactured.

2. Local/Regional Materials:
  a. Sourcing location(s): Indicate location of extraction, harvesting, and recovery; indicate distance between extraction, harvesting, and recovery include the project site.
  b. Manufacturing location(s): Indicate location of manufacturing facility; indicate distance between manufacturing facility and the project site.
  c. Product value: Indicate dollar value of product containing local/regional materials; include materials cost only.
  d. Product component(s) value: Where product components are sourced or manufactured in separate locations, provide location information for each component. Indicate the percentage by weight of each component per unit of product.

B. Submit environmental data in accordance with Table 1 of ASTM E2129 for products provided under work of this section.

C. Documentation of manufacturer's take-back program for (units, full and partial) (packaging) (xxxx); include the following:

1. Appropriate contact information
2. Indicate manufacturer's commitment to reclaim materials for recycling and/or reuse.
3. Limitations and conditions, if any, applicable to the project

Part 2: Products

Specifier Note

EO 13423 includes requirements for federal agencies to use "sustainable environmental practices, including acquisition of bio-based, environmentally preferable, energy-efficient, water-efficient, and recycled-content products."

Specifically, under the sustainable-building requirements per Guiding Principle #5: Reduce Environmental Impact of Materials, EO13423 directs federal agencies to "use products meeting or exceeding EPA's recycled-content recommendations" for EPA-designated products and for other products to "use materials with recycled content such that the sum of postconsumer recycled content plus one-half of the preconsumer content constitutes at least 10 percent (based on cost) of the total value of the materials in the project."

2.1 Materials

Load-bearing and non-load-bearing AAC elements: Comply with ASTM C1386.

Part 3: Execution

3. X Site Environmental Procedures

Waste management: As specified in Section 01 74 19: Construction Waste Management and as follows:

1. Broken, waste AAC units: May be used as nonstructural fill (if approved by architect/engineer)
2. Coordinate with manufacturer for take-back program; set aside (scrap) (packaging) (xxxx) to be returned to manufacturer for recycling into new product.

4 LEED™ Professional Accreditation, Standards, and Codes

4.1 Overview

From the outset, LEED™ Professional Accreditation was designed to distinguish building professionals with the knowledge and skills to successfully steward the LEED™ certification process and sustainable design in general. LEED™ has indeed made some dramatic advances in recent years; toward this end the LEED™ Professional Accreditation program was transitioned in 2008 to the Green Building Certification Institute (GBCI) to manage. The GBCI, with the support of the USGBC, now handles exam development and delivery to ensure objective, balanced management of the credentialing program (Figure 4.1).

In an effort to capitalize on the existing market momentum the USGBC has significantly revamped and reorganized the existing LEED™ Rating Systems. LEED™ v3 is the latest version of the LEED™ green-building certification system. It was launched on April 27, 2009, and builds on the fundamental structure and familiarity of the existing rating system while providing a new structure to ensure that LEED™ v3 incorporates new technology and addresses the most urgent priorities. The USGBC states that LEED™ 2009 has been greatly improved, particularly with the inclusion of three major enhancements to the LEED™ rating system:

1. Harmonization and alignment: LEED™ credits across the rating systems will bring the core elements of LEED™ into one elegant rating system. According to USGBC, this is intended to synchronize the development and deployment of LEED™ rating systems while creating the capacity to respond to previously underserved markets. Credits and prerequisites from all LEED™ rating systems have now been updated, consolidated, and aligned, drawing on their most effective common denominators, so that credits and prerequisites are consistent. Necessary precedent-setting and clarifying information from Credit Interpretation Rulings (CIRs) were incorporated into the rating systems. LEED™ for Homes and LEED™ for Neighborhood Development have not changed under LEED™ 2009. USGBC also announced that "a scrub of the existing Credit Interpretation Rulings (CIRs) was conducted, and necessary precedent-setting and clarifying language has been incorporated into the prerequisites/credits."
2. Credit weightings: The second major advancement that comes with LEED™ 2009 is that credits now consist of different weightings that rely on their ability to impact different environmental and human health concerns. Penny Bonda, co-author of *Sustainable Commercial Interiors*, describes the new credit weightings as "scientifically grounded reevaluations that place an increased emphasis on energy use and carbon-emission reductions." The recently

DOI: 10.1016/B978-1-85617-691-0.00004-7

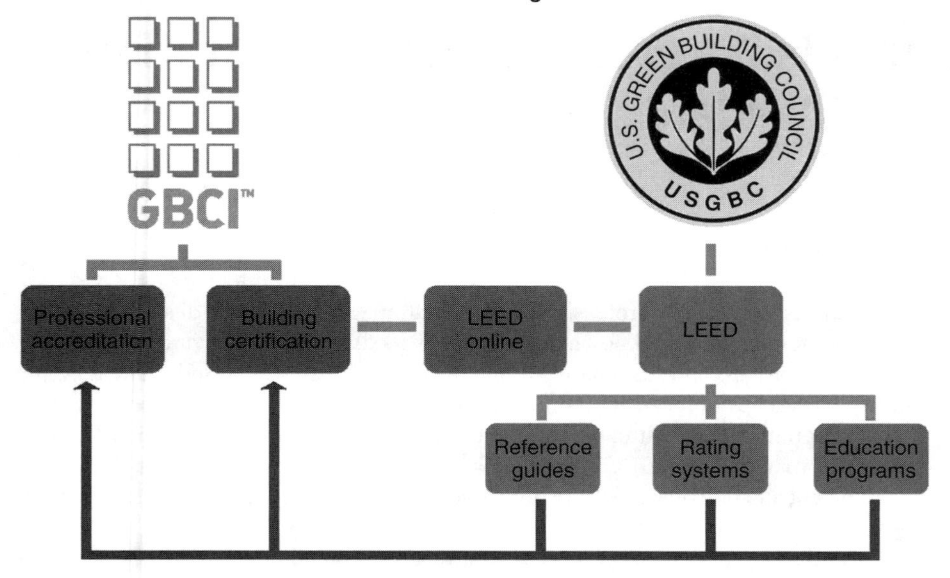

Figure 4.1 Diagram showing the relationship between USGBC and GBCI and how they relate to Project Certification and LEED™ Accreditation.
Source: USGBC.

revised LEED™ credit weightings therefore award more points to green strategies and practices that reflect the most positive impact on critical issues such as energy efficiency, CO_2 reductions, and life-cycle performance of assemblies. For the revised weightings credits were evaluated individually against a list of 13 environmental-impact categories that include climate change, indoor environmental quality, water intake, resource depletion, and others (Figure 4.2). LEED™ prioritized the impact categories, and credits were assigned a value based on the size of their contribution to mitigating each impact. This resulted in giving the most value to credits that offer the greatest potential for making the biggest change. And while the credits are all intact, they nevertheless are now worth different amounts. In LEED™ 2009 the evaluation of each credit against the set of impact categories produces a point total for each credit; the point totals for all credits come to 100, and these final point totals are reflected in the rating-system scorecard. In previous LEED™ versions, for example, the total points for the New Construction category came to 69 points and for the Commercial Interiors category to 57 points.

3. Regionalization: One of the most significant changes in the LEED™ 2009 Rating System is the incorporation of Regional Priority Credits. This was done to enhance the flexibility of LEED™ and provide a more effective method of addressing the need for regional adaptation. Moreover, USGBC's introduction of Regional Bonus Credits is intended to increase the value of pursuing credits that address environmental areas of concern in a project's region. This is in contrast to the previous rating system in which LEED™ was applied uniformly across the United States and point values were equal across various regions. Six credits existing

Figure 4.2 Diagram illustrating the 13 LEED™ Weightings Categories on which much of the LEED™ 2009 Rating System is based.
Source: USGBC.

> in each rating system will be identified as regional credits, similar to current exemplary performance points, and are meant to address regionally prioritized environmental issues that have been identified as regionally critical within a project's environmental zone by the regional USGBC Board. A maximum of four points are available for project teams to pursue. The implication is that in the new rating system, a project can earn up to four extra points, or one point for each of four such Regional Priority Credits. USGBC Chapters and Regional Councils played a crucial role in this effort, based on their knowledge of issues of concern in their locales.

In 2008 the final version of LEED™ for Homes was rolled out on a national basis. LEED™ for Homes is a voluntary initiative to actively promote more sustainable building practices in the homebuilding industry. The LEED™ for Homes rating system is targeting the top 25 percent of homes with best-practice environmental features. It should also be noted that the LEED™ for Homes format differs somewhat from the other certification categories such as New Construction and Commercial Interiors.

4.2 LEED™ Exam Sections/Certification Categories

The Green Building Certification Institute (GBCI) states that "The specific credits in the rating system provide guidance for the design and construction of buildings of all sizes in both private and public sectors." Jerry Yudelson author of *The Green Building Revolution*, comments that "The essence of LEED™, and its particular genius, is that it is a point-based rating system that allows vastly different green buildings to be compared in the aggregate." Yudelson goes on to say that "The LEED™ rating is a form of 'eco-label' that describes the environmental attributes of a project." Below are found checklists for the main categories for LEED™ certification upon which much of the LEED™ v3 exams are based.

4.2.1 LEED™ Professional Accreditation Requirements for New Construction and Major Renovations Project Checklist

Following are checklists governing the sections in the New Construction and Major Renovations category:

Sustainable Sites (SS)	26 Possible Points
❑ **SS Prerequisite 1 Construction Activity Pollution Prevention**	**Required**
❑ SS Credit 1 Site Selection	1
❑ SS Credit 2 Development Density and Community Connectivity	5
❑ SS Credit 3 Brownfield Redevelopment	1
❑ SS Credit 4.1 Alternative Transportation – Public Transportation Access	6
❑ SS Credit 4.2 Alternative Transportation – Bicycle Storage and Changing Rooms	1
❑ SS Credit 4.3 Alternative Transportation – Low-Emitting and Fuel-Efficient Vehicles	3
❑ SS Credit 4.4 Alternative Transportation – Parking Capacity	2
❑ SS Credit 5.1 Site Development – Protect or Restore Habitat	1
❑ SS Credit 5.2 Site Development – Maximize Open Space	1
❑ SS Credit 6.1 Stormwater Design – Quantity Control	1
❑ SS Credit 6.2 Stormwater Design – Quality Control	1
❑ SS Credit 7.1 Heat-Island Effect – Nonroof	1
❑ SS Credit 7.2 Heat-Island Effect – Roof	1
❑ SS Credit 8 Light-Pollution Reduction	1

Water Efficiency (WE)	10 Possible Points
❑ **WE Prerequisite 1 Water-Use Reduction**	**Required**
❑ WE Credit 1 Water-Efficient Landscaping	2–4
❑ WE Credit 2 Innovative Wastewater Technologies	2
❑ WE Credit 3 Water-Use Reduction	2–4

Energy and Atmosphere (EA)	35 Possible Points
❑ **EA Prerequisite 1 Fundamental Commissioning of Building Energy Systems**	**Required**
❑ **EA Prerequisite 2 Minimum Energy Performance**	**Required**
❑ **EA Prerequisite 3 Fundamental Refrigerant Management**	**Required**

❑ EA Credit 1 Optimize Energy Performance	1–19
❑ EA Credit 2 On-site Renewable Energy	1–7
❑ EA Credit 3 Enhanced Commissioning	2
❑ EA Credit 4 Enhanced Refrigerant Management	2
❑ EA Credit 5 Measurement and Verification	3
❑ EA Credit 6 Green Power	2

Materials and Resources (MR) — 14 Possible Points

❑ **MR Prerequisite 1 Storage and Collection of Recyclables**	**Required**
❑ MR Credit 1.1 Building Reuse – Maintain Existing Walls, Floors, and Roof	1–3
❑ MR Credit 1.2 Building Reuse – Maintain Existing Interior Nonstructural Elements	1
❑ MR Credit 2 Construction Waste Management	1–2
❑ MR Credit 3 Materials Reuse	1–2
❑ MR Credit 4 Recycled Content	1–2
❑ MR Credit 5 Regional Materials	1–2
❑ MR Credit 6 Rapidly Renewable Materials	1
❑ MR Credit 7 Certified Wood	1

Indoor Environmental Quality (IEQ) — 15 Possible Points

❑ **IEQ Prerequisite 1 Minimum Indoor-Air-Quality Performance**	**Required**
❑ **IEQ Prerequisite 2 Environmental Tobacco Smoke (ETS) Control**	**Required**
❑ IEQ Credit 1 Outdoor-Air Delivery Monitoring	1
❑ IEQ Credit 2 Increased Ventilation	1
❑ IEQ Credit 3.1 Construction Indoor-Air-Quality Management Plan – During Construction	1
❑ IEQ Credit 3.2 Construction Indoor-Air-Quality Management Plan – Before Occupancy	1
❑ IEQ Credit 4.1 Low-Emitting Materials – Adhesives and Sealants	1
❑ IEQ Credit 4.2 Low-Emitting Materials – Paints and Coatings	1
❑ IEQ Credit 4.3 Low-Emitting Materials – Flooring Systems	1
❑ IEQ Credit 4.4 Low-Emitting Materials – Composite Wood and Agrifiber Products	1
❑ IEQ Credit 5 Indoor Chemical and Pollutant Source Control	1
❑ IEQ Credit 6.1 Controllability of Systems – Lighting	1
❑ IEQ Credit 6.2 Controllability of Systems – Thermal Comfort	1
❑ IEQ Credit 7.1 Thermal Comfort – Design	1
❑ IEQ Credit 7.2 Thermal Comfort – Verification	1
❑ IEQ Credit 8.1 Daylight and Views – Daylight	1
❑ IEQ Credit 8.2 Daylight and Views – Views	1

Innovation in Design (ID)	**6 Possible Points**
❑ ID Credit 1 Innovation in Design	1–5
❑ ID Credit 2 LEED™ Accredited Professional	1

Regional Priority (RP)	**4 Possible Points**
❑ RP Credit 1 Regional Priority	1–4

4.2.2 LEED™ Professional Accreditation Requirements for Core and Shell Development Project Checklist

Following are checklists governing the sections in the Core and Shell Development category:

Sustainable Sites (SS)	**28 Possible Points**
❑ **SS Prerequisite 1 Construction Activity Pollution Prevention**	**Required**
❑ SS Credit 1 Site Selection	1
❑ SS Credit 2 Development Density and Community Connectivity	5
❑ SS Credit 3 Brownfield Redevelopment	1
❑ SS Credit 4.1 Alternative Transportation – Public-Transportation Access	6
❑ SS Credit 4.2 Alternative Transportation – Bicycle Storage and Changing Rooms	2
❑ SS Credit 4.3 Alternative Transportation – Low-Emitting and Fuel-Efficient Vehicles	3
❑ SS Credit 4.4 Alternative Transportation – Parking Capacity	2
❑ SS Credit 5.1 Site Development – Protect or Restore Habitat	1
❑ SS Credit 5.2 Site Development – Maximize Open Space	1
❑ SS Credit 6.1 Stormwater Design – Quantity Control	1
❑ SS Credit 6.2 Stormwater Design – Quality Control	1
❑ SS Credit 7.1 Heat-Island Effect – Nonroof	1
❑ SS Credit 7.2 H eat-Island Effect – Roof	1
❑ SS Credit 8 Light-Pollution Reduction	1
❑ SS Credit 9 Tenant Design and Construction Guidelines	1

Water Efficiency (WE)	**10 Possible Points**
❑ **WE Prerequisite 1 Water Use Reduction**	**Required**
❑ WE Credit 1 Water-Efficient Landscaping	2–4
❑ WE Credit 2 Innovative Wastewater Technologies	2
❑ WE Credit 3 Water-Use Reduction	2–4

Energy and Atmosphere (EA)	**37 Possible Points**
❑ **EA Prerequisite 1 Fundamental Commissioning of Building Energy Systems**	**Required**

❑ **EA Prerequisite 2 Minimum Energy Performance**	**Required**
❑ **EA Prerequisite 3 Fundamental Refrigerant Management**	**Required**
❑ EA Credit 1 Optimize Energy Performance	3–21
❑ EA Credit 2 On-site Renewable Energy	4
❑ EA Credit 3 Enhanced Commissioning	2
❑ EA Credit 4 Enhanced Refrigerant Management	2
❑ EA Credit 5.1 Measurement and Verification – Base Building	3
❑ EA Credit 5.2 Measurement and Verification – Tenant Submetering	3
❑ EA Credit 6 Green Power	2

Materials and Resources (MR) — 13 Possible Points

❑ **MA Prerequisite 1 Storage and Collection of Recyclables**	**Required**
❑ MA Credit 1 Building Reuse – Maintain Existing Walls, Floors, and Roof	1–5
❑ MA Credit 2 Construction Waste Management	1–2
❑ MA Credit 3 Materials Reuse	1
❑ MA Credit 4 Recycled Content	1–2
❑ MA Credit 5 Regional Materials	1–2
❑ MA Credit 6 Certified Wood	1

Indoor Environmental Quality (IEQ) — 12 Possible Points

❑ **IEQ Prerequisite 1 Minimum Indoor-Air-Quality Performance**	**Required**
❑ **IEQ Prerequisite 2 Environmental Tobacco Smoke (ETS) Control**	**Required**
❑ IEQ Credit 1 Outdoor-Air Delivery Monitoring	1
❑ IEQ Credit 2 Increased Ventilation	1
❑ IEQ Credit 3 Construction Indoor-Air-Quality Management Plan – During Construction	1
❑ IEQ Credit 4.1 Low-Emitting Materials – Adhesives and Sealants	1
❑ IEQ Credit 4.2 Low-Emitting Materials – Paints and Coatings	1
❑ IEQ Credit 4.3 Low-Emitting Materials – Flooring Systems	1
❑ IEQ Credit 4.4 Low-Emitting Materials – Composite Wood and Agrifiber Products	1
❑ IEQ Credit 5 Indoor Chemical and Pollutant Source Control	1
❑ IEQ Credit 6 Controllability of Systems – Thermal Comfort	1
❑ IEQ Credit 7 Thermal Comfort – Design	1
❑ IEQ Credit 8.1 Daylight and Views – Daylight	1
❑ IEQ Credit 8.2 Daylight and Views – Views	1

Innovation in Design (ID) **6 Possible Points**

- ❑ ID Credit 1 Innovation in Design 1–5
- ❑ ID Credit 2 LEED™ Accredited Professional 1

Regional Priority (RP) **4 Possible Points**

- ❑ RP Credit 1 Regional Priority 1–4

4.2.3 LEED™ Professional Accreditation Requirements for Schools: New Construction and Major Renovations

Following are checklists governing the sections in the Schools category:

Sustainable Sites (SS) **24 Possible Points**

- ❑ **SS Prerequisite 1 Construction Activity Pollution Prevention** **Required**
- ❑ **SS Prerequisite 2 Environmental Site Assessment** **Required**
  - ❑ SS Credit 1 Site Selection 1
  - ❑ SS Credit 2 Development Density and Community Connectivity 4
  - ❑ SS Credit 3 Brownfield Redevelopment 1
  - ❑ SS Credit 4.1 Alternative Transportation – Public Transportation Access 4
  - ❑ SS Credit 4.2 Alternative Transportation – Bicycle Storage and Changing Rooms 1
  - ❑ SS Credit 4.3 Alternative Transportation – Low-Emitting and Fuel-Efficient Vehicles 2
  - ❑ SS Credit 4.4 Alternative Transportation – Parking Capacity 2
  - ❑ SS Credit 5.1 Site Development – Protect or Restore Habitat 1
  - ❑ SS Credit 5.2 Site Development – Maximize Open Space 1
  - ❑ SS Credit 6.1 Stormwater Design – Quantity Control 1
  - ❑ SS Credit 6.2 Stormwater Design – Quality Control 1
  - ❑ SS Credit 7.1 H eat-Island Effect – Nonroof 1
  - ❑ SS Credit 7.2 H eat-Island Effect – Roof 1
  - ❑ SS Credit 8 Light-Pollution Reduction 1
  - ❑ SS Credit 9 Site Master Plan 1
  - ❑ SS Credit 10 Joint Use of Facilities 1

Water Efficiency (WE) **11 Possible Points**

- ❑ **WE Prerequisite 1 Water Use Reduction** **Required**
  - ❑ WE Credit 1 Water-Efficient Landscaping 2–4
  - ❑ WE Credit 2 Innovative Wastewater Technologies 2
  - ❑ WE Credit 3 Water-Use Reduction 2–4
  - ❑ WE Credit 4 Process Water-Use Reduction 1

Energy and Atmosphere (EA) — 33 Possible Points

Item	Points
❑ **EA Prerequisite 1 Fundamental Commissioning of Building Energy Systems**	**Required**
❑ **EA Prerequisite 2 Minimum Energy Performance**	**Required**
❑ **EA Prerequisite 3 Fundamental Refrigerant Management**	**Required**
❑ EA Credit 1 Optimize Energy Performance	1–19
❑ EA Credit 2 On-site Renewable Energy	1–7
❑ EA Credit 3 Enhanced Commissioning	2
❑ EA Credit 4 Enhanced Refrigerant Management	1
❑ EA Credit 5 Measurement and Verification	2
❑ EA Credit 6 Green Power	2

Materials and Resources (MR) — 13 Possible Points

Item	Points
❑ **MR Prerequisite 1 Storage and Collection of Recyclables**	**Required**
❑ MR Credit 1.1 Building Reuse – Maintain Existing Walls, Floors, and Roof	1–2
❑ MR Credit 1.2 Building Reuse – Maintain Existing Interior Nonstructural Elements	1
❑ MR Credit 2 Construction Waste Management	1–2
❑ MR Credit 3 Materials Reuse	1–2
❑ MR Credit 4 Recycled Content	1–2
❑ MR Credit 5 Regional Materials	1–2
❑ MR Credit 6 Rapidly Renewable Materials	1
❑ MR Credit 7 Certified Wood	1

Indoor Environmental Quality (IEQ) — 19 Possible Points

Item	Points
❑ **IEQ Prerequisite 1 Minimum Indoor-Air-Quality Performance**	**Required**
❑ **IEQ Prerequisite 2 Environmental Tobacco Smoke (ETS) Control**	**Required**
❑ **IEQ Prerequisite 3 Minimum Acoustical Performance**	**Required**
❑ IEQ Credit 1 Outdoor-Air Delivery Monitoring	1
❑ IEQ Credit 2 Increased Ventilation	1
❑ IEQ Credit 3.1 Construction Indoor-Air-Quality Management Plan – During Construction	1
❑ IEQ Credit 3.2 Construction Indoor-Air-Quality Management Plan – Before Occupancy	1
❑ IEQ Credit 4 Low-Emitting Materials	1–4
❑ IEQ Credit 5 Indoor Chemical and Pollutant Source Control	1
❑ IEQ Credit 6.1 Controllability of Systems – Lighting	1
❑ IEQ Credit 6.2 Controllability of Systems – Thermal Comfort	1
❑ IEQ Credit 7.1 Thermal Comfort – Design	1
❑ IEQ Credit 7.2 Thermal Comfort – Verification	1

- ❑ IEQ Credit 8.1 Daylight and Views – Daylight 1–3
- ❑ IEQ Credit 8.2 Daylight and Views – Views 1
- ❑ IEQ Credit 9 Enhanced Acoustical Performance 1
- ❑ IEQ Credit 10 Mold Prevention 1

Innovation in Design (ID) 6 Possible Points

- ❑ ID Credit 1 Innovation in Design 1–4
- ❑ ID Credit 2 LEED™ Accredited Professional 1
- ❑ ID Credit 3 School as a Teaching Tool 1

Regional Priority (RP) 4 Possible Points

- ❑ RP Credit 1 Regional Priority 1–4

4.2.4 LEED™ Professional Accreditation Requirements for Existing Buildings: Opertions and Maintenance

Following are checklists governing the sections in the Existing Buildings: Operations and Maintenance category:

Sustainable Sites (SS) 26 Possible Points

- ❑ SS Credit 1 LEED™ Certified Design and Construction 4
- ❑ SS Credit 2 Building Exterior and Hardscape Management Plan 1
- ❑ SS Credit 3 Integrated Pest Management, Erosion Control, and Landscape Management Plan 1
- ❑ SS Credit 4 Alternative Commuting Transportation 3–15
- ❑ SS Credit 5 Site Disturbance – Protect or Restore Open Habitat 1
- ❑ SS Credit 6 Stormwater Quantity Control 1
- ❑ SS Credit 7.1 Heat-Island Reduction – Nonroof 1
- ❑ SS Credit 7.2 Heat-Island Reduction – Roof 1
- ❑ SS Credit 8 Light-Pollution Reduction 1

Water Efficiency (WE) 14 Possible Points

❑ **WE Prerequisite 1 Minimum Indoor Plumbing Fixture and Fitting Efficiency Required**

- ❑ WE Credit 1 Water Performance Measurement 1–2
- ❑ WE Credit 2 Additional Indoor Plumbing Fixture and Fitting Efficiency 1–5
- ❑ WE Credit 3 Water-Efficient Landscaping 1–5
- ❑ WE Credit 4 Cooling-Tower Water Management 1–2

Energy and Atmosphere (EA) 35 Possible Points

❑ **EA Prerequisite 1 Energy Efficiency Best Management Practices – Planning, Documentation, and Opportunity Assessment Required**

❑ **EA Prerequisite 2 Minimum Energy-Efficiency Performance**	**Required**
❑ **EA Prerequisite 3 Fundamental Refrigerant Management**	**Required**
❑ EA Credit 1 Optimize Energy-Efficiency Performance	1–18
❑ EA Credit 2.1 Existing Building Commissioning – Investigation and Analysis	2
❑ EA Credit 2.2 Existing Building Commissioning – Implementation	2
❑ EA Credit 2.3 Existing Building Commissioning – Ongoing Commissioning	2
❑ EA Credit 3.1 Performance Measurement – Building Automation System	1
❑ EA Credit 3.2 Performance Measurement – System Level Metering	1–2
❑ EA Credit 4 On-site and Off-site Renewable Energy	1–6
❑ EA Credit 5 Enhanced Refrigerant Management	1
❑ EA Credit 6 Emissions Reduction Reporting	1

Materials and Resources (MR) — 10 Possible Points

❑ **MR Prerequisite 1 Sustainable Purchasing Policy**	**Required**
❑ **MR Prerequisite 2 Solid-Waste Management Policy**	**Required**
❑ MR Credit 1 Sustainable Purchasing – Ongoing Consumables	1
❑ MR Credit 2 Sustainable Purchasing – Durable Goods	1–2
❑ MR Credit 3 Sustainable Purchasing – Facility Alterations and Additions	1
❑ MR Credit 4 Sustainable Purchasing – Reduced Mercury in Lamps	1
❑ MR Credit 5 Sustainable Purchasing – Food	1
❑ MR Credit 6 Solid-Waste Management – Waste Stream Audit	1
❑ MR Credit 7 Solid-Waste Management – Ongoing Consumables	1
❑ MR Credit 8 Solid-Waste Management – Durable Goods	1
❑ MR Credit 9 Solid-Waste Management – Facility Alterations and Additions	1

Indoor Environmental Quality (IEQ) — 15 Possible Points

❑ **IEQ Prerequisite 1 Minimum Indoor-Air-Quality Performance**	**Required**
❑ **IEQ Prerequisite 2 Environmental Tobacco Smoke (ETS) Control**	**Required**
❑ **IEQ Prerequisite 3 Green Cleaning Policy**	**Required**
❑ IEQ Credit 1.1 Indoor-Air-Quality Best Management Practices – Indoor-Air-Quality Management Program	1

- ❑ IEQ Credit 1.2 Indoor-Air-Quality Best Management Practices – Outdoor-Air Delivery Monitoring 1
- ❑ IEQ Credit 1.3 Indoor-Air-Quality Best Management Practices – Increased Ventilation 1
- ❑ IEQ Credit 1.4 Indoor-Air-Quality Best Management Practices – Reduce Particulates in Air Distribution 1
- ❑ IEQ Credit 1.5 Indoor-Air-Quality Best Management Practices – Indoor-Air-Quality Management for Facility Alterations and Additions 1
- ❑ IEQ Credit 2.1 Occupant Comfort – Occupant Survey 1
- ❑ IEQ Credit 2.2 Controllability of Systems – Lighting 1
- ❑ IEQ Credit 2.3 Occupant Comfort – Thermal Comfort Monitoring 1
- ❑ IEQ Credit 2.4 Daylight and Views 1
- ❑ IEQ Credit 3.1 Green Cleaning – High-Performance Cleaning Program 1
- ❑ IEQ Credit 3.2 Green Cleaning – Custodial Effectiveness Assessment 1
- ❑ IEQ Credit 3.3 Green Cleaning – Purchase of Sustainable Cleaning Products and Materials 1
- ❑ IEQ Credit 3.4 Green Cleaning – Sustainable Cleaning Equipment 1
- ❑ IEQ Credit 3.5 Green Cleaning – Indoor Chemical and Pollutant Source Control 1
- ❑ IEQ Credit 3.6 Green Cleaning – Indoor Integrated Pest Management 1

Innovation in Operations (IO) **6 Possible Points**

- ❑ ID Credit 1 Innovation in Operations 1–4
- ❑ ID Credit 2 LEED™ Accredited Professional 1
- ❑ ID Credit 3 Documenting Sustainable Building Cost Impacts 1

Regional Priority (RP) **4 Possible Points**

- ❑ RP Credit 1 Regional Priority 1–4

4.2.5 LEED™ Professional Accreditation Requirements for Commercial Interiors

Following are checklists governing the sections in the Commercial Interiors category:

Sustainable Sites (SS) **21 Possible Points**

- ❑ SS Credit 1 Site Selection 1–5
- ❑ SS Credit 2 Development Density and Community Connectivity 6

❑ SS Credit 3.1 Alternative Transportation – Public Transportation Access	6
❑ SS Credit 3.2 Alternative Transportation – Bicycle Storage and Changing Rooms	2
❑ SS Credit 3.3 Alternative Transportation – Parking Availability	2

Water Efficiency (WE) — 11 Possible Points

❑ **WE Prerequisite 1 Water Use Reduction**	**Required**
❑ WE Credit 1 Water-Use Reduction	6–11

Energy and Atmosphere (EA) — 37 Possible Points

❑ **EA Prerequisite 1 Fundamental Commissioning of Building Energy Systems**	**Required**
❑ **EA Prerequisite 2 Minimum Energy Performance**	**Required**
❑ **EA Prerequisite 3 Fundamental Refrigerant Management**	**Required**
❑ EA Credit 1.1 Optimize Energy Performance – Lighting Power	1–5
❑ EA Credit 1.2 Optimize Energy Performance – Lighting Controls	1–3
❑ EA Credit 1.3 Optimize Energy Performance – HVAC	5–10
❑ EA Credit 1.4 Optimize Energy Performance – Equipment and Appliances	1–4
❑ EA Credit 2 Enhanced Commissioning	5
❑ EA Credit 3 Measurement and Verification	2–5
❑ EA Credit 4 Green Power	5

Materials and Resources (MR) — 14 Possible Points

❑ **MR Prerequisite 1 Storage and Collection of Recyclables**	**Required**
❑ MR Credit 1.1 Tenant Space – Long-Term Commitment	1
❑ MR Credit 1.2 Building Reuse – Maintain Interior Nonstructural Components	1–2
❑ MR Credit 2 Construction-Waste Management	1–2
❑ MR Credit 3.1 Materials Reuse	1–2
❑ MR Credit 3.2 Materials Reuse – Furniture and Furnishings	1
❑ MR Credit 4 Recycled Content	1–2
❑ MR Credit 5 Regional Materials	1–2
❑ MR Credit 6 Rapidly Renewable Materials	1
❑ MR Credit 7 Certified Wood	1

Indoor Environmental Quality (IEQ) — 17 Possible Points

❑ **IEQ Prerequisite 1 Minimum Indoor-Air-Quality Performance**	**Required**

- ❑ **IEQ Prerequisite 2 Environmental Tobacco Smoke (ETS) Control** — **Required**
  - ❑ IEQ Credit 1 Outdoor-Air Delivery Monitoring — 1
  - ❑ IEQ Credit 2 Increased Ventilation — 1
  - ❑ IEQ Credit 3.1 Construction Indoor-Air-Quality Management Plan – During Construction — 1
  - ❑ IEQ Credit 3.2 Construction Indoor-Air-Quality Management Plan – Before Occupancy — 1
  - ❑ IEQ Credit 4.1 Low-Emitting Materials – Adhesives and Sealants — 1
  - ❑ IEQ Credit 4.2 Low-Emitting Materials – Paints and Coatings — 1
  - ❑ IEQ Credit 4.3 Low-Emitting Materials – Flooring Systems — 1
  - ❑ IEQ Credit 4.4 Low-Emitting Materials – Composite Wood and Agrifiber Products — 1
  - ❑ IEQ Credit 4.5 Low-Emitting Materials – Systems Furniture and Seating — 1
  - ❑ IEQ Credit 5 Indoor Chemical and Pollutant Source Control — 1
  - ❑ IEQ Credit 6.1 Controllability of Systems – Lighting — 1
  - ❑ IEQ Credit 6.2 Controllability of Systems – Thermal Comfort — 1
  - ❑ IEQ Credit 7.1 Thermal Comfort – Design — 1
  - ❑ IEQ Credit 7.2 Thermal Comfort – Verification — 1
  - ❑ IEQ Credit 8.1 Daylight and Views – Daylight — 1–2
  - ❑ IEQ Credit 8.2 Daylight and Views – Views for Seated Spaces — 1

Innovation in Design (ID) — 6 Possible Points

- ❑ ID Credit 1 Innovation in Design — 1–5
- ❑ ID Credit 2 LEED™ Accredited Professional — 1

Regional Priority (RP) — 4 Possible Points

- ❑ RP Credit 1 Regional Priority — 1–4

4.2.6 LEED™ Professional Accreditation Requirements for Retail (Proposed)

Following are checklists governing the sections in the proposed Retail category:

Sustainable Sites (SS) — 26 Possible Points

- ❑ **SS Prerequisite 1 Construction Activity Pollution Prevention** — **Required**
  - ❑ SS Credit 1 Site Selection — 1
  - ❑ SS Credit 2 Development Density and Community Connectivity — 5
  - ❑ SS Credit 3 Brownfield Redevelopment — 1
  - ❑ SS Credit 4 Alternative Transportation – Public Transportation Access — 1–10

Credit	Points
❑ SS Credit 5.1 Site Development – Protect or Restore Habitat	1
❑ SS Credit 5.2 Site Development – Maximize Open Space	1
❑ SS Credit 6.1 Stormwater Design – Quantity Control	1
❑ SS Credit 6.2 Stormwater Design – Quality Control	1
❑ SS Credit 7.1 Heat-Island Effect – Nonroof	1–2
❑ SS Credit 7.2 Heat-Island Effect – Roof	1
❑ SS Credit 8 Light-Pollution Reduction	2
Water Efficiency (WE)	**10 Possible Points**
❑ **WE Prerequisite 1 Water-Use Reduction**	**Required**
❑ WE Credit 1 Water-Efficient Landscaping	2–4
❑ WE Credit 2 Innovative Wastewater Technologies	2
❑ WE Credit 3 Water-Use Reduction	2–4
Energy and Atmosphere (EA)	**35 Possible Points**
❑ **EA Prerequisite 1 Fundamental Commissioning of Building Energy Systems**	**Required**
❑ **EA Prerequisite 2 Minimum Energy Performance**	**Required**
❑ **EA Prerequisite 3 Fundamental Refrigerant Management**	**Required**
❑ EA Credit 1 Optimize Energy Performance	1–19
❑ EA Credit 2 On-site Renewable Energy	1–7
❑ EA Credit 3 Enhanced Commissioning	2
❑ EA Credit 4 Enhanced Refrigerant Management	2
❑ EA Credit 5 Measurement and Verification	3
❑ EA Credit 6 Green Power	2
Materials and Resources (MR)	**14 Possible Points**
❑ **MR Prerequisite 1 Storage and Collection of Recyclables**	**Required**
❑ MR Credit 1.1 Building Reuse – Maintain Existing Walls, Floors, and Roof	1–3
❑ MR Credit 1.2 Building Reuse – Maintain Existing Interior Nonstructural Elements	1
❑ MR Credit 2 Construction Waste Management	1–2
❑ MR Credit 3 Materials Reuse	1–2
❑ MR Credit 4 Recycled Content	1–2
❑ MR Credit 5 Regional Materials	1–2
❑ MR Credit 6 Rapidly Renewable Materials	1
❑ MR Credit 7 Certified Wood	1
Indoor Environmental Quality (IEQ)	**15 Possible Points**
❑ **IEQ Prerequisite 1 Minimum Indoor-Air-Quality Performance**	**Required**
❑ **IEQ Prerequisite 2 Environmental Tobacco Smoke (ETS) Control**	**Required**

❑ IEQ Credit 1 Outdoor-Air Delivery Monitoring	1
❑ IEQ Credit 2 Increased Ventilation	1
❑ IEQ Credit 3.1 Construction Indoor-Air-Quality Management Plan – During Construction	1
❑ IEQ Credit 3.2 Construction Indoor-Air-Quality	
❑ Management Plan – Before Occupancy	1
❑ IEQ Credit 4 Low-Emitting Materials – Adhesives and Sealants	1–5
❑ IEQ Credit 5 Indoor Chemical and Pollutant Source Control	1
❑ IEQ Credit 6 Controllability of Systems – Lighting and Thermal Comfort	1
❑ IEQ Credit 7.1 Thermal Comfort – Design	1
❑ IEQ Credit 7.2 Thermal Comfort – Employee Verification	1*
❑ IEQ Credit 8.1 Daylight and Views – Daylight	1
❑ IEQ Credit 8.2 Daylight and Views – Views	1

*In addition to IEQ Credit 7.1

Innovation in Design (ID) — 6 Possible Points

❑ ID Credit 1 Innovation in Design	1–5
❑ ID Credit 2 LEED™ Accredited Professional	1

Regional Priority (RP) — 4 Possible Points

❑ RP Credit 1 Regional Priority	1–4

4.2.7 LEED™ Professional Accreditation Requirements for Homes

The USGBC designed the LEED™ for Homes rating system to promote the design and construction of high-performance green homes because such homes use less energy, water, and natural resources; create less waste; and are healthier and more comfortable for the occupants. One example of a LEED™ certified home is in Port St. Lucie, Florida; the home which boasts hurricane-resistant wall panels that provide superior energy efficiency, ENERGY STAR®, appliances, and a drought-tolerant and low-maintenance landscape (Figure 4.3).

The LEED™ for Homes Rating System has 35 topic areas, each with a specific intent or goal. The piloting program has only recently been completed, and LEED™ for Homes was not officially launched until February 2008.

The Rating System differs somewhat from, say, New Construction in that there is no Regional Priority category in LEED™ Homes, and, as seen below, the accreditation requirements for the Innovation in Design Process (ID) category was moved to the front of the LEED™ for Homes Rating System to emphasize the importance of design in a LEED™ home. In LEED™ Homes there is also an added Awareness and Education category. Likewise, the various category contents in LEED™ Homes differ from the other Rating Systems. There are a total of 18 measures in LEED™ for Homes, which must be completed during the design or construction phase.

Figure 4.3 Photo of a Port St. Lucie (Florida) home, which became the first in the state to earn a certification through the U.S. Green Building Council's LEED™ for Homes pilot program after several inspections and performance tests were completed by the Florida Solar Energy Center. The home completed the certification process on March 19, 2007 (Courtesy: Florida Solar Energy Center).

Following are checklists governing the sections in LEED™ for Homes:

Innovation & Design Process (ID)	**11 Possible Points**
❑ **ID Prerequisite 1.1 Integrated Project Planning – Preliminary Rating**	**Required**
❑ ID Credit 1.2 Integrated Project Planning – Integrated Project Team	1
❑ ID Credit 1.3 Integrated Project Planning – Professional Credentialed for LEED™ for Homes	1
❑ ID Credit 1.4 Integrated Project Planning – Design Charrette	1
❑ ID Credit 1.5 Integrated Project Planning – Building Orientation for Solar Design	1
❑ **ID Prerequisite 2.1 Durability Management Process – Durable Planning**	**Required**
❑ **ID Prerequisite 2.2 Durability Management Process – Durability Management**	**Required**
❑ ID Credit 2.3 Durability Management Process – Third-Party Durability Management Verification	3
❑ ID Credit 3.1 Innovative or Regional Design – Innovation #1	1
❑ ID Credit 3.2 Innovative or Regional Design – Innovation #2	1
❑ ID Credit 3.3 Innovative or Regional Design – Innovation #3	1
❑ ID Credit 3.4 Innovative or Regional Design – Innovation #4	1

Location and Linkages (LL)	**10 Possible Points**
❑ LL Credit 1 LEED™ ND – LEED™ for Neighborhood Development	10
❑ LL Credit 2 Site Selection	2
❑ LL Credit 3.1 Preferred Locations – Edge Development	1
❑ LL Credit 3.2 Preferred Locations – Infill	2
❑ LL Credit 3.3 Preferred Locations – Previously Developed	1
❑ LL Credit 4 Infrastructure – Existing Infrastructure	1
❑ LL Credit 5.1 Community Resources – Basic Community Resources	1
❑ LL Credit 5.2 Community Resources – Extensive Community Resources	2
❑ LL Credit 5.3 Community Resources – Outstanding Community Resources	3
❑ LL Credit 6 Access to Open Space	1

Sustainable Sites (SS) (Minimum 5 SS Points Required)	**22 Possible Points**
❑ **SS Prerequisite 1.1 Site Stewardship – Erosion**	**Required**
❑ SS Credit 1.2 Site Stewardship – Minimize Disturbed Area of Site	1
❑ **SS Prerequisite 2.1 Landscaping – No Invasive Plants**	**Required**
❑ SS Credit 2.2 Landscaping – Basic Landscape Design	2
❑ SS Credit 2.3 Landscaping – Limit Conventional Turf	3
❑ SS Credit 2.4 Landscaping – Drought-Tolerant Plants	2
❑ SS Credit 2.5 Landscaping – Reduce Overall Irrigation Demand by at least 20 percent	6
❑ SS Credit 3 Local Heat-Island Effects – Reduce Local-Heat Island Effects	1
❑ SS Credit 4.1 Surface Water Management – Permeable Lot	4
❑ SS Credit 4.2 Surface Water Management – Permanent Erosion Controls	1
❑ SS Credit 4.3 Surface Water Management – Management of Runoff from Roof	2
❑ SS Credit 5 Nontoxic Pest Control – Pest-Control Alternatives	2
❑ SS Credit 6.1 Compact Development – Moderate Density	2
❑ SS Credit 6.2 Compact Development – High Density	3
❑ SS Credit 6.3 Compact Development – Very High Density	4

Water Efficiency (WE) – (Minimum of 3 WE Points Required)	**15 Possible Points**
❑ WE Credit 1.1 Water Reuse – Rainwater Harvesting System	4
❑ WE Credit 1.2 Water Reuse – Graywater Reuse System	1

- ❑ WE Credit 1.3 Water Reuse – Use of Municipal Recycled Water System — 3
- ❑ WE Credit 2.1 Irrigation System – High-Efficiency Irrigation System — 3
- ❑ WE Credit 2.2 Irrigation System – Third-Party Inspection — 1
- ❑ WE Credit 2.3 Irrigation System – Reduce Overall Irrigation Demand by at least 45 percent — 4
- ❑ WE Credit 3.1 Indoor Water Use – High-Efficiency Fixtures and Fittings — 3
- ❑ WE Credit 3.2 Indoor Water Use – Very High-Efficiency Fixtures and Fittings — 6

Energy and Atmosphere (EA) **38 Possible Points**

- ❑ **EA Prerequisite 1.1 Optimize Energy Performance – Performance of ENERGY STAR® for Homes** — **Required**
  - ❑ EA Prerequisite 1.2 Optimize Energy Performance – Exceptional Energy Performance — 34
  - ❑ EA Credit 7.1 Water Heating – Efficient Hot-Water Distribution — 2
  - ❑ EA Credit 7.2 Water Heating – Pipe Insulation — 1
- ❑ **EA Prerequisite 11.1 Residential Refrigerant Management – Refrigerant Charge Test** — **Required**
  - ❑ EA Credit 11.2 Residential Refrigerant Management – Appropriate HVAC Refrigerants — 1

Materials and Resources (MR) – (Minimum of 2 MR Points Required) **16 Possible Points**

- ❑ **MR Prerequisite 1.1 Material-Efficient Framing – Framing Order Waste Factor Limit** — **Required**
  - ❑ MR Credit 1.2 Material-Efficient Framing – Detailed Framing Documents — 1
  - ❑ MR Credit 1.3 Material-Efficient Framing – Detailed Cut List and Lumber Order — 1
  - ❑ MR Credit 1.4 Material-Efficient Framing – Framing Efficiencies — 3
  - ❑ MR Credit 1.5 Material-Efficient Framing – Off-site Fabrication — 4
- ❑ **MR Prerequisite 2.1 Environmental Preferable Products – FSC-Certified Tropical Wood** — **Required**
  - ❑ MR Credit 2.2 Environmentally Preferable Products — 8
- ❑ **MR Prerequisite 3.1 Waste Management – Construction Waste Management Planning** — **Required**
  - ❑ MR Credit 3.2 Waste Management – Construction Waste Reduction — 3

Indoor Environmental Quality (IEQ) (Minimum of 6 IEQ Points Required)	**21 Possible Points**
❑ IEQ Credit 1 ENERGY STAR® with Indoor-Air Package	13
❑ **IEQ Prerequisite 2.1 Combustion Venting – Basic Combustion Venting Measures**	**Required**
❑ IEQ Credit 2.2 Combustion Venting – Enhanced Combustion Venting Measures	2
❑ IEQ Credit 3 Moisture Control – Moisture Load Control	1
❑ **IEQ Prerequisite 4.1 Outdoor-Air Ventilation – Basic Outdoor-Air Ventilation**	**Required**
❑ IEQ Credit 4.2 Outdoor-Air Ventilation – Enhanced Outdoor-Air Ventilation	2
❑ IEQ Credit 4.3 Outdoor-Air Ventilation – Third-Party Performance Testing	1
❑ **IEQ Prerequisite 5.1 Local Exhaust – Basic Local Exhaust**	**Required**
❑ IEQ Credit 5.2 Local Exhaust – Enhanced Local Exhaust	1
❑ IEQ Credit 5.3 Local Exhaust – Third-Party Performance Testing	1
❑ **IEQ Prerequisite 6.1 Distribution of Space Heating and Cooling – Room-by-Room Load Calculations**	**Required**
❑ IEQ Credit 6.2 Distribution of Space Heating and Cooling – Return-Air Flow/Room-by-Room Controls	1
❑ IEQ Credit 6.3 Distribution of Space Heating and Cooling – Third-Party Performance Testing/Multiple Zones	2
❑ **IEQ Prerequisite 7.1 Air Filtering – Good Filters**	**Required**
❑ IEQ Credit 7.2 Air Filtering – Better Filters	1
❑ IEQ Credit 7.3 Air Filtering – Best Filters	2
❑ IEQ Credit 8.1 Contaminant Control – Indoor Contaminant Control During Construction	1
❑ IEQ Credit 8.2 Contaminant Control – Indoor Contaminant Control	2
❑ IEQ Credit 8.3 Contaminant Control – Preoccupancy Flush	1
❑ **IEQ Prerequisite 9.1 Radon Protection – Radon-Resistant Construction in High-Risk Areas**	**Required**
❑ IEQ Credit 9.2 Radon Protection – Radon-Resistant Construction in Moderate-Risk Areas	1
❑ **IEQ Prerequisite 10.1 Garage Pollutant Protection – No HVAC in Garage**	**Required**
❑ IEQ Credit 10.2 Garage Pollutant Protection – Minimize Pollutants from Garage	2
❑ IEQ Credit 10.3 Garage Pollutant Protection – Exhaust Fan in Garage	1

- ❑ IEQ Credit 10.4 Garage Pollutant Protection – Detached Garage or No Garage — 3

Awareness and Education — 3 Possible Points

- ❑ **AE Prerequisite 1.1 Education of the Homeowner or Tenant – Basic Operations Training** — **Required**
  - ❑ AE Credit 1.2 Education of the Homeowner or Tenant – Enhanced Training — 1
  - ❑ AE Credit 1.3 Education of the Homeowner or Tenant – Public Awareness — 1
  - ❑ AE Credit 2 Education of Building Manager — 1

Project Checklist – Prescriptive Approach for Energy and Atmosphere (EA) Credits (Points cannot be earned in both the Prescriptive (below) and the Performance approaches of the EA section)

Energy and Atmosphere (EA) — 38 Possible Points

- ❑ **EA Prerequisite 2.1 Insulation – Basic Insulation** — **Required**
  - ❑ EA Credit 2.2 Insulation – Enhanced Insulation — 2
- ❑ **EA Prerequisite 3.1 Air Infiltration – Reduced Envelope Leakage** — **Required**
  - ❑ EA Credit 3.2 Air Infiltration – Greatly Reduced Envelope Leakage — 2
  - ❑ EA Credit 3.2 Air Infiltration – Minimal Envelope Leakage — 3
- ❑ **EA Prerequisite 4.1 Windows – Good Windows** — **Required**
  - ❑ EA Credit 4.2 Windows – Enhanced Windows — 2
  - ❑ EA Credit 4.3 Windows – Exceptional Windows — 3
- ❑ **EA Prerequisite 5.1 Heating and Cooling Distribution System – Reduced Distribution Losses** — **Required**
  - ❑ EA Credit 5.2 Heating and Cooling Distribution System – Greatly Reduced Distribution Losses — 2
  - ❑ EA Credit 5.3 Heating and Cooling Distribution System – Minimal Distribution Losses — 3
- ❑ **EA Prerequisite 6.1 Space Heating and Cooling Equipment – Good HVAC Design and Installation** — **Required**
  - ❑ EA Credit 6.2 Space Heating and Cooling Equipment – High-Efficiency HVAC — 2
  - ❑ EA Credit 6.3 Space Heating and Cooling Equipment – Very High-Efficiency HVAC — 4
  - ❑ EA Credit 7.1 Water Heating – Efficient Hot-Water Distribution — 2
  - ❑ EA Credit 7.2 Water Heating – Pipe Insulation — 1
  - ❑ EA Credit 7.3 Water Heating – Efficient Domestic Hot-Water Equipment — 3

- ❑ **EA Prerequisite 8.1 Lighting – ENERGY STAR® Lights** **Required**
  - ❑ EA Credit 8.2 Lighting – Improved Lighting 2
  - ❑ EA Credit 8.3 Lighting – Advanced Lighting Package 3
  - ❑ EA Credit 9.1 Appliances – High-Efficiency Appliances 2
  - ❑ EA Credit 9.2 Appliances – Water-Efficient Clothes Washer 1
  - ❑ EA Credit 10 Renewable Energy – Renewable Energy System 10
- ❑ **EA Prerequisite 11.1 Residential Refrigerant Management – Refrigerant Charge Test** **Required**
  - ❑ EA Credit 11.2 Residential Refrigerant Management – Appropriate HVAC Refrigerants 1

4.2.8 LEED™ Professional Accreditation Requirements for Neighborhood Development (Pilot)

The LEED™ for Neighborhood Development pilot program is in its final stages and is not affected by the launch of LEED™ v3. However, upon the launching of LEED™ for Neighborhood Development, USGBC says that the new LEED™ Online v3 platform will be used for submittals and a stand-alone reference guide will be released.

4.3 Identifying Standards that Support LEED™ Credits

There are an increasing number of organizations today writing and maintaining standards. Most of these standards are developed by trade associations, government agencies, or standards-writing organizations. Moreover, there is a long-standing relationship between construction codes and standards that address design, installation, testing, and materials related to the building industry. The pivotal role standards play (including green related standards) in the building regulatory process is that they represent an extension of the code requirements and are therefore equally enforceable. However, standards only have legal standing when stipulated by a particular code that is accepted by a jurisdiction.

Building standards function as a valuable design guideline for architects and engineers while establishing a framework of acceptable practices from which codes are frequently taken. When a standard is stipulated, the acronym of the standard organization and a standard number are called out. National and international standards are a critical aspect of supporting and achieving LEED™ credits. The most relevant organizations and standards relating to green issues are discussed below.

4.3.1 Organizations and Agencies

American Institute of Architects (AIA): Based in Washington, D.C., the AIA has been the leading professional membership association for licensed architects, emerging professionals, and allied partners since 1857. The goal of the AIA is to serve as the voice of the architecture profession and the resource for members in terms of service to society. The AIA carries out its goal through advocacy, information, and community.

American National Standards Institute (ANSI): ANSI approves standards as American National Standards and provides information and access to the world's standards. It is also the official U.S. representative to the world's leading standards bodies, including the International Organization for Standardization (ISO). It provides and administers the only recognized system in the United States for establishing standards.

American Society of Heating, Refrigerating, and Air-Conditioning Engineers (ASHRAE): ASHRAE is an international organization whose purpose is to advance the arts and sciences of heating, ventilation, air conditioning, and refrigeration for the public's benefit. ASHRAE's stated purpose is to write "standards and guidelines in its fields of expertise to guide industry in the delivery of goods and services to the public." The HVACR industry is making great strides in the areas of indoor-air quality, protection of the ozone layer, and energy conservation. ASHRAE leads this effort through its standards and guidelines. There are currently more than 80 ASHRAE standards and guidelines available.

American Wind Energy Association (AWEA): AWEA is a national trade association representing wind-power project developers, equipment suppliers, services providers, parts manufacturers, utilities, researchers, and others involved in the wind industry – one of the world's fastest-growing energy industries. In addition, AWEA represents hundreds of wind-energy advocates from around the world. The mission of the American Wind Energy Association is to promote wind-power growth through advocacy, communication, and education.

ASTM International (previously known as American Society of Testing and Materials): ASTM is one of the largest voluntary standards-development organizations in the world, and it provides a global forum for the development and publication of voluntary consensus standards for materials, products, systems, and services with internationally recognized quality and applicability.

BIFMA International (formerly known as the Business and Institutional Furniture Manufacturers Association): BIFMA is a nonprofit trade association of furniture manufacturers and suppliers addressing issues of common concern. The association was established in 1973 and serves as the industry voice for workplace solutions by providing standards development, statistical-data generation, government relations, industry promotion, and education.

California Energy Commission (CEC): The California Energy Commission is the state's primary energy policy and planning agency; it was created by the California Legislature in 1974. It is located in Sacramento and has responsibilities that include forecasting future energy needs and keeping historical energy data; licensing thermal power plants 50 megawatts or larger; promoting energy efficiency by setting the state's appliance and building efficiency standards and working with local government to enforce those standards; supporting public-interest energy research that advances energy science and technology through research, development, and demonstration programs; supporting renewable energy by providing market support to existing, new, and emerging renewable technologies; providing incentives for small wind and fuel-cell electricity systems; providing incentives for solar electricity systems in new home construction; implementing the state's Alternative and Renewable Fuel and Vehicle Technology Program; and planning for and directing state response to energy emergencies.

Carpet and Rug Institute (CRI): CRI is a nonprofit trade association representing the manufacturers of more than 95 percent of all carpet made in the United States, as well as their suppliers and service providers. The CRI coordinates with other segments of the industry, such as distributors, retailers and installers, to help increase consumers' satisfaction with carpet and to show how carpet creates a better environment.

Center for Resource Solutions (CRS): CRS is a national nonprofit with global impact. CRS brings forth expert responses to climate-change issues with the speed and effectiveness necessary to provide real-time solutions. Its leadership through collaboration and environmental innovation builds policies and consumer-protection mechanisms in renewable energy, greenhouse gas reductions, and energy efficiency that foster healthy and sustained growth in national and international markets.

Chartered Institution of Building Services Engineers (CIBSE): CIBSE is a professional body that received its Royal Charter in 1976 and exists to support the science, art, and practice of building-services engineering by providing its members and the public with first-class information and education services and promoting a spirit of fellowship.

Code of Federal Regulation (CFR): CFR is the codification of the general and permanent rules published in the Federal Register by the executive departments and agencies of the federal government. It is divided into 50 titles that represent broad areas subject to federal regulation. Each volume of the CFR is updated once each calendar year and is issued on a quarterly basis.

Department of Energy (DOE): DOE's primary mission is to advance the national, economic, and energy security of the United States; to promote scientific and technological innovation in support of that mission; and to ensure the environmental cleanup of the national nuclear-weapons complex. In addition, part of its mission is protecting the environment by providing a responsible resolution to the environmental legacy of nuclear-weapons production.

Energy Information Administration (EIA): The EIA was created by Congress in 1977; it is the independent statistical agency within the U.S. Department of Energy. EIA's mission is to provide policy-independent data, forecasts, and analyses to promote sound policy making, efficient markets, and public understanding regarding energy and its interaction with the economy and the environment. The agency collects data on energy reserves, production, consumption, distribution, prices, technology, and related international, economic, and financial matters. This information is disseminated as policy-independent data, forecasts, and analyses. EIA programs cover data on coal, petroleum, natural gas, electric, renewable, and nuclear energy.

Environmental Protection Agency (EPA): The EPA was founded in July 1970 by the White House and Congress working together in response to the growing public demand for cleaner water, air, and land. Prior to the establishment of the EPA, the federal government was not structured to make a coordinated attack on the pollutants that harm human health and degrade the environment. The EPA was assigned the daunting task of repairing the damage already done to the natural environment and to establish new criteria to guide Americans in making a cleaner environment a reality.

Federal Emergency Management Agency (FEMA): FEMA is an independent agency of the federal government, reporting to the President. Since its founding in 1979, FEMA's mission has been clear: to reduce loss of life and property and protect our nation's critical infrastructure from all types of hazards through a comprehensive, risk-based, emergency-management program of mitigation, preparedness, response, and recovery.

Forest Stewardship Council (FSC): The FSC was created in 1993 to establish international forest-management standards (known as the FSC Principles and Criteria) to assure that forestry practices are environmentally responsible, socially beneficial, and economically viable. Certification is a "seal of approval" awarded to forest managers who adopt environmentally and socially responsible forest-management practices and to companies that manufacture and sell products made from certified wood. Certification enables consumers, including architects, designers, and specifiers, to identify and procure wood products from well-managed sources and thereby use their purchasing power to influence and reward improved forest-management activities around the world. LEED™ accepts certification established by the internationally recognized FSC. Of note, the FSC itself does not issue certifications but rather accredits certification bodies in order to maintain its independence from those seeking certification. The FSC offers two types of certificates. The holder of a Forest Management Certificate can maintain that its operations comply with FSC standards. However, before selling products as FSC-certified, a Chain of Custody Certificate must also be obtained.

Green Building Certification Institute (GBCI): The GBCI was established as a separately incorporated entity with the support of the U.S. Green Building Council. GBCI administers credentialing programs related to green-building practice. These programs support the application of proven strategies for increasing and measuring the performance of buildings and communities as defined by industry systems such as the Leadership in Energy and Environmental Design (LEED™®) Green Building Rating Systems™.

GREENGUARD Environmental Institute: GREENGUARD is a nonprofit organization that oversees the GREENGUARD Certification Program, a program that seeks to establish acceptable indoor-air standards for indoor products, environments, and buildings. This third-party testing program evaluates low-emitting products and materials. The GREENGUARD Indoor Air Quality Certification Program certifies products based on their measured chemical emissions levels. This program incorporates the Standard for Cleaning Products and Systems and the Standard for Children and Schools. (The latter standard requires reduced chemical levels compared to the former.) Chemical emissions are measured and reviewed across a broad range of exposure levels established by the U.S. Environmental Protection Agency (EPA) and the Agency for Toxic Substances and Disease Registry (ATSDR).

Green Seal (GS): Green Seal is a Washington D.C. – based independent, nonprofit, standard-setting organization certifying a range of products and services. Green Seal (www.greenseal.org) is an independent organization that provides third-party "green" certification to various products and services. Approved products carry a Green Seal logo that is well recognized throughout industry and government as a leading environmental standard. Manufacturers pay Green Seal a fee for each product that is reviewed for certification. In addition to complying with Green Seal standards, manufacturers of approved products are subject to ongoing factory inspections (at the manufacturer's expense), product testing, and annual maintenance fees.

Illuminating Engineering Society of North America (IESNA): IESNA is the recognized technical authority on illumination. Its principal objective is to communicate information on all aspects of good lighting practice to its members, to the lighting community, and to consumers, through a variety of programs, publications, and services. IES is a forum for the exchange of ideas and information and a vehicle for its members' professional development and recognition. Through technical committees, with hundreds of qualified individuals from the lighting and user communities, IES correlates research, investigations, and discussions to guide lighting professionals and laypersons via consensus-based lighting recommendations.

Industrial Material Exchange (IMEX): IMEX is a regional program of local governments working together to protect public health and environmental quality by reducing the threat posed by the production, use, storage, and disposal of hazardous materials.

International Organization for Standardization (ISO): ISO is the world's largest developer and publisher of international standards. ISO is a network of the national standards institutes of 159 countries, one member per country, with a Central Secretariat in Geneva, Switzerland, that coordinates the system. It is a non-governmental organization that forms a bridge between the public and private sectors. On the one hand, many of its member institutes are part of the governmental structure of their countries or are mandated by their government. On the other hand, other members have their roots uniquely in the private sector, having been set up by national partnerships of industry associations.

National Fenestration Rating Council (NFRC): NFRC is a nonprofit organization that administers the only uniform, independent rating and labeling system for the energy performance of windows, doors, skylights, and attachment products. NFRC's goal is to provide fair, accurate, and reliable energy performance ratings.

New Building Institute (NBI): NBI was incorporated as a 501(c)(3) nonprofit in 1997. NBI works with national, regional, state, and utility groups to promote improved energy performance in commercial new construction, managing projects involving building research, design guidelines, and code activities. Additionally, NBI serves as a carrier of ideas between states and regions, researchers and the market.

Office of Solid Waste and Emergency Response (OSWER): OSWER provides policy, guidance and direction for the Environmental Protection Agency's solid-waste and emergency-response programs. It develops guidelines for the land disposal of hazardous waste and underground storage tanks and provides technical assistance to all levels of government to establish safe practices in waste management. The Office administers the brownfields program, which supports state and local governments in redeveloping and reusing potentially contaminated sites. The Office also manages the

Superfund program, which responds to abandoned and active hazardous-waste sites and accidental oil and chemical releases as well as encouraging innovative technologies to address contaminated soil and groundwater.

Sheet Metal and Air Conditioning Contractors' National Association (SMACNA): This standard-granting body provides an overview of air pollutants associated with construction, control measures, construction process management, quality control, communications with occupants, and case studies. Consult the referenced standard for measures to protect the building HVAC system and maintain acceptable indoor-air quality during construction and demolition activities.

South Coast Air Quality Management District (SCAQMD): SCAQMD is a governmental organization in southern California with the mission to maintain healthful air quality for its residents. The organization established source-specific standards to reduce air-quality impacts, including South Coast Rule #1168.

Sustainable Building Industry Council (SBIC): SBIC is an independent, nonprofit trade association that seeks to dramatically improve the long-term performance and value of buildings through our outreach, advocacy, and education programs. It was founded in 1980 as the Passive Solar Industries Council (PSIC) and served as a clearinghouse for passive-solar design for the major building trade groups, including those representing architects, homebuilders, the glass industry, and building owners. In the 1990s, PSIC's scope expanded to include the other major aspects of sustainable design and construction including optimizing site potential, minimizing energy use, using renewable energy sources, conserving and protecting water, using environmentally preferable products, enhancing indoor environmental quality, and optimizing operations and maintenance practices.

U.S. Green Building Council (USGBC): USGBC is a nonprofit organization, founded in 1993 to transform the way buildings and communities are designed, built, and operated, enabling an environmentally and socially responsible, healthy, and prosperous environment that improves the quality of life. A steering committee of the USGBC developed the Leadership in Energy and Environmental Design (LEED™) Green Building Rating System™ to provide universally understood and accepted tools and performance criteria that encourage and accelerate global adoption of sustainable green building and development practices. LEED™ encourages construction practices that meet specified standards, resolving much of the negative impact of buildings on their occupants and on the environment. Green buildings in the United States are certified with this voluntary, consensus-based rating system.

4.3.2 Referenced Standards and Legislation

Advanced Buildings Energy Benchmark for High-Performance Buildings (E-Benchmark): For Option A, install HVAC systems that comply with the efficiency requirements outlined in the New Buildings Institute, Inc., publication "Advanced Buildings: Energy Benchmark for High Performance Buildings (E-Benchmark)," which provides prescriptive criteria for mechanical-equipment efficiency requirements; see sections 2.4 (less ASHRAE Standard 55), 2.5, and 2.6.

ANSI/ASHRAE 52.2-1999: A method of testing general ventilation air-cleaning devices for removal efficiency by particle size. This standard presents methods for testing air cleaners for two performance characteristics: the ability of the device to remove particles from the airstream and the device's resistance to airflow. The minimum efficiency reporting value (MERV) is based on three composite average particle size removal efficiency (PSE) points. Consult the standard for a complete explanation of MERV value calculations.

ANSI/ASHRAE 55-2004: A standard for thermal environmental conditions for human occupancy. This standard specifies the combinations of indoor thermal environmental factors and personal factors that produce thermal environmental conditions acceptable to a predicted percentage of the occupants within a defined space and provides a methodology to be used for most applications including naturally ventilated spaces.

ANSI/ASHRAE 62.1-2004: A standard for ventilation for acceptable indoor-air quality. This standard specifies minimum ventilation rates and indoor-air quality (IAQ) levels to reduce the potential for adverse health effects. The standard specifies that mechanical or natural ventilation systems be designed to prevent uptake of contaminants, minimize the opportunity for growth and dissemination of microorganisms, and filter particulates if necessary. Makeup air inlets should be located away from contaminant sources such as cooling towers, sanitary vents, and vehicular exhaust from parking garages, loading docks, and street traffic.

ANSI/ASHRAE/IESNA 90–1999: An energy standard for buildings except low-rise Residential. On-site renewable or site-recovered energy that might be used to capture EA Credit 2 is handled as a special case in the modeling process. If either renewable or recovered energy is produced at the site, the ECB method considers it free energy, and it is not included in the design energy cost. See the calculation section for details.

ANSI/ASHRAE/IESNA 90.1-2004: An energy standard for buildings except low-rise residential buildings. This standard was formulated by the American Society of Heating, Refrigerating, and Air-Conditioning Engineers (ASHRAE) under an American National Standards Institute (ANSI) consensus process. It establishes minimum requirements for the energy-efficient design of buildings except low-rise residential buildings. Of note, the new LEED™ v3 references ASHRAE 90.1-2007, which is the latest updated version. The provisions of this standard do not apply to single-family houses, multifamily structures of three habitable stories or fewer above grade, manufactured houses (mobile and modular homes), buildings that do not use either electricity or fossil fuel, or equipment and portions of building systems that use energy primarily for industrial, manufacturing, or commercial processes. Building-envelope requirements are provided for semiheated spaces, such as warehouses.

ANSI/ASTM 779-03: A standard test method for determining air leakage rate by fan pressurization. Acceptable sealing of residential units must be demonstrated by blower-door tests conducted in accordance with this standard, using the progressive sampling methodology defined in Chapter 7, "Home Energy Rating Systems (HERS) Required Verification and Diagnostic Testing" of the California Low-Rise Residential Alternative Calculation Method Approval Manual.

ASHRAE 52.2: Standardized method of testing building ventilation filters for removal efficiency by particle size.

ASHRAE 55: Standard describing thermal and humidity conditions for human occupancy of buildings.

ASHRAE 62: Standard that defines minimum levels of ventilation performance for acceptable indoor-air quality.

ASHRAE 90.1: Building energy standard covering design, construction, operation, and maintenance.

ASHRAE 192: Standard for measuring air-change effectiveness.

ASTM E408: Standard of inspection-meter test methods for normal emittance of surfaces.

ASTM E408-71 (1996)e1: This provides standard test methods for total normal emittance of surfaces using inspection-meter techniques. It describes how to measure total normal emittance of surfaces using a portable inspection-meter instrument. The test methods are intended for large surfaces where nondestructive testing is required. See the standard for testing steps and a discussion of thermal emittance theory.

ASTM E903: Standard of integrated-sphere test method for solar absorptance, reflectance, and transmittance.

ASTM E903-96: This provides a standard test method for solar absorptance, reflectance, and transmittance of materials using integrating spheres. Referenced in the ENERGY STAR® roofing standard, this test method uses spectrophotometers and need only be applied to initial reflectance measurement. Methods of computing solar-weighted properties from the measured spectral values are specified. This test method is applicable to materials having both specular and diffuse optical properties. Except for transmitting sheet materials that are inhomogeneous, patterned, or corrugated, this test method is preferred over E1084. The ENERGY STAR® roofing standard also allows the use of reflectometers to measure solar reflectance of roofing materials. See the roofing standard for more details.

ASTM E1903-97 Phase II Environmental Site Assessment: This guide covers a framework for employing good commercial and customary practices in conducting a Phase II environmental site assessment of a parcel of commercial property. It covers the potential presence of a range of contaminants that are within the scope of CERCLA, as well as petroleum products.

ASTM E1980-01: This covers standard practice for calculating the solar reflectance index of horizontal and low-sloped opaque surfaces. The standard describes how surface reflectivity and emissivity are combined to calculate a solar reflectance index (SRI) for a roofing material or other surface. The standard also describes a laboratory and field-testing protocol that can be used to determine SRI.

California Title 24-2001: Though Title 24 is recognized as being more stringent for EA Prerequisite 2, for consistency and fairness projects in California must use Standard 90.1-2007 in determining performance in EA Credit 1.1.

CIBSE Applications Manual 10-2005: This provides for natural ventilation in nondomestic buildings. This manual sets out the various approaches to ventilation and cooling of buildings, summarizes the relative advantages and disadvantages of those approaches, and gives guidance on the overall approach to design. It provides detailed information on how to implement a decision to adopt natural ventilation, either as the sole servicing strategy for a building or as an element in a mixed-mode design.

"Compendium of Methods for the Determination of Air Pollutants in Indoor Air": This provides information to conduct baseline IAQ testing, after construction ends and prior to occupancy, using testing protocols of the Environmental Protection Agency and additionally detailed in the CI Reference Guide.

Comprehensive Environmental Response, Compensation, and Liability Act (CERCLA): Also known as Superfund, the Act provides a federal fund to clean up uncontrolled or abandoned hazardous-waste sites as well as accidents, spills, and other emergency releases of pollutants and contaminants into the environment. Through CERCLA, EPA was given power to seek out those parties responsible for any release and assure their cooperation in the cleanup.

EcoLogo is the standards-based labeling system in the Environmental Choice Program (ECP), Environment Canada's ecolabeling program. The program certifies a range of products, including hand cleaners, window and glass cleaners, boat and bilge cleaners, vehicle cleaners, degreasers, cooking-appliance cleaners, cleaning products with low potential for environmental illness and endocrine disruption, bathroom cleaners, dish cleaners, carpet cleaners, and disinfectants.

Energy Policy Act of 1992: This Act was promulgated by the U.S. government and addresses energy and water use in commercial, institutional, and residential facilities.

ENERGY STAR® is a government-backed program administered by the U.S. Environmental Protection Agency (EPA) and the U.S. Department of Energy (DOE). Its goal is to protect the environment by identifying and promoting the use of energy-efficient products and services. The program labels more than 35 product categories, including residential and light commercial HVAC equipment, as well as new homes and commercial buildings. For labeling commercial buildings, ENERGY STAR® evaluates conformance to energy efficiency and indoor environmental standards, primarily ASHRAE and IES standards. It uses a statistical-analysis data set (the DOE's Commercial Buildings Energy Consumption Survey) to compare energy intensity of similar buildings across the country. The blue ENERGY STAR® label can make it easier to choose energy-efficient appliances, since it can only be placed on products that meet their strict requirements. ENERGY STAR® appliances often cost more to buy than their less efficient counterparts but cost less to operate because of the lower energy requirements, which also benefit the environment. Some utilities and state governments even offer financial incentives to use these energy-efficient appliances.

EPA ENERGY STAR® Roofing Guidelines: In addition to several other building product categories, the ENERGY STAR® program identifies roofing products that reduce the amount of air conditioning needed in buildings and can reduce energy bills. Roofing products with the ENERGY STAR® logo meet EPA criteria for reflectivity and reliability. Roofing products that meet these criteria are a good starting point for achievement of this credit, but note that the requirements are not as stringent as LEED™ credit requirements; LEED™ also accounts for good emissivity in the SRI calculation. An ENERGY STAR® rating alone does not necessarily meet LEED™ credit requirements.

Environmental Technology Verification (ETV): This covers large chamber test protocols for measuring emissions of VOCs and aldehydes (September 1999). The standards referenced were developed by a testing-protocol committee under the leadership of the EPA. The protocol uses a climatically controlled test chamber in which the seating product or furniture assembly being tested is placed. A controlled quantity of conditioned air is drawn through the chamber with emission concentrations measured at set intervals over a four-day period.

FTC Guides for the Use of Environmental Marketing Claims, 16 CFR 260.7(e): According to the guide: "A recycled-content claim may be made only for materials that have been recovered or otherwise diverted from the solid-waste stream, either during the manufacturing process (pre-consumer) or after consumer use (post-consumer)." To the extent the source of recycled content includes pre-consumer materials, the manufacturer or advertiser must be able to substantiate that the pre-consumer material would otherwise have entered the solid-waste stream. In asserting a recycled-content claim, distinctions may be made between pre-consumer and post-consumer materials. In such cases any express or implied claim about the specific pre-consumer or post-consumer content of a product or package must be substantiated.

Green-e Renewable Electricity Certification Program: The Green-e Program is a voluntary certification and verification program for green electricity products. Those products exhibiting the Green-e logo are greener and cleaner than the average retail electricity product sold in that particular region. To be eligible for the Green-e logo, companies must meet certain threshold criteria for their products. Criteria include qualified sources of renewable-energy content such as solar electric, wind, geothermal, biomass, and small or certified low-impact hydro facilities; "new" renewable-energy content (to support new generation capacity); emissions criteria for the nonrenewable portion of the energy product; absence of nuclear power; and other criteria regarding renewable portfolio standards and block products. Criteria are often specific per state or region of the United States. Refer to the standard for further details.

Green Label: The Carpet and Rug Institute is a trade organization representing the carpet and rug industry. Green Label Plus is an independent testing program that identifies carpets with very low emissions of volatile organic compounds (VOCs). The carpet pad must meet or exceed CRI Green Label testing and product requirements.

Green Label Plus: Green Label Plus is an independent testing program that identifies carpets with very low emissions of volatile organic compounds (VOCs). Carpet must meet or exceed the Carpet and Rug Institute's Green Label Plus testing and product requirements. Green Label Plus does not address backer or adhesive.

Green Seal Standard GS-03: This is a standard that sets VOC limits for anticorrosive and anti-rust paints.

Green Seal Standard GS-11: This is a standard that sets VOC limits for commercial flat and nonflat paints.

Green Seal Standard 36: This is a standard that sets VOC limits for commercial adhesives.

HERS: This refers to Home Energy Rating Systems required verification and diagnostic testing. See California Low Rise Residential Alternative Calculation Method

Approval Manual. Acceptable sealing of residential units must be demonstrated by blower-door tests conducted in accordance with ANSI/ASTM-779-03, Standard Test Method for Determining Air Leakage Rate by Fan Pressurization and using the progressive sampling methodology defined in Chapter 7.

IESNA Recommended Practice Manual Lighting for Exterior Environments (IESNA RP-33-99): This standard provides general exterior lighting-design guidance and acts as a link to other IESNA outdoor lighting recommended practices (RPs). IESNA RP documents address the lighting of different types of environments. RP-33 was developed to augment other RPs with subjects not otherwise covered and is especially helpful in the establishment of community lighting themes and complements LEED™ in defining appropriate light-trespass limitations based on environmental area classifications. RP-33 addresses visual issues such as glare, luminance, visual acuity, and illuminance. Also covered are exterior-lighting design issues including community-responsive design, lighting ordinances, luminaire classification, structure lighting, and hardscape and softscape lighting. Light-level recommendations in RP-33 are lower than in many other RPs, since RP-33 was written to address environmentally sensitive lighting.

International Performance Measurement and Verification Protocol, Volume 1, 2001 Version: The IPMVP presents best practices for verifying savings produced by energy- and water-efficiency projects. While the emphasis is on a methodology geared toward performance contracting for retrofits, the protocol identifies the required steps for new building design in Section 6.0. Section 3.0 provides a general approach, procedures, and issues, while Section 4.0 provides guidance on retrofit projects.

Management Measures for Sources of Non-Point Pollution in Coastal Waters, January 1993: This document discusses a variety of management practices that can be incorporated to remove pollutants from storm-water volumes. Chapter 4, Part II, addresses urban runoff and suggests a variety of strategies for treating and infiltrating storm-water volumes after construction is completed.

Natural Ventilation in Non-Domestic Buildings: A Guide for Designers, Developers, and Owners (Good Practice Guide G237): The Guide is available for no charge, but registration (also free) is required to get access. Under the energy section of the website (http://www.carbontrust.co.uk/publications/publicationdetail.htm?productid=GPG237), search for "natural ventilation" to find the Guide. Also see: http://products.ihs.com/Ohsis-SEO/892814.html. It is based on an earlier version of the CIBSE AM10.

Resource Conservation and Recovery Act (RCRA): Congress enacted RCRA to address the increasing problems the nation faced from its growing volume of municipal and industrial waste.

South Coast Rule #1113: This Rule sets VOC limits for architectural coatings.

South Coast Rule #1168: This Rule sets VOC limits for adhesives.

U.S. Energy Policy Act of 1992 (EPACT): This Act covers efficiency levels of general-purpose industrial motors. The basic intent of the law is to reduce the rate of energy consumption in the U.S. by requiring the use of energy-efficient products. In an attempt to accomplish this goal, the Act mandates that most industrial AC motors imported or produced for sale in the United States must meet energy-efficient requirements as defined by Table 12-10 from NEMA Standard MG 1. The Act consists of 27 titles detailing various measures designed to lessen the nation's dependence on imported energy, provide incentives for clean and renewable energy, and promote energy conservation in buildings.

4.4 LEED™ Education and the LEED™ AP Exam

Nearly 75,000 people have earned their LEEDTM APs since LEEDTM first launched the professional accreditation exam in 2001. LEEDTM APs work in all sectors of the building industry and are recognized experts in green-building practices and principles and the LEEDTM Rating System. The new version of the LEEDTM exam is now formally called LEEDTM v3. Eligibility requirements to take the LEEDTM v3 exam include agreeing to disciplinary policy and credential maintenance guidelines, demonstrating or documenting involvement in support of LEEDTM projects, employment in a sustainable field of work engaged in an education program in green building principles and LEEDTM, and submitting to an application audit.

4.4.1 Preparing for the LEED™ AP Exam

The LEEDTM AP exam, while not particularly difficult, does require a good understanding of the integrated approach to green building in addition to the ability to recall a host of facts related to the rating system. The retirement date for the "old" LEEDTM AP exams (e.g., LEEDTM NC 2.2 and CI 2.0 AP exams) was June 30, 2009. Eligibility standards, exam content, exam standards, fees, and guidelines are subject to change.

In terms of the LEEDTM Green Associate exam, Beth Holst, Vice President of Credentialing for the GBCI, states that the exam will focus on the green-building principles and information that are consistent and present across all of the various LEEDTM specialties, such as New Construction, Commercial Interiors, Existing Buildings Operations and Maintenance, etc. While the LEEDTM Green Associate exam is similar in principle and length to the existing LEEDTM AP exam, it will reportedly better test a candidate's general knowledge of good environmental practices and his or her understanding of green-building design, construction, and operations.

The issue of whether a project is able to receive the Innovation Credit for utilizing a LEEDTM AP on a project using the LEEDTM Green Associate credential requires clarification. The assumption is that this benefit will remain limited to Legacy LEEDTM APs and LEEDTM AP+'s only.

The LEEDTM Professional Accreditation Exam has four sections. Each question on the exam will typically be from one of these four sections. The exam sections are listed as follows:

1. Green Building Design and Construction Industry Knowledge
2. LEEDTM Rating System Knowledge

3. LEED™ Resources and Processes
4. Green Design Strategies

It should be mentioned that LEED™ AP exam subject areas overlap somewhat, and some questions will fit into more than one area. There are four areas of competency to master for the LEED™ AP Exam. The most important subjects covered will include Implement LEED™ Process, Coordinate the Project and the Team, Knowledge of LEED™ Credit Intents and Requirements, and being able to Verify, Participate in, and Perform Technical Analyses Required for LEED™ Credits.

4.4.2 New LEED™ Reference Guides

It is encouraging to note that the new LEED™ reference guides have been consolidated into three books that address buildings by type and phase:

1. Green Building Design and Construction (for rating systems that address LEED™ for New Construction; LEED™ for Core & Shell; LEED™ for Schools; LEED™ for Healthcare; LEED™ for Retail): The USGBC says: "The LEED™ 2009 Reference Guide for Green Building Design and Construction is a user's manual that guides a LEED™ project from registration to certification of the design and construction of new or substantially renovated commercial or institutional buildings and high-rise residential buildings of all sizes, both public and private. The reference guide is a unique tool for all new construction projects, as it incorporates guidance for the three design- and construction-based LEED™ rating systems: LEED™ 2009 for New Construction and Major Renovations, LEED™ 2009 for Schools New Construction and Major Renovations and LEED™ 2009 for Core and Shell Development. This new consolidated LEED™ 2009 Reference Guide for Green Building Design and Construction includes detailed information on the process for achieving LEED™ certification, detailed credit and prerequisite information, and resources and standards for the LEED™ 2009 Design and Construction Rating Systems. For each credit or prerequisite, the guide provides: intent, requirements, point values, environmental and economic issues, related credits, summary of reference standards, credit-implementation discussion, timeline and team recommendations, calculation methods and formulas, documentation guidance, examples, operations and maintenance considerations, regional variations, resources, and definitions."
2. Green Interior Design and Construction (for rating systems that address LEED™ for Commercial Interior; LEED™ for Retail Interiors*): According to the USGBC: "The LEED™ 2009 Reference Guide for Green Interior Design and Construction is a user's manual that guides a LEED™ project from registration to certification of high-performance building interiors. This guide is specifically designed to provide the tools necessary for sustainable choices to be made by tenants and designers, who may not have control over whole building operations. The LEED™ 2009 Reference Guide for Green Interior Design and Construction includes detailed information on the process for achieving LEED™ certification, detailed credit and prerequisite information, and resources and standards for the LEED™ 2009 for Commercial Interiors Green Building Rating System. For each credit or prerequisite, the guide provides: intent, requirements, point values, environmental and economic issues, related credits, summary of reference standards, credit-implementation discussion, timeline and team recommendations, calculation methods and formulas, documentation guidance,

examples, operations and maintenance considerations, regional variations, resources, and definitions."

3. Green Building Operations & Maintenance (for rating systems that address LEED™ for Existing Buildings; LEED™ for Existing Schools): USGBC says: "This is a user's manual that guides a LEED™ project from registration to certification of a high-performance existing building. This guide is specifically designed to provide the tools necessary for building owners, managers, and practitioners to maximize the operational efficiency of their buildings while minimizing environmental impacts. The LEED™ 2009 Reference Guide for Green Building Operations and Maintenance includes detailed information on the process for achieving LEED™ certification, detailed credit and prerequisite information, and resources and standards for the LEED™ 2009 for Existing Buildings: Operations and Maintenance Rating System. For each credit or prerequisite, the guide provides: intent, requirements, point values, environmental and economic issues, related credits, summary of reference standards, credit-implementation discussion, timeline and team recommendations, and calculation methods. The LEED™ 2009 Reference Guide for Green Building Operations and Maintenance expands upon the information presented in the LEED™ for Existing Buildings 2008 edition. LEED™ for Existing Buildings: 2008 was aligned with the other long-standing LEED™ rating systems; the LEED™ 2009 Reference Guide for Green Building Operations and Maintenance has also been aligned."

The new LEED™ v3 Reference Guides can be purchased from the USGBC website, which should be checked for prices and information.

4.4.3 Procedure for Taking the LEED™ AP Exam

Below are the main points to follow in order to take the LEED™ AP exam:

- Register online to take the test through www.gbci.org (GBCI will ask you to pay the required fee) and schedule your test date and local test-center location online, utilizing the Prometric website (www.prometric.com/gbci).
- Arrive at the test center at least 30 minutes before your scheduled appointment to complete the required check-in process before testing begins. Make sure you have correct identification ready when you check in. Ensure that the name on your ID exactly matches the name you provided when you registered for the exam.
- You will not be permitted to take anything into the testing room, and you will be required to put all personal belongings (books, notes, etc.) in a locker provided at the test center.
- You will be given paper and pencil before you enter the exam for notes and calculations. Use these to write down important details that are fresh in your mind just before you start the computer-based exam.
- Upon being seated, you will have the option of taking a quick tutorial on how to use the testing software and program. Once you complete this tutorial, you can begin taking the exam.
- You will have exactly two hours to complete the exam.
- The exam will consist of 80 multiple-choice questions, the majority of which will have multiple answers. The system allows you to skip by questions and/or to mark questions you want to come back to later. You can review all of your answers at the end, or you can do so at any time by pressing the review answers button. Unanswered questions will show an "I" for incomplete. Marked questions will be flagged with a red flag to make it easier to find

them. You can call up any question to review your answer and make any changes you wish prior to completing the exam.

- The maximum passing score is 200 points; the minimum passing score is 170 points. All unanswered or incomplete exam questions will be scored as incorrect when time expires.
- Upon completing the exam, click "end exam," and your score will appear on the computer screen. If you have not completed the exam and you are reviewing questions, do not click "finish," as this will exit you from the exam and you will not be able to get back in.
- Unsuccessful candidates can schedule to retake the exam by repeating the previously described process. There is no limit to the number of times a failing candidate may retake the exam, but the full examination fee must be paid each time.
- Candidates who successfully pass the Tier I exam will be awarded LEED™ Green Associate certificates through the U.S. mail and listed as LEED™ Green Associate Professionals on the USGBC website. Candidates who successfully pass the Tier II exam will be awarded LEED™ Accredited Professional certificates through the U.S. mail and listed as LEED™ Accredited Professionals on the USGBC website.

There are three tiers of credentials. The exam for Tier 1, the LEED™ Green Associate covers basic green-building knowledge and will not require the technical details of the individual rating systems (e.g., you are not required to know the difference between ASHRAE 55 and ASHRAE 62.1) nor the in-depth knowledge that it takes to actually get LEED™ buildings built. It is the one exam that is the same for all of the different LEED™ tracks.

Beth Holst has provided guidance and clarification on the new tiered LEED™ Green Associate credentials under LEED™ 2009. Ms. Holst indicated that the LEED™ Green Associate credential will provide a level of distinction for those who are looking to establish basic green-building knowledge while working to implement the processes and principles of the LEED™ certification system and is ideal for students, teachers, marketers, manufacturers, program managers, and anyone who has ever wanted to understand what LEED™ is and why it is important to the green-building industry without actually needing to know all the technical details related to its implementation. To be eligible, a candidate is required to agree to disciplinary policy and credential maintenance guidelines.

Tier II is the LEED™ Accredited Professional. This tier builds on the Green Associate exam and tests credit knowledge of specific details and regulations. It goes into in-depth knowledge of the rating system for the area that you work in. The LEED™ AP test is more appropriate for practitioners actively working on LEED™ projects at the certification level. Here is a list of the LEED™ Rating Systems under LEED™ 2009:

- LEED™ AP BD+C: New Construction
- LEED™ AP ID+C: Interiors
- LEED™ AP Homes: Homes
- LEED™ AP O+M: Operations and Maintenance (previously EB)
- LEED™ AP ND: Neighborhood Development (pilot)

Tier III, the LEED™ AP Fellow distinction, is expected to encompass an elite class of leading professionals who are distinguished by their years of experience and who

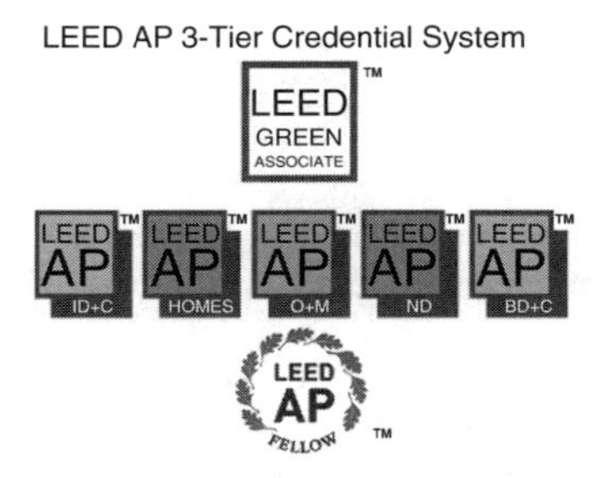

Figure 4.4 Diagram showing the new three-tier LEED™ Credential System.
Source: US Green Building Council.

have demonstrated a large track record of work in green building. Requirements for this tier remain under discussion and have yet to be determined.

Existing LEED™ APs, will become **Legacy LEED™ APs** but may opt in or upgrade to the new LEED™ AP+ by enrolling in the new tiered 2009 system. To do this, the LEED™ AP must agree to the LEED™ 2009 Credentialing Maintenance Program (CMP), sign the Disciplinary Policy, and complete the prescriptive CMP requirements for the initial two-year reporting period. Upon becoming a LEED™ AP+, you will use one of the new specialty suffixes (O&M, BD&C, ID&C) after your name. Enrollment must occur before June 2011. In choosing to make the transition to become a LEED™ AP+, there are no fees if you enroll in the CMP between June 30, 2009, and June 30, 2011. The transition to LEED™ AP+ may also be made by passing one of the new specialty examinations, called the Tier II exam. The Tier II LEED™ AP+ exam will only be required if testing takes place by June 2011. The ame rules apply here as described in the paragraph above. A current LEED™ AP also has the option to do nothing, in which case he/she will continue to be designated a LEED™ AP without a specialty title in the LEED™ Professional Directory.

4.4.4 *Tips for Passing the Exam*

Here are some tips for passing the exam:

- Success really does require a detailed study of the reference manual and the USGBC/GBCI website in addition to this book. Pay particular attention to the introductory sections on green-building design/construction and the introductions to each individual credit area (introductory pages to SS, WE, EA, MR, and IEQ). Many questions are drawn from these areas.
- For the exam the candidate should understand:
  - How to register a project
  - The steps that are involved in the certification process
  - The Credit Interpretation Process
  - The Appeal Process
  - Who is (and is not) part of a project team and project review
- Know and memorize which items are prerequisites and which are credits. There are usually several questions that will ask you to differentiate.
- Know the various industry publications (*Green Spec*, *Building Green*, etc.)

- Understand and memorize each of the opportunities for exemplary performance, innovation and design, and regional priority credit thresholds and percentages.
- Know which credits have calculations and ensure that you know how to perform them – especially credits dealing with FTE or number of plumbing fixtures.
- Know all of the SMACNA standards and what LEED™ points they relate to.
- Practice tests and then, more practice tests! Sample practice tests can be found on various Internet sites and in the Exhibits section of this book.
- Know all of the sample questions in USGBC's study guide (on the website). There are often a number of these word for word on the test.
- Many questions focus on the Energy and Atmosphere credits, especially strategies to optimize energy performance.
- Understand the three types of renewable energy discussed in the reference guide and in Chapter 9 of this book (wind, biomass, PVs)
- Understand the differences between the Fundamental Commissioning prerequisite and the Enhanced Commissioning credit.
- Understand the differences between the IAQ prerequisite and the IAQ credit; know the steps for Construction IAQ.
- Toilets use the most water of any fixture in commercial buildings.
- Know who the commissioning agent should be for Fundamental Commissioning and for Enhanced Commissioning.
- Know that for brownfields, the letter must be from the local regulatory agency or the regional EPA office.
- There are two free credit interpretations. Know how you apply for an CIR interpretation.
- Green Seal is the standard by which paints are measured in VOCs. Know the difference between GS-36, GS-11, and GS-3.
- Make sure you scan through all of the Summary or Referenced Standards. There are a couple of questions that ask which of several is not a reference standard.
- Know the definitions for emissivity and albedo as well as how they are used within the different credits (heat islands, energy efficiency).
- Know the definition and uses of composting toilets, waterless urinals, low-flow fixtures, etc.
- Know the definition of Rapidly Renewable Resources and the various materials.
- Understand the difference between Stormwater Management, Erosion and Sedimentation Control, and Wastewater Technologies.
- Be very familiar with where ASHRAE 55 and ASHRAE 62 apply.
- Try to memorize each prerequisite and credit number, name, intent, requirements, and any referenced standards, strategies, exemplary performance possibilities, point value, submittals, and decision makers.
- Some of the test questions may throw you. Don't let them; just skip the ones you don't think you know and come back to them later. Concentrate first on completing the questions you know. Mark any questions you are not sure of and come back to them toward the end. You should have plenty of time for review.

Finally, if you decide to sit the exam and you have adequately prepared for it, you will pass. Good luck!

5 Design Strategies and the Green Design Process

5.1 Conventional versus Green Delivery Systems

Although most of our time is spent inside buildings, we seem to take for granted the shelter, protection, comfort, air, and light that buildings provide, and rarely do we give much thought to the systems that allow us to enjoy these services unless there's a power interruption or other problem. Moreover, few Americans understand the environmental consequences of maintaining indoor comfort levels. This may be partly because buildings are often deceptively complex. While connecting us with the past, they reflect our greatest legacy for the future. Buildings also provide us with shelter to live and work, embody our culture, and play a pivotal part in our daily life on this earth. But the role of buildings continues to change as they become increasingly costly to build and maintain, as well as requiring constant adjustment to function effectively over their life cycle.

The term "green or sustainable architecture" basically means smart construction and design; it is smart for the consumer and smart for the planet. This is particularly relevant today with the supply of the world's fossil fuel rapidly dwindling, the concerns for energy security increasing, and the impact of greenhouse gases on our climate rising; there is a dire need to find ways to reduce energy loads, increase efficiency, and employ renewable energy resources in our facilities. Green construction is environmentally friendly because it uses sustainable, location-appropriate building materials and employs building techniques that reduce energy consumption. Indeed, the primary objectives of sustainable design are to avoid resource depletion of energy, water, and raw materials and prevent environmental degradation caused by facilities and infrastructure throughout their life cycle. Likewise, it places a high priority on health, which is why green buildings are also safe, aesthetically pleasing, and comfortable.

In addition to the classical building design concerns – economy, utility, durability, and delight – green design strategies underline emerging concerns regarding occupant health, the environment, and resource depletion. There are many green design strategies and measures that can be employed to help address these concerns, including:

- Reduce human exposure to hazardous materials.
- Maximize the use of renewable energy and materials that are sustainably harvested.
- Conserve nonrenewable energy and scarce materials.
- Use water efficiently and minimize waste water and runoff.
- Use energy as efficiently as possible and minimize the ecological impact of energy and materials used.

DOI: 10.1016/B978-1-85617-691-0.00005-9

- Protect and restore local air, water, soils, flora, and fauna.
- Encourage occupants to use bicycles, mass transit, and other alternatives to fossil-fueled vehicles.
- Optimize site selection in order to preserve green space and minimize transportation impacts.
- Building orientation should take maximum advantage of sunlight and microclimate.
- Minimize materials impacts by employing green products.

The majority of conventionally designed and constructed buildings existing today are having a negative impact on the environment as well as on occupant health and productivity. Moreover, these buildings are expensive to operate and maintain, and they contribute to excessive resource consumption, waste generation, and pollution. Reducing negative impacts on our environment and adopting new environmentally friendly goals and guidelines that facilitate the development of green/sustainable buildings such as are outlined by the USGBC and similar organizations must be a priority for our generation. USGBC's LEED™ guidelines include required and recommended practices that are intended to reduce life-cycle environmental impacts associated with the construction and operation of both commercial and municipal developments and major remodeling projects.

Green construction, unfortunately, is still not considered mainstream in the United States. However, rising costs of energy and building materials, coupled with warnings from the EPA about the toxicity of today's treated and human-made materials, are prompting engineers and architects to take a new look at building techniques that use native resources as construction materials and utilize nature in the heating and cooling process. The majority of green buildings are efficient and high-quality; they last longer and cost less to operate and maintain. They also provide greater occupant satisfaction than conventional developments, which is why most sophisticated buyers and tenants prefer them, and would consider paying a premium for their advantages.

Green building takes into account renewable, recycled, and native building materials to minimize the adverse impact on the environment and to harmonize with their surroundings. A green building takes advantage of the sun's seasonal position to heat the interior in winter and often incorporates design features such as light shelves, overhanging eaves, or landscaping to mitigate the sun's heat in summer. Room orientation is designed to improve ventilation. Naturally insulative materials such as earth and straw are often employed in the construction of walls.

Some green buildings also incorporate solar, wind, or other alternative energy sources into the HVAC system to reduce operational costs and minimize the use of fossil fuels. Green buildings also generally avoid using potentially toxic materials such as treated woods, plastics, and petroleum-based adhesives, which can degrade air and water quality and cause health problems. Incorporating operating and maintenance factors into the design process of a building project will contribute significantly to healthy working environments, higher productivity, and reduced energy and resource costs. Whenever possible, designers should always specify materials and systems that simplify and reduce maintenance and life-cycle costs, use less water and energy, and are cost-effective.

The vast majority of today's buildings use mechanical equipment powered by electricity or fossil fuels for heating, cooling, lighting, and maintaining indoor-air quality. According to the U.S. Department of Energy (DOE), buildings in the U.S. annually consume more than one-third of the nation's energy and contribute roughly 36 percent of the carbon dioxide (CO_2) emissions released into the atmosphere. The fossil fuels used to generate electricity and condition buildings cause a heavy toll on the environment; they emit a plethora of hazardous pollutants (e.g., volatile organic compounds) that cost building occupants and insurance companies millions of dollars annually in healthcare costs. Mining and extraction of fossil fuels add to the environmental impacts in addition to the instability in pricing, which causes concern among both business people and homeowners. Buildings that use less energy both reduce and stabilizes cost as well as having a positive impact on the environment.

The DOE had the foresight to realize early on the need for buildings that were more energy-efficient and in 1998 took the initiative to work with the commercial-buildings industry to develop a 20-year plan for research and development on energy-efficient commercial buildings. The primary mission of DOE's High-Performance Buildings Program is to help create better buildings that save energy and provide a quality, comfortable environment for workers. The principal target of the program is the building community, especially building owners/developers, engineers, and architects. Today we have the knowledge and technologies required to reduce energy use in our homes and workplaces without compromising comfort and aesthetics. Unfortunately, however, the building industry is not taking full advantage of these advances, and to this day many buildings are still being designed and operated without considering all the environmental impacts.

Careful consideration of building performance is a valuable prerequisite for designers. Once performance measures are selected, it is necessary to follow up and establish performance targets and the metrics to be used for each measure. Minimum expectations, or baselines, are generally defined by codes and standards. Otherwise, performance baselines can be set to exceed the average performance of a building type. Alternatively, performance can be measured against the most recent building built or against the performance of the best building of a particular type that has been documented.

A number of green-building rating systems have been developed to set standards for the evaluation of high performance. The most widely recognized system for rating building performance in the United States is LEED™ (Leadership in Energy and Environmental Design), which provides various consensus-based criteria to measure performance, along with useful reference to baseline standards and performance criteria. Nevertheless, LEED™ certification by itself does not ensure high performance in terms of energy efficiency, as certification may have been achieved by succeeding in other non-energy-related LEED™ categories such as Materials and Resources or Sustainable Sites. For this reason specific energy-related goals must still be set.

Integrated design is enhanced by the use of computer energy-modeling tools such as the Department of Energy's DOE 2.1E and other computer programs, which can inform the building team of energy-use implications very early in the design process by factoring in relevant information such as climate data, seasonal changes, building massing and orientation, and daylighting. Modeling facilitates prompt exploration of cost-effective design alternatives for the building envelope and mechanical systems by forecasting energy use of various alternatives in combination.

High-performance building projects are more effectively initiated with a green-design "charrette," or multidisciplinary kickoff meeting to formulate a clear roadmap for the project team to follow. A charrette is useful because it offers team professionals, often with the assistance of green-design experts and facilitators, the opportunity to brainstorm design goals and alternatives. This goal-setting approach helps identify green strategies for members of the design team and facilitate the group's ability to reach a consensus on performance targets for the project.

As the world of buildings continues to grow in complexity, additional programs and information will have an impact on the entire design, planning, and construction community. Among them is the introduction of building-information modeling (BIM) software, which is the newest trend in computer-aided design and is being touted by many industry professionals as a lifesaver for complicated projects because of its ability to correct errors at the design stage and accurately schedule construction. Moreover, some industry professionals forecast that buildings in the near future will be built directly from the electronic models that BIM creates and that the design role of architects will dramatically change (Figure 5.1). BIM is gradually changing the role of drawings in the construction process, improving architectural productivity, and making it easier to consider and evaluate design alternatives. BIM is also helping to facilitate the process of integrating the various design teams' work, thus adding to the urgency of utilizing an integrated team process.

BIM consists of 3D-modeling concepts, information-database technology, and interoperable software in a computer-application environment that architects, engineers, and contractors can use to design a facility and simulate construction. This modeling technology enables members of the project team to create a virtual model of the structure and all of its systems in 3D in a format that can be shared with the entire team. All drawings, specifications, and construction details are integral to the model, which incorporates building geometry, spatial relationships, geographic information, and quantity properties of building components. This allows team members to identify design issues and construction conflicts and resolve them in a virtual environment long before construction actually commences.

As BIM increases in popularity, it will rapidly become the foundation of building design, visualization studies, contract documents, cost analysis, 3D simulation, and facilities management. Autodesk Revit® is making considerable headway in its market penetration into architectural firms, and it is anticipated that within the next few years the majority of major projects designed in the United States will be modeled in Revit®.

5.2 Green Design Strategies

Clear design objectives are critical to any project's success, yet to achieve maximum benefit the project goals must be identified early on and held in proper balance during

Figure 5.1 Highlands Lodge Resort and Spa project, a joint venture of Q&D Construction and Swinerton Builders, Inc., in which Vico, a BIM software package, was used. The multi-use five-star hotel and high-end luxury condominiums have a total gross floor area of 406,500 square. feet/37,720 square meters and are built on a roughly 20-acre site at the Northstar-at-Tahoe Ski Resort in northern California.
Source: Vico Software, Inc.

the design process. An integrated approach to design and construction must be employed to achieve a high-performance building. By working together, for example, the civil engineer, landscape architect, and design architect can maneuver and direct the ground plane, building shape, section, and planting scheme to provide increased thermal protection and reduce wind loads and heat loss/gain. This allows them to reduce heating and cooling loads and, with the participation of the mechanical engineer, reduce the size of mechanical equipment necessary to achieve comfort. The architectural, lighting, and mechanical engineers can work together to design, for example, an effective interior/exterior light shelf that can serve not only as an architectural feature but can also provide sunscreening, and reducing summer cooling load while at the same time letting daylight penetrate deep into the interior. The resultant outcome is almost always more efficient environmental performance at the equivalent first cost, followed by a stream of ongoing operational savings. Should first cost exceed initial estimates, it should be easily offset by decreased life-cycle dollar and environmental costs.

Another feature of green design is "smart growth," which is an issue that concerns many communities around the country. It relates to the ability to control sprawl, reuse existing infrastructure, and create walkable neighborhoods. The locating of places to live and work near public transportation is an obvious plus toward reducing energy. Likewise, it is more resource-efficient and logical to reuse existing roads and utilities than build new ones in rural areas at considerable distances from cities. Smart growth is needed to preserve open spaces, farmlands, and undeveloped land and strengthen the evolution of existing communities and their quality of life.

However, before delving too deeply into green design and the integrated design process, it may be prudent to first describe the more conventional design process. The traditional approach to design, simply put, usually begins with the architect and the client agreeing on a budget and design concept, which consists of a general massing scheme, typical floor plan, schematic elevations, and, usually, the general exterior appearance as determined by these characteristics as well as by the basic materials. The mechanical and electrical engineers are then asked to implement the design and to suggest appropriate systems.

This conventional design approach, while greatly oversimplified, in fact continues to be employed by the majority of general-purpose design consultant firms, which in turn unfortunately tends to limit the achievable performance to conventional levels. The traditional design process consists mainly of a linear structure, due to the sequential contributions of the members of the design team. The opportunity for optimization is limited during the traditional process, while optimization in the later stages of the process is often difficult if at all viable. Thus, while this process may seem adequate and even appropriate, the actual results often tell a different story, with high operating costs and a substandard interior environment. These factors can have a significant impact on a property's ability to attract quality tenants or achieve desirable long-term rentals in addition to reducing the asset value of the property. Moreover, since conventional design methods do not typically use computer simulations of predicted energy performance, the consequential dismal performance and steep operating costs will often come as an unpleasant surprise to the owners, operators, or users.

5.2.1 The Integrated Design Process (IDP)

The IDP approach differs from the conventional design process in a number of ways. For example, the owner takes on a more direct role in the process, and the architect takes on the role of team leader rather than sole decision maker. Other consultants including the structural, electrical, mechanical, lighting, and other relevant players become part of the team from the outset, not after completion of the initial design, and participate in the project's decision-making process (Figure 5.2).

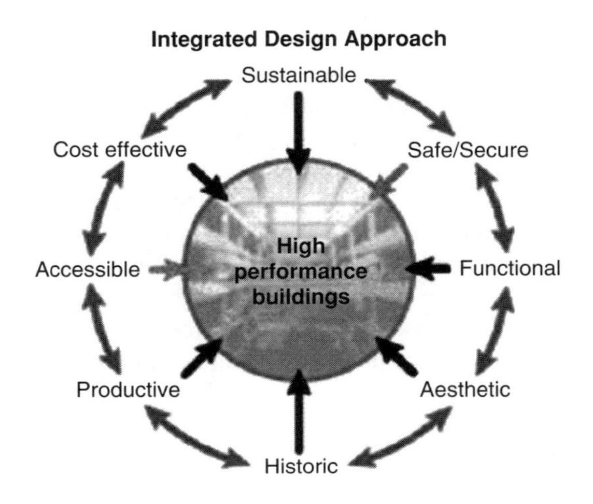

Figure 5.2 Diagram showing the various elements that impact the design of high-performance buildings using IDP.
Source: Don Prowler, FAIA – Donald Prowler & Associates/Stephanie Vierra – Steven Winter Associates, Inc.

To ensure success of any building project, the following strategies should be carefully considered:

- Establish a vision that embraces sustainable principles in order to apply an integrated design approach to a project.
- Articulate a clear statement of the project's requirements, objectives, design criteria, and priorities.
- Develop a budget for the project that encompasses green-building measures. Allocate contingencies for possible incorporation of additional options.
- Establish performance goals for siting, energy, water, materials, and indoor environmental quality along with other sustainable-design goals and ensure the incorporation of these goals throughout the design and life cycle of the building.
- Incorporate building-information modeling (BIM) concepts to enable members of the project team to create a virtual model of the project and its various systems in 3D in a format that can be shared with the entire project team.
- Form a design and construction team that is committed to the project vision. During the selection process ensure that contractors are appropriately qualified and capable of implementing an integrated system of green-building measures.
- Include in the contract documents a project schedule that allows for systems testing and commissioning.
- Develop contract plans and specifications that ensure that the building design is at an appropriate level of building performance.
- Implement a building management plan that ensures that operating decisions and tenant education are implemented in a satisfactory manner with regard to integrated, sustainable building operations and maintenance.
- Consider all stages of the building's life cycle, including deconstruction.

Depending on a project's size, type, and complexity, the decision to identify a leader for the project usually rests with the user/client, who sees the need for a person who is proficient and able to lead a team to design and build the project on the basis of quantifiable requirements for space and budgetary capacity to undertake the activity. A project brief should accompany this planning activity that describes existing-space use; develops realistic estimates of both spatial and technical requirements, and arrives at a space program around which design activity can develop. In the case of larger projects, a construction manager or a general contractor may come on board at this point.

Upon completion of the predesign activities, the architect, designer of record (DOR), and other prime consultants, in collaboration with the other team members or subconsultants, may produce initial graphic suggestions for the project or portions of it via a 3D modeling program (e.g., BIM) or manually. Such suggestions are meant more to stimulate thought and discussion than to describe the final outcome, although normally the fewer changes made before bidding the project, the more cost-effective the project will be. Involvement of subconsultants is a critical part of the process at this stage, and their individual insights can prevent costly changes further along in the process. A final design gradually emerges that embodies the interests and requirements of all participants including the owner while also meeting the overall area requirements and project budget that was established during predesign activities.

The resulting schematic design produced at this stage should show site location and organization, a 3D model of the project, space allocation, and an outline specification including an initial list of components and systems to be designed and/or specified for the final result. A preliminary cost estimate should also be made, and, depending on the size of the project, it may be performed by a professional cost estimator or computer program at this point. For smaller projects, one or more possible builders may perform this service as part of a preliminary bidding arrangement. On larger projects, a cost estimate can be linked to the selection process for a builder, assuming that other prerequisites such as bonding capacity, experience, and satisfactory references are met.

The design-development phase entails going into greater detail for all aspects of the building, such as systems, materials, etc. The collaborative process continues, with the architect working hand in hand with the owner and the various contributors. The conclusion of this phase is a detailed design on which a consensus exists among all players, who may be asked to sign off. When the design is developed in a spirit of collaboration among the main contributors, the end result is usually a design that is highly efficient with minimal, if not zero, incremental capital costs, and reduced long-term operating and maintenance costs.

The development and production of contract documents involves converting the design-development data into formats that can be used for pricing and bidding, permitting, and construction. While no set of construction documents can ever be perfect, an efficient set can be achieved by careful scrutiny and accountability to the initial program needs as outlined by the design team and the client in addition to careful coordination and collaboration with the technical consultants on the design team. Design, budgetary, and other decisions continue to be made with the appropriate contributions of the various players. Any changes in scope during this phase can significantly impact the project and, once pricing has begun, can invite confusion, errors, and added costs. Cost estimates may be made at this point, prior to or simultaneously with bidding, in order to assure compliance with the budget and to check the bids. Accepted bids may be used as a basis for selecting a contractor.

Upon selection of the general contractor and during the construction phase, members of the project team must remain fully involved. Previous decisions may require clarification; supplier samples and information must be reviewed for compliance with the contract documents; and proposed substitutions need evaluation. Whenever proposed changes affect the operation of the building, it is imperative to inform and involve the user/client. Any changes in user requirements may necessitate changes in the building; such changes will require broad consultation among the consultants and subconsultants, revised pricing, as well as incorporation into the contract documents and the building.

The design team is ultimately responsible for ensuring that upon the building's completion it meets the requirements of the contract documents. The building's success or otherwise of meeting the program performance requirements can be assessed by an independent third party in a process known as commissioning, where the full range of functions in the building are evaluated and the design and construction team can be called upon to make any required changes and adjustments. This is discussed in greater detail in Section 5.5 of this chapter.

After the building is fully operational, a postoccupancy evaluation is sometimes conducted to assess how well the building meets the original and emerging requirements for its use. Such information is especially useful if the owner/developer anticipates further construction of the same type to assist in avoiding mistakes and repeating successes.

As buildings consist of a number of complex systems of interacting elements, careful combinations of design strategies are necessary to produce a successful and sustainable project. Intelligent green design takes into account the effects of the various elements on one another and on the building as a whole. For example, the need for mechanical and electrical systems is greatly influenced by a building's form, orientation, and envelope design. Allied strategies such as daylighting, natural cooling, solar-load control, and natural ventilation can all work in unity to reduce lighting, heating, and cooling loads, and the cost of necessary equipment to meet them.

Buildings that are designed to adapt to changing uses over long periods of time reduce life-cycle resource consumption. Adaptable structural elements that provide generous service space and accommodate open plans with non-load-bearing partitions can last centuries and easily outlast buildings that cannot adapt to changing times and functions. Durable, well-designed envelope assemblies reduce life-cycle maintenance and energy costs while improving indoor-air quality and comfort. Mechanical and electrical systems that are readily accessible and that allow easy modifications save materials and money whenever tenant improvements or renovations are needed. Durability and use remain valuable criteria in evaluating aspects of sustainability and green design.

Preventing problems from the outset instead of fixing them after the fact always makes practical and economic sense. Thus, the use of low-toxicity building materials and installation practices is much more effective than diluting indoor-air pollution from toxic sources by the use of increased air ventilation. Likewise, using efficient designs that minimize heating, cooling, and lighting loads is far more cost-effective than installing bulky or larger HVAC and electrical equipment to address the problem.

Climate-responsive design has come of age as it rediscovers the powerful relationship of buildings to place. Buildings that respond to local topography, microclimate, vegetation, and water resources are typically better suited to meet today's challenges; they are more comfortable and efficient than conventional designs that ignore their surroundings by relying on technological fixes to address the various environmental and other issues that continuously crop up. Many regions have excellent solar and wind resources for passive-solar heating, natural cooling, ventilation, and daylighting. Local water supplies, on the other hand, are rapidly being depleted or polluted. Taking advantage of available natural resources and conserving commodities that are being rapidly depleted are two of the best ways to reduce costs and connect occupants to their surroundings.

5.2.2 The Green-Building Project Delivery Process

For conventional (non-green) buildings, the various specialties associated with project delivery, from design and construction through building occupancy, are responsive in nature and utilize restricted approaches to address particular problems. Each of these specialties typically has wide-ranging knowledge in a specific field and seeks solutions to any problems that arise by solely using that knowledge. For example, an air-conditioning specialist, when asked to address the problem of an unduly warm room, will immediately suggest increasing the cooling capacity of the HVAC system servicing that room rather than investigating why this room is unduly warm. The excessive heat gain could, for example, be mitigated by incorporating external louvers or operable windows. The end result of the conventional process is often a functional but highly inefficient building comprised of different materials and systems with little integration among them.

From a design perspective, the key process difference between green-building design and conventional design is the concept of integration. Green buildings typically use an integrated design process (IDP), which relies on a multidisciplinary team of building professionals who typically collectively work together from the predesign phase through postoccupancy to produce a superior, cost-effective, environmentally friendly, high-performance building. The essence of integrated design is that buildings consist of interconnected or interdependent systems, all of which impact one another to some degree.

Properly engineered systems help to ensure the comfort and safety of all building occupants. They also enable designers to create environments that are healthy, efficient, and cost-effective. Integrated design is a critical and consistent component in the design and construction of green buildings. The brief description above highlights the benefits of integrated design and the main differences between conventional and integrated design.

5.2.3 The Integrated Multidisciplinary Project Team

Some of the professions required to join the project team in an integrated process may include the following.

The owner's representative (OR) is the person who represents the owner and speaks on his/her behalf. The OR must be prepared to devote the time needed to advocate, defend, clarify, and develop the owner's interests. The OR may come from within the organization commissioning the project or be hired by the owner as a consultant.

The construction manager (CM) is normally hired on a fee basis to represent the logistics and costs of the construction process and can be an architect, a general contractor, or specifically a consulting construction manager. It is important for this person to be involved from the beginning of the project.

In most building projects it is the architect who typically leads the design team and who coordinates with subconsultants and other experts to ensure compliance with the project brief and budget. In some cases the architect has the authority to hire some or all of the sub-consultants; in larger projects the owner may decide to contract directly with some or all of them. The architect usually manages the production of the contract documents and oversees the construction phase of the project, ensuring compliance with the contract documents by conducting appropriate inspections and managing submissions approvals, and evaluations by the subconsultants. The architect also oversees the evaluation of requests for payment by the builder and other professionals.

Early involvement of the civil engineer is essential for understanding the land, soil, and regulatory aspects of a construction project; the civil engineer may be hired directly by the owner or by the architect. The role of the civil engineer usually includes preparation of the civil-engineering section of the contract documents in addition to assessing compliance of the work with the contract documents.

Consulting structural, mechanical, and electrical engineers may be engaged by the architect or on larger or more complex projects directly by the owner. They are responsible for designing the structural, heating, ventilation, air-conditioning, power, and illumination aspects of the project. Each produces his or her own portions of the contract documents, and all participate in assessing their part of the work for compliance with those documents.

The landscape architect, depending on the type and size of the project, is often hired as an independent consultant. He/she should be involved early in the design process to assess existing natural systems, how they will be impacted by the project, and the best ways to accommodate the project to those systems. The landscape architect should have extensive experience in sustainable landscaping including erosion control, green roofs, indigenous plant species, managing stormwater runoff, and other factors.

Specialized consultants are adopted into the design team as needed by the requirements of the project. They may include lighting consultants, specifications writers, materials and components specialists, sustainability consultants, and technical specialists such audiovisual, materials handling, and parking engineers. The size, complexity, and specialization of the project will suggest the kinds of additional experts that may be needed. As with all contributors to the integrated design process, they should be involved early on to include their suggestions and requirements for the design so as to ensure that their contributions are taken into account and do not become remedial.

5.3 Design Process for High-Performance Buildings

The world is rapidly changing, and building-construction practices and advances in architectural modeling technologies have reached a unique crossroad in history. And with many successful new building projects taking shape throughout the country today, they call into question the performance level of our more typical construction endeavors and make us wonder just how far our conventional buildings are falling short of the mark and what needs to be done to meet the new challenges.

The process of green-building design and construction differs fundamentally from current standard practice. The features that are pivotal to successful green-building design require a stringent commitment to health, the environment, and resource-use performance targets by developers, designers, and builders. Measurable targets challenge the design and construction team and allow progress to be tracked and managed throughout development and beyond. Post-occupancy evaluations are often used as a marketing tool as well as to demonstrate performance achievement of ambitious targets.

High-performance outcomes also necessitate a far more integrated team approach to the design process and mark a departure from traditional practices, in which emerging designs are handed sequentially from architect to engineer to subconsultant. Close collaboration by multi-disciplinary teams, from the commencement of the conceptual design process throughout design and construction, is critical for success. Typically, the design team is expanded to include additional members, such as energy analysts, materials consultants, environmentalists, lighting designers, and cost consultants (Figure 5.3).

An integrated design process requires the inclusion of contractors, operating staff, and prospective occupants to increase the likelihood of success. The enlarged design team offers fresh perspectives and approaches and feedback on performance and cost. This results in a typically more unified, more team-driven design and construction process that encompasses different experts early in the design setting process. This in turn increases the likelihood of creating high-performance buildings that achieve significantly higher targets for energy efficiency and environmental performance. The design process becomes one continuous, sustained team effort from conceptual design through commissioning and final occupancy.

A team-driven approach is considered to be a "front-loading" of expertise. The process should typically begin with the consultant and owner leading a green-design charrette with all stakeholders (design professionals, operators, and contractors) in a brainstorming session or "partnering" approach that encourages collaboration in achieving high performance green goals for the new building while breaking down

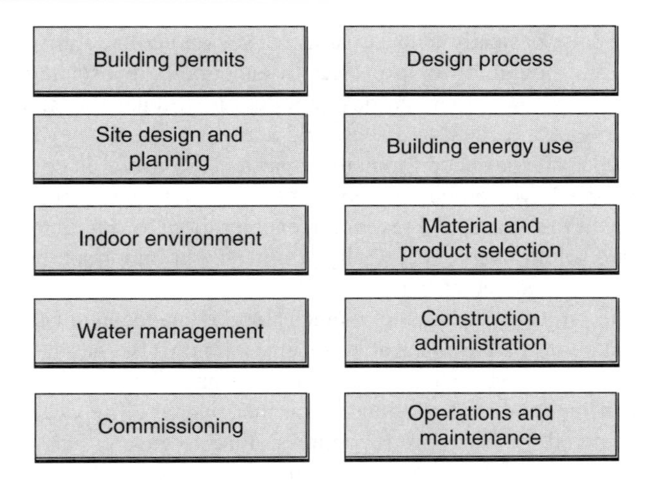

Figure 5.3 Elements of high-performance building design.

traditional adversarial roles. During the design development phase, continuous input from users and operators can advance progress, instigate commitment to decision making, minimize errors, and identify opportunities for collaboration.

Superior results can be achieved in building design and construction through the application of best practices and an integrated team-driven approach. It is the application of integrated design methods that elevates mere energy- and resource-efficiency practices into the realm of high performance. This approach differs from the typical planning and design process of relying on the expertise of various specialists, who work in their respective specialties in relative isolation from one another. The Integrated design process encourages designers from all the disciplines – architectural, civil, electrical, mechanical, landscape, hydrology, interior design, and others—to collectively be involved in the design decision-making process. Coming together at the beginning of the project's preliminary-design phase and at key stages in the project-development process allows professionals to work together to achieve exceptional and creative design solutions that yield multiple benefits and beneficial synergies at no extra cost.

In the absence of an interactive approach to the design process it would be extremely difficult to achieve a successful high-performance building. The successful design of buildings requires the integration of many kinds of information into a syncretic whole. The process draws its strength from the knowledge pool of all the stakeholders (including the owner) across the life cycle of the project. They are collaborative involved from defining the need for a building through planning, design, construction, operation, maintenance of the facility, and building occupancy. The best buildings result from active, consistent, organized collaboration among all players, which is why the stakeholders need to fully understand the issues and concerns of all the parties and be able to interact closely throughout the various phases of the project.

A design charrette basically consists of a focused and collaborative brainstorming session held at the beginning of a project to encourage an exchange of ideas and information and to allow integrated design solutions to take shape. Team members (the stakeholders) are expected to discuss and address problems beyond their field of expertise. The charrette has been found to be particularly useful in complex situations where many people have conflicting needs, interests, and constituencies. Participants are educated about the issues, and resolution enables them to "buy into" the schematic solutions. A final solution isn't necessarily produced, but important, often interdependent, issues are explored and clarified.

Designing the project in a holistic manner alone is inadequate, because the effectiveness and outcome of the integrated design solution also has to be determined. Conducting a facility performance evaluation to ensure that the high-performance goals have been met and will continue to be met over the life cycle of the project should be considered. Likewise, retrocommissioning to ensure that the building will continue to optimally perform through continual adjustments and modifications should be considered.

Computer energy simulations should be conducted as early as possible in the design process and continue until the design is complete to enable proper assessment of energy-conservation measures. The collaboration of the expanded design team early in conceptual design allows the team to generate several alternative concepts for the building's form, envelope, and landscaping and to concentrate on reduction of peak energy loads, demand, and consumption. Computer energy simulation is an excellent tool to assess the project's effectiveness in energy conservation as well as its construction costs. Employing sustainable approaches that reduce heating and cooling loads, such as more efficient glazing, insulation, lighting systems, daylighting, and similar measures, allows the mechanical engineer to follow suit by designing a more appropriate, more efficient, and less expensive HVAC equipment and systems that in turn would result in minimal if any increase in construction cost compared to conventional designs.

Simulations are most often used to see how a design can be improved so that energy-conservation and capital-cost goals are met and to ensure that a design complies with regulatory requirements. Design alternatives are typically evaluated on the basis of either capital cost or the reduced life-cycle cost. In either case the principal aim of producing alternative designs is to simultaneously minimize both a building's life-cycle cost and its construction cost. But in order to more accurately estimate these costs, a comprehensive approach is required that includes costs and environmental impacts of everything from resource extraction; materials and assembly manufacture; construction; operation, and maintenance in use to final reuse, recycling, or disposal. Computer energy simulation is but one of several tools employed to include operational costs into the analysis; other computer tools facilitate life-cycle cost analysis.

The integrated design process reflects a deeper analysis of the project than would be typical of a conventional design practice and requires greater effort from design consultants. Design fees for this additional work typically reflects the increased work being implemented, but such an investment is insignificant when compared to the environmental and cost impacts over the life of a typical building.

High-performance, sustainable/green buildings are gradually entering the mainstream as an important market sector in the United States and globally. At the same time this increased demand for high-performance buildings is encouraging facility owners and design professionals to rethink the design process. This reevaluation of emerging patterns and key processes on successful high-performance building projects is having a substantial impact on both the private and public sectors.

For example, a Federal Leadership in High-Performance and Sustainable Buildings Memorandum of Understanding (MOU) was signed in January 2006, in which the signatory agencies committed to federal leadership in the design, construction, and operation of high-performance and sustainable buildings. An important element of this strategy is the implementation of common strategies for planning, acquiring, siting, designing, building, operating, and maintaining the buildings. Included in the MOU are a number of guiding principles, including optimizing energy performance, conserving water, improved IEQ, integrated design, and reduction of the impact of materials. Since the signing of the MOU many federal facilities have succeeded in creating high-performance buildings that save energy and reduce the environmental impact on our lives.

On December 5, 2008, a new memorandum on high-performance federal buildings was issued. It includes revised guiding principles for new construction as well as for existing buildings. It also includes clarification of reporting guidelines for entering information on the sustainability data element (#25) in the Federal Real Property Profile and on the method to calculate the percentage of buildings and square footage that are compliant with the principles for agencies' scorecard input.

5.4 Sustainable Site Selection

This section should be read in conjunction with Chapter 3, Section 3.2.1, on the LEED™ Sustainable Sites (SS) Credit Category.

5.4.1 General Considerations

Site considerations are critical to creating an environmentally friendly and aesthetically pleasing project. Siting of green buildings should take into account the natural environmental features of the site and minimize any potential adverse impacts on these features. Attentive design and construction of a green building within an undeveloped natural setting such as a greenfield site can mitigate the damage inflicted on natural systems, wildlife habitat, and biodiversity. Exploiting previously developed brownfield sites that have been disturbed and contaminated by past uses is also a good strategy both because such sites are usually less expensive to purchase and also, Because development benefits the environment, averting impacts on natural resources, providing for site remediation, and taking advantage of existing infrastructure.

Site design and building placement should also encourage sustainable transportation alternatives such as walking and bicycling, thereby minimizing automobile use.

The implementation of sustainable landscaping practices that use hardy, indigenous plant species that do not require potable water for irrigation or the use of pesticides is another strategy to be encouraged.

Creation of sustainable buildings begins typically with the selection of a suitable site. A building's location and orientation on a site impact a wide range of environmental factors in addition to other important factors such as security and accessibility. Likewise, a building's location can impact the energy likely to be consumed by occupants for commuting and their choice of transportation methods and can affect local ecosystems and the degree to which existing structures and infrastructures are employed. This is why it is preferable to locate buildings in areas of existing development and whenever possible, to consider reusing or retrofitting existing buildings and rehabilitating historic properties. Whether the project consists of a single building or group of buildings, a campus or a military base, it is important to incorporate smart-growth principles in the project's development process.

Physical security has recently taken center stage to become a critical issue in optimizing site design; it should not be an afterthought. Site security requires that the determination and location of access roads, parking, vehicle barriers, and perimeter lighting all be integrated into the design at the earliest phase of the design, along with sustainable site considerations.

Generally to achieve LEED™ credits, sustainable site planning should employ an integrated-system approach that seeks to:

- Minimize the development of open space through the selection of disturbed land, brownfields, or building retrofits.
- Consider energy implications in site selection and building orientation to take advantage of natural ventilation and maximum daylighting. Buildings should be sited to allow integration of passive and active solar strategies. Investigate the potential impact of possible future developments surrounding the site including daylighting, ventilation, solar, etc.
- Control/prevent erosion through improved landscaping practices and stabilization techniques (e.g., grading, seeding and mulching, installing pervious paving) and/or retaining sediment after erosion has occurred (e.g., earth dikes and sediment basins). This mitigates the negative impacts on water and air quality and reduces the potential damage to a building's foundation and structural system that may be caused by natural hazards such as floods, mudslides, and torrential rainstorms. Installing retention ponds and berms should also be considered to control erosion, manage storm water runoff on site, and reduce heat islands while also serving as physical barriers to control access to a building. Use of vegetated swales and depressions will facilitate reducing runoff.
- Reduce heat islands using landscaping techniques such as the use of existing trees and other vegetation to shade walkways, parking lots, and other open areas. However, the sitework and landscaping should be integrated with security and design safety. Landforms and landscaping should be integrated into the site-planning process to enhance resource protection. Employing light-colored materials to cover the facility's roof helps reduce energy loads and extend the life of the roof. In hot, dry climates, consider creating shading by covering walkways, parking lots, and other open areas that are paved or made with low-reflectivity materials. For security reasons any shading devices in place must not block critical ground-level sight lines.

- In warm climates consider incorporating green roofs into the project as well as using roofing products that meet or exceed ENERGY STAR® standards. The main site benefits of employing green/vegetated roofs include helping to control storm-water runoff, improving water quality, and mitigating urban heat-island effects.
- Minimize habitat disturbance by limiting the site to a minimal area around the building perimeter; locate buildings adjacent to existing infrastructure and retain prime vegetation features to the extent possible. Habitat disturbance can also be positively impacted by reducing building and paving footprints, and planning construction staging areas with the environment in mind.
- Restore the health of degraded areas by increasing and improving the presence of healthy habitat for indigenous species through native plants and closed-loop water systems in addition to conserving water use through xeriscaping with native plants and native or climate-tolerant trees to improve the quality of the site as well as providing visual protection by obscuring assets and people.
- Consider the installation of retention ponds and berms to control erosion, manage storm water, and reduce heat islands; they can also serve as physical barriers to control access to a building.
- Incorporate transportation solutions along with site plans that acknowledge the need for bicycle parking, carpool staging (Figure 5.4), and proximity to mass transit. Alternatives to traditional commuting can be encouraged by siting the proposed building to take into account availability and access of public transport and to limit on-site parking. Adequate provisions for bicycling, walking, carpool parking, and telecommuting should be provided. Facilities for refueling/recharging alternative fuel/electric vehicles are also a good idea (Figure 5.5).

Figure 5.4 Drawing illustrating the placement of carpool and vanpool parking spaces; they should be closer to the building entrance than other automobile parking. Prominent signage may be required to draw attention to the location of car- and vanpool parking and pickup areas. Attractive and comfortable waiting areas should be provided to encourage car- and vanpool commuters.
Source: City of Santa Monica Green Building Design and Construction Guidelines.

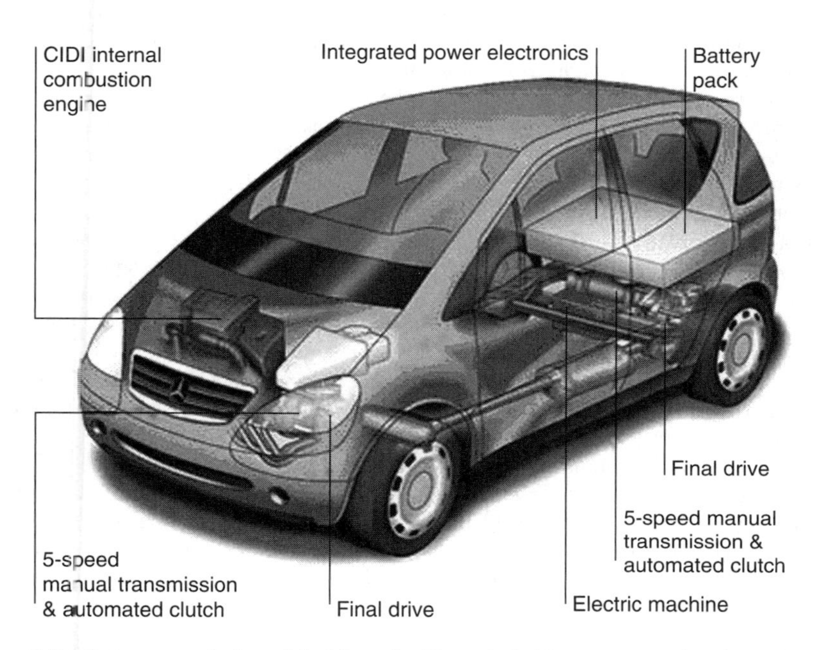

Figure 5.5 Cutaway rendering of the Mercedes Hyper hybrid-car concept showing major components. It is an example of an energy-efficient car designed to run on biodiesel fuels. EPA studies suggest that using biodiesel reduces many of the toxic substances present in diesel exhaust – including many cancer-causing compounds.
Source: PhysOrg.com.

- Site security should be addressed concurrently with other key sustainable site issues such as location of access roads, parking, vehicle barriers, and perimeter lighting, among others.

5.4.2 Site Selection

The LEED™ Sustainable Site section consists of various credits depending on which category is being considered (e.g., New Construction, Schools, Commercial Interiors, etc.). Discussed below are some of the major issues to be considered.

The general intent of selecting a sustainable site is essentially to reduce environmental impacts by avoiding inappropriate sites and by selecting an appropriate building location. The intent for LEED™ Commercial Interior differs somewhat from, say, LEED™ New Construction and LEED™ Schools in that the purpose here is mainly to encourage tenants to select buildings with best-practices systems and green strategies.

The main LEED™ requirements have the intent to avoid developing building projects, hardscape areas, roads, or parking areas on segments of sites that meet any one of the following criteria:

- Prime farmland (as defined by the United States Department of Agriculture)
- Previously undeveloped land with an elevation lower than five feet above FEMA's 100-year flood elevation
- Land that is specifically identified as habitat for any endangered or threatened species (species on federal or state threatened or endangered lists)
- Within 100 feet of any wetlands, isolated wetlands, or areas of special concern identified by state or local rule or within setback distances from wetlands prescribed in state or local regulations, as defined by local or state rule or law, whichever is more stringent
- Previously undeveloped land that is within 50 feet of a water body, defined as seas, lakes, rivers, streams, and tributaries that support or could support fish, recreation, or industrial use, consistent with the terminology of the Clean Water Act
- Land that prior to acquisition for the project was public parkland, unless the public landowner accepts in trade parkland that is of equal or greater value; Park Authority projects are exempt
- Consider reuse of existing building space whenever possible

During the site-selection process, give preference to those sites that do not include sensitive elements and restrictive land types and formulate during the design phase an erosion- and sedimentation-control plan for the project's development. Select a suitable building location and design the building with a minimal footprint to minimize site disruption of those environmentally sensitive areas identified above. Once a site is selected, consider employing programs that encourage temporary and permanent seeding, mulching, earthen dikes, sediment basins, etc.

During the site-selection process preference should be given to urban and urbanized sites. Direct development of urban areas with existing infrastructure should be encouraged. Greenfields should be protected, and habitat and natural resources should be preserved. Buildings should be sited near mass transit whenever possible to minimize transportation impacts, and walking and/or biking should be encouraged. The building should be programmed to be used by the community by integrating the building function with a variety of uses.

During the site-selection process, give preference to brownfields redevelopment sites and also to the rehabilitation of previously utilized and/or damaged sites with suspected or confirmed environmental contamination, thereby reducing pressure on undeveloped land. Develop and implement a site-remediation plan.

5.4.3 Development Density and Community Connectivity

The LEED™ intent here is to preserve natural resources by developing urban areas with existing infrastructure, and protecting greenfields.

There are two LEED™ requirements. Option 1, Development Density, seeks to construct/renovate on previously developed land in a community with a development density of at least 60,000 square feet per acre net. The calculation is to be based on two-story downtown development. Option 2, Community Connectivity, seeks to construct/renovate on previously developed land which is within a half mile of a residential area or neighborhood with an average density of 10 units per acre, is within a half mile of at least 10 basic services, and has pedestrian access between the building and the services.

For Commercial Interiors the LEED™ requirement is to find a space in a building that is located in an existing, walkable community with a development density of at least 60,000 square feet per acre net.

Basic services as per LEED™ requirements include but are not limited to:

1. Bank
2. Place of worship
3. Convenience grocery
4. Daycare
5. Cleaner
6. Fire station
7. Beauty salon
8. Hardware store
9. Laundry
10. Library
11. Medical/dental facilities
12. Senior-care facility
13. Park
14. Pharmacy
15. Post office
16. Restaurant
17. School
18. Supermarket
19. Theater
20. Community center
21. Fitness center
22. Museum.

For Site selection preference should be given to urban sites with pedestrian access to a variety of services.

5.4.4 Brownfield Redevelopment

The LEED™ intent is to protect greenfields by developing on environmentally contaminated or damaged sites.

There are two LEED™ requirements. Option 1 is to develop on a contaminated site as determined by ASTM E1903-97 Phase II Environmental Site Assessment or a local voluntary cleanup program. Option 2 is to develop on a brownfield site determined by a local, state, or federal government agency.

Site selection should give preference to brownfield sites and identify all tax incentives and property cost savings. Coordinate site-development plans with remediation activity as required.

5.4.5 Alternative Transportation

5.4.5.1 Public-Transport Access

The LEED™ intent is to reduce the impacts of pollution and land development on the environment by reducing automobile use.

LEED™ requirements are to develop within a half-mile walking distance of existing, planned, or funded commuter rail, light rail, or subway station, and/or to develop within a quarter-mile walking distance of at least one stop for at least two public or private/campus bus lines that can be used by building occupants.

To achieve exemplary performance, institute a comprehensive transportation management plan that quadruples subway, commuter rails, light rails, or bus service and doubles ridership.

Potential technologies and strategies are to identify transportation needs of future building occupants by performing a transportation survey. Site the proposed building near mass transit.

5.4.5.2 Bicycle Storage and Changing Rooms

The LEED™ intent is to reduce the impacts of pollution and land development on the environment by reducing automobile use.

LEED™ requirements for commercial or institutional buildings are to provide secure bicycle racks and/or storage within 200 yards of a building entrance for at least 5 percent or more of FTE occupants and transient occupants during peak building hours. For residential buildings, provide covered bicycle storage, for at least 15 percent of the building occupants in lieu of changing/shower facilities.

To achieve exemplary performance, institute a comprehensive transportation management plan that shows a quadrupling of subway, commuter rails, light rails, or bus service and a doubling of potential ridership.

Potential technologies and strategies include planning on transportation amenities such as secured bicycle racks and showering/changing facilities.

5.4.5.3 Low-Emitting and Fuel-Efficient Vehicles

The LEED™ intent is to reduce pollution and land-development impacts from automobile use.

LEED™ requirements have four options. Option 1 is to provide low-emitting and fuel-efficient vehicles for 3 percent of FTE occupants and preferred parking for these vehicles. Option 2 is to provide preferred parking for low-emitting and fuel-efficient vehicles for 5 percent of the total vehicle parking capacity of the site and/or a discounted parking rate for low-emitting/fuel efficient vehicles. Discount the parking rate by at least 20 percent in order to establish a meaningful incentive in all potential markets. Option 3 is to provide alternative-fuel refueling stations for 3 percent of the site's total vehicle parking capacity. Option 4 is to provide building occupants access to a low-emitting/fuel-efficient vehicle sharing program. LEED™ requires the following conditions to be met:

1. One low-emitting/fuel-efficient vehicle is to be provided for every 3 percent of FTE, assuming that one shared vehicle can service 8 people, which equates to one vehicle per 267 FTE occupants. For buildings with fewer than 267 FTE occupants, at least one low-emitting/fuel-efficient vehicle must be provided.
2. Provide a vehicle-sharing contract valid for at least two years.

3. Provide documentation of the vehicle-sharing program's customers served/vehicle estimates and an explanation of the program and its administration.
4. Locate the low-emitting/fuel-efficient vehicle parking in the nearest available spaces in the nearest available parking area and provide a site plan or area map clearly highlighting the walking path from the parking area to the project site and noting the distance.

For credit purposes, LEED™ defines low-emitting and fuel-efficient vehicles as "vehicles that are either classified as zero-emission vehicles (ZEV) by the California Air Resources Board or have achieved a minimum green score of 40 on the American Council for an Energy-Efficient Economy (ACEEE) annual vehicle rating guide."

To achieve exemplary performance, institute a comprehensive transportation management plan demonstrating a reduction in personal automobile use through the use of alternative options.

Potential technologies and strategies include providing transportation amenities such as alternative-fuel refueling stations. Consider sharing the costs and benefits of refueling stations with neighbors.

5.4.5.4 Parking Capacity

The LEED™ intent is to reduce pollution and land-development impacts from automobile use.

LEED™ requirements have five options. Option 1, for nonresidential buildings, is to have parking capacity meet but not exceed minimum local zoning requirements and to reserve 5 percent of the preferred parking spaces for carpools or vanpools. Option 2, for nonresidential building projects providing parking for less than 5 percent of FTE building occupants, is to reserve 5 percent of the preferred parking spaces for carpools or vanpools. Providing a discounted parking rate is an acceptable substitute for preferred parking for carpools or vanpools or for low-emitting/fuel efficient vehicles, providing that the discounted parking rate is at least 20 percent in order to establish a meaningful incentive. Option 3, for residential buildings, is to have parking capacity not to exceed minimum local zoning requirements and to provide infrastructure and support programs to facilitate shared vehicle usage, designated parking for vanpools, car-share services, ride boards, and shuttle services to mass transit. Option 4, for all building types, is not to provide any new parking. Option 5 relates to mixed-use (commercial and residential) buildings. For mixed-use buildings in which the commercial area is less than 10 percent, the entire building should be considered residential and adhere to the residential requirements of Option 3. For mixed-use buildings in which the commercial area is more than 10 percent, the commercial space is to adhere to nonresidential requirements, while the residential component is to adhere to residential requirements. This option only applies to mixed-use projects that have residential and commercial/retail components.

Potential technologies and strategies are to minimize the size of parking lots/garages and consider sharing parking facilities with adjacent buildings as well as alternatives that will limit the use of single-occupancy vehicles.

5.4.6 Site Development

5.4.6.1 Protect or Restore Habitat

The LEED™ intent is to promote biodiversity and provide habitat by conserving existing natural areas and restoring damaged areas.

LEED™ requirements have two options. Option 1, for greenfield sites, is to limit all site disturbance to 40 feet beyond the building perimeter; 10 feet beyond surface walkways, patios, parking, and utilities less than 12 inches in diameter; 15 feet beyond primary roadway curbs and main utility-branch trenches; and 25 feet beyond constructed areas with permeable surfaces. Option 2, for previously developed/graded sites, restore or protect with native or adapted vegetation a minimum of 50 percent of the site area, not to include the building footprint, or 20 percent of the total site area including the building footprint, whichever is greater. Native/adapted plants are plants indigenous to a locality or cultivars of native plants that are adapted to the local climate and are not considered invasive species or noxious weeds.

Potential technologies and strategies for greenfield sites are to perform a site survey to identify site elements and adopt a master plan for development of the project site. Minimize any disruption to existing ecosystems by carefully siting the building and designing it to minimize its building footprint. Establish clearly marked construction boundaries and restore previously degraded areas to their natural state.

5.4.6.2 Maximize Open Space

The LEED™ intent is to provide a high ratio of open space to development footprint to promote biodiversity.

LEED™ requirements have three options. Option 1 is to reduce the development footprint (which is defined as the total area of the building footprint, hardscape, access roads, and parking) and/or provide more than 25 percent vegetated open space within the project boundary than required by the local zoning's open-space requirement for the site. Option 2 is for areas where there are no local zoning ordinances in place (e.g., military bases); provide vegetated open-space area that is equal to the building footprint. Option 3 is for areas where a zoning ordinance exists, but there is no requirement for open space; provide a vegetated open space equal to 20 percent of the project's site area.

Potential technologies and strategies begin with a site survey to identify site elements and adopt a master plan for development of the project site. To minimize site disruption, select a suitable building location and design the building with a minimal footprint.

5.4.7 Stormwater Design

5.4.7.1 Quantity Control

The LEED™ intent is to support natural water hydrology by increasing porosity, increasing on-site infiltration, and reducing or eliminating pollution and contaminants from storm-water runoff.

LEED™ requirements include two cases. In Case 1 existing cover is less than or equal to 50 percent pervious. - Option 1 is to implement a storm-water management plan in which the post-development peak discharge rate and quantity must not exceed the predevelopment peak discharge rate and quantity for one- and two-year, 24-hour design storms. Option 2 is to put in place a storm-water management plan that protects receiving stream channels from excessive erosion by implementing stream-channel protection and quantity-control strategies. In Case 2 existing cover is greater than 50 percent pervious. Implement a storm-water management plan that results in a 25 percent decrease in the volume of storm-water runoff from the two-year, 24-hour design storm.

Potential technologies and strategies during the project site design are to maintain natural storm-water flows by promoting infiltration and to specify vegetated roofs, pervious paving, and other measures to minimize impervious surfaces. Encourage reuse of stormwater for nonpotable uses including landscape irrigation, toilet and urinal flushing, and custodial uses.

5.4.7.2 Quality Control

The LEED™ intent is to reduce water pollution by increasing pervious areas and on-site infiltration, eliminating contaminants, and removing pollutants from storm-water runoff.

LEED™ requirements are to implement a storm-water management plan that reduces impervious cover, promotes infiltration, and captures and treats 90 percent of the average annual rainfall using best management practices (BMPs).

BMPs used to treat runoff must be capable of removing 80 percent of the average annual post-development total suspended solids (TSS) load. BMPs are considered to be acceptable if they are designed in accordance with standards and specifications from a state or local program that has adopted these performance standards or if there exists in-field performance monitoring data that demonstrates compliance with the criteria. Data must conform to accepted protocols for BMP monitoring.

Potential technologies and strategies are to use alternative sustainable surfaces and nonstructural techniques such as rain gardens, vegetated swales, and rain-water recycling to reduce imperviousness and promote infiltration so as to reduce pollutant loadings. Use sustainable-design strategies to design integrated natural and mechanical treatment systems such as constructed wetlands, vegetated filters, and open channels to treat storm-water runoff.

5.4.8 Heat-Island Effect

5.4.8.1 Nonroof Applications

The LEED™ intent is to minimize impact on microclimates and human and wildlife habitats by reducing heat islands (thermal-gradient differences between developed and undeveloped areas).

LEED™ requirements have two options. Option 1 is to provide for 50 percent of the site hardscape, including roads, sidewalks, courtyards, and parking lots, using any combination of the strategies listed below:

- Provide shade by utilizing tree canopies, either existing or within five years of landscape installation; trees must be in place at time of occupancy.
- Provide shade from structures covered by solar panels that produce energy used to offset some nonrenewable-resource use.
- Provide shade from architectural devices or structures that have a solar-reflectance-index (SRI) value of at least 29.
- Employ hardscape materials with an SRI of at least 29.
- Employ an open-grid pavement system that is at least 50 percent pervious.

Option 2 is to cover a minimum of 50 percent of the parking spaces (underground, under deck, under roof, or under a building). Roofs used to shade or cover parking must have a minimum SRI of 29, consist of green vegetatation, or be covered by solar panels that produce energy used to offset some nonrenewable-resource use (Figure 5.6).

Employ strategies, materials, and landscaping techniques that reduce heat absorption of exterior materials. Shade is to be provided by using native or adapted trees and large shrubs, vegetated trellises, or other exterior structures supporting vegetation. Photovoltaic cells should be positioned to shade impervious surfaces.

Figure 5.6 Sketch showing the mitigation of urban heat-island effect.
Source: Bruce Hendler.

5.4.8.2 Roof Applications

The LEED™ intent is to minimize impact on microclimate and human and wildlife habitat by reducing heat islands (thermal-gradient differences between developed and undeveloped areas).

LEED™ requirements have three options. Option 1 for low-slope (less than 2:12) roofs is to use roofing materials having a SRI equal to or greater than 78 for a minimum of 75 percent of the roof surface.

For steep slopes (greater or equal to 2:12) 75 percent of the roof surface area must be covered with materials having an SRI value greater than 29. Roofing materials with a lower SRI value than those listed above may also be used if the weighted rooftop SRI average meets the following criteria:

$$(\text{area SRI roof/total roof area})\ (\text{SRI of installed roof/required SRI}) \geq 75\ \text{percent}$$

Option 2 is to install a vegetated roof for at least 50 percent of the roof area. Option 3 is to install high-albedo and vegetated roof surfaces that in combination meet the following criteria:

$$(\text{area SRI roof}/0.75) + (\text{area of vegetated roof}/0.5) \geq \text{total roof area}$$

where the SRI for low-sloped roofs is 78 or greater and for steep-sloped roofs is 29 or greater (Figure 5.7).

Potential technologies and strategies include the installation of high-albedo and vegetated roofs to reduce heat absorption. SRI is calculated according to ASTM E 1980.

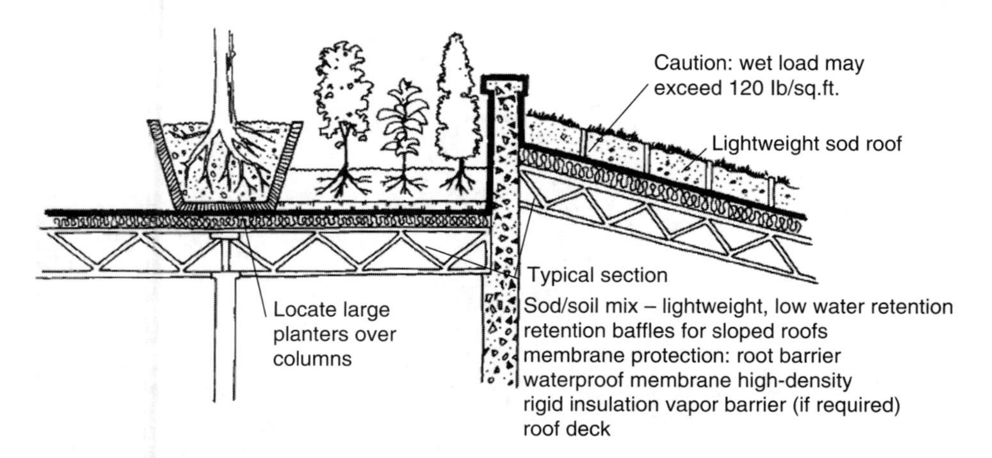

Figure 5.7 Sketch showing the use of planting on balconies, terraces, and roofs, which can significantly enhance the environment and provide habitat for wildlife and opportunities for food and decorative gardens. Moderately sloped and flat roofs can be planted with drought-tolerant perennial grasses and groundcovers that require minimal maintenance.
Source: City of Santa Monica Green Building Design and Construction Guidelines.

Reflectance is measured according to ASTM E 903, ASTM E 1918, or ASTM C 1549. Emittance is measured according to ASTM E 408 or ASTM C 1371. For default values consult the LEED™ 2009 Reference Guide (www.usgbc.org). Product information is available from the Cool Roof Rating Council website at www.coolroofs.org and the ENERGY STAR® website at www.energystar.gov.

5.4.9 Light-Pollution Reduction

The LEED™ intent is to minimize light trespass from the building and site, reduce sky glow to increase night-sky access, improve nighttime visibility through glare reduction, and reduce development impact from lighting on nocturnal environments.

LEED™ requirements for interior lighting **have two options.** Option 1 is for all nonemergency interior luminaires with a direct line of sight to any openings in the envelope (translucent or transparent) to have their input power automatically reduced by at least 50 percent between the hours of 11 PM and 5 AM. After-hours override may be provided by a manual or occupant sensing device, provided that the override does not last in excess of 30 minutes. Option 2 is for all openings in the envelope with a direct line of sight to any nonemergency luminaires to have shielding that is controlled/closed by automatic device between the hours of 11 PM and 5 AM. For exterior lighting only light areas as required for safety and comfort. Lighting power densities should not exceed ASHRAE/IESNA Standard 90.1-2007 for the classified zone. Meet exterior-lighting control requirements from ASHRAE/IESNA Standard 90.1-2007, Exterior Lighting Section, without amendments. For zone classification as defined in IESNA RP-33, see the USGBC LEED™ reference guide for the relevant category.

To maintain safe light levels while avoiding off-site lighting and night-sky pollution, site-lighting criteria should be adopted and computer modeling of the site lighting considered. Technologies to reduce light pollution include full cutoff luminaires, low-reflectance surfaces, and low-angle spotlights.

5.5 Commissioning Process

5.5.1 Overview

Commissioning is a systematic process that ensures that all building systems perform interactively according to the design intent and the owner's operational needs. Ideally it starts at a project's inception – i.e., the beginning of the design process – and culminates in the inspection, testing, balancing, and actual verification that new building systems are operating properly and efficiently within an agreed warranty period. Commissioning is designed to make sure that the heating, ventilating, air conditioning, and refrigeration (HVACR) systems (mechanical and passive) and associated controls, lighting controls (including daylighting), domestic hot-water systems, renewable energy systems (PV, wind, solar, etc.), and other energy-using building systems meet the owner's performance requirements, function the way they were intended, and operate at maximum efficiency. Commissioning results can identify changes that will dramatically reduce operating and maintenance costs, provide better occupant conditions, and facilitate upgrades, as well as fulfilling LEED™ requirements.

Many buildings today contain highly sophisticated conservation and environmental control technologies that, in order to function correctly, require careful supervision of installations, testing, and calibration and instruction of building operators. Some buildings possess unusual electrical or air-conditioning systems or employ sustainable features that may require particular attention to ensure that they operate as designed. Commissioning can reduce operating and maintenance costs, improve the comfort of a building's occupants, and extend the useful life of equipment. In addition, commissioning new construction projects will help reduce construction delays, ensure that the correct equipment is installed, reduce future maintenance costs, and reduce employee absenteeism. It is important that complete as-built information and operating and maintenance information are passed on to owners and operating staff upon completion of the project. Returns for these commissioning services often pay for themselves in energy savings within a year of completion.

Commissioning is a fairly recent concept that includes what was formerly referred to as "testing, adjusting, and balancing," but it goes much further. Commissioning is particularly important when complex mechanical and electrical systems are involved to ensure that these systems operate as intended and to realize energy savings and a quality building environment and thereby justify the incorporation and installation of more complex systems. When special building features are installed to generate renewable energy, recycle waste, or reduce other environmental impacts, commissioning is often necessary to ensure optimum performance. More importantly, commissioning practices should be tailored to the size and complexity of the building and its systems and components in order to verify their performance and to ensure that all requirements are met as per the construction documents and specifications. In addition to verification of the installation and performance of systems, commissioning will producing a commissioning report.

Few experiences can be more frustrating than finding out that essential systems in a new building don't operate according to specifications. To prevent this sort of outcome, total building commissioning should be included as part of the design, construction, and operation process.The degree of commissioning required should be appropriate to the complexity of the project and its systems, the owner's need for assurances, and the budget and time available. HVAC commissioning costs vary but are usually from one to four percent of the value of the mechanical contract.

Commissioning of building systems will vary from one project to another but will generally encompass and coordinate equipment startup; HVAC, electrical, plumbing, communications, security, and fire-management systems; and controls and calibration. Other systems and components may be included, particularly in large or complex projects. Commissioning begins by checking documentation and design intent for future reference. This is followed by performance testing of components when they first arrive on the job site and again after installation is complete. Balancing of air- and water-distribution systems to deliver services as designed is followed by checking and adjusting control systems to ensure that energy savings and environmental conditions meet design intent. Providing maintenance training and manuals is typically the final step of commissioning.

A final complete commissioning report must be prepared and submitted to the owners along with drawings and equipment manuals. The commissioning report should contain all documentation pertaining to the commissioning procedures and testing results in addition to deficiency notices and records of satisfactory corrections of deficiencies. System commissioning is typically conducted by a mechanical consultant with appropriate experience and training, ideally hired by and responsible directly to the project owner and independent of the mechanical consulting firm and general contractor. In cases where commissioning is required for very complex projects, a special commissioning coordinator may be designated. The architect or designer of record (DOR) typically has the responsibility of overseeing the completion of commissioning.

Formal commissioning is strongly recommended for buildings incorporating complex and digitally controlled HVAC systems or natural ventilation systems integrated with HVAC systems, those with renewable-energy, on-site water treatment systems, occupancy sensor lighting controls, or those incorporating other high technology systems. Commissioning is rarely required for projects with minimal mechanical or electrical complexity, such as typical residential projects. In addition, commissioning also serves an important construction quality-control function for all building types and helps consultants track the progress of contracts.

For U.S. Green Building Council (USGBC) LEED™ certification, commissioning is an integral and prerequisite component. For the New Construction, Commercial Interiors, Schools, and Core and Shell categories, LEED™ has two commissioning components, Fundamental Commissioning of Building Systems, which is a prerequisite (i.e., obligatory), and Enhanced Commissioning, which receives a credit but is not a prerequisite.

5.5.2 Fundamental Commissioning

The LEED™ intent of fundamental commissioning is to ensure that installation, calibration, and performance of energy systems meet project requirements, basis of design, and construction documents.

The following commissioning process activities should be implemented by the commissioning team:

1. The owner or project team must designate an individual as the commissioning authority (CxA) to lead, review, and oversee the commissioning process activities until completion. This individual should be independent of the project's design or construction management unless the project is less than 50,000 square feet.
2. The designated CxA should have documented commissioning-authority experience in at least two building projects. Additionally, he or she should ideally meet the minimum qualifications and have an appropriate level of experience in energy-systems design, installation, and operation, as well as commissioning planning and process management. LEED™ also recommends that a designated CxA have hands-on field experience with energy-systems performance, startup, balancing, testing, troubleshooting, operation, and maintenance procedures and energy-systems automation-control knowledge.

3. The CxA should clearly document and review the owner's project requirements (OPR) and the basis of design (BOD) for the building's energy-related systems (usually prepared by A/E). Updates to these documents should be made during design and construction by the design team. The commissioning process does not absolve or diminish the responsibility of the contractor to meet the contract-documents requirements.

Commissioning during the design phase is intended to achieve the following specific objectives:

- The design and operational intent are clearly documented and fully understood.
- The recommendations are communicated to the design team during the design process to facilitate the development of commissioning and avoid later contract modifications.
- The commissioning process for the construction phase is appropriately reflected in the construction documents.

Commissioning during the construction phase is intended to achieve the following objectives in line with the contract documents:

- The commissioning requirements should be incorporated into the construction documents.
- A commissioning plan should be developed and put in place.
- Verify and document that the installation and performance of energy-consuming equipment and systems meet the owner's project requirements and basis of design.
- Verify that applicable equipment and systems are installed according to the manufacturer's recommendations and to industry-accepted minimum standards.
- A complete summary commissioning report should be executed.
- Verify that operations and maintenance (O&M) documentation left on site is complete.
- Verify that training of the owner's operating personnel is adequate.

Commissioning is a LEED™ prerequisite and should be conducted for both new construction and major retrofits, as well as for medium or large energy-management control systems that incorporate more than 50 control points. Commissioning is also necessary when large or very complex mechanical or electrical systems are in place and where on-site renewable-energy generation systems, such as solar hot-water heaters or photovoltaic arrays, are in place. Likewise, commissioning is useful where innovative water-conservation strategies, such as gray-water irrigation systems or composting toilets, are in place.

Most building systems over the years become less efficient, due to changing occupant needs, building renovations, and obsolete systems, which cause significant occupant discomfort and complaints. These issues drive up a facility's operating costs and make it less attractive to new and existing tenants. But these problems can be mitigated by investing in a commissioning process that typically pays for itself in less than a year. The following checked systems are to be commissioned:

HVAC Equipment and System

- ❑ Variable-speed drives
- ❑ Hydronic piping systems
- ❑ HVAC pumps
- ❑ Boilers

- ❑ Chemical-treatment system
- ❑ Air-cooled condensing units
- ❑ Makeup-air systems
- ❑ Air-handling units
- ❑ Underfloor air distribution
- ❑ Centrifugal fans
- ❑ Ductwork
- ❑ Fire/smoke dampers
- ❑ Automatic temperature controls
- ❑ Laboratory fume hoods
- ❑ Testing, adjusting, and balancing
- ❑ Building/space pressurization
- ❑ Ceiling radiant heating
- ❑ Underfloor radiant heating

Electrical Equipment and System

- ❑ Power-distribution system
- ❑ Lighting-control systems
- ❑ Lighting-control programs
- ❑ Engine generators
- ❑ Transfer switches
- ❑ Switchboard
- ❑ Panelboards
- ❑ Grounding
- ❑ Fire alarm and interface items with HVAC
- ❑ Security system

Plumbing System

- ❑ Domestic water heater
- ❑ Air compressor and dryer
- ❑ Storm-water oil /grit separators

Building Envelope

- ❑ Building insulation installation
- ❑ Building roof installation methods
- ❑ Door and window installation methods
- ❑ Water infiltration/shell drainage plain

Commissioning needs may differ by project; however, commissioning the building-envelope systems, domestic water heating, power distribution, ductwork, and any hydronic piping systems is strongly recommended for any project (*Source*: BuildingGreen, Inc.).

5.5.3 Retrocommissioning

While building commissioning is an important aspect of new construction projects as a means of ensuring that all installed systems perform as intended, we find that most buildings have never encountered the commissioning or quality-assurance process and not surprisingly are therefore performing well below their potential. Retrocommissioning (RCx) is simply the commissioning of existing buildings. It is witnessing increasing prominence as a cost-effective strategy for improving energy performance. Retrocommissioning is an independent systematic process designed to be used with buildings that have never been previously commissioned in order to improve the current conditions and operations, to check the functionality of equipment and systems, and to optimize how they operate and function together in order to reduce energy waste and improve building operation and comfort.

The inefficiencies found in most existing buildings are confirmed by a Lawrence Berkeley National Laboratory study of 60 different types of buildings, which showed (*Source*: Association of State Energy Research Technology Internships and US Department of Energy) that:

- Over 50 percent had control problems.
- 40 percent had HVAC-equipment problems.
- 15 percent had missing equipment.
- 25 percent had building automation systems (BAS) with economizers, variable-frequency drives (VFD), and advanced applications that were simply not operating correctly.

Other advantages of retrocommissioning, in addition to energy savings, are to address issues such as modifications made to system components, space changes from original design intent, building systems not operating efficiently against benchmarks, complaints regarding IAQ and temperature, and others.

Retrocommisioning requires an assessment of BAS capabilities, potential building opportunities, and investigation approach. A determination is needed on how systems are intended to operate; measure and monitor how they operate, and prepare a prioritized list of the operating opportunities. Retrocommissioning is designed to implement opportunities and verify proper operation. Operating improvements made should be recorded and the building operator trained on how to sustain efficient operation and implement capital improvements.

Recommissioning (ReCx) applies only to buildings that have previously been commissioned or retrocommissioned. The original commissioning process documented that the building systems performed as intended at one point in time. The intent of recommissioning is to help ensure that the benefits of the initial commissioning or retrocommissioning process endure. In some cases, ongoing commissioning (ongoing Cx) may be necessary to resolve operating problems, improve comfort, optimize energy use, and identify energy and operational retrofits for existing buildings.

5.5.4 Enhanced Commissioning

For a LEED™ credit, enhanced commissioning is required in addition to the fundamental commissioning prerequisite. The intent of enhanced commissioning is to start the commissioning process early during the design process and execute additional activities upon completion of the systems performance verification. For the Commercial

Interiors category the intent is to verify and ensure that the tenant space is designed, constructed, and calibrated to operate as intended.

LEED™ requirements include implementation, or a contract in place requiring implementation, of further commissioning process activities in addition to the fundamental commissioning prerequisite requirements and in accordance with the relevant LEED™ Reference Guide (e.g., *Green Building Design and Construction*, 2009 edition). Duties of the commissioning authority (CxA) include:

1. Prior to the end of design development and commencement of the construction-documents phase, a commissioning authority (CxA) independent of the firms represented on the design and construction team must be designated to lead, review, and oversee the completion of all commissioning process activities. This person can be an employee or consultant of the owner. Furthermore, this person must:
   - Have documented commissioning-authority experience in at least two building projects.
   - Be independent of the project's design and construction management
   - Not be an employee of the design firm, though the individual may be contracted through the firm
   - Not be an employee of or contracted through a contractor or construction manager of the construction project
2. The CxA must report all results, findings, and recommendations directly to the owner.
3. The CxA must conduct a minimum of one commissioning design review of the owner's project requirements (OPR), basis of design (BOD), and design documents prior to the mid-construction-documents phase and backcheck the review comments in the subsequent design submission.
4. The CxA must review contractor submittals and confirm that they comply with the owner's project requirements and basis of design for systems being commissioned. This review must be conducted in parallel with the review of the architect or engineer of record and submitted to the design team and the owner.
5. The CxA and/or other members of the project team are required to develop a systems manual that provides future operating staff with the necessary information to understand and optimally operate the commissioned systems. For Commercial Interiors the manual must contain the information required for recommissioning the tenant-space energy-related systems.
6. The task of verifying that the requirements for training operating personnel and building occupants have been completed may be performed by either the CxA or other members of the project team.
7. The CxA must be involved in reviewing the operation of the building with operations and maintenance (O&M) staff and occupants and having a plan in place for resolving outstanding commissioning-related issues within 10 months after substantial completion. For Commercial Interiors there must also be a contract in place to review tenant-space operation with O&M staff and occupants.

5.6 Cost Management

For optimum results integrate the project's activities. For example, manage costs collaboratively, with the integrated project team engaged wherever possible at the earliest phases of design – using target costing, value management, and risk management. The temptation to put in place a guaranteed maximum price on the project before the design stage is complete should be resisted to ensure quality and functionality for the building owner or stakeholder; if the price has to be fixed at an earlier stage, agree on an incentive scheme for sharing benefits. A clear understanding of actual construction costs should be aimed for in terms of labor, plant, and materials. Identify and separate underlying costs from risk allowances and distinguish between profit and overhead margins. All parties in the supply chain, including material and component suppliers, should have at their disposal reliable data on the operational costs of their products, including running and maintenance costs.

Management of the overall cost of the project is the responsibility of the project manager, who in turn reports to the owner. Limits of authority for these two roles should be agreed upon at the start of the project so that everyone knows exactly what they are empowered to do in managing project costs.

The UK Office of Government Commerce says that the main ingredients for successful project cost management are:

- To manage the base estimate and risk allowance
- To operate change control procedures
- To produce cost reports, estimates, and forecasts; the project manager is directly responsible for understanding and reporting the cost consequences of any decisions and for initiating corrective actions if necessary
- To maintain an up-to-date estimated out-turn cost and cash flow
- To manage expenditure of the risk allowance
- To initiate action to avoid overspending
- To issue a monthly financial status report.

The cost-management objectives during the construction phase include:

- Delivering the project at the appropriate capital cost using the value criteria established at the commencement of the project
- Making sure that throughout the project full and proper accounts are monitored of all transactions, payments, and changes

The key areas that cost-management teams should consider during the design and execution of a project are:

- Identifying elements and components to be included in the project and constricting expenditures accordingly
- Defining the project program from inception to completion
- Making sure that designs meet the scope and budget of the project and that delivered quality is appropriate and conforms to the brief
- Checking that orders are properly authorized
- Certifying that the contracts provide full and proper control and that all incurred costs are as authorized; all materials are to be appropriately specified to meet the project's scope and design criteria and to be procured effectively
- Monitoring all expenditures relating to risks to ensure that they are appropriately allocated from the risk allowance and properly authorized; monitoring use of risk allowance to assess impact on overall out-turn cost

- Maintaining strict planning and control of both commitments and expenditures within budgets to help prevent any unexpected cost over-/underruns; all transactions are to be properly recorded and authorized and, where appropriate, decisions justified

The project owner or owner's representative should manage the risk allowance, with support and advice from the project manager. Management of the risk allowance consists essentially of a procedure to transfer costs out of the risk allowance into the base estimate for the project work as risks materialize or actions are taken to manage the risks. Formal procedures are required to be put in place for controlling quality, cost, time, and changes. Risk allowances should not be disbursed unless the identified risks to which they relate actually occur. When risks occur that have not previously been identified, they should be treated as changes (variation orders) to the project. Likewise, risks that materialize but have insufficient risk allowance allocated for them also need to be treated as changes.

Risks and risk allowances should be reviewed on a regular basis, particularly when formal estimates are prepared but also throughout the design, construction, and equipping stages. With the increase in firm commitments being entered into and work being carried out, the risks to future commitments and work are naturally reduced. Any risk-allowance estimate should reflect this.

Every effort should be made to avoid introducing changes after the briefing and outline design stages are complete. Changes can be minimized by ensuring that from the very beginning the project brief is as complete and comprehensive as possible and that the stakeholders have approved it. This may involve early discussions with planning authorities to anticipate their requirements and to ensure that designs are adequately developed and coordinated before construction begins. In the case of existing buildings to be renovated, site investigations or condition surveys may be required.

In an elemental cost plan the estimate is broken down into a series of elements that can then be compared with later estimates or with actual costs, in line with the project's progress. In using this approach each element is treated as a separate cost center, although money can be transferred between elements as long as a reasonable balance is maintained and the overall target budget is not exceeded. Often the initial cost plan is based on estimates, which provide a fair basis for determining the validity of future assessments. Control by the project manager is achieved by an ongoing review of estimates for each cost center against its target budget. As the design develops and is priced, any differential in cost from the cost plan is identified. Decisions are then taken on whether that element can be authorized to increase in cost, which would require a corresponding reduction elsewhere, or whether the element needs to be redesigned in order to keep within the budget.

One of the project manager's responsibilities is to maintain ongoing reviews of designs as they develop and provide advice on costs to the integrated project team. This continuous cost oversight is of particular benefit in assessing individual decisions and is especially useful on large and complex schemes. It may also prove useful to schedule periodic formal assessments and budget estimates of the whole scheme at each phase of the project.

The owner's designated representative normally has overall responsibility for the project, including the estimated cost, and as such has to be satisfied that appropriate systems are in place for controlling the project's cost. Where substantial monies are earmarked to a design, these costs must be assessed against the budget decision and appropriately authorized. The owner's representative will often delegate a level of financial authority for design-development decisions to the integrated project team, as appropriate to the project. For complex projects the owner may prefer to have different delegated levels for each cost center.

During the construction process, instructions issued to the integrated project team, whether for change via a formal change-order procedure or for clarification of detail, will have an immediate impact on the project's cost. For this reason, the project team needs to establish procedures for issuance of instructions and information that ensure that they are issued within delegated authority and that before being issued their cost is estimated and their impact assessed. Any issued instruction should be justified in terms of value for money and overall impact on the project. Furthermore, the cost of all issued instructions is to be constantly monitored, and, where costs are estimated to be outside the delegated authority, specific approval is required.

The normal payment procedure is for the client, who is the contracting party and therefore obligated for paying the integrated project team, to make the interim and final payments as per the contract. These payments usually take the form of milestone/stage payments that are due during the course of the work. The client organization's finance division (e.g., bank) should be kept aware of future payment requirements by means of updated reports and cash-flow forecasts.

The terms of the contract may include clauses that allow the integrated project team to claim additional payments in certain circumstances as defined in the contract conditions. The justification for these additional payment claims may be the result of or caused by occurring risks that are essentially client risks under the contract, requesting additional/varied work, or failure by the client to comply with its contract obligations – often expressed as disruption to the integrated project team's work program due to modifications or other reasons.

The payment process should be managed in the same manner as the design/construction process. All payments should be made on time and as per the contract agreement. Payments for variations, provisional sums, etc., should be discharged after formal approval is given and as the work is carried out.

6 Green Materials and Products

6.1 General

Building material is any material that is used for the purpose of construction. The USGBC states that building-materials choices are important in sustainable design due to the extensive network of extraction, processing, and transportation steps required to process them, and the activities needed to create building materials may pollute the air and water, destroy natural habitats, and deplete natural resources. Incorporating green products into a project does not imply sacrifice in performance or aesthetics and does not necessarily entail higher cost.

Many naturally occurring substances, such as clay, sand, wood, and stone, have been used to construct buildings. Apart from naturally occurring materials, many man-made products are also in use, some more and some less synthetic. The manufacture of building materials has long been an established industry in many countries around the world, and their use is typically segmented into specific specialty trades, such as carpentry, plumbing, roofing, and flooring.

Repair and reuse of a building instead of tearing it down and building new structure is one of the most compelling strategies for minimizing environmental impacts. And rehabilitation of existing buildings components saves natural resources, including the raw materials, energy, and water resources required to build new; prevents pollution that might take place as a byproduct of extraction, manufacturing, and transportation of virgin materials; and minimizes the creation of solid waste that could end up in landfills.

Some states, such as North Carolina, provide grants to renovate vacant buildings in rural counties or in economically distressed urban areas. The USGBC's LEED™ Rating System likewise recognizes the importance of building reuse. Reusing a building can contribute to earning points under LEED™ Materials Resource Credits on Building Reuse.

6.1.1 What is a Green Building Material?

Green building materials defy easy definition. In general, building materials are called "green" because they are good for the environment. The ideal building material would have no negative environmental impacts and might even have positive environmental impacts, including air, land, and water purification. There is no perfect green material, but in practice there are a growing number of materials that reduce or eliminate negative impacts on people and the environment.

Green materials may be made from recycled materials, which puts waste to good use and reduces the energy required to make them. Good examples include counter

DOI: 10.1016/B978-1-85617-691-0.00006-0

tiles made from recycled automobile windshields and carpeting made from recycled soda bottles. Building materials are also classified as green because they can be recycled once their useful lifespan is over – for example, aluminum roofing shingles. If that product itself is made from recycled materials, the benefit of recyclability is multiplied.

Building materials are also considered green when they are made from renewable resources that are sustainably harvested. Flooring made from sustainably grown and harvested lumber or bamboo is a good example.

Some building materials are even considered green because they are durable. A durable form of siding, for instance, outlasts less durable products, resulting in a significant savings in energy and materials over the lifetime of a property. If a durable product is made from environmentally friendly materials, such as recycled waste, it offers even greater benefits. Siding made from cement and recycled wood fiber (wood waste) is a good example of a green building material that offers at least two significant advantages over less environmentally friendly materials.

6.1.2 Natural versus Synthetic

Building materials can be generally categorized into two categories, natural and synthetic. Natural building materials are those that are unprocessed or minimally processed by industry, such as timber or glass. Synthetic materials, on the other hand, are made in industrial settings after considerable human manipulations; they include plastics and petroleum-based paints. Both have their advantages and disadvantages.

Mud, stone, and fibrous plants are among the most basic natural building materials. These materials are being used together by people all over the world to create shelters and other structures to suit their local weather conditions. In general stone and/or brush are used as basic structural components in these buildings, while mud is generally used to fill in the space between, acting as a form of concrete and insulation.

Plastic is an excellent example of a typical synthetic material. The term "plastic" covers a range of synthetic or semisynthetic organic condensation or polymerization products that can be molded or extruded into objects, films, or fibers. The term is derived from the material's malleable nature when in a semiliquid state. Plastics vary immensely in heat tolerance, hardness, and resiliency. Combined with this adaptability, their general uniformity of composition and lightness facilitate their use in almost all industrial applications.

The evaluation of a material's greenness is generally based on certain criteria, such as whether the material is renewable and resource-efficient in its manufacture, installation, use, and disposal. Other considerations are whether the material supports the health and wellbeing of occupants, construction personnel, the public, and the environment; whether the material is appropriate for the application; and what the environmental and economic tradeoffs among alternatives are.

Much research remains to be conducted to satisfactorily evaluate alternatives and select the best material for a project. Material selection ideally considers the impacts of a product throughout its life cycle (from raw-material extraction to use and then to reuse, recycling, or disposal). The areas of impact to consider at each stage in the life cycle of a material include:

- Natural-resource depletion; air and water pollution; hazardous- and solid-waste disposal; durability
- Energy required for extraction, manufacturing, and transport
- Energy performance in useful life of material; durability
- Effect on indoor-air quality; exposure of occupant, manufacturer, or installer to harmful substances; moisture and mold resistance; cleaning and maintenance methods

Among the properties that typical green building materials and techniques may have include:

- Durability
- Readily recyclable or reusable when no longer needed
- Sustainably harvested from renewable resources
- Can be salvaged for reuse, refurbished, remanufactured, or recycled
- Manufactured from a waste material such as straw or fly ash or a waste-reducing process
- Minimally packaged and/or wrapped with recyclable packaging
- Locally extracted and processed
- Energy-efficient in use
- Less energy used in extraction, processing, and transport to the job site
- Generates renewable energy
- Water-efficient
- Manufactured with a water-efficient process

6.1.3 Storage and Collection of Recyclables

A Materials and Resources prerequisite in most of the LEED™ Rating System categories is Storage and Collection of Recyclables. The intent of this prerequisite is essentially to reduce waste sent to landfills by encouraging storage and collection of recyclables. While LEED™ has not set any specific standards or requirements for this area, the USGBC guidelines (Figure 6.1) note that the area should be easily accessible, serve the entire building(s), and be dedicated to the storage and collection of nonhazardous materials for recycling – including paper, corrugated cardboard, glass, plastics, and metals.

Commercial Building Square Footage (sf)	Minimum Recycling Area (sf)
0 to 5,000	82
5,001 to 15,000	125
15,001 to 50,000	175
50,001 to 100,000	225
100,001 to 200,000	275
200,001 or greater	500

Figure 6.1 Recycling storage-area guidelines based on overall building square footage. *Source*: USGBC.

6.2 Low-Emitting Materials

LEED™ addresses low-emitting materials within the Indoor Environmental Quality section; they are outlined in Chapters 3 and 4 of this book.

6.2.1 Adhesives, Finishes, and Sealants

For LEED™ credits, all adhesives and sealants with volatile-organic-compound (VOC) content must not exceed the VOC content limits of South Coast Air Quality Management District (SCAQMD) Rule #1168 as of 2007, as indicated in Figure 6.2. Aerosol adhesives not covered by Rule 1168 must meet Green Seal Standard GS-36 requirements. All indoor-air contaminants that are odorous, potentially irritating, and/or harmful to the comfort and wellbeing of installers and occupants should be minimized.

Sealants increase the resistance of materials to water or other chemical exposure, while caulks and other adhesives can help control vibration and strengthen assemblies by spreading loads beyond the immediate vicinity of fasteners. Both properties enhance durability of surfaces and structures, although they do so at a cost, as they are

Architectural Applications (SCAQMD 1168)	VOC Limit (g/L less water)	Specialty Applications (SCAQMD 1168)	VOC Limit (g/L less water)
Ceramic tile	65	Welding: ABS (avoid)	325
Contact	80	Welding: CPVC (avoid)	490
Drywall and panel	50	Welding: plastic cement	250
Metal to metal	30	Welding: PVC (avoid)	510
Multipurpose construction	70	Plastic primer (avoid)	650
Rubber floor	60	Special-purpose contact	250
Wood: structural member	140		
Wood: flooring	100	**Sealants and Primers (SCAQMD 1168)**	
Wood: all other	30	Architectural porous primers (avoid)	775
All other adhesives	50	Sealants and nonporous primers	250
Carpet pad	50	Other primers (avoid)	750
Structural glazing	100		
		Aerosol Adhesives (GS-36)	**VOC Limit**
Substrate-Specific Applications		General-purpose mist spray	65%
Fiberglass	80	General-purpose web spray	55%
Metal to metal	30	Special-purpose aerosol adhesives	70%

VOC weight limit is based on grams/liter of VOC minus water. Percentage is by total weight.

Figure 6.2 Adhesives, sealants, and sealant primers: South Coast Air Quality Management District (SCAQMD) Rule #1168. VOC limits are listed in the table and correspond to an effective date of July 1, 2005, and rule-amendment date of January 7, 2005.
Source: USGBC.

often hazardous in manufacture and application. Many construction-adhesive formulas contain more than 30 percent volatile petroleum-derived solvents, such as hexane, to maintain liquidity until application. Because of this, workers become exposed to toxic solvents; and, as the materials continue to outgas during curing, the occupants can also be potentially exposed to emissions for extended periods.

Water-based adhesives are available from a number of different manufacturers. Industry tests indicate that these products work as well as or better than solvent-based adhesives, can pass all relevant ASTM and APA performance tests, and are available at comparable costs. When adhesives are purchased in bulk, larger containers can often be returned to vendors for refill.

Stains and sealants also normally emit potentially toxic volatile organic compounds (VOCs) into indoor air. The most efficient method to manage this problem is to employ materials that do not require additional sealing, such as stone, ceramic and glass tile, and clay plasters. The toxicity and the air and water pollution generated through the manufacture of chlorinated hydrocarbons such as methylene chloride emphasize strongly the preferability of responsible, effective alternatives, such as plant-based, non-toxic, or low-toxicity sealant formulations.

According to LEED™ requirements, all adhesives and sealants used on the interior of the building (defined as inside the weatherproofing system and applied on-site) should comply with the reference standards shown in Figure 6.2.

Environmentally preferable cleaning methods and products can reduce indoor-air pollution and solid/liquid waste generation. Safe cleansers are readily available and are competitively priced and environmentally friendly. The improper use and disposal of some common cleaning and maintenance products can contribute to indoor-air contamination, water pollution, and toxic waste.

Biodegradability is a key factor for surfactants, which are the active ingredients in cleaners. Even low surfactant concentrations in runoff can pose greater risks to the environment. Generally, petroleum-derived surfactants break down more slowly than vegetable-oil-derived fatty acids; some materials are even resistant to municipal sewage treatment.

To minimize the harmful effects of toxins, the following should be followed:

- Select materials that have a durable finish and that do not require frequent stripping, waxing, or oiling (e.g., colored concrete, linoleum, or cork.).
- Hazardous materials should be stored outside the building envelope.
- Select whenever possible, biodegradable, nontoxic cleansers.
- Avoid selecting cleansers, waxes, and oils that are labeled as toxic or highly toxic, poisonous, harmful or fatal if swallowed, corrosive, flammable, explosive, volatile, causing cancer or reproductive harm, or requiring "adequate ventilation" or safety equipment.
- Select products that have approved third-party or government-agency certification: Green Seal, Scientific Certification Systems (SCS), EPA Environmentally Preferable Purchasing Program, General Services Agency, CIWMB Recycled Content Product Directory.
- Minimize stripper use by placing mats at all building entrances and clean regularly; dust-mop and/or vacuum frequently and wet mop with a liquid cleaner; refinish only areas where the finish surface is wearing.

6.2.2 Paints and Coatings

Paints basically consist of a mixture of solid pigment suspended in a liquid vehicle and applied as a thin, usually opaque coating to a surface for protection and/or decoration. Primers are basecoats applied to a surface to increase the adhesion of subsequent coats of paint or varnish. Sealers are also basecoats but are applied to a surface to help reduce the absorption of subsequent coats of paint or varnish or to prevent bleeding through the finish coat.

The USGBC requires that paints and coatings applied on-site and used on the interior of the building (defined as inside the weatherproofing system) must comply with the following referenced standards:

1. Architectural paints, coatings, and primers applied to interior walls and ceilings: Do not exceed the VOC content limits established in Green Seal Standard GS-11, Paints, second edition. May 12, 2008:
   - Flats 50 g/L
   - Nonflats: 100 g/L
2. Anticorrosive and antirust paints applied to interior ferrous metal substrates: Do not exceed the VOC content limit of 250 g/L established in Green Seal Standard GC-03, Anti-Corrosive Paints, Second Edition, January 7, 1997.
3. Clear wood finishes, floor coatings, stains, and shellacs applied to interior elements: Do not exceed the VOC content limits established in South Coast Air Quality Management District (SCAQMD) Rule 1113, Architectural Coatings, January 1, 2004.
   - Clear wood finishes: varnish 350 g/L, lacquer 550 g/L
   - Floor coatings: 100 g/L
   - Sealers: waterproofing sealers 250 g/L, sanding sealers 275 g/L, all other sealers 200 g/L
   - Shellacs: clear 730 g/L, pigmented 550 g/L
   - Stains: 250 g/L

Figure 6.3 shows the allowable VOC levels stipulated by SCAQMD.

According to Santa Cruz County, paint has significant environmental and health implications in its manufacture, application, and disposal. Most paint, even water-based "latex," is derived from petroleum. Its manufacture requires substantial energy and water and creates air pollution and solid/liquid waste. Volatile organic compounds (VOCs) are typically the pollutants of greatest concern in paints. VOCs from the solvents found in most paints (including latex paints) are released into the atmosphere during manufacture and application and for weeks or months after application. VOCs emitted from paint and other building materials are associated with eye, lung, and skin irritation, headaches, nausea, respiratory problems, and liver and kidney damage.

Paint Type	VOC Limit (grams/liter)
Flat	50 g/L
Nonflat	50 g/L
Primers, sealers, and undercoats	100 g/L
Quick-dry enamels	50 g/L

Figure 6.3 Allowable VOC levels (grams/liter less water and exempt compounds) according to the South Coast Air Quality Management District (SCAQMD).

Exposure to solvents emitted by finish products can be significant. However, renewable alternatives, such as milk paint, address many of these concerns but often at a premium price, and some products are only suitable for indoor applications. Reformulated low- and zero-VOC latex paints with excellent performance in both indoor and outdoor applications are now available at the same or lower price than older high-VOC products. Paints that meet GS-11 standard are low in VOCs and aromatic solvents, do not contain heavy metals, formaldehyde, or chlorinated solvents, and meet stringent performance criteria.

Silicate paints provide an alternative type of paint that is solvent-free and may be used on concrete, stone, and stucco. Silicate paint has many advantages; it is odorless, nontoxic, vapor-permeable, naturally resistant to fungi and algae, noncombustible, colorfast, light-reflective, and even resists acid rain. These paints cannot spall or flake off and will only crack if the substrate cracks. Though silicate paints are expensive, their extraordinary durability can be a significant compensation.

6.2.3 Flooring Systems

6.2.3.1 Carpet

Carpet manufacture, use, and disposal have significant environmental and health implications. Most carpet products are synthetic, usually derivatives of nonrenewable petroleum products; their manufacture requires substantial energy and water and creates harmful air and solid/liquid waste. However, many carpets are now becoming available with recycled content, and a growing number of carpet manufacturers are refurbishing and recycling used carpets into new carpet. At the end of its useful life most carpet tends to end up in landfills; in California, for example, 840,000 tons of carpet – roughly two percent of its waste stream – ends up in landfills every year.

Leasing arrangements in which the manufacturer will recycle worn or stained carpets also help to reduce waste. Resources are required for recycling, and, unless it can be recycled indefinitely, the carpet will still end up in a landfill after a finite number of uses. Carpet tiles limit waste because only worn or stained tiles need to be replaced and they are available for both commercial and residential applications.

Synthetic carpets, backings, and adhesives typically offgas volatile organic compounds (VOCs), all of which pollute indoor and outdoor air. Redesigned carpets, new adhesives, and natural fibers are available that emit low or zero amounts of VOCs. For improved air quality selected carpets and adhesives should meet a third-party standard, such as the Carpet and Rug Institute (CRI) Green Label Plus or the State of California's Indoor Air Emission Standard 1350. VOCs often have an odor and are often characterized as "new carpet smell."

Natural fibers are an environmentally preferable carpeting option because they are renewable and biodegradable. Options include jute, sisal, coir, and wool floor coverings. Biodegradable carpets made from plant extracts and plant-derived chemicals are also available. One of the disadvantages of carpets is that they tend to harbor more dust, allergens, and contaminants than many other materials (Figure 6.4). Durable flooring,

Figure 6.4 Photo of FLOR carpet squares laid in a flexible and practical "tile" format. The tiles are made from renewable and recyclable materials and are available in a range of colors, textures, and patterns. More than two million tons of carpets are landfilled in the U.S. each year (Courtesy: FLOR Inc.).

such as a concrete-finish floor, linoleum, cork, or reclaimed hardwoods, are generally preferable in helping to improve indoor-air quality.

The intent of low-Emitting carpet systems, according to LEED™, is to reduce the quantity of indoor-air contaminants that are odorous and potentially irritating and or harmful to the comfort and wellbeing of installers and occupants. In addition, for LEED™ credits, all carpet installed within a building's interior must meet or exceed testing and product requirements of the Carpet and Rug Institute (CRI) Green Label Plus program for VOC emission limits. Carpet pads installed within a building's interior must meet or exceed CRI's Green Label program for VOC emission limits. Adhesives and sealants used in carpet-system installation must comply with South Coast Air Quality Management District (SCAQMD) Rule #1168.

It should be noted that in LEED™ 2009 NC, C&S, and CI the title of IEQ 4.3 has been changed from Carpet to Flooring Systems, and this credit has been substantially expanded. From LEED™ 2009 and forward, all hard-surface flooring must be certified as compliant with the FloorScore standard by an independent third party. FloorScore is a program developed by the Resilient Floor Covering Institute (RFCI) and Scientific Certification Systems (SCS) that tests and certifies hard-surface flooring and flooring adhesive products for compliance with indoor-air quality (IAQ) emissions targets. Flooring products covered by FloorScore include vinyl, linoleum, laminate flooring, wood flooring, ceramic flooring, rubber flooring, wall base, and associated sundries.

6.2.3.2 Vinyl/PVC

Polyvinyl chloride (PVC, often referred to as "vinyl") deserves special attention because it accounts for almost 50 percent of total plastic use in construction and because it is increasingly recognized as problematic. Vinyl is ubiquitous—some 14 billion pounds are produced every year in North America—and artificially cheap, and not all of its alternatives have yet worked out all their negative issues. Moreover, as the U.S. Green Building Council (USGBC) suggested in its long-awaited report on PVC, all materials have potential pitfalls, from indoor-air quality to disposal. In addition to flooring, PVC is common in pipes, siding, wire insulation, conduit, window frames, wallcovering, and roofing, among other areas.

Vinyl has been extensively used because of its benefits: good strength relative to its weight, durability, water resistance, and adaptability. Vinyl tends to be inexpensive, in part because production typically requires roughly half the energy required to produce other plastics. Products made from vinyl can be resistant to biodegradation and weather and are effective insulators. The physical properties of vinyl can be tailored for a wide variety of applications. Many firms are increasingly concerned about the dramatic environmental liabilities of vinyl but are struggling to find appropriate substitutes.

Vinyl's life cycle begins and ends with hazards, most stemming from chlorine, its primary component. Chlorine makes PVC more fire-resistant than other plastics. Vinyl chloride, the building block of PVC, causes cancer. Lead, cadmium, and other heavy metals are sometimes added to vinyl as stabilizers; and phthalate plasticizers, which give PVC its flexibility, pose potential reproductive risks. Also, over time phthalates can leach out or off-gas, exposing building occupants to materials linked to reproductive-system damage and cancer in laboratory animals. Manufacturing vinyl or burning it in incinerators produces dioxins, which are among the most toxic chemicals known to man. Research has shown that the health effects of dioxin, even in minute quantities, include cancer and birth defects.

Polyvinyl chloride is produced from vinyl chloride monomer (VCM) and ethylene dichloride (EDC), which are carcinogens and acutely toxic. During the production of PVC, VCM and EDC are released to the environment, and there is no safe exposure level. Clean-air regulation and liability concerns have been effective in reducing total

VCM releases since 1980, while PVC use has roughly tripled. Although PVC resin is inert in normal use, older PVC products are frequently contaminated with traces of VCM, which can leach into the surrounding environment and contaminate drinking water.

Substitutes may cost more or require different maintenance, but many will also outlast plastics with proper care. Moreover, for many applications, particularly indoors where occupants can be directly exposed to off-gassing plasticizers, substitution of vinyl would clearly be prudent for the sake of health and wellbeing. Examples of possible alternatives include:

- Flooring made from natural linoleum, cork, tile, finished concrete, or earth
- Windows framed with fiberglass, FSC-certified wood, or possibly wood-based composites utilizing formaldehyde-free binders
- Stucco, lime plaster, reclaimed wood, fiber-cement, and FSC-certified wood siding
- Glass shower doors instead of vinyl curtains.
- Natural wallcoverings instead of vinyl wallpaper

6.2.3.3 Tile

Tile is made primarily from fired clay (porcelain and other ceramics), glass, or stone; it provides a useful option for flooring, countertops, and wall applications whose principal environmental requirement is durability. Tile can last indefinitely even in high-traffic areas, eliminating the waste and expense of replacements. Tile production, however, is energy-intensive, although tile made from recycled glass requires less energy than tile made from virgin materials. Among tiles' attributes is that they do not burn, will not retain liquids, and do not absorb fumes, odors, or smoke and, when installed with low- or zero-VOC mortar, can contribute to good indoor-air quality.

Tile can offer such performance only if it has the appropriate surface hardness for the location. Tile hardness is measured on the PEI (Porcelain Enamel Institute) scale of 0 to 5, with 0 being the least hard, indicating a tile that should not be used as flooring, and 5 signifying a surface designed for very heavy foot traffic and abrasion. If kept relatively free of sand and grit, floor tile can easily last as long as the building it is installed in.

The impacts of mining, producing, and delivering a unit of tile are important considerations. Approximately 75 percent of the tiles used in the U.S. are imported. The remaining amount represents about 650 million square feet of ceramic tile produced by U.S. factories each year, together with the billions of square feet manufactured throughout the world. This requires mining millions of tons of clay and other minerals and substantial energy to fire material into hardened tile. Once the raw materials are processed, a number of steps take place to obtain the finished product. These steps include batching, mixing and grinding, spray-drying, forming, drying, glazing, and firing. Many of these steps are now accomplished using automated equipment. Stone requires relatively little energy to process but significant energy to quarry and ship. Selecting tile produced regionally may dramatically reduce the energy use and pollution of transport.

Figure 6.5 Interior showing installation of Crystal Micro double-loading polished tile. *Source*: Foshan YeShengYuan Ceramics Co.

Nearly 99 percent of the tile industry's product consists of glazed or unglazed floor tile and wall tile, including quarry tile and ceramic mosaic tile (Figure 6.5). Due to the industry's focus on decorative tiles, it has become completely dependent on the economic health of the construction and remodeling industries.

The production process for ceramic glazed tile is the same as for ordinary ceramic tile, except that it includes a step known as glazing. Glazing ceramic tile requires a liquid made from colored dyes and a glass derivative known as flirt that is applied to the tile, either using a high-pressure spray or by direct pouring. This in turn gives a glazed look to the ceramic tile. Tile in low-traffic areas, particularly roofing, may use lower-impact water-based glazes. Though they can be more slippery, glazed tiles are practically stainproof. In addition to their slip resistance, the integral color and generally greater thickness of unglazed tiles tend to make them more durable than glazed tiles. However, unglazed tiles may require a sealant. Factory-sealed tiles can help minimize or eliminate a source of indoor VOC emissions. Glass floor tile can also offer a non-skid surface appropriate for ADA compliance. When installing stone tile, especially for countertop applications, use a nontoxic sealer for the grout and tile surface.

6.2.4 Earthen Building Materials

The use of earthen building materials goes back to before the invention of writing. Types include adobe bricks, made from clay, sand, and straw; rammed earth, compressed with fibers for stabilization; and cob, made of clay, sand, and straw that are stacked and shaped while wet. Provided they are obtained locally, earthen building materials can reduce or eliminate many of the environmental problems posed by conventional building materials since they are plentiful, nontoxic, reusable, and biodegradable.

Well-built earthen buildings are known to be durable and long-lasting with little maintenance. For thousands of years, people throughout the world have crafted cozy homes and communities with earthen materials that provide excellent shelter after centuries of use. A key element of American architectural vernacular, the Great Plains are home to sod and straw-bale construction, and in the Southwest adobe construction provides protection from extremes in summer and winter. Though the domestic popularity of earthen materials waned during the 20th century, a revival has emerged since the 1970s. By contrast, modern "stick-frame" construction, which requires specialized skills and tools, has been standard practice in the U.S. only since the end of World War II and remains uncommon in many parts of the world.

Considerations in regard to earthen construction include:

- Earthen walls are thick and may comprise a high percentage of floor area on a small site.
- Construction is generally labor-intensive, although minimal skill is required.
- Multistory and cob structures require post-and-beam designs.
- It may be more difficult to obtain necessary permits, although code recognition and structural testing are available
- If labor is done primarily by building professionals, the square-foot cost of earthen construction is comparable to conventional building methods.

Benefits of earthen materials include:

- Environmental impact is minimal, provided materials come from local sources.
- They are durable and low-maintenance.
- Thermal mass helps keep indoor temperatures stable, particularly in mild to warm climates.
- They are biodegradable or reusable.
- Structures can be easy to build, requiring few special skills or tools.
- If well designed, they provide pleasing aesthetics.
- They are highly resistant to fire and insects.
- They require no toxic treatments and do not off-gas hazardous fumes, thus are good for chemically sensitive individuals.

Earthen flooring, also called adobe floor, is a durable, inexpensive, environmentally friendly, and uniquely aesthetic complement to a home or office. Because "dirt" is

Figure 6.6 Illustration showing the method for installing an earth floor.
Source: Dancing Rabbit.

plentiful and locally available, earthen flooring virtually eliminates the waste, pollution, and energy necessary to manufacture a floor and can save money. In the United States earth floors are still most often used in outbuildings and sheds, but, if properly installed, can also be used in interior spaces (Figure 6.6). For interior use, earth floors must be properly insulated, moisture-sealed, and protected from capillary action of water by sealing with a watertight membrane underlayment.

Construction preparation includes removal of any vegetation under the floor area, followed by ramming of the area. The ground must be dry before installation of the floor. After the surface is moistureproofed, a foundation of stone, gravel, or sand is installed, 8 to 10 inches (20 to 25 cm) deep. Then an insulating layer is installed, such as a straw-clay mixture. The key to a good earthen floor is the proper mixture of dirt, clay, and straw. (Stabilizers such as starch paste, casein, glues, or Portland cement may be added for a harder floor.) Earthen floors are first troweled to a smooth finish and then usually sealed with an oxidizing oil such as linseed or hemp oil, which hardens it and allows it to be swept and mopped.

Earthen-floor considerations include:

- Elimination of construction waste; any excess earth can be reincorporated into the landscape.
- Materials are generally inexpensive when found locally.
- Minimal to zero pollution – earthen materials require only simple processing and little or no transport. Even when produced by a machine, a finished earthen slab is estimated to have 90 percent lower embodied energy than finished concrete.
- The floor is durable with proper care and repairable.

- The floor is low-maintenance and when properly sealed can be swept or moist-mopped; stabilized earthen flooring is not dusty.
- It is labor-intensive to install
- High-traffic areas such as entries or workspaces may require flagstones or other protective materials.
- It is more vulnerable to scratching and gouging than hard tile or cement but more durable than vinyl because it is repairable.
- Few local contractors are experienced in installing earthen flooring.

6.2.5 Windows

Windows are an essential element in construction because they provide ventilation, light, views, and a connection to the outside world. Drafty, inefficient, poorly insulated, or simply poorly chosen windows can compromise the energy efficiency of a building's envelope. The fabrication of windows, whether made of wood, aluminum, plastic, or steel, as with any other manufactured product, will require energy and will likely generate air pollution. Energy efficiency is one of the main considerations in reducing the environmental impacts of a window, followed by waste generated in manufacturing and general durability. Figure 6.7 shows the various components of a window.

Windows are available in a variety of glazing options. Each option offers a different thermal resistance or R-value. R-values are approximate and vary with temperature, type of coating, type of glass, and distance between glazings. From lowest resistance to greatest:

1. Single glazing and acrylic single glazing are similar; $R = 1.0$.
2. Single glazing with a storm window and double glazing are similar: $R = 2.0$.
3. Double glazing with a low-E coating and triple glazing are similar: $R = 3.0$.
4. Triple glazing with a low-E coating: $R = 4.0$.

By comparison, for a conventional insulated stud wall $R = 14.0$.

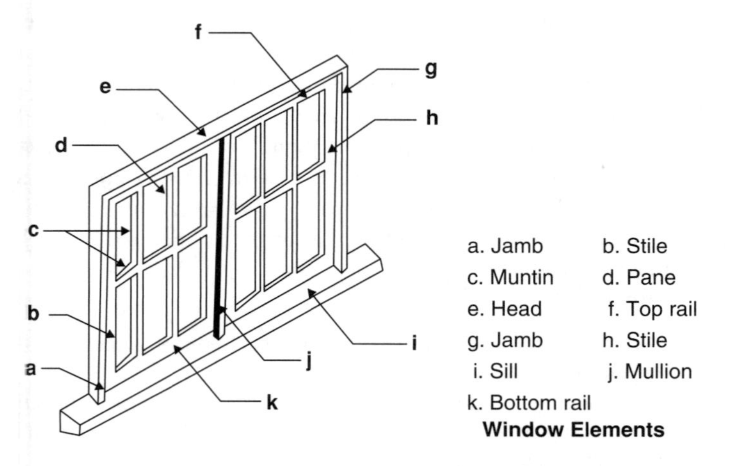

Figure 6.7 Drawing of a window showing individual elements.

Older, single-pane windows are very unlikely to perform as well as new windows and should preferably be reused only in unheated structures such as greenhouses. Residential window frames are typically made from wood, vinyl, aluminum, or fiberglass, or combinations of wood and aluminum or vinyl (i.e., "clad"). Each has a different cost, insulating ability, and durability, as shown below:

- Wood requires continuous maintenance for durability. The wood source should be certified by an accredited organization such as the FSC.
- Fiberglass is energy-intensive to manufacture, but is strong and durable and has excellent insulating value.
- Aluminum and steel are poor insulators and very energy-intensive to manufacture. When using metal-framed windows, look for recycled content and frames with "thermal breaks" to limit the loss of heat to outdoors.
- Vinyl offers good insulation but is highly toxic in its manufacture and if burned. High-efficiency windows typically utilize dual or triple panes with low-E (low emissivity) coatings and gas fill (typically argon) between panes to help control heat gain and loss. Factory-applied low-E coatings on internal glass surfaces are more durable and effective than films. High-quality, efficient windows are widely available from local retailers. To make an informed choice, consider only windows that have NFRC ratings (see sidebar). The EPA ENERGY STAR® label for windows can be a useful summary of these factors.

NFRC (National Fenestration Rating Council) ratings include:

- The U-factor summarizes a window's ability to keep heat inside or outside a building. The lower the U-factor, the better its insulating value; look for values of 0.4 or lower.
- The solar heat-gain coefficient (SHGC) summarizes a window's capability to block heat from sunlight. Seek out SHGC values of less than 0.4.
- Visible-light transmittance (VLT) is a measure of how much light gets through a window. Desired VLT varies with taste and application.
- Low values for air leakage are best.
- The higher the condensation resistance, the better; values range from 0 to 100. Condensation can contribute to mold growth, although new, high-quality windows (with a low U-factor) are generally better equipped to resist condensation than older windows.

6.2.6 Miscellaneous Building Elements

6.2.6.1 Gypsum Wall Board

In the United States and Canada gypsum board is manufactured to comply with ASTM Specification C 1396. This standard must be met whether the core is made of natural ore or synthetic gypsum.

Gypsum wall board, also known as plasterboard or drywall, is a plaster-based wall finish that is available in a variety of sizes; 4 feet wide by 8 feet high is the most common. Thicknesses vary in 1/8-inch increments from 1/4 to 3/4 inch.

The vast majority of the synthetic gypsum used by the industry is a byproduct of the process used to remove pollutants from the exhaust created by the burning of fossil fuels for power generation. Nearly 100 percent of the fiber used in the production of gypsum-board face and back paper comes from newsprint and postconsumer waste materials.

Advantages of gypsum board include: low cost, ease of installation and finishing, fire resistance, sound control, and availability. Disadvantages include: difficulty in curved-surface application and low durability when subject to damage from impact or abrasion.

Due to its ease of installation, familiarity, fire resistance, nontoxicity, and sound attenuation, gypsum wall board, known by its proprietary names Dywall® and Sheetrock®, is ubiquitous in construction. Gypsum wall board is a benign substance (basically paper-covered calcium sulfate), but it has significant environmental impacts because it is used on a vast scale; domestic construction uses an estimated 30 billion square feet per year.

The primary environmental impacts of gypsum are habitat disruption from mining, energy use and associated emissions in processing and shipment, and solid waste from disposal. Using "synthetic" or recycled gypsum board can significantly reduce several of these impacts. Synthetic gypsum accounts for approximately 20 percent of U.S. raw gypsum use and is made from the byproducts of manufacturing and energy-generating processes, primarily from desulfurization of coal-power-plant exhaust gases. In excess of 80 percent of coal fly ash sold in the U.S. is used in gypsum board.

Though synthetic gypsum-board use is growing in popularity, diverting drywall from the waste stream is proving more challenging. Reclaimed gypsum board can easily be recycled into new gypsum panels that conform to the same quality standards as natural and synthetic gypsum, but doing this may not be practical because gypsum is an inexpensive material that can require significant labor to separate and prepare for recycling. Gypsum-board face paper is commonly 100 percent recycled from newsprint, cardboard, and other postconsumer waste streams, but most recycled gypsum in wall-board products is postindustrial, made from gypsum-board manufacture. Gypsum board should be purchased in sizes that minimize the need for trimming (saving time and waste). Working crushed gypsum off-cuts (that have not been painted, glued, or otherwise contaminated) into soil helps reduce waste while improving the workability and calcium availability of many soils.

6.2.6.2 Siding

Siding is an external element that provides protection for wall systems from moisture and the heat and ultraviolet radiation of the sun. Selecting siding that is reclaimed, recyclable, or incorporating recycled material will reduce waste and pollution. However, there are many types of siding, and the environmental impacts of siding products vary considerably.

Earth or lime plasters last a long time and require relatively little maintenance. Cement or lime is commonly added for improved hardening and durability, but the relatively (or zero) overall cement content of natural plasters means that the material requires relatively small amounts of pollution and energy use to prepare and install. Deep eaves or overhangs that protect the siding from extended moisture exposure are critical to the longevity of natural plasters.

Fiber-cement siding has proven to be very durable, and many products are backed by 50-year or lifetime warranties. It is fire- and pest-resistant and emits no pollutants in use. However, it possesses a high embodied energy due to its cement content and because it is manufactured with wood fiber from overseas.

Cement stucco is another extremely durable material, which helps minimize long-term waste, but it is also energy-intensive to manufacture. Cement substitutes such as fly ash or rice-hull ash can mitigate the environmental cost of stuccos. In coastal regions salt spray can accelerate corrosion of reinforcing meshes.

Metal siding is very durable and recyclable, and it typically contains significant postconsumer recycled content. It is energy-intensive to manufacture, but recycled steel and aluminum require far less energy than virgin ore. Some types of metal siding are prone to damage.

Composite siding (hardboard) is made of newspaper or wood fiber mixed with recycled plastic or binding agents. It is highly durable, resists moisture and decay, often has significant recycled content, and is not prone to warping or cracking like wood. Composites require less frequent repainting, and some need not be painted at all, saving waste and resources.

Wood siding requires more maintenance than many of the other siding options, but it is renewable and requires relatively little energy to harvest and process. If it is not well maintained, wood can easily be the least durable option, generating significant waste. The most durable solid wood siding, unfortunately, comes from old growth and tropical forests.

Siding selection considerations include the following:

- The most durable siding product that is appropriate should be selected. Siding failures that allow water into the wall cavity can lead to expensive repairs, the waste of damaged components, and the environmental costs of replacement materials. Fire resistance is a feature that helps reduce the financial and environmental impact of rebuilding, particularly in high-risk areas.
- For existing buildings, consider refinishing existing siding to minimize waste, pollution, and energy use.
- Select materials that are biodegradable, have recycled content, and/or are recyclable.
- Reclaimed or remilled wood siding should be used to minimize demand for virgin wood and reduce waste. Painted wood should be tested for lead contamination prior to use.
- New wood siding should display an FSC-certified label.
- Vinyl is somewhat durable, but it is not a green building material. Attributed disadvantages include pollution generated in manufacturing, air emissions, human health hazards of manufacturing and installation, the release of dioxin and other toxic persistent organic pollutants in the event of fire, and the difficulty in recycling.

6.2.7 Roofing

A roof's main role is that of keeping the weather outside a structure and protecting the structural members and interior materials from deterioration. Moisture, dependability, and durability are the most critical characteristics of roofing materials.

The extraction, manufacture, transport, and disposal of roofing materials pollutes air and water, depletes resources, and damages natural habitats. Roofing materials comprise an estimated 12 to 15 percent of construction and demolition waste.

An environmentally sustainable roof must first be durable and long-lasting but ideally should also contain recycled or low-impact materials. Roofs that are environmentally friendly can provide aesthetically pleasing design options, reduced life-cycle costs, and environmental benefits such as reduced landfill waste, energy use, and impacts from harvesting or mining of virgin materials. It takes roughly the same materials, energy, and labor to manufacture and install a 50-year-warranted roof as a 30-year roof, yet disposal and replacement are delayed. A well-installed 50+ rated roof can reduce roofing waste by 80 to 90 percent over its lifetime, relative to a roof with a 20-year warranty.

Mild climates are well suited for passive temperature controls that reduce winter heating and reduce or eliminate the need for mechanical cooling. Air conditioning is used less in mild climates because operable windows and skylights are often employed that can easily provide ventilation and cooling, particularly in smaller buildings. Larger commercial buildings can also be effectively cooled in mild climates without air conditioning, but more careful design is required, and roofing that minimizes heat gain is a key consideration.

Ideally, roofing should renew natural resources. For example, the potential habitat for birds and native plants on a green or living roof can be an island of safety in the urban environment and can help provide pathways for migration through fragmented ecosystems. Similarly, electricity from solar photovoltaic panels reduces demand for fossil fuels. Cool roofing does not renew resources but is often a highly cost-effective way to conserve them.

6.2.7.1 Roofing Materials

Roofing systems should be chosen that are durable and reduce waste, liability, and frustration. Some of the considerations that will impact the type of roof chosen include:

- Its capacity to reflect sunlight and re-emit surface heat. Cool roofs can reduce cooling loads and urban heat-island effects while providing longer roof life.
- Its ability to resist the flow of heat from the roof into the interior, whether through insulation, radiant barriers, or both.
- Its ability to reduce ambient roof air temperatures through evaporation and shading, as in the case of vegetated green roofs.
- Its recyclability and/or capability of being reusable to reduce waste, pollution, and resource use. Options with high postconsumer recycled content—up to 30 percent – are preferable.
- Nonhalogenated roofing membranes (i.e., materials that do not contain bromine or chlorine.) are preferable. In the event of fire burning polyvinyl chloride (PVC) and thermoplastic olefin (TPO) produce strong acids and persistent toxic organic pollution, including dioxin.

Roofs may require protective ballast such as concrete tile to comply with local fire codes. Existing PVC and TPO (thermoplastic polyolefin) roofing membranes, as well as underlying polystyrene insulation, can sometimes be recycled, and this practice is expected to become more prevalent as federal construction-specification requirements generate increased demand.

Residential and commercial roofing options include:

- Clay or cement tiles are very durable and made from abundant materials, but they are heavy and expensive.
- Recycled plastic, rubber, or wood composite shingles are durable, lightweight, and sometimes recyclable but not biodegradable.
- Composition shingles can have 50-year warranties, which are better than 20- to 40 year-products; they can be recycled but are typically landfilled.
- Fiber cement is durable and fire- and insect-proof, but it is heavy and not renewable or biodegradable. They may be ground up and used as inert fill at demolition.
- Metal is a durable and fire- and insect-proof, recyclable material. It typically contains recycled content, but manufacture is energy-intensive and causes pollution and habitat destruction.
- Built-up roofing's durability depends largely on the structure, installation, flashing, and membrane chosen. Most membranes are not made from renewable resources, but some may contain recycled content. High-VOC products emit air pollution during installation.
- Vegetated green roofs are most commonly installed on roofs with slopes less than 30° degrees.
- Wood shakes are biodegradable but flammable and not very durable. They are not typically considered to be a "green" option for fire-prone areas.

Emerging new technologies are helping to promote green building, and there are increasing efforts to make usable space on existing rooftops and/or new roofs to allow additional living space. The key to creating these spaces is the need to use lightweight and recycled materials and to help with storm-water management. Traditional drainage systems using pipe and stone are not plausible. Green roof systems are a natural way of providing additional clean air through the transference of CO_2 and oxygen between the plants and vegetation and the atmosphere.

Green or "living" roofing involves the use of vegetation as the weathering surface. It has proved very successful because it reduces extremes in rooftop temperature, saves energy, and extends the useful life of the roof. High temperatures shorten the life of a roof (leading to increased construction and demolition waste) and increase summer cooling costs.

Green roofing has proven to be effective for three principal reasons: the large surface area of soil and plants helps to reradiate heat; it provides shade and insulation for the waterproof roof membrane; and the plants' transpiration provides cooling. The net result is a 25- to 80-degree decrease in peak roof temperature and up to 75 percent reduction in cooling-energy demand.

Green roofs provide additional environmental and aesthetic benefits. The soil and vegetation in many common designs can detain up to 75 percent of a one-inch rainfall and will filter the remainder. This on-site storm-water management helps reduce demand on storm-water infrastructure, saving resources and money for the entire community. Green roofs provide urban wildlife microhabitat. Although not a replacement for wild land, a vegetated roof can accommodate birds, beneficial insects, and native plants far better than tar and gravel. Cooler roof temperatures also reduce the urban heat-island effect, helping to reduce the cooling load for surrounding buildings. It can also benefit local property values.

Figure 6.8 Example of a 70,000-square-foot vegetated roof on the LDS Assembly Hall building, Salt Lake City, UT.
Source: American Hydrotech, Inc.

Contemporary green-roof designs generally contain a mixture of hard and soft landscaping (Figure 6.8). It is very important that the selected drainage/retention layer is capable of supporting any type of landscape – from roadways and paths to soil and trees so as to permit excess water to drain unobstructed underneath (Figure 6.9).

- Green roof systems generally contain certain essential components, though the location of these components, the materials they are made from, and whether they are present or not vary widely from one installation to another. AWD, a leader in roof garden and prefabricated drainage systems, notes that the main components, materials, and locations to take into consideration if you are contemplating the installation of a green or vegetated roof are:
    - Structural support: Roof and roof garden systems are required to have an underlying structural system in place to support additional weights resulting from use of normal building materials such as concrete, wood, etc. The structural engineer must consider the load of the roof-garden system in the initial design.
    - Roofing membrane: Various roofing membrane options are available to the design engineer. The final membrane choice may be affected by loads imposed by the rooftop garden, by available membrane protection elements in the rooftop-garden system, by root penetration properties of the membrane, and by membrane drainage and aeration requirements.

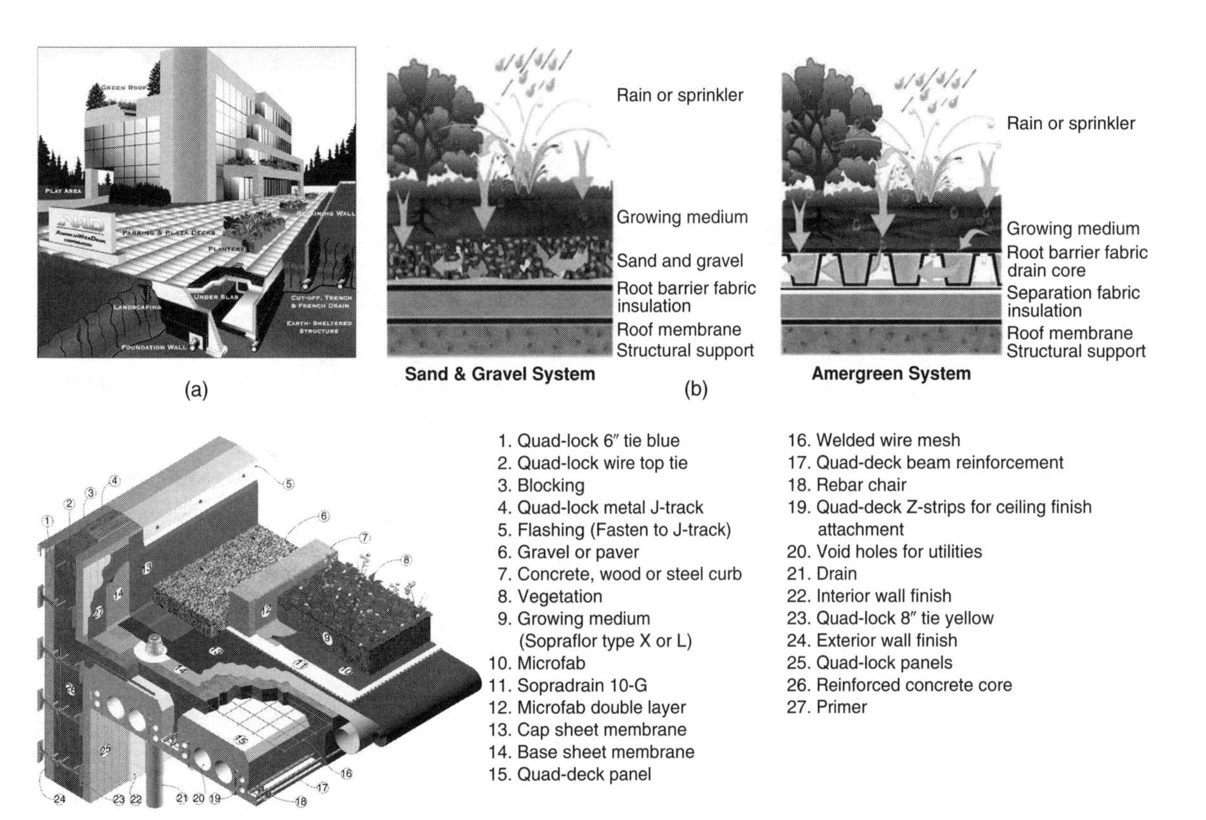

Figure 6.9 (a) Rendering showing commercial application of green elements. (b) Drawing showing two types of roofing systems: a sand and gravel system and the Amergreen roofing system.
Source: American Wick Drain Corporation.

  - Membrane protection: The roofing membrane may require protection from installation damage, long-term water exposure, UV exposure, drainage-medium loads, growing-medium loads, or chemical properties.
- Root barrier: A root barrier may be required between the growing medium and the lower components to eliminate or mitigate potential root penetration into the roofing membrane, drainage medium, or water-storage medium. The optimum location is above the membrane, drainage medium, and water-storage medium so that all three components are protected. Alternate locations may require multiple roof-guard elements to avoid long-term root penetration.
  - Insulation: Insulation may be installed above the structural support, depending on the thermal design of the structure. It may be installed above or below the roofing membrane.
- Drainage medium: Some sort of drainage medium is necessary to ensure the proper range of water content in the growing medium. Too much water will have an adverse affect and cause root rot. Too little water will result in poor vegetation growth. Drainage-medium options range from gravel to various materials designed specifically for this purpose. Plastic materials usually combine the drainage function with water storage and aeration and also protect the membrane from roots and potentially damaging materials in the growing medium.
  - Aeration medium: Aeration is an essential element to promote optimal vegetation growth. The aeration medium usually serves as both drainage and aeration medium. Open channels in the prefabricated drain core are designed to provide air to the plant roots.
  - Water-storage medium: Although the growing medium will store a certain amount of water, additional water storage may be required to provide more successful growth of vegetation. Most prefabricated plastic roof-garden products have water-storage capability. The plastic cones of the prefabricated core provide positive water-storage reservoirs. Sand and gravel may also hold some water.
  - Growing medium: Various natural and manufactured materials may be used as a growing medium. Soil is often mixed with other materials to reduce weight, to provide better structure for roots, and to provide essential nutrients, water, and oxygen.
  - Vegetation: The upper and most visible layer of the roof-garden system is the vegetation. A wide variety of typical landscaping and garden plants may be employed for a rooftop garden. An experienced landscape architect should be involved in plant selection to ensure that vegetation is appropriate both for the geographic region and for a rooftop-garden environment. The type of vegetation affects the selection of the other components of the roof-garden system. Such issues as root protection, water and aeration needs, and drainage requirements are affected by plant selection and vice versa.

6.2.7.2 Extensive and Intensive Green Roofs

Extensive green roofs typically contain a layer of soil medium that is relatively thin (two to six inches) and lightweight (10 to 50 pounds per square foot for the entire system when saturated with water). They are lightweight, relatively easy to install, durable, and cost-effective.

Intensive green roofs are designed to accommodate trees and gardens. Soil can be as deep as needed to accommodate the desired tree or plant species, but deeper, denser soil dramatically increases dead load, requiring a stronger and more expensive structure, greater maintenance, and either terracing or a relatively flat roof. But green-roof options are available for virtually any building type or location.

Some of the benefits attributed to green roofs include:

- They provide enhanced insulation and more moderate rooftop temperatures, which reduce cooling and heating requirements, saving energy and money.
- They facilitate filtration and detention of storm water, reducing pollution and the cost of new and expanded infrastructure as paved areas increase.
- They absorb dust and airborne pollutants.
- They reduce ambient air temperatures, lowering urban heat-island effects and helping to enhance microclimate of surrounding areas.
- They extend the life of roof membranes, which are protected from ultraviolet radiation, extreme temperatures, and mechanical damage. (Plant species, soil depth, and root-resistant layers are carefully matched to ensure that the roof membrane is not damaged by the roots themselves.).
- Lightweight extensive systems can be designed with dead loads comparable to standard low-slope roofing ballast. Structural reinforcement may not be necessary, and cost can be comparable to conventional high-quality roofing options.

6.2.8 Wood

6.2.8.1 Wood Types

The popularity over the decades of pressure-treated lumber has been partially due to its resistance to rot and insects. The familiar copper chromated arsenate (CCA) was largely phased out in a cooperative effort between manufacturers and the Environmental Protection Agency. Arsenic is acutely toxic and carcinogenic and was shown to be leaching into surrounding soils, which prompted the cessation of its production for residential use at the end of 2003. It is unfortunate that these problems were not identified before it was put into widespread use.

Existing CCA-treated lumber poses a challenge. It remains resistant to rot and insects, and its reuse would help conserve forest resources and keep a potentially useful resource out of landfills. However, permitting its reuse would allow it to continue to leach arsenic into soils. CCA-treated wood should not be composted or disposed of in green-waste or wood-waste bins. Burning CCA-treated wood is highly toxic. Disposal of CCA-treated wood is now mandated to be in a lined landfill or as class I hazardous waste. The newer, much less toxic wood treatments, copper azole (CA) and alkaline copper quaternary (ACQ), are more corrosive than CCA. Manufacturer-recommended fasteners should be employed to minimize rust and prevent staining.

Points to consider include the following:

- Repair and/or refinish existing decks, railing, or fencing.
- Reuse wood in good condition.
- Build with durable materials such as plastic lumber.
- For structural elements that will be in contact with soil and water, consider:
  1. Heartwood from decay-resistant species such as redwood or cedar that has been FSC-certified as harvested from a responsibly managed forest.
  2. If pressure-treated lumber is required, the two water-resistant preservatives currently used are CA and ACQ, which are significantly less toxic than CCA.
  3. Avoid the few remaining stocks of CCA.

- For fencing, consider friendly alternatives such as a living fence of bushes, shrubs, or live bamboo in urban settings or fencing made of cut bamboo (a rapidly renewable material).

Wood is a renewable material and requires less energy than most materials to process into finished products. However, logging, manufacture, transport, and disposal of wood products have substantial environmental impacts. Standard logging practices cause erosion, pollute streams and waterways with sediments, damage sensitive ecosystems, reduce biodiversity, and lead to a loss of soil carbon. The key to reducing these impacts is the minimization of wood use by substituting suitable alternatives, reusing salvaged wood, selecting wood from responsibly managed forests, controlling waste, and minimizing redundant components.

Where salvaged or reclaimed wood is not available or applicable (i.e., for structural applications) specify products that are certified by an approved and accredited organization such as the Forest Stewardship Council (FSC).

Engineered lumber, also known as composite wood or man-made wood, consists of a range of derivative wood products that are manufactured by pressing or laminating together the strands, particles, fibers, or veneers of wood with a binding agent to form a composite material. The superior strength and durability of engineered lumber allows it to displace the use of large (and increasingly unavailable), mature timber. It is also less susceptible to humidity-induced warping than equivalent solid woods, although the majority of particle- and fiber-based boards require treatment with sealant or paint to prevent water penetration.

These products are engineered to meet precise application-specific design specifications and are tested to meet national or international standards. Using engineered lumber instead of large-dimension rafters, joists, trusses, and posts can save money and reduce total wood use by as much as 35 percent. The wider spacing of members possible with engineered lumber also has the advantage of increasing the insulated portion of walls. Other advantages include its ability to form large panels from fibers taken from small-diameter trees, small pieces of wood, and wood that has defects; these panels can be used in many engineered-wood products, especially particle- and fiber-based boards. Engineered-wood products are used in a number of ways, usually in applications similar to solid wood products, but many builders prefer engineered products because they are economical and typically longer, stronger, straighter, more durable, and lighter than comparable solid lumber.

Engineered-wood products also have some disadvantages; for example, they require more primary energy for their manufacture than solid lumber. Furthermore, the adhesives used may be toxic. A concern with some resins is the release of formaldehyde in the finished product, often seen with urea-formaldehyde-bonded products. Cutting and otherwise working with engineered-wood products can therefore expose workers to toxic constituents that could potentially cause harm.

The primary types of adhesives used in engineered wood include urea-formaldehyde resin (UF), which is not waterproof but is popular because it is inexpensive; melamine-formaldehyde resin (MF), which is white in color, heat- and water-resistant, and preferred for exposed surfaces in more costly designs; phenol-formaldehyde resin (PF), which has a yellow/brown color and is commonly used for exterior exposure products;

and methylene diphenyl diisocyanate (MDI) or polyurethane (PU) resins, which are expensive, generally waterproof, and do not contain formaldehyde.

Sheathing is the structural covering of plywood or oriented-strand board (OSB) that is applied to studs and roof/floor joists to provide shear strength and serve as a base for finish flooring or a building's weatherproof exterior. OSB relies on smaller (aspen and poplar) trees, which are a more rapidly renewable resource than the mature timber required for plywood. Nonetheless, sheathing is considered to be the second most wood-intensive element of wood-frame construction.

Wood is a renewable product and requires less energy than most materials to process into finished products. However, the logging, transport, and disposal of wood have substantial environmental impacts. Standard logging techniques lead to erosion and fouling of streams and waterways with sediments. They result in loss of carbon stored in soil, damage sensitive ecosystems, and reduce biodiversity. Minimizing wood use through a combination of substitute materials, selection of wood from responsibly managed forests, and design and construction practices that control waste and minimize redundant components is the key to reducing these impacts. Salvaged, remanufactured, or certified wood products (by accredited organizations such as the FSC) should be considered.

Engineered-wood sheathing materials do have some environmental tradeoffs because the wood fibers are typically bound with formaldehyde-based resins. Interior-grade plywood typically contains urea formaldehyde (UF), which is less chemically stable than the phenol formaldehyde (PF) found in water-resistant exterior-grade plywood and OSB. This advantage makes exterior-grade plywood preferable for indoor applications, as it emits less toxic and suspected carcinogenic compounds.

There are other alternatives to these wood-intensive conventional and engineered materials. Fiberboard products rated for structural applications (such as Homasote®'s 100-percent-recycled nailable structural board) are alternatives to plywood and OSB. Structural-grade fiber-cement composite siding combines sheathing and cladding, providing shear strength and protection from the elements while reducing labor costs for installation. Water-resistant exterior-grade gypsum sheathing can be one of the options employed under brick and stucco exterior finishes to reduce wood requirements.

Designs that combine bracing with nonstructural sheathing can provide necessary strength while enhancing insulation and reducing wood requirements. Structural insulated-panel construction provides interior and exterior sheathing as well as insulation in precut, factory-made panels. And by designing for disassembly, sheathing materials can be readily reused or recycled. For example, while adhesives distribute loads over larger areas than fasteners alone, materials attached with removable fasteners (in appropriate designs) are more easily deconstructed than materials installed with adhesives.

Medium-density fiberboard (MDF) is a composite panel product typically consisting of low-value wood byproducts such as sawdust combined with a synthetic resin such as urea formaldehyde (UF) or other suitable bonding system and joined together under heat and pressure. Additives may be introduced during manufacturing to impart additional characteristics. MDF is widely used in the manufacturing of furniture, kitchen cabinets, door parts, moldings, millwork, and laminate flooring. MDF panels are manufactured with a variety of physical properties and dimensions, providing the opportunity to design the end product with the specific characteristics and density needed.

Although MDF is highly toxic to manufacture, it does not emit volatile organic compounds (VOCs) in use. MDF will accept a wide range of sealers, primers, and coatings to produce a hard, durable tool surface, but it is not suitable for high-temperature applications. It is also entirely possible to make MDF-type panels using waste wood fiber from demolition wood and waste paper.

The properties of MDF can vary based on the country or region where it is produced. It also comes in a variety of densities depending on the application. The surface of MDF is flat, smooth, uniform, dense, and free of knots and grain patterns. The homogenous density profile of MDF allows intricate and precise machining and finishing techniques for superior finished products. Trim waste is significantly reduced when using MDF compared to other substrates. Stability and strength are important assets of MDF, which can be machined into complex patterns that require precise tolerances.

One of the more recent developments in North American composites has been the introduction of boards made from agricultural residues. Increasing constraints on residue burning have been a prime motivator for their introduction. Their manufacture entails compressing agricultural residue materials with nonformaldehyde glues; the panels provide an excellent alternative to plywood sheets 3/8 inch and thicker and can be used in much the same way as medium-density fiberboard. They can also replace OSB and MDF for interior walls and partitions work. Agricultural-residue (ag-res) boards are made from waste wheat straw, rice straw, or even sunflower seed husks. Ag boards are both aesthetically pleasing, often stronger than MDF, and just as functional. Under heat and pressure, microscopic "hooks" on the straws link together, reducing or eliminating the need for binders. The use of new soy adhesives promises both improved performance and economics to the ag-composites industry. They are also expected to be safer to handle and to reduce volatile-organic-compound emissions.

Homasote® is a panel product made of 100-percent postconsumer recycled newspaper fiber and has actually been in production longer than plywood and OSB. It has many potential applications including walls, structural roof decking, paintable interior panels, and concrete forms. At the time of writing, agricultural-straw panels are difficult to find due to a combination of intense demand, lack of supply, and the time required to bring a manufacturing facility online. Homasote® is weather-resistant, structural, insulating, and extremely durable, and it has two to three times the strength of typical light-density wood fiberboards.

6.2.8.2 Framing

Advanced framing, also known as optimum-value engineering (OVE), refers to a variety of framing techniques designed to reduce the amount of lumber used and waste generated in the construction of a wood-framed house, thereby reducing material cost and use of natural resources while at the same time increasing energy efficiency through increased space for insulation. The extraction, manufacture, transport, and disposal of lumber deplete resources, damage natural habitats, and pollute air and water. Dimensional lumber supplies depend upon larger trees that require decades to mature.

OVE advanced framing techniques include studs spaced at 24-inch on center (oc); 2-foot modular designs that reduce cutoff waste from standard-sized building materials (Figure 6.10); in-line framing that reduces the need for double top plates; building

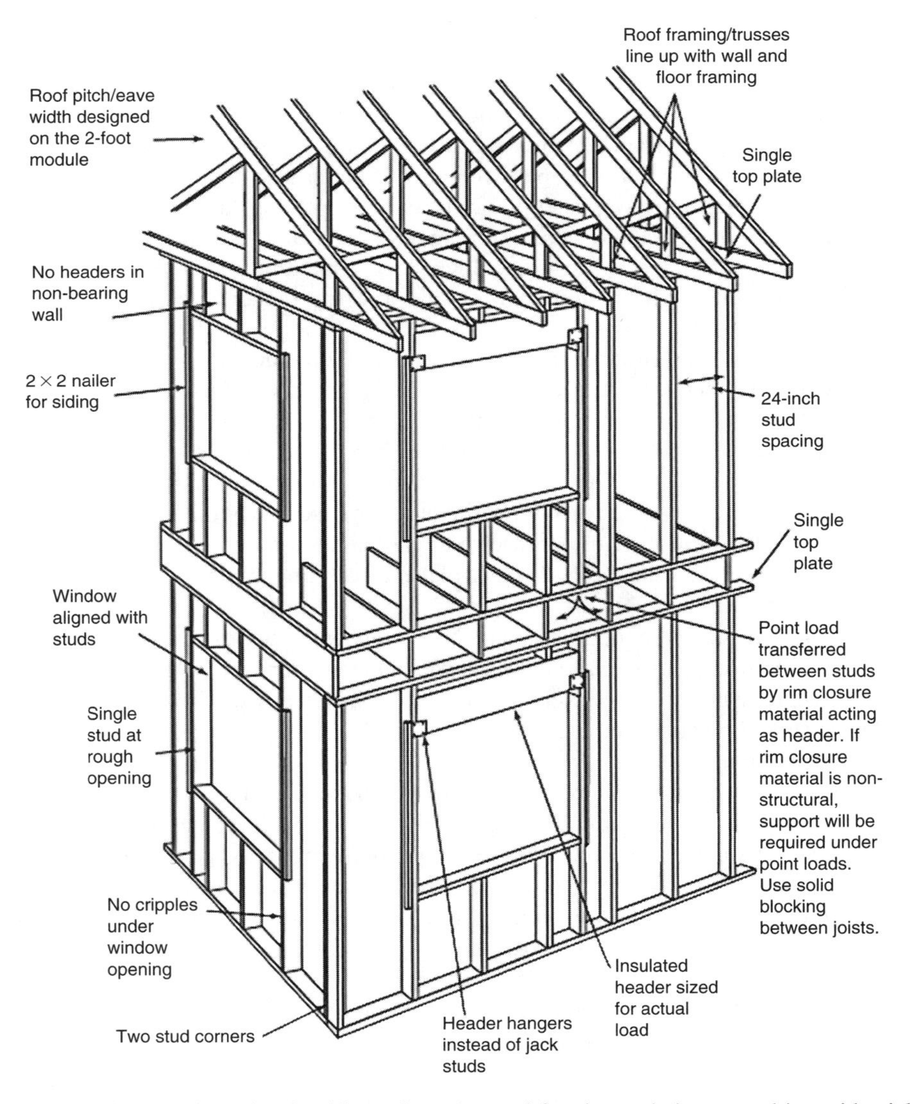

Figure 6.10 Isometric drawing illustrating advanced framing techniques used in residential construction.
Source: Adapted from Building Science Corporation.

corners with two studs; and insulated headers over exterior building openings (or no headers for non-load-bearing walls). The advantage of spacing studs at 24-inch oc rather than 16-inch oc is that it reduces the amount of framing lumber required to construct a home and replaces framing members with insulation. This allows the wall to achieve a higher overall insulating value and costs less to construct than a conventionally framed wall while still meeting the structural requirements of the home. However, to be successful, advanced framing techniques must be considered early in the design process.

Both builders and homeowners can benefit from advanced framing. Advanced framing techniques create a structurally sound home that has lower material and labor costs than a conventionally framed house. Additional construction cost savings result from the generation of less waste to be disposed of, which also helps the environment. Advanced framing improves energy efficiency by replacing lumber with insulation material. The whole-wall R-value is improved and insulation values are enhanced because fewer studs means the insulated wall area is maximized, the deeper wall cavity allows for thicker insulation, and thermal bridging (conduction of heat through framing) is reduced.

In many cases conventional framing is structurally redundant, using wood unnecessarily for convenience. The Department of Energy's Office of Building Technology notes that with advanced framing techniques $500 in material costs per 1200 square feet can be saved in addition to 3 to 5 percent of labor costs and an annual heating and cooling cost saving of 5 percent. While advanced framing is more wood-efficient than conventional framing, keep in mind that alternative structural technologies such as insulated structural systems, straw bale or earthen construction, and high-recycled-content steel framing with thermal breaks place fewer demands on our forest resources than stick framing.

Although advanced framing techniques have been researched extensively and proven effective, some techniques may not be allowed under certain circumstances (i.e., areas prone to high winds or with seismic potential) or in some jurisdictions. Local building officials should be consulted early in the design phase to verify or obtain acceptance of these techniques.

Structural insulated panels (SIPs) are factory-built walls and roof panels that consist of two sheets of rigid structural facing board – e.g., oriented-strand board (OSB) or plywood that is applied to both sides of a rigid EPS (expanded polystyrene) insulating foam core that is four or more inches thick. The result of this simple sandwich is a strong structural panel for building walls, roofs, and floors that is significantly more energy-efficient, yielding increased R-values compared to traditional framing. In addition to SIP's excellent insulation properties, it offers airtight assembly, noise attenuation, and superior structural strength. Though SIP panels cost more per square foot than conventional construction, total construction costs are often minimized due to reduced labor and faster completion.

SIP's reduction in construction waste and superior insulation are especially significant when compared to conventional stick or steel stud systems. They can be delivered precut to the precise dimensions required, and each panel contains the structure, insulation, and moisture barrier of the wall system. OSB is the most common sheathing and facing material in SIP, reducing wood use by as much as 35 percent and reducing pressure on mature forests by allowing the use of smaller farm-grown trees for structural applications. The exterior-grade plywood used in some products requires more mature timber.

SIP wall assemblies tend to be well sealed, enhancing energy efficiency. As with any tightly sealed structure, moisture control and well-designed ventilation are critical. SIP construction can contribute to very good indoor-air quality; the plastic insulating foams (expanded polystyrene or polyurethane/polyisocyanurate) are very chemically stable, and OSB is a low-emitting material.

Structural insulated panels are engineered and custom-manufactured to give the designer greater control over the project, including materials and costs. SIPs are usually chosen for their versatility, strength, cost effectiveness, and energy efficiency. They are suitable for a wide range of residential and commercial projects and are custom-made according to specifications and drawings. Another advantage of insulated structural systems is that they integrate a building's structure and insulation into a single component.

There are several materials used for SIP cores. Expanded polystyrene (EPS) is the most common core material. It requires less energy to manufacture than other options and is more recyclable than polyurethane or polyisocyanurate. Many products offer a one-hour fire rating when installed with 5/8-inch or thicker gypsum sheathing. EPS foam is expanded with pentane, which does not contribute to ozone depletion or global warming and is often recaptured at the factory for reuse.

Polyurethane and polyisocyanurate possess greater insulation properties per inch of foam than EPS and offer greater resistance to thermal breakdown. However, polyurethane and polyisocyanurate are unlikely to be recycled. Polyurethane and polyisocyanurate use HCFC blowing agents, which contribute to global warming and ozone depletion (though to a lesser degree than CFCs). Research is currently underway on another alternative: new resins (polyurethane/polyisocyanurate) derived from soy that may soon be available for use in SIPs.

Straw-core SIPs are made from waste agricultural straw. They are renewable and recyclable, and the pressed-straw core does not require a binding agent. However, straw-core SIPs offer less insulation per inch of thickness and are considerably heavier than other options; energy use in shipping is a significant consideration.

Some of the factors to consider when building with SIPs include:

- Design to minimize waste by ordering SIP panels precut to meet project requirements, including window and door openings. Likewise, design should be to standardized panel dimensions.
- Enquire whether the SIP supplier or manufacturer will take back any offcuts for recycling.
- When sizing your heating system, consider the thermal performance of SIPs to save money up front and energy over time. Oversized heating and cooling systems are inefficient.
- SIP roofs do not necessarily require ventilation, making them appropriate for low-slope roofs. If local jurisdiction mandates ventilated roofs, consider SIPs with integrated air channels or upgrading from composition roofing.

At their most basic level, integrated concrete forms (ICFs) serve as a forming system for poured-concrete walls and consist of stay-in-place formwork for energy-efficient, cast-in-place, reinforced-concrete walls. The forms are interlocking modular units that are dry-stacked (without mortar) and filled with concrete. The forms lock together somewhat like LEGO® bricks and serve to create a form for the structural walls of a building. Concrete is pumped into the cavity to form the structural element of the walls. ICFs use an insulating material as permanent formwork that becomes a part of the finished wall. All ICFs can be considered "green" materials; they are durable, produce little or no waste during construction, and dramatically improve the thermal performance of concrete walls.

ICFs usually employ reinforcing steel (rebar) before concrete placement to give the resulting walls flexural strength, similar to that of bridges and high-rise buildings made of concrete. The forms are filled with concrete every several feet in order to reduce the risk of blowouts. After the concrete has cured or firmed up, the forms are left in place permanently to increase thermal and acoustic insulation, increase fire protection, and provide backing for gypsum boards on the interior and stucco, brick, or other siding on the exterior. It can also accommodate electrical and plumbing installations (Figure 6.11).

Figure 6.11 Photo depicting application of insulating concrete formwork Quad-Lock system. *Source*: Quad-Lock Building Systems, Ltd.

There are several different ICF systems on the market that vary in the shape of the resulting concrete within the wall. Examples are:

- The flat system forms an even thickness of concrete throughout the walls like a conventionally poured wall.
- The waffle-grid system creates a waffle pattern where the concrete is thicker at some points than others.
- The post-and-beam system forms detached horizontal and vertical columns of concrete.

One of the attributes of standard concrete is that it is a dense material with a high heat capacity that can be used as thermal mass, reducing the energy required to maintain comfortable interior temperatures. However, concrete is not a good insulator; standard formwork is waste-intensive; and toxic materials are frequently needed to separate formwork from the hardened product. ICF address these weaknesses by reducing solid waste, air and water pollution, and (potentially) construction cost. ICF wall systems are thermally superior, which enhances their usefulness for passive heating and cooling; comfort is also enhanced and energy costs are reduced, while first costs can be minimized through the resulting reduced heating/cooling system size.

ICF systems can be manufactured from a variety of materials, such as lightweight foamed-concrete panels, rigid foams such as expanded polystyrene and polyurethane, and composites that combine concrete with mineral wool, wood waste, paper pulp, or expanded polystyrene beads. There are also ICF systems, such as BaleBlock™ and Faswall™, that substitute straw bales or fiber cement for polystyrene. Rigid foams used in ICFs generally do not have significant recycled content and are less likely to be recyclable at the end of their life but may be reused in fill or other composite concrete products.

ICFs offer the structural and fire-resistance benefits of reinforced concrete; structural failure due to fire is rare to nonexistent. Due to the addition of flame-retardant additives, polystyrene ICFs tend to melt rather than burn, and interior ICF walls tend to contain fires much better than wood frame walls, improving overall fire safety. Some ICFs such as RASTRA are made from recycled postconsumer polystyrene (foam) waste products.

As is the case in most heated structures, moisture control is a key design consideration for ICF walls. Solid concrete walls sandwiched in polystyrene blocks tend to be very well sealed to enhance energy efficiency, but they consequently also tend to seal water vapor within the structure. Potential mold growth and impaired indoor-air quality are serious health concerns that must be addressed. The easiest method to resolve this is by incorporating mechanical ventilation. Certain systems such as straw bale and RASTRA tend to be more vapor-permeable, thereby reducing this concern.

6.2.9 Concrete

Concrete is a composite building material made from a combination of aggregates such as sand and crushed stone and a binder or paste such as cement. The most common form of concrete consists of mineral aggregate such as stones, gravel, sand, cement, and water. The cement hydrates after mixing and hardens into a stonelike material.

Concrete has a low tensile strength and is generally strengthened by the addition of steel reinforcing bars; this is commonly referred to as reinforced concrete.

Concrete is a strong, durable, and inexpensive material that is the most widely used structural building material in the United States. Due to the vast scale of concrete demand, the impacts of its manufacture, use, and demolition are widespread. Habitats are disturbed from materials extraction; significant energy is used to extract, produce, and ship cement; and toxic air and water emissions result from cement manufacturing. Cement manufacture in particular is energy-intensive.

Estimates indicate that approximately one ton of carbon dioxide is released for each ton of cement produced, resulting in 7 to 8 percent of man-made CO_2 emissions. And although concrete is typically only 9 to 13 percent cement, it accounts for 92 percent of concrete's embodied energy. Cement dust contains free silicon-dioxide crystals, the trace element chromium, and lime, all of which can adversely impact worker health. Mixing concrete requires a great deal of water and generates alkaline waste water and runoff that can contaminate waterways and vegetation.

Incorporating local and/or recycled aggregate (such as ground concrete from demolition) is an excellent way to reduce the impacts of solid waste, transit emissions, and habitat disturbance. Environmental impacts can also be reduced substantially by substituting alternative pozzolan ash (industrial byproducts such as fly ash, silica fume, rice husk ash, furnace slag, and volcanic tuff) for Portland cement. Fly ash, a residue from coal combustion, is quite popular as a cement substitute that generally decreases porosity, increases durability, and improves workability and compressive strength, although the curing time is increased. Fly ash generally constitutes 10 to 15 percent of standard mixes, but many applications will allow substituting of up to 35 to 60 percent of cement, and with certain types of fly ash (Class C) cement can be completely replaced in some projects.

In nonstructural applications concrete use may be reduced by trapping air in the finished product or through the use of low-density aggregates. Trapped air displaces concrete while enhancing insulation value and reducing weight and material costs without compromising the durability and fire resistance of standard concrete. Low-density aggregates such as pumice, vermiculite, perlite, shale, polystyrene beads, or mineral fiber provide similar insulation and weight-reduction benefits.

Cast-in-place or precast concrete and concrete-masonry unit (CMU) considerations include:

- Design for the reuse of portions of existing structures, such as slabs or walls that are in satisfactory condition.
- Recycle demolished concrete at local landfills or use on site for use as aggregate or fill material for new projects.
- Incorporate the maximum amount of fly ash, blast-furnace slag, silica fume, and/or rice-husk slag appropriate to the project, thereby reducing cement use by 15 to 100 percent.
- Use of precast systems will minimize waste of forming material and reduce the impact of wash water on soils.
- Consider alternative material substitutes for concrete such as insulating concrete forms (ICF) to reduce waste, enhance thermal performance, and reduce construction schedules.

Likewise, cellular, foamed, autoclaved-aerated (AAC), and other lightweight concretes add insulation value while reducing weight and concrete required. The use of earthen and rapidly renewable materials, such as rammed earth, cob, or straw bale, reduce the need for insulation and finish materials in both residential and commercial projects.

- Use nontoxic form-release agents.
- Waste can be minimized by carefully planning concrete material quantities.
- For footings, consider fabric-based form systems for fast installation and wood savings.
- Reduce wood waste and material costs by employing steel or aluminum concrete forms, which can be reused many more times than wood forms.

Up to 75 percent of urban surface area is covered by impermeable pavement, which inhibits groundwater recharge, contributes to erosion and flooding, conveys pollution to local waters, and increases the complexity and expense of storm-water treatment. One of the main characteristics of pervious paving is that it contains voids that allow water to percolate through to the base materials below. It also reduces peak storm-water flow and water pollution and promotes ground-water recharge. Pervious paving may incorporate recycled aggregate and fly ash, which help reduce waste and embodied energy. Pervious paving is suitable for use in parking and access areas, as it has a compressive strength of up to 4000 psi. It also mitigates problems with tree roots; percolation areas encourage roots to grow deeper. Enhanced heat exchange with the underlying soil can decrease summer ambient-air temperature by two to four degrees F.

Poured-in-place applications require on-site formwork to give shape to walls, slabs, and other project elements as they cure (Figure 6.12). Plywood and milled lumber are the most common form materials, contributing to construction waste and the impacts of timber harvesting and processing. Wooden formwork can be made from salvaged wood and typically be disassembled and reused several times. Disassembling construction-grade lumber and exterior-grade plywood forms should be considered for reuse within the project.

Form releasers or parting agents are materials that facilitate the separation of forms from hardened concrete. Such materials prevent concrete from bonding to the form, which can mar the surface when forms are disassembled. Traditional form releasers such as diesel fuel, motor oil, and home heating oil are carcinogenic, which limits the potential for reuse of wood formwork because it exposes construction personnel (and potentially occupants as well) to volatile organic compounds. They are now prohibited by a variety of state and federal regulations, including the Clean Air Act. Low- and zero-VOC water-based form-release compounds that incorporate soy or other biologically derived oils dramatically reduce health risks to construction staff and occupants and often make it easier to apply finishes or sealants when necessary. Many soy-based options are less expensive than their petroleum-based counterparts.

When designing concrete formwork, one should also consider factors that will adversely affect concrete-formwork pressure. These factors include the rate of placement, concrete mix, and temperature. The rate of placement should generally be lower in the winter than in the summer. It doesn't matter how many cubic yards are actually placed per hour or how large the project is. What does matter is the rate of placement per height and time (height of wall poured per hour).

Figure 6.12 Photo of carpenters setting concrete formwork for high-level-waste facility pit walls.
Source: Bechtel Corporation.

6.3 Building and Material Reuse

6.3.1 Building Reuse

6.3.1.1 Maintain Existing Walls, Floors, and Roof

The LEED™ intent is to protect virgin resources by reusing building materials and products, thereby reducing impacts associated with extraction and processing, and to extend the life cycle of existing building stock, conserve resources, retain cultural resources, reduce waste, and reduce environmental impacts of new buildings as they relate to materials' manufacturing and transport.

LEED™ requirements for New Construction are to maintain a minimum of 50, 75, or 95 percent (for up to three points) of the existing building structure, including structural floor and roof decking as well as the envelope (exterior skin and framing but excluding window assemblies and nonstructural roofing material). It is possible to achieve a credit by maintaining a minimum of 50 percent (by area) of interior nonstructural elements, such as interior walls, doors, floor coverings, and ceiling systems. Hazardous materials that are remediated as a part of the project scope are to be excluded from the calculation of the percentage maintained. The credit will not apply if the project includes an addition to an existing building where the square footage of the addition is more than two times the square footage of the existing building.

The table below describes the minimum percent of building-structure reuse requirements for credit achievement for New Construction:

% Building Reuse	Points
55	1
75	2
95	3

However, for Core and Shell you are required to maintain a minimum of 25, 33, 42, 50, or 75 percent of existing walls, floors, and roof for up to five credits. Schools must maintain 55 or 75 percent of existing walls, floors, and roof for up to two credits. It is strongly advised to check with LEED™ for the updated requirements of a particular category.

Potential technologies and strategies: Consider the use of salvaged, refurbished, or reused materials from previously occupied buildings, including structure, envelope, and elements. Remove elements that pose contamination risk to building occupants and upgrade components that would improve energy and water efficiency, such as windows, mechanical systems, and plumbing fixtures. However, mechanical, electrical, plumbing, or specialty items and components should not be included for this credit. Furniture may be included only if it is included in the other MR Credits.

6.3.1.2 Interior Nonstructural Elements

Maintain 50 percent of interior nonstructural elements for New Construction and Schools and 40 and 60 percent for Commercial Interiors. The intent here according to LEED™ is to extend the life cycle of existing building stock, conserve resources, retain cultural resources, reduce waste, and reduce environmental impacts of new buildings as they relate to materials' manufacturing and transport.

LEED™ requirements: Maintain at least 50 percent (by area) of existing interior nonshell, non-structural elements (interior walls, doors, floor coverings, and ceiling systems) of the completed building (including additions). If the project includes an addition to an existing building, this credit is not applicable if the square footage of the addition is more than two times the square footage of the existing building.

In terms of potential technologies and strategies, LEED™ requires that consideration be given to the reuse of existing buildings, including structure, envelope, and interior nonstructural elements. Elements that pose contamination risk to building occupants are to be removed, and components that would improve energy and water efficiency, such as mechanical systems and plumbing fixtures, should be upgraded. For the LEED™ credit the extent of building reuse needs to be quantified.

The owner/developer must provide a report prepared by a suitably qualified person outlining the extent to which major building elements from a previous building were incorporated into the existing building. The report should include preconstruction and postconstruction details highlighting and quantifying the reused elements, be it foundations, structural elements, or facades. Windows, doors, and similar assemblies may be excluded.

The rehabilitation of old buildings highlights the many successful commercial redevelopments being executed in many cities around the world. There is potential to lower building costs and provide a mix of desirable building characteristics. However, the reuse of existing structural elements depends on many factors, not least of which are fire safety, energy efficiency, and regulatory requirements.

6.3.2 Materials Reuse

Materials reuse should be 5 and 10 percent for New Construction, Schools, and Commercial Interiors (30 percent for furniture and furnishing),and 5 percent for Core and Shell. The intent is to reuse building materials and products in order to protect and reduce demand for virgin resources and to reduce waste, thereby reducing impacts associated with the extraction and processing of virgin resources.

LEED™ requirements: Use salvaged, refurbished, or reused materials such that the sum of these materials constitutes at least 5, 10, or 30 percent (for Commercial Interiors, Furniture and Furnishings), based on cost, of the total value of materials on the project. Mechanical, electrical, and plumbing components and specialty items such as elevators and equipment are not included in this calculation. Include only materials permanently installed in the project. Furniture may be included, providing it is included consistently in MR Credits 3–7. Most credits in the Material and Resource category are calculated using a percentage of total building materials.

LEED™ potential technologies and strategies include the identification of opportunities to incorporate salvaged materials into the building design and research potential material suppliers. Salvaged materials such as beams and posts, flooring, paneling, doors and frames, cabinetry and furniture, brick, and decorative items should be considered.

Reuse is essentially the salvage and reinstallation of materials in their original form. Recycling is the collection and remanufacture of materials into a new material or product, typically different from the original. Biodegradable material breaks down organically and may be returned to the earth with none of the damage associated with the generation of typical waste materials.

Reusing materials slated for the landfill has become one of the most environmentally sound ways to build because the extraction, manufacture, transport, and disposal of virgin building materials pollutes air and water, depletes resources, and damages natural habitats. Construction and demolition are estimated to be responsible for about 30 percent of the U.S. solid-waste stream. Real-world case studies by the Alameda County Waste Management Authority, for example, have concluded that more than 85 percent of that material, from flooring to roofing to packaging, is reusable or recyclable.

By salvaging materials from renovation projects and specifying salvaged materials, material costs can be reduced while adding character to projects and maximizing environmental benefits, such as reduced landfill waste, reduced embodied energy, and reduced impacts from harvesting/mining of virgin materials (e.g., logging old-growth or tropical hardwood trees, mining metals, etc.).

It should be noted that some materials require remediation or should not be reused at all. For example, materials contaminated by hazardous substances such as asbestos, arsenic, and lead paint must be treated and/or disposed of properly. Avoiding materials that will cause future problems is critical to long-term waste reduction as well as the health of communities and the planet.

A system should be developed for tracking the volume or weight of all materials salvaged, recycled, and disposed of during the project. At the end of the project, or more frequently if stipulated in the specifications, submit a calculation documenting that the project achieved a 50 or 75 percent diversion rate. This rate will depend on the targeted goals set for the project.

Considerations for selecting reusable building materials include:

- Reusing existing building shells, when appropriate, can yield the greatest overall reduction in project impacts. Additionally, for remodels/renovations, materials may be reused on site.
- For remodels and redevelopment, adequate time should be allowed in the construction schedule for deconstruction and recycling.
- Replace inefficient fixtures, components, and appliances (e.g., toilets using more than 1.6 gallons per flush, single-pane windows, and refrigerators or other appliances over five years old).
- Products containing hazardous materials such as asbestos, lead, or arsenic should either be disposed of properly or remediated prior to reuse.
- Note that salvaged materials can vary in availability, quality, and uniformity. Ensure that materials are readily available to meet project needs before specifying them.

- Building materials composed of one substance (e.g., steel, concrete, wood, etc.), or that are readily disassembled are generally easiest to reuse or recycle.
- Materials composed of many ingredients, such as vinyl siding, OSB, or particleboard are generally not recyclable or biodegradable.
- Materials should be carefully evaluated to ensure that they offer the best choice for the application. They need to be durable and can preferably be readily disassembled for reuse, recycling, or biodegrading at the end of the useful life of the building.

6.4 Construction Waste Management

The broad intent of the Construction Waste Management credit is to avoid materials going to landfills during construction by diverting the construction waste, demolition, and land-clearing debris from landfill disposal; redirecting recyclable recovered resources back to the manufacturing process; and redirecting reusable materials to appropriate sites.

LEED™ requirement: Recycle and/or salvage at least 50 or 75 percent of nonhazardous construction, demolition, and packaging debris. Develop and implement a construction waste-management plan that at a minimum identifies and quantifies the materials generated during construction that is to be recycled or diverted from disposal and notes whether such materials will be sorted on-site or comingled. Typical items would include brick debris, concrete, steel, ductwork, clean dimensional wood, paperboard an plastic used in packing, etc. (Figure 6.13). Excavated soil and land-clearing debris do not contribute to this credit. Calculations can be done by weight or by volume but must be consistent throughout.

Figure 6.13 Photo of typical landfill showing debris.

Each credit a project attempts to achieve using the LEED™ system requires documentation to prove the activity was completed. LEED™ letter templates are to be used to certify that requirements are met for each prerequisite and credit. Additional documentation may also be required. As the contractor is responsible for construction waste management, he or she will generally be responsible for completing the required LEED™ documentation for these two credits.

Provide the LEED™ Letter Template, signed by the architect, owner, or other responsible party, tabulating the total waste material, quantities diverted and the means by which they were diverted, and declaring that the credit requirements have been met. A portion of the credits in each application may be audited, and the contractor should be prepared with backup documentation.

LEED™ projects will generally require a waste-management plan and regular submittals tracking progress. The plan should show how the required recycling rate is to be achieved, including materials to be recycled or salvaged, cost estimates comparing recycling to disposal fees, materials-handling requirements, and how the plan will be communicated to the crew and subcontractors. All subcontractors are required to adhere to the plan in their contracts.

Design considerations relating to construction waste reduction include:

- Smaller projects use less material, reducing both solid waste and operating costs.
- Assemblies should be designed to match the standard dimensions of the materials to be used.
- Disassembly design should be considered so that materials can be readily reused or recycled.
- Clips and stops should be employed to support drywall or wood paneling at top plates, end walls, and corners. Clips can provide the potential for two-stud corners, reducing wood use, easing electrical and plumbing rough-in, and improving thermal performance.
- Materials attached with removable fasteners are generally quicker, cheaper, and more feasible to deconstruct than materials installed with adhesives. However, adhesives distribute loads over larger areas than fasteners used alone.
- Remodel to make use of existing foundations and structures in good condition, reducing waste, material requirements, and possibly labor costs.
- Design for flexibility and changing use of spaces.
- Specify materials such as structural insulated panels, panelized wood framing, and precast concrete that can be delivered precut for rapid, almost waste-free installation.
- For wood construction, consider 24-inch on center framing with insulated headers, trusses for roofs and floors, finger-jointed studs, and engineered-wood framing and sheathing materials.
- Whenever practical specify materials with high recycled content.

It is estimated that commercial construction typically generates between 2 and 2.5 pounds of solid waste per square foot, the majority of which can be recycled. Salvaging and recycling construction and demolition waste reduces demand for virgin resources and the associated environmental impacts. Effective construction-waste management, including appropriate handling of nonrecyclables, can reduce contamination from and extend the life of existing landfills. Whenever feasible, reducing initial waste generation is environmentally preferable to reuse or recycling.

The construction waste-management plan should from the outset recognize project waste as an integral part of overall materials management. The premise that waste management is a part of materials management and the recognition that one project's wastes are materials available for another project facilitate efficient and effective waste management.

Waste-management requirements should be taken into account early in the design process and be a topic of discussion at both preconstruction and ongoing regular job meetings to ensure that contractors and appropriate subcontractors are fully informed of the implications of these requirements on their work prior to and throughout construction.

Waste management should also be coordinated with or be part of the standard quality-assurance program, and waste-management requirements should be addressed regularly at each phase of the project. Any topical application of processed clean wood waste and ground gypsum board as a soil amendment must be done in accordance with local and state regulations. If possible, adherence to the plan is facilitated by tying completion of recycling documentation to one of the payments for each trade contractor.

6.5 Recycled Materials

Recycling is the practice of recovering used materials from the waste stream and then incorporating those same materials into the manufacturing process. Recycled content refers to that portion of material used in a product that has been diverted from the solid-waste stream and is the most widely cited attribute of green building products. The LEED™ intent is to protect virgin resources by increasing demand for building products with recycled content. If the materials are diverted during the manufacturing process, they are referred to as preconsumer recycled content (sometimes also known as postindustrial). If they are diverted after consumer use, they are termed postconsumer. Postconsumer content is generally viewed as offering greater environmental benefit than preconsumer content. While preconsumer waste is much greater, it is also more likely to be diverted from the waste stream. Postconsumer waste is more likely to fill limited space in municipal landfills and is typically mixed, making recovery more difficult.

LEED™ requirements: Use materials with recycled content such that the total material costs of postconsumer recycled content plus one-half of the preconsumer content represents at least 10 to 20 percent (based on cost) of the total value of the materials used in the project. The recycled-content value of the material used is determined by weight. The recycled-content value is arrived at by multiplying the recycled fraction of the assembly by the cost of the assembly. Mechanical, electrical, and plumbing components and specialty items such as elevators are not included in this calculation. Furniture may be included, providing it is included consistently in MR Credits 3 to 7.

For a company to claim that it is using preconsumer recycled content, it must be able to substantiate that the material it is using would have become garbage had it not been purchased from another company's waste stream. However, if a manufacturer routinely collects scraps and feeds them back into its own process, that material does not qualify as recycled.

Numerous federal, state, and local government agencies have "buy recycled" programs aimed at increasing markets for recycled materials. These programs usually have a specific goal of supporting recycling programs to reduce solid-waste disposal, and many communities in the United States now offer curbside collection or dropoff sites for certain recyclable materials. Collecting materials, however, is only the first step toward making the recycling process work. Successful recycling also depends on manufacturers producing products from recovered materials and, in turn, consumers purchasing products made of recycled materials. The LEED™ Rating System applies credits for using recycled materials.

Recyclability as the ability of a product to be recycled can only describe products that can be collected and recycled through an established process. This definition has often been stretched beyond credibility by numerous manufacturers who make the claim based on a laboratory process. Sometimes they attempt to justify this lax approach with new products by noting that they aren't expected to enter the waste stream in large quantities for years. In fact, many national and international companies seek an environmental marketing edge by advertising the recycled content of their products, which is often undocumented or certified.

Such claims come under the jurisdiction of the Federal Trade Commission (FTC), which first published definitions for common environmental terms in its Green Guides in 1992. The LEED™ Rating System offers credit for recycled-content materials, referencing definitions from ISO 14021. These definitions can still leave a lot of gray areas, which many manufacturers tend to interpret in their own favor. Third-party certification of recycled content is useful in maintaining a high standard and offering the ability to verify any claims that are made.

In some industries, recycling pathways have become very reliable. Products such as engineered lumber and fuel pellets have created a reliable demand for waste from the forest-products industry, including sawdust and wood chips. Although these materials may at one time have been part of the waste stream, the active market for them means that they are nearly always diverted. That does not change their status as preconsumer recycled content, however.

Postconsumer recycled content does not always refer to the average individual consumer; the consumer may be a large business or manufacturer. For example, Timbron International, Inc., an interior-molding company, offers interior decorative wood-alternative moldings (crowns, bases, quarters, and rounds) that are 90 percent recycled expanded polystyrene. Timbron's recycled content has been certified by Scientific Certification Systems (SCS) as 75 percent postconsumer and 15 percent preconsumer. Most of that comes from large industrial facilities that receive shipments packed in polystyrene. Since those facilities are the end users of that packaging, it can qualify as postconsumer recycled content.

Waste has a cost, and we all bear it. Moreover, the extraction, manufacture/transport, and disposal of building materials clogs our landfills, pollutes our air and water, depletes resources, and damages natural habitats. The California Integrated Waste Management Board (CIWMB) notes that construction and demolition (C & D) are responsible for about 28 percent of California's solid-waste stream. In excess of 85 percent of that material, from flooring to roofing, is either reusable or recyclable. In addition to C & D waste, the material in our recycling bins, our used bottles, paper, cans, and cardboard, is also considered to be suitable raw materials for recycled-content products.

Keeping a substance out of the landfill is only the first step to putting "waste" back into productive use. The material must be processed into a new, high-quality item, and that product must be sold to a builder or homeowner who recognizes its benefits. By mining our "waste" as the raw material for new products, we increase demand for recycling, and convince manufacturers to use more recycled material, continuously strengthening this cycle.

Benefits of recycled-content materials include reduced solid waste, reduced energy and water use, reduced pollution, reduced greenhouse-gas emissions, and a healthier economy. The following materials are readily recyclable and generally cost less to recycle than to dispose of as garbage:

- Cardboard
- Clean wood (Includes engineered products; nails OK)
- Land-clearing debris
- Metals
- Window glass
- Plastic film (sheeting, shrink wrap, packaging)
- Acoustical ceiling tiles
- Asphalt roofing
- Plastic and wood-plastic composite lumber from plastic and wood chips, ideal for outdoor decking and railings
- Fluorescent lights and ballasts
- Insulation, such as cotton made from denim, newspaper processed into cellulose, or fiberglass with some recycled-glass content
- Carpet and carpet pad made of plastic bottles (or sometimes from used carpet); up to half of all polyester carpet made in the U.S. contains recycled plastic
- Tile containing recycled glass
- Concrete containing ground-up concrete as aggregate, fly ash – a cementious waste product from coal-burning power plants – asphalt, brick, and other cementitious materials
- Countertops made with everything from recycled glass to sunflower-seed shells
- Drywall made with recycled gypsum and Homasote™ wall board made from recycled paper

For maximum benefit, when selecting a recycled-content building material:

- Choose material with the highest recycled content available. For example, some recycled products may only be five percent recycled and 95 percent virgin material.
- Seek high postconsumer recycled content. Some "recycled" content is waste from manufacturing processes. Reducing manufacturing waste is the first step, but recycling postconsumer material is necessary to close the loop.
- Check that the materials are appropriate for the application in hand.
- Salvaging (reusing) whole materials is preferable to recycling. Reuse all but eliminates waste, energy, water use, and pollution.
- Seek out materials that are not only recycled but also recyclable or biodegradable at the end of their useful life. Ideally, a material may be recycled again and again back into the same product.

Reclaimed wood has many applications, including flooring, trim, siding, furniture, and in some cases structural members. Consider reusing wood from an existing building on site, or look to salvage yards and on-site deconstruction sales for a portion of your material needs.

Reclaimed wood flooring is manufactured from timbers salvaged from old buildings, bridges, or other timber structures, or it may be manufactured from logs retrieved from river bottoms or from trees being removed in urban and suburban areas. The character and aesthetic of reclaimed flooring can be particularly attractive. But equally important, salvaging or reusing wood can reduce solid waste, save forest resources, and save money. Moreover, reclaimed wood is often available in dimensions, species, and old-growth quality that is no longer obtainable from virgin forests at any price. However, planning and research are necessary, as available species, dimensions, and lumber quality can vary considerably from location to location.

Examples of reusable (RU), recyclable (RC), and biodegradable (B) building materials include:

- Asphalt (RC)
- Bricks (RU, RC)
- Concrete, ground and used as aggregate (RC)
- Steel, aluminum, iron, copper (RU, RC)
- Gypsum wall board (RU, B)
- Earthen materials (RU, B)
- Wood and dimensional lumber, including beams, studs, plywood, and trusses (RU, RC, B)
- Straw bales (B)
- Wool carpet (B)
- Linoleum flooring (B)
- Doors and windows (RU)
- Plumbing and lighting fixtures (RU)
- Unique and antique products that may no longer be available (RU)

Deconstruction consists of the dismantling of a building to preserve the useful value of its component materials. Deconstruction is preferable to demolishing; the combination of tax breaks, new tools, and increasing local expertise are making it easier to keep materials out of the landfill and money in owner's wallets. Although deconstruction takes longer and may initially cost more than demolition, it is nevertheless likely to reduce the overall project cost.

Best Construction practices include the following:

- Basic but true: Measure twice and cut once. Protect materials from the elements.
- Consider deconstructing and salvaging existing materials.
- Develop a waste-reduction plan that includes waste prevention. This should be followed by assigning responsibility for its implementation to a motivated individual on the construction team. Post the plan and set up on-site locations for recycling, with color coding for separation. Be sure to include time in the schedule for salvage and recycling. Require participation of all team members, including subcontractors.
- Delineate and limit the construction footprint and coordinate construction with a landscape professional to minimize grading and retain native soils and vegetation.

- Drywall clips should be used to fasten drywall. Recycled-content polyethylene clips are available as an alternative to metal.
- Surplus materials can be donated to organizations like Habitat for Humanity.

From an environmental and financial point of view avoiding waste in the first place is a far better practice than recycling. Waste reduction has the benefits of minimizing energy use, conserving resources, and easing pressure on landfill capacity.

6.6 Regional Materials

The main LEED™ intent here is to reduce material transport by increasing demand for building products made in the region where the project is located. This is achieved by increasing demand for building materials and products that are extracted and manufactured within the designated region. It will also support the regional economy and reduce the environmental impacts emanating from transportation.

LEED™ Requirement: Use a minimum of 10 or 20 percent (based on cost) of total building materials and products that are extracted, harvested, recovered, or manufactured regionally within a radius of 500 miles of the site (Figure 6.14). Either the default 45-percent rule or actual materials cost may be used. All mechanical, electrical, plumbing, and specialty items such as elevator equipment should be excluded. If only

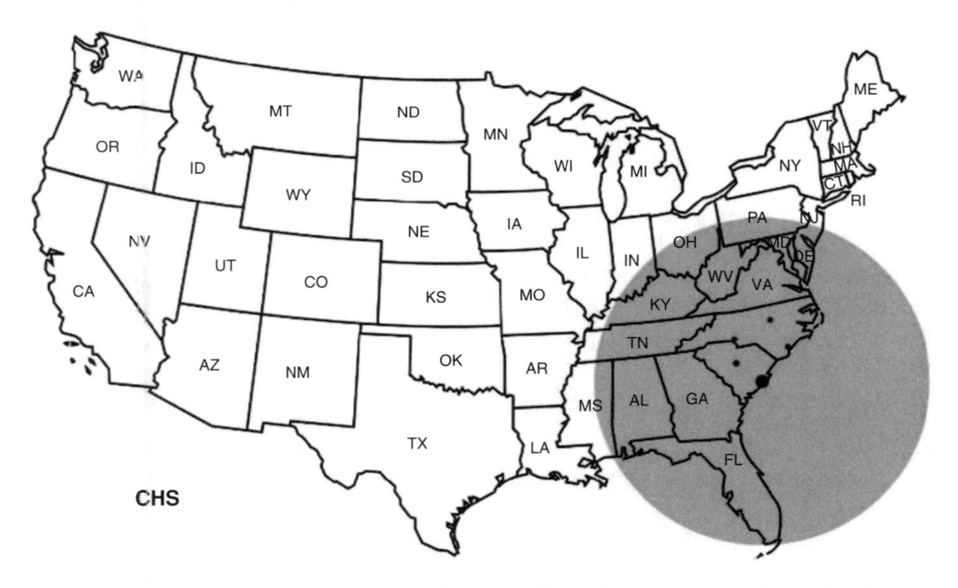

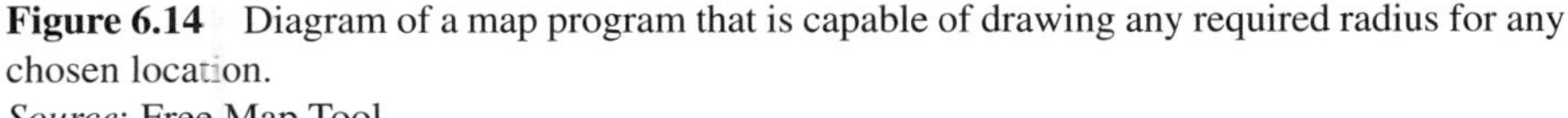
Figure 6.14 Diagram of a map program that is capable of drawing any required radius for any chosen location.
Source: Free Map Tool.

a fraction of the product/material is extracted, harvested, recovered, or manufactured within 500 miles of the site, then only that percentage (based on weight) will contribute to the regional value. Furniture may be included only if it is included throughout MR Credits 3 to 7. (Note that "manufacturing" alludes to the final assembly of components into the building product that is furnished and installed by the contractor. For example, if the hardware comes from Los Angeles, California, the lumber from Vancouver, British Columbia, and the joist is assembled in Fairfax, Virginia, then the location of the final assembly is Fairfax, Virginia.)

The 20-percent-credit threshold can often be attained with no cost impact simply by tracking the materials that are normally produced and supplied within 500 miles of a project site. In some cases, however, the 20-percent threshold can only be achieved by targeting certain materials (e.g., specific types of stone or brick) or limiting the number of manufacturers whose products will be considered in the project bids. In these cases, cost premiums may be incurred.

In order to verify that the extraction/harvest/recovery site is located within a 500-mile radius of the project site, project teams are required to indicate the actual mileage between the project site and the manufacturer and, similarly, the distance between the project site and the extraction site for each raw material in the submittal template. Alternatively, a statement on the manufacturer's letterhead indicating that the point of manufacture is within 500 miles of the LEED™ project site will also be accepted as part of the documentation and credit submittals.

6.7 Rapidly Renewable Materials

LEED™ states that the intent of using rapidly renewable materials is to "reduce the use and depletion of finite raw materials and long-cycle renewable materials by replacing them with rapidly renewable materials."

6.7.1 Bamboo

Bamboo is emerging as an alternative resource to other types of wood commonly used in the United States and abroad. It was previously intuitively used as a basic material for making household objects and small structures. But ongoing research and engineering efforts are enabling bamboo's true value to be realized as a renewable, versatile, and readily available economic resource.

Bamboo is a giant grass; it comes in 1500 varieties that produces hard, strong, dimensionally stable wood. It is commonly found in the tropical regions of Asia, Africa, and South America. It has been used as both a building material and for furniture construction for thousands of years. Bamboo is considered to be the fastest-growing woody plant on earth and is also one of the most versatile and sustainable building materials available. It grows remarkably fast – itcan reach maturity in months in a wide range of climates – and is exceedingly strong for its weight. It can be used both structurally and as a finish material. Bamboo is characterized as a totally renewable resource; you can clear-cut it, and it grows right back.

The swift growth, strength, and durability of bamboo make it an environmentally superior alternative to conventional hardwood flooring. Some hardwood trees may require decades to reach maturity, whereas bamboo is typically harvested on a four- or five-year cycle, and the mature forest will continue to send up new shoots for decades. Pine forests have the most rapid growth among tree species, but bamboo grass species used in flooring can grow more than three feet per day and produce almost twice as much harvestable fiber per year. Though bamboo grows rapidly, it can nonetheless yield a product that is 13 percent harder than rock maple, with durability comparable to red oak. Bamboo canes are beautiful when exposed and may also be used for paneling, furnishings, and cabinetry (Figure 6.15). Vertically laminated flooring and plywood products consist of layers of bamboo compressed with a binder, creating a durable, resilient finish material.

There is a long vernacular tradition for the use of bamboo in structures in many parts of the world, especially in more tropical climates, where it grows into larger-diameter canes. When using bamboo, care should be given to joinery details, since its strength comes from its integral structure, which means that it cannot be joined with many of the traditional methods used with wood. In this respect, the old ways of building with bamboo can be especially informative.

Bamboo is an extremely strong fiber that has twice the compressive strength of concrete and also roughly the same strength-to-weight ratio of steel in tension. The strongest bamboo fibers also have greater shear strength than structural woods, and they take much longer to come to ultimate failure. However, this ability of bamboo to bend without breaking makes it unsuitable for building floor structures due to our low tolerance for deflection and unwillingness to accept a floor that feels "alive." However, bamboo as a $^{3}/_{4}$-inch-thick finish floor is an appropriate substitute for the standard oak because

(a) (b)

Figure 6.15 (a) Kitchen cabinets made of bamboo. (b) Photo showing use of bamboo as exterior siding.
Source: Bamboo Technologies.

it installs the same way, is harder, and expands less. Like most interior-grade hardwood plywood, bamboo flooring is typically made with a urea-formaldehyde binder, which can emit tiny amounts of formaldehyde. Choosing high-quality products, particularly from manufacturers that provide independent air-quality testing data, can help to minimize this source of indoor-air pollution. And when well maintained, bamboo floors can last decades.

6.7.2 Cork

Cork is a natural, sustainable product harvested from the bark of the cork oak, *Quercus suber*, which grows in the sunny Mediterranean. Cork oak can be first harvested when it is 25 years old, when the virgin bark is carefully cut from the tree. From this point on, the tree can be "stripped" of its cork every 9 years for nearly 200 years without harm coming to the tree, which helps to encourage long-term management of this renewable resource. An 80-year-old cork tree can produce up to about 500 pounds of cork.

Cork is becoming increasingly popular due to its extraordinary combination of beauty, durability, insulation, and renewability. Modern cork floors are durable and fire-resistant, provide thermal and acoustic insulation, and are soft on the feet. They are typically covered with an acrylic finish but may be covered with polyurethane for bathroom or kitchen applications. Cork floors can last for decades, and the material is biodegradable at the end of its useful life (Figure 6.16).

By contrast, the extraction, manufacture, and transport, and disposal of synthetic flooring materials pollutes air and water, depletes resources, damages natural habitats, and can have negative health impacts. Hardwood flooring requires logging slow-maturing trees that require decades to centuries to mature.

(a) (b)

Figure 6.16 (a) Residential interior using cork flooring. (b) Cork pattern detail.
Source: Globus Cork.

While cork is mainly considered to be a resilient flooring option, it is also a natural alternative to carpet. While carpet can attract and hold indoor pollutants in its fibers, cork is easier to thoroughly clean, is inherently resistant to mold and mildew, sheds no dust or fibers, and is naturally antistatic. These hypoallergenic properties, combined with thermal and acoustic insulation, allow cork floors to provide the majority of the benefits of carpet without its liabilities. Furthermore, the benefits of cork extend beyond human health; they include reduced landfill waste, low embodied energy, local availability of many products, excellent aesthetics, and reduced impacts from the harvest or mining of virgin materials.

6.7.3 Insulation

Well-insulated building envelopes are primary considerations in comfort and sustainability. Insulation helps to protect a building's occupants from heat, cold, and noise; in addition, it reduces pollution while conserving the energy needed to heat and cool a building. Environmentally preferable insulation options offer additional benefits, such as reduced waste and pollution in manufacture and installation, more efficient resource use, recyclability, enhanced R-value, and reduced or eliminated health risks for installers and occupants.

The comfort and energy efficiency of a home and, to a lesser degree, an office depend on the R-value of the entire wall, roof, or floor (i.e., whole-wall R-value), not just the R-value of the insulation.

Techniques such as advanced framing increase the wall area covered by insulation, thereby increasing the whole wall's effectiveness. Framing conducts far more heat than insulation, in the same manner that most window frames conduct more heat than double-paned glass. An additional layer of rigid insulation between framing and exterior sheathing can help improve the whole-wall R-value by insulating the entire wall, and not just the clear space. In non-"breathable" wall designs, closed-cell rigid foam with taped seams can provide an effective vapor barrier.

For economic reasons, fiberglass is often the insulation of choice. Any fiberglass insulation used should be a formaldehyde-free product with a minimum 50 percent total recycled content (minimum 25 percent postconsumer). Some products are manufactured with heavier, intertwined glass fibers to reduce airborne fibers and mitigate the fraction of fibers that can enter the lungs. Like all glass products, fiberglass insulation is made primarily from silica heated to high temperatures, requiring significant energy and releasing formaldehyde. Short-term effects include irritation to eyes, nose, throat, lungs, and skin during installation or other contact. Longer-term effects are controversial, but OSHA requires fiberglass insulation to carry a cancer warning label. Binders in most fiberglass batts contain toxic formaldehyde that is slowly emitted for months or years after installation, potentially contaminating indoor air.

Some of the environmentally preferable insulation options include the use of recycled cotton, which insulates as well as fiberglass and offers superior noise reduction. Cotton insulation poses no health risk and is not irritating during installation. Likewise, cellulose (recycled newspaper) insulation is an acceptable alternative. It is sprayed in wet or dry, poses no health risk, and offers superior R-value per inch. Both cotton and cellulose are treated with borate, which is not toxic to humans and makes both materials more resistant to fire and insects than fiberglass. Sprayed polyurethane foams expand to fill cracks, providing insulation, vapor barrier, and additional shear strength. Sprayed cementious foams such as Air-Krete have similar properties. Finally, structural insulating systems integrate a building's structure and insulation into a single component. They produce little or no waste during construction and provide excellent thermal performance.

6.7.4 Linoleum

Linoleum is a highly durable, resilient material used mainly for flooring. It is made from natural materials – a mixture of linseed oil, wood flour, powdered cork, and pine resin – which are pressed onto a jute-fiber backing.

Flexible vinyl flooring, which displaced linoleum from the marketplace in the 1960s, is often incorrectly referred to as linoleum. The two materials are quite different. First costs of linoleum are higher, but linoleum offers performance that is in many ways superior to vinyl: linoleum lasts longer, is inherently antistatic, and is antibacterial. All-natural, linoleum requires less energy and creates less waste in its manufacture, and it can be chipped and composted at the end of its useful life. Maintenance of linoleum is also less labor-intensive and less costly because it does not require sealing, waxing, or polishing as often as vinyl.

By contrast, flexible vinyl flooring is a more prolific generator of solid waste because it is manufactured from toxic materials and typically lasts less than 10 years; it is neither biodegradable nor recyclable. Linoleum emits far fewer volatile organic compounds (VOCs) when installed with a low-VOC adhesive than flexible vinyl and does not exude the phthalate plasticizers that are an increasing concern for human health.

The durability of hard vinyl composition tile (VCT) is comparable to linoleum, but recycling it is impractical, which is why VCT tile will ultimately end up in a landfill. Moreover, vinyl products can be harmful because they involve toxic manufacturing chemistry that generates hazardous wastes and air pollution, and vinyl manufacturing also consumes petroleum.

Linoleum attributes include:

- Very durable and can often last for decades; this helps reduce waste associated with the frequent replacement of flexible vinyl flooring.
- It is quiet and comfortable.
- It is made from natural, nontoxic components; it does not contain formaldehyde, asbestos, or plasticizers.
- It is biodegradable at the end of its useful life.
- It is easy to clean and maintain, using minimal water and gentle detergent.
- It is resistant to temporary water exposure, making it an appropriate for use in kitchens. However, its sensitivity to standing water can be a concern in bathrooms.
- Being naturally antistatic, it helps control dust.

- It has very low VOC emissions when installed with appropriate adhesives, although the scent of curing linseed oil may not agree with chemically sensitive persons.
- Square-foot cost can be comparable to high-quality flexible vinyl flooring. However, flexible vinyl is commonly replaced within 10 years and is toxic to manufacture; it is neither biodegradable nor recyclable.
- As linoleum is the same color all the way through, gouges and scratches can be buffed out, reducing long-term costs and waste.

6.7.5 Straw-Bale Construction

Straw-bale construction consists of using compressed blocks (bales) of straw, either as fill for a wall cavity (non-load-bearing) or as a structural component of a wall (load-bearing.). A post-and-beam framework that supports the basic structure of the building, with the bales of straw used as infill, is the most common non-load-bearing approach. This is also the primary method that is permitted in many jurisdictions, although many localities have specific codes for straw-bale construction, and some banks are even willing to lend on this technique. Straw is a renewable resource that acts as excellent insulation and is fairly easy to build with.

The interior and exterior sides of the bale wall are typically covered by stucco, plaster, clay, or other treatment. This type of construction can offer structural properties superior to the sum of its parts. Both load- and non-load-bearing straw-bale design divert agricultural waste from the landfill for use as a building material with many exceptional qualities.

Building with bales of straw has become popular in many parts of the country, and there are now thousands of straw-bale homes in the U.S. However, with load-bearing straw-bale buildings care must be taken to consider the possible settling of the straw bales as the weight of the roof and other elements compresses them. Care should also be taken to assure that the straw is kept dry, or it will eventually rot. For this reason it is generally best to allow a straw-bale wall to remain breathable; any moisture barrier will invite condensation to collect and undermine the structure. Other possible concerns with straw-bale walls are infestation of rodents or insects, so the skin on the straw should be treated to resist them.

Straw bale houses typically save about 15 percent of the wood used in a conventionally framed house. Because of the specialized work that goes into plastering both sides of straw-bale walls, the cost of finishing a straw-bale house sometimes exceeds that of standard construction. Nevertheless, the result is usually worth it because of the superior insulation and wall depth that is achieved. In many cases, straw bales provide an excellent “alternative” building material that helps to reduce or eliminate many environmental problems (Figure 6.17).

Issues to consider when using straw-bale construction include:

- Straw-bale walls are thick and may constitute a high percentage of the floor area on a small site.
- Straw bales may be plastered on both sides to provide thermal mass, and, similar to standard construction, the walls must be protected from moisture.

Figure 6.17 An interior of a residence built of straw bale.
Source: StrawBale Innovations, LLC.

- When labor is executed primarily by building professionals, the square-foot cost of straw-bale construction may be equivalent to conventional building methods.
- Straw bale can be more resistant to termites and vermin than stick construction, but, as is the case with other types of construction, elimination of cracks and holes is critical.
- Straw bales do not hold nails as well as wood, and thus nailing surfaces need to be provided.

Benefits of using straw-bale construction methods include:

- Very good thermal and acoustic insulation, thereby enhancing occupant comfort throughout the year. The interior plaster of straw-bale houses increase the "thermal mass" of the home, which helps to stabilize interior temperature fluctuations.
- A traditional "stick frame" home of 2-x-6 construction usually has an insulating value of R-14, whereas with a properly insulated roof straw bale enables a warm winter home with an R factor of R-35 to R-50.
- Straw bale does not require toxic treatment, which helps chemically sensitive individuals.
- It can be economical. Straw bales are inexpensive (or free), and owners, builders, and volunteers can contribute significantly to labor.
- It reduces construction waste: the main building material is a waste product, and excess straw can be used on-site in compost or as soil-protecting ground cover.
- Material is biodegradable or reusable at the end of its useful life.
- It has the potential for major reductions in wood and cement use, particularly in load-bearing straw-bale designs.
- Conventional foundations and roofs can be used with straw bale buildings.

- It is highly resistant to vermin (including termites).
- It offers aesthetic flexibility from conventional linearity to organic undulation.
- It can be used in new construction, additions, and remodels.

6.7.6 Wheatboard

Environmentally friendly furniture comes from a variety of resources. One of the more popular renewable materials is wheatboard, which is a byproduct of wheat straw. This material has no formaldehyde and can be used to create, among other things, quality furniture and cabinets. Produced in sheets, this durable substance can be filled, sealed, painted, stained, or varnished. It can also be shaped in a wide variety of designs.

Wheatboard has traditionally been burned or added to landfills. A number of manufacturers produce wheatboard both in the U.S. and Canada. It is a viable substitute for wood and benefits the environment by reducing deforestation and also lessening both air pollution and landfill use. Along with its versatility, using wheatboard can help a building project earn crucial LEED™ credits for rapidly renewable materials.

6.8 Use and Selection of Green Office Equipment

Wherever you turn, people are searching for ways to reduce their environmental footprint and act greener, both at home and at the office. With buildings contributing nearly 40 percent of U.S. CO_2 emissions, even small changes at the office can potentially have a dramatic impact. In fact the impact of energy costs directly affects the bottom line of both building owners and tenants alike. Energy represents 30 percent of operating expenses in a typical office building, which makes it the single largest and most manageable operating expense in the provision of office space. The cost of energy to power an appliance over time is typically many times the original price of the equipment. By choosing the most energy-efficient models available, a positive impact on the environment is achieved while at the same time saving money.

The energy and water that appliances and office equipment consume directly translates into more fuel being burned at power plants, which in turn contributes to air pollution, global climate change, and waste of our limited natural resources. The good news is that efficient new appliances and office equipment can use only one-half to one-third as much energy as those purchased a decade ago. The Lawrence Berkeley Laboratory reports that replacing older refrigerators, clothes washers, dishwashers, thermostats, heating equipment, and incandescent lighting with ENERGY STAR® equipment can save enough energy and water to provide an average aftertax return on investment over 16 percent, which is substantially better than the stock market. When appliances are well maintained and in good condition, one should weigh the waste, energy, and pollution required to make a new piece of equipment against gains in efficiency. There are no tangible savings or environmental benefits if older appliances remain in use, as is common with refrigerators.

Retailers typically carry efficient, durable appliances and office equipment. Appliances and equipment with the ENERGY STAR® label are preferred, although they are not necessarily the most efficient of all available models. Nevertheless, ENERGY STAR® products do usually perform significantly better than federal minimum efficiency standards. According to a 2002 EPA report, ENERGY STAR® labeled office buildings generate utility bills 40 percent less than the average office building. There are often rebates and/or incentives for the purchase of energy- and water-saving appliances.

Factors to consider when selecting equipment:

- Appliances and equipment that use the least energy and/or water
- Appliances and equipment that have the ENERGY STAR® label
- Durable appliances and equipment that meet long-term needs
- Natural-gas appliances for space and water heating; gas is often more cost-effective and can reduce overall energy ubut, like other fossil fuels, it is not a renewable resource
- Sealed combustion and direct-vent furnaces and water heaters to increase indoor-air quality
- Occupancy sensors in offices to minimize unnecessary lighting, as well as "smart" power strips that combine an occupancy sensor with a surge protector; smart power strips will shut down devices (such as monitors, task lights, space heaters, and printers) that can be safely turned off when space is unoccupied

With relation to residential construction ENERGY STAR® offers homebuyers the features they want in a new home, plus energy-efficient improvements that deliver better performance, greater comfort, and lower utility bills. To earn the ENERGY STAR® label, a home must meet strict guidelines for energy efficiency set by the U.S. Environmental Protection Agency. Such homes are typically 20 to 30 percent more efficient than standard homes.

There is also the potential to achieve LEED™ credit points for using green office equipment. For example, for existing buildings the requirements for electronics can be found in the Materials and Resources (MR 2.1) section under "Sustainable Purchasing: Durable Goods, Electric," which states:

- One point is awarded to projects that achieve sustainable purchases of at least 40 percent of total purchases of electric-powered equipment (by cost) over the performance period. Examples of electric-powered equipment include, but are not limited to, office equipment (computers, monitors, copiers, printers, scanners, fax machines), appliances (refrigerators, dishwashers, water coolers), external power adapters, and televisions and other audiovisual equipment.

Sustainable purchases are those that meet one of the following criteria:

- The equipment is ENERGY STAR® labeled (for product categories with developed specifications). This will reduce the carbon footprint of the equipment as well as the office's operating costs.
- The equipment (either battery or corded) replaces conventional gas-powered equipment.

When applying for project certification, you should constantly consult the USGBC website (http://www.usgbc.org).

6.9 Certified Wood

Forest certification is a means for independent organizations to develop standards of good forest management, and independent auditors issue certificates to forest operations that comply with those standards. This certification verifies that forests are well-managed, as defined by a particular standard, and ensures that certain wood and paper products come from responsibly managed forests. The Intent of the LEED™ Certified Wood credit is to encourage such environmentally responsible forest-management programs.

After more than two years of studying the issue, the U.S. Green Building Council (USGBC) has recently proposed a major change for certified wood in its LEED™ Rating System. Previously, LEED™ awarded credit to projects that used wood certified to the standards of the Forest Stewardship Council (FSC) for at least half of their wood-based materials. The USGBC has now broadened the credit to recognize any forest-certification program that meets its criteria. The change is partly in response to criticism that LEED™ favors one forest-certification program, FSC, over others – particularly the Sustainable Forestry Initiative (SFI), a rival to the FSC that is portrayed by some environmentalists as less rigorous. Nevertheless, the revision brings the credit into line with a trend in LEED™ toward using transparent criteria to decide which third-party certification programs to reference.

Under the newly proposed credit language, wood certification systems would be evaluated for eligibility to earn points towards LEED™ certification against a measurable benchmark that includes:

- Governance
- Technical/standards substance
- Accreditation and auditing
- Chain of custody and labeling

LEED™ says that it will recognize wood certification programs that are, after thorough objective analysis, deemed compliant with the benchmark. On the other hand, wood certification programs that are not found to be in alignment with the benchmark would have a clear and transparent understanding of what modifications are necessary to receive recognition under LEED™. The FSC appears likely to meet the criteria, but a number of other programs, including SFI and the Canadian Standards Association (CSA), may face some difficulty with parts of the benchmark system, such as the ban on genetically modified organisms, the emphasis on integrated pest management and bans on certain chemicals. All of the benchmark criteria must be met for recognition in LEED™.

The implementation requirements to achieve the LEED™ credit include:

1. Use of a minimum of 50 percent of wood materials and products certified in accordance with an approved certifier (e.g., Forest Stewardship Council) meeting these principles and criteria for wood building components:
   - Framing
   - Flooring and subflooring

- Wood doors
- Other

2. Based on cost of certified wood products compared to total wood material cost
3. Exclude MEP and elevator equipment
4. Furniture may be included in this calculation
5. Contractor doesn't need the certification number but supplier does

The concept of forest certification was launched over a decade ago to help protect forests from destructive logging practices. Forest certification was intended as a seal of approval and a means of notifying consumers that a wood or paper product comes from forests managed in accordance with strict environmental and social standards. For example, a person shopping for wall paneling or furniture would seek a certified forest product to be sure that the wood was harvested in a sustainable manner from a healthy forest and not procured from a tropical rainforest or the ancestral homelands of forest-dependent indigenous people.

Forest certification involves the green labeling of companies and wood products that meet standards of "sustainable" or "responsible" forestry. The primary purpose of forest certification is to provide market recognition for forest producers who meet a set of agreed-upon environmental and social standards.

The SCS (Scientific Certification Systems) first developed its Forest Conservation Program in 1991 and has since emerged as a global leader in certifying forest-management operations and wood-product manufacturers. The Forest Stewardship Council (FSC) in 1996 accredited SCS as a certification body, enabling it to evaluate forests according to the FSC Principles and Criteria for Forest Stewardship. Through a well-developed network of regional representatives and contractors, SCS offers timely and cost-effective certification services around the world.

6.10 Life-Cycle Assessment of Building Materials and Products

Life-cycle cost analysis (LCCA) is discussed in greater detail in Chapter 10.

6.10.1 Life-Cycle Cost Analysis

The National Institute of Standards and Technology (NIST) defines Life-cycle cost (LCC) as "the total discounted dollar cost of owning, operating, maintaining, and disposing of a building or a building system" over a period of time. Life-cycle costing (LCC) is therefore an economic evaluation technique that determines the total cost of owning and operating a facility over a period of time and takes into consideration relevant costs of alternative building designs, systems, components, materials, or practices in addition to the multiple impacts on the environment (both positive and negative) that building materials and certain products have.

This method takes into account the impacts of every stage in the life of a material, from first costs, including the cost of planning, design, extraction/harvest, and installation, as well as future costs, including costs of fuel, operation, maintenance, repair, replacement, recycling, or ultimate disposal. It also takes into account any resale or salvage value recovered during or at the end of the time period examined. This type of analysis allows for comprehensive and multidimensional product comparisons. Potential health and safety issues that emerge during the construction, occupation, maintenance, alteration, and disposal of the facility should be taken into consideration. Initial failure to address the ease with which the built environment can be safely maintained can lead to unnecessary costs and risks to health and safety at a later date.

Life-cycle cost analysis (LCCA) can be employed on both large and small buildings as well as on isolated building systems. LCCA is especially useful when applied to project alternatives that fulfill the same performance requirements but differ with respect to initial costs and operating costs and have to be compared in order to select the one that maximizes net savings. For example, LCCA will help determine whether the incorporation of a high-performance HVAC or glazing system, which may increase initial cost but result in dramatically reduced operating and maintenance costs, is cost-effective or not. LCCA is not useful for budget allocation.

Many building owners apply the principles of life-cycle cost analysis in decisions they make regarding construction or improvements to a facility. From the building owner who opts for aluminum windows in lieu of wood windows to the strip-mall developer who chooses paving blocks over asphalt, owners are taking into consideration the future maintenance and replacement costs in their selections. While initial cost has always been a factor in the decision-making process, it is not the only factor.

Related to LCA, life-cycle costing (LCC) is the systematic evaluation of financial ramifications of a material, a design decision, or a whole building. LCC tools can help calculate payback, cash flow, present value, internal rate of return, and other financial measures. Such criteria can go a long way toward comprehending how a modest up-front cost for environmentally preferable materials or design features can provide a very sound investment over the life of a building.

Lowest life-cycle cost (LLCC) is considered the most straightforward and easy-to-interpret system of economic evaluation. There are other commonly used measures available, such as net savings (or net benefits), savings-to-investment ratio (or savings benefit-to-cost ratio), internal rate of return, and payback period. All of these systems are consistent with the lowest LCC measure of evaluation if they use the same parameters and length of study period. Building economists, architects, cost engineers, quantity surveyors, and others might use any or several of these techniques to evaluate a project. The approach to making cost-effective choices for building-related projects can be quite similar whether it is called cost estimating, value engineering, or economic analysis. Open-book accounting, when shared across the whole project team, helps everyone to see the actual costs of the project.

It is worth reiterating that the purpose of an LCCA is to estimate the overall costs of project alternatives and to select the design that ensures that the facility will provide the lowest overall cost of ownership consistent with its quality and function. The LCCA should be performed early in the design process while there remains an opportunity to refine the design in a manner that results in a reduction in life-cycle costs (LCC).

The first and most challenging task of an LCCA, or any economic evaluation method for that matter, is to determine the economic effects of alternative designs of buildings and building systems and to quantify these effects and express them in dollar amounts. When viewed over a 30-year period, initial building costs have been shown to generally account for approximately just 2 percent of the total, while operations and maintenance costs equal 6 percent and personnel costs reflect the lion's share of 92 percent (*Source*: Sustainable Building Technical Manual).

There are various costs associated with acquiring, operating, maintaining, and disposing of a building or building system. Building-related costs usually fall into one of the following categories (see Chapter 10):

- First costs: purchase, acquisition, construction costs
- Operation, maintenance, and repair costs
- Fuel costs
- Replacement costs
- Residual values: resale or salvage values or disposal costs
- Finance charges: loan interest payments
- Nonmonetary benefits or costs

Only costs within each category that are relevant to the decision and significant in amount are needed to make a valid investment decision. To be relevant, costs must differ from one alternative to another; they achieve significance when they are large enough to make a credible difference in the LCC of a project alternative. All costs are entered as base-year amounts in today's dollars; the LCCA method accelerates all amounts to their future year of occurrence and discounts them back to the base date to convert them to present values.

Detailed estimates of construction costs are not required for preliminary economic analyses of alternative building designs or systems. Such estimates are usually not available until the design is relatively advanced and the opportunities for cost-reducing modifications have been missed. LCCA can be repeated throughout the design process whenever more detailed cost information becomes available. Initially, construction costs are estimated by reference to historical data from similar facilities. They can also be determined from government or private-sector cost-estimating guides and databases.

Detailed cost estimates are prepared at the submittal stages of design (typically at 30, 60, and 90 percent) based on quantity takeoff calculations. These estimates rely mainly on cost databases such as the *R. S. Means Building Construction Cost Database*. Testing organizations such as ASTM International and trade organizations also have reference data for materials and products they test or represent.

The most astute and logical way for the owner/developer to avoid cost overruns is to have:

- Objectives that are reasonable and not subject to modification during the course of the project
- A complete design that meets planning and statutory requirements and will not require later modification
- Clear leadership and appropriate management controls in place
- Project estimates that are realistic and not unduly optimistic

- A project brief that is comprehensive, clear, and consistent
- A coordinated design that takes into account maintenance, health and safety, and sustainability
- An appropriate risk allocation and sufficient contingency that is unambiguous and clear
- A payment mechanism that is simple and that incentives the parties to achieve a common and agreed goal

6.10.2 Third Party Certification

The LEED™ third-party certification program is recognized as an internationally accepted benchmark for the design, construction and operation of high-performance green buildings. According to Alice Soulek, VP of LEED™ development, "Third-party certification is the hallmark of the LEED™ program," and "Moving the administration of LEED™ certification under GBCI will continue to support market transformation by delivering auditable third-party certification. Importantly, it also allows UGSBC to stick to the knitting of advancing the technical and scientific basis of LEED™."

Moving administration of the LEED™ certification process to the Green Building Certification Institute (GBCI), a nonprofit organization established in 2007 with the support of USGBC, will likely have far-reaching ramifications for the U.S. Green Building Council (USGBC) and its influential LEED™ rating systems. Working together with the selected certification bodies, GBCI will be in a position to deliver a substantially improved, ISO-compliant certification process that will be able to grow with the green-building movement.

USGBC has essentially outsourced LEED™ certification to independent, accredited certifiers overseen by GBCI. In that respect, the U.S. Green Building Council's (USGBC) Leadership in Energy and Environmental Design (LEED™) v3 has identified the certification bodies for the updated LEED™ Green Building Rating System. The companies are well-known and respected for their role in certifying organizations, processes, and products to ISO and other standards. These members include:

- ABS Quality Evaluations, Inc., Houston, TX
- BSI Management Systems America, Inc., Reston, VA
- Bureau Veritas North America, Inc., San Diego, CA
- DNV Certification, Houston, TX
- Intertek, Houston, TX
- KEMA-Registered Quality, Inc., Chalfont, PA
- Lloyd's Register Quality Assurance Inc., Houston, TX
- NSF-International Strategic Registrations, Ann Arbor, MI
- SRI Quality System Registrar, Inc., Pittsburgh, PA
- Underwriters Laboratories-DQS Inc., Northbrook, IL

In the spirit of the many successful LEED™ projects, this development in the certification process has been undertaken as an integrated part of a major update to the technical rating system that was put in place as LEED™ 2009. The update will also include a comprehensive technology upgrade to LEED™ Online aimed at improving the user experience and expanding its portfolio-management capabilities.

Third-party testing and certification are required for LEED™ to provide an independent analysis of manufacturers' environmental performance claims, based upon established standards. They provide building owners and operators with the tools necessary to have an immediate and measurable impact on their buildings' performance. Sustainable building strategies should be considered early in the development cycle. An integrated project team will include the major stakeholders of the project, such as the developer/owner, architect, engineer, landscape architect, contractor, and asset and property-management staff. Making choices based on third-party analysis is simpler than life-cycle analysis (LCA), but care should be taken to evaluate the independence, credibility, and testing protocols of the third-party certifiers.

Michelle Moore, senior vice president of policy and public affairs at USGBC, says, "We believe in third-party certification," and "the USGBC provides independent third-party verification to ensure that a building meets these high performance standards. As part of this process, USGBC requires technically rigorous documentation that includes information such as project drawings and renderings, product manufacturer specifications, energy calculations, and actual utility bills. This process is facilitated through a comprehensive online system that guides project teams through the certification process. All certification submittals are audited by third-party reviewers." Moore also believes that separating LEED™ from the certification process will bring LEED™ into alignment with norms established by the International Organization for Standardization (ISO) for certification programs.

To be healthy, green-building materials and techniques should typically have zero or low emissions of toxic or irritating chemicals and be moisture- and mold-resistant. They are typically manufactured with a low-pollution process from nontoxic components, have low maintenance requirements, and do not require the use of toxic cleansers. Most green materials do not emit volatile organic compounds (VOCs), particularly indoors, and are free of toxic materials such as chlorine, lead, mercury, and arsenic. It should be noted that individual products do not carry LEED™ points; instead they can contribute to LEED™ points. Green-building techniques include the monitoring of indoor pollutants and poor ventilation with the use of radon and carbon-monoxide detectors. The use of ozone-depleting gases is avoided (e.g., free of HCFCs and halons).

7 Indoor Environmental Quality

7.1 General Overview

During the past few decades the general public has become increasingly aware of the health hazards related to contaminated air. A variety of factors have been found to contribute to poor indoor-air quality in buildings, the primary one being indoor pollution sources that release gases or particles into the air. Other major sources include outdoor pollutants near the building; pollution carried by faulty or inadequate ventilation systems; and a variety of combustion sources such as tobacco smoke, gas, oil, kerosene, coal, wood, and emissions from building materials, furnishings, and various types of equipment. The relative importance of a particular source depends mostly on the amount of a given pollutant it emits and how hazardous those emissions are. The people who are affected most by poor indoor-air quality are those who are exposed to it for the longest periods of time. These groups typically consist of the young, the elderly, and the chronically ill, especially those suffering from respiratory or cardiovascular disease.

Poor indoor environmental quality has become a major concern in homes, schools, and workplaces; it can lead to poor health, learning difficulties, and productivity problems. Since the majority of us spend up to 90 percent of our time indoors (especially in the United States), it is not surprising that we should expect our indoor environment to be healthy and free from the plethora of hazardous pollutants. Yet studies by the American College of Allergies show that roughly 50 percent of all illness is aggravated or caused by polluted indoor air. Moreover, cases of building-related illness (BRI) and sick-building syndrome (SBS) continue to rise. The main reason is that the indoor environment we live in is often contaminated by various toxic or hazardous substances as well as pollutants of biological origin (Figure 7.1). In fact, recent studies point to the presence of more than 900 possible contaminants, from thousands of different sources, in a given indoor environment.

Indoor-air pollution is now generally recognized as having a greater potential impact on public health than most types of outdoor-air pollution, causing numerous health problems from respiratory distress to cancer. Furthermore, a building interior's air quality is one of the most pivotal factors in maintaining building occupants' safety, productivity, and wellbeing.

This heightened public awareness has led to a sudden surge of demands from building occupants for compensation for their illnesses. Tenants are not only suing building owners but also architects, engineers, and others involved in the building's construction. To shift the blame, building owners have made claims against the consultant, the contractor, and others involved in the construction of the facility.

DOI: 10.1016/B978-1-85617-691-0.00007-2

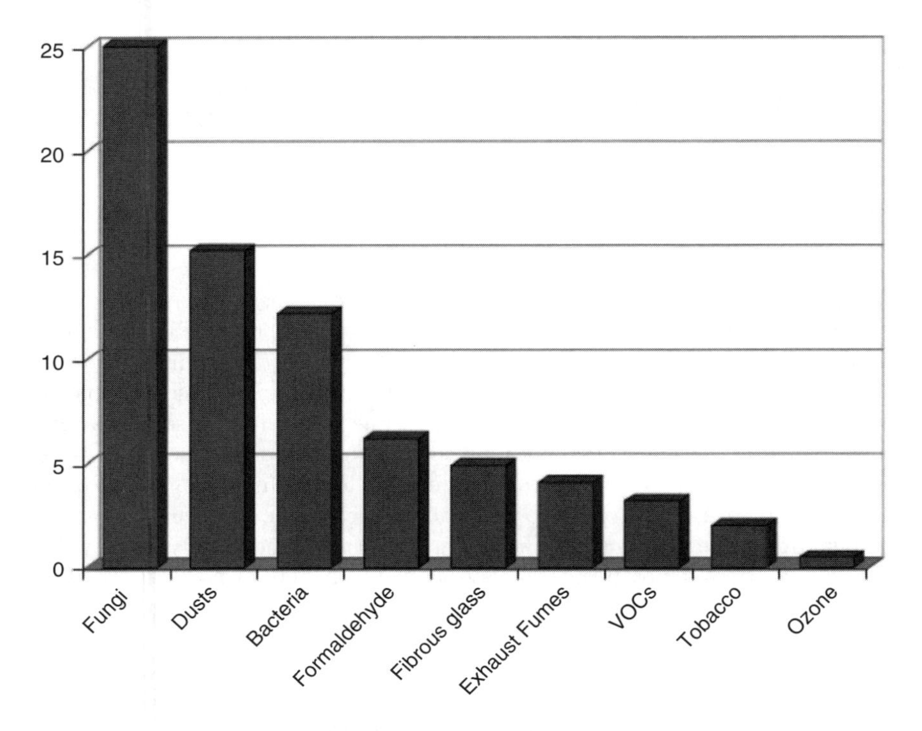

Figure 7.1 Percentage of buildings with inappropriate concentrations of contaminants. *Source*: HBI database of 2,500 buildings.

While architects and engineers to date have not been a major target of publicity or litigation arising out of IAQ issues, nevertheless the potential scope and cost of some of the incidents have resulted in everyone associated with a project being blamed when the inside air of a building appears to be the cause of occupants becoming sick. This is causing great concern among design professionals, because it can result in a loss of reputation as well as time and money.

Due to the intense competition to maintain high occupancy rates, forward-thinking owners and managers of offices and public buildings find themselves under increasing pressure to meet or exceed the demands of the marketplace in attracting and retaining tenants. Moreover, in today's increasingly litigious society, they add another factor to be addressed – protecting their investment from liability due to air-quality issues.

Technological breakthroughs are bringing down the cost of facility monitoring systems and making them more affordable for a wider range of building types. By reducing the cost of facility monitoring many financial and maintenance obstacles are removed, making permanent monitoring systems an appropriate consideration for a broader range of facility managers. Schools, healthcare facilities, and general office buildings can benefit from measuring many of the environmental conditions and using that information to respond to occupant complaints, optimize facility performance, and keep energy costs in check. In addition, feedback from the indoor environment can be used to establish baselines for building performance and document improvements to indoor-air quality.

Facility monitoring systems can be valuable instruments for improving indoor-air quality, identifying energy-savings opportunities, and validating facility performance. Automating the process of recording and analyzing relevant data and providing facility managers adequate access to this information can improve their ability to meet the challenge of maintaining healthy, productive environments.

Buildings are dynamic environments – but sometimes the original design calculations, which may have been sound, need modifying after they are operational. Experience tells us that a building's functions and occupancy rates may change. These changes can have a significant impact on the HVAC system's ability (as originally designed) to maintain a balance between occupant comfort, health, productivity, and operating costs.

Indoor pollution is found to exist under many diverse conditions from dust and bacterial buildup in ductwork to secondhand smoke and the offgassing of paint solvents, all of which are potential health hazards.

The primary causes of poor indoor-air quality are sources that release gases or particles into the air. Inadequate ventilation is considered the single most common cause of pollutant buildup (Figure 7.2a) because it can increase indoor pollution levels by not bringing enough outdoor air in to dilute emissions from indoor sources and by not removing indoor-air pollutants to the outside. High temperature and humidity levels can also increase concentrations of some pollutants. The second most common cause of pollutant buildup is inefficient filtration (Figure 7.2b). But despite fundamental improvements in air-filter technology, far too many buildings persist in relying on inefficient filters or continue to be negligent in the maintenance of filters.

An investigation into indoor-air contamination can be triggered by several factors including the presence of biological growth (mold), unusual odors, and adverse health concerns of occupants, including respiratory problems, headaches, nausea, irritation of eyes, nose, or throat, and fatigue.

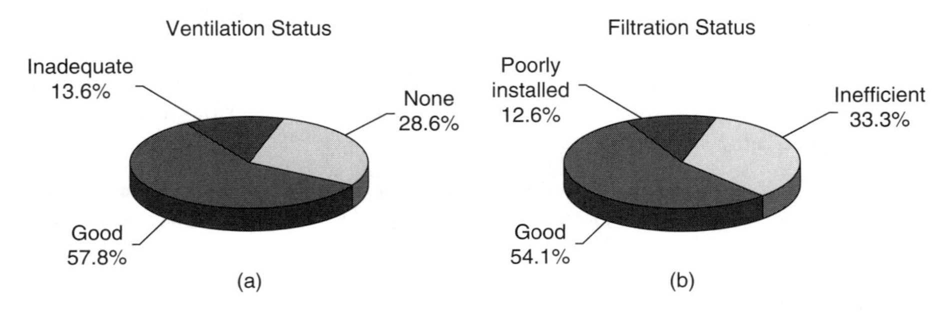

Figure 7.2 (a) Inadequate ventilation is the single most common cause of pollutant buildup. (b) Inefficient filtration is the second most important factor in indoor pollution.
Source: HBI Database.

Indoor environmental quality (IEQ) and energy efficiency may be classified into three basic categories: comfort and ventilation, air cleanliness, and building pollutants. Within these basic categories, facility-wide monitoring systems are available that can provide independent measurement of a range of parameters, such as temperature, humidity, total volatile organic compounds (TVOCs), carbon dioxide (CO_2), carbon monoxide (CO), and airborne particulates. Any information that is extracted from continuous monitoring can help minimize the total investigative time and expense needed to respond to occupant complaints; the information can also be used proactively in the optimization of building performance.

Unfortunately, to this day there has been insufficient federal legislation governing indoor-air quality. This has led several engineering societies such as the American Society of Heating, Refrigerating, and Air-Conditioning Engineers (ASHRAE) to establish guidelines, which are now generally accepted by designers as minimum design requirements for commercial buildings. ASHRAE has established two procedures for determining minimum acceptable ventilation rates: the ventilation-rate procedure, which stipulates a minimum ventilation rate based on space functions within a specified building type and respiration rates resulting from occupants' activities; and the indoor-air quality procedure, which requires the monitoring of certain indoor-air contaminants below specified values.

Air sampling techniques require the use of a device to impinge organisms from a specific volume of air and place it onto a sterile agar growth medium. The sample is then incubated for a specified period of time (say, seven days). The colonies are then counted and the results recorded. When testing the air of a potentially contaminated area, it is best practice to have comparative samples of air from both the contaminated area and the air outside the potentially contaminated building.

Sometimes building occupants complain of symptoms that do not appear to fit the pattern of any specific illness and are difficult to trace to any specific source. This has been labeled sick building syndrome (SBS) and is a fairly recent phenomenon. It is a term used to describe situations in which building occupants experience acute health and discomfort effects that appear to be attributed to time spent in a building, but often no specific illness or cause can be identified. Complaints may be localized in a particular room or zone or may be widespread throughout the building. Factors that may impact SBS include, noise, poor lighting, thermal discomfort, and psychological stress. The EPA's indoor-air-quality website contains pertinent information regarding strategies for identifying the causes of SBS as well as finding possible solutions to the problem. SBS indicators include:

- Building occupants complaining of symptoms associated with acute discomfort, such as eye, nose, or throat irritation; headaches; respiratory infections and dry cough; dry or itchy skin; erythema (redness of the skin, inflammation); dizziness and nausea; irritability and forgetfulness; difficulty in concentrating; fatigue or lethargy; and sensitivity to odors.
- Cause of symptoms not identified.
- Most complainants find relief shortly after leaving building.

In contrast, building-related illness (BRI) is the general term used to describe symptoms of a medically diagnosable illness that is caused by or related to occupancy of a building and that can be attributed directly to airborne building contaminants. The causes of BRIs can be determined and are typically related to allergic reactions and infections. BRI indicators of include:

- Building occupants complain of symptoms such as cough, fever, chills, chest tightness, and muscle aches.
- Symptoms can be clinically defined and have clearly identifiable causes.
- Complainants may require extended recovery times after leaving the building.

It should be noted that there is no particular manner in which these health problems appear. Sometimes, they begin to appear as workers enter their offices and diminish or disappear as workers leave; at other times symptoms persist until the illness is treated. On occasion there are outbreaks of illness among many workers in a single building; in other cases, they show up only in individual workers.

Complaints may also result from other causes. Since SBS symptoms are quite diverse and nonspecific, it can seem that they could arise from numerous other ailments completely unrelated to indoor air quality. In light of these types of complaints, technical and medical scholars have searched for a satisfactory definition of SBS, especially since it is the most common indoor-air-quality problem in commercial buildings today. To this end, some World Health Organization (WHO) experts estimated that 30 percent of all new and renovated commercial buildings in the U.S. suffer from SBS, and furthermore the expenses in terms of medical costs and lost production may be in the tens of billions of dollars per year in the U.S. alone.

According to industry IAQ standards, SBS is diagnosed if significantly more than 20 percent of a building's occupants complain of adverse health effects such as headaches, eye irritation, fatigue, and dizziness over a period of two weeks or more, but no clinically diagnosable disease is identified, and the SBS symptoms disappear or are diminished when the complainants leave the building.

We are surrounded by microorganisms both indoors and outdoors. Some microorganisms are good for our health, while others can harm us; most microorganisms are benign. It is important to prevent buildings from becoming breeding grounds for harmful microbes, and to do that one needs to deprive them of the essentials they need to live. Water is one of microbes' main survival ingredients. It can come from leaky roofs, burst water pipes, or warm, moist air, as when water is absorbed by ceiling tiles, drywall, and carpet padding.

The greatest amount of our time is spent indoors, and more than half of today's work force spends much of it sitting in front of a computer screen. It is not surprising that we should witness a steady increase in chronic work-related illness, including repetitive stress injuries, asthma, and cardiovascular disease – all suggesting that much of our artificial environment is hazardous to workplace productivity.

It has been known for some time that indoor environments strongly affect human health. The EPA, for example, has estimated that pollutant concentration levels (such as volatile organic compounds) inside a building may be two to five times higher than outside levels. A 1997 study by W.J. Fisk and A.H. Rosenfeld ("Estimates of Improved Productivity and Health from Better Indoor Environments," *Indoor Air,* vol. 7, pp. 158–172) estimates that the cost to the nation's workforce of upper-respiratory diseases in

1995 reached $35 billion in lost work, which doesn't include an estimated additional $29 billion in healthcare costs. The report suggests that these costs could be reduced by 10 to 30 percent by healthier and more efficient indoor environments.

Certain attributes of indoor environments can also have a strong impact on occupant wellbeing and functioning; these include personal control over environmental conditions, the amount and quality of light and color, the sense of privacy, access to window views, connection to nature, and sensory variety. Proper design that takes into consideration occupants' psychological well-being will have a positive impact on worker productivity and effectiveness as well as on other high-value issues, such as stress reduction, job satisfaction, and organizational loyalty.

In order to fully profit from the fiscal, physical, and psychological benefits of healthy buildings, projects need to incorporate a comprehensive, integrated design and development process that seeks to:

- Ensure adequate ventilation
- Provide maximum access to natural daylight and views to the outdoors
- Eliminate or control sources of indoor-air contamination
- Prevent water leaks and unwanted moisture accumulation
- Improve the psychological and social aspects of space

There are currently several environmental rating methods for buildings, but it is not always clear whether these methods assess the most relevant environmental aspects or whether other considerations lie behind the specific methods chosen. The Swedish rating method bases the selection on a number of factors such as the severity and extent of problems, mandatory rules, official objectives, and current practice. Upon identifying the presence of building-related health problems, possible indicators for monitoring these problems can be tested with respect to the theoretical and practical criteria applied in order to better understand the strengths and limitations of different indicators.

Concern for occupant health continues to increase with increased awareness, and this has translated into public demand for more exacting performance requirements for materials selection and installation, improved ventilation practices, and better commissioning and monitoring protocols.

Many of the building products used today contain chemicals that evaporate or off-gas for significant periods of time after installation. When substantial quantities of these products are utilized inside a building or products are used that have particularly strong emissions, they pollute the indoor air and can be hazardous. Some products readily trap dust and odors and release them over time. Building materials, particularly when damp, can also support growth of mold and bacteria, which can cause allergic reactions, respiratory problems, and persistent odors – i.e., SBS symptoms.

The Insurance Information Institute (III) reports a dramatic increase in IAQ-related lawsuits within the United States; there are currently more than 10,000 IAQ-related cases pending. Several lawsuits have resulted in large damage awards to building occupants suffering from health problems linked to chemicals off-gassed from building materials, setting legal precedents across the country.

The flood of IAQ-related lawsuits has prompted insurance companies to reexamine their policies and their clients' design and building methods. An effective way to reduce health risks and thus minimize potential liability is to follow a rigorous selection procedure for construction materials aimed at minimizing harmful effects to occupants.

7.2 Indoor Environmental Quality and Factors Affecting the Indoor Environment

According to a report on indoor environmental quality released by the United States Access Board in July 2005, there is a growing number of people suffering a range of debilitating physical reactions from exposures to everyday materials and chemicals found in building products, floor coverings, cleaning products, and fragrances, among others. In addition, there are those who have developed an acute sensitivity to various types of chemicals, a condition known as multiple chemical sensitivity (MCS). The range and severity of these reactions vary as do the potential triggering agents.

7.2.1 Indoor-Air Quality

The health and productivity of employees and tenants are greatly influenced by the quality of the indoor environment, and studies consistently reinforce the correlation between improved indoor environmental quality and occupants' health and wellbeing. The adverse effects to building occupants caused by poor air quality and lighting levels, the growth of molds and bacteria, and off-gassing of chemicals from building materials can be significant. One of the chief characteristics of sustainable design is to support the wellbeing of building occupants by reducing indoor-air pollution. This can best be achieved through the selection of materials with low off-gassing potential, appropriate ventilation strategies, adequate access to daylight and views, and optimum comfort through control of lighting, humidity, and temperature levels.

7.2.1.1 Inorganic Contaminants

Inorganic substances such as asbestos, radon, and lead are among the leading indoor contaminants whose exposure can create significant health risks.

7.2.1.1.1 Asbestos

Asbestos is a generic term given to a variety of naturally occurring, hydrated fibrous silicate minerals that posses unique physical and chemical properties that distinguish them from other silicate minerals. Such properties include thermal, electric, and acoustic insulation; chemical and thermal stability; and high tensile strength, all of which have contributed to their wide use by the construction industry (Figure 7.3).

High concentrations of airborne asbestos can occur during demolition and after asbestos-containing materials are disturbed by cutting, sanding, and other activities. Inadequate attempts to remove these materials could release asbestos fibers into the air,

Sample Asbestos-Containing Material List

- Acoustical ceiling texture
- Asphalt flooring
- Base flashing
- Blown-in insulation
- Boiler/tank insulation
- Breaching insulation
- Brick mortar
- Built-up roofing
- Caulking/putties
- Ceiling tiles/panels/mastic
- Cement board
- Cement pipes
- Cement roofing shingles
- Chalkboards
- Construction mastics
- Duct tape/paper
- Ductwork flexible connections
- Electrical cloth
- Electrical panel partitions
- Electrical wiring insulation
- Elevator brake shoes
- Fire blankets
- Fire curtains/hose
- Fire doors
- Fireproofing
- Furnace insulation
- Gray roofing paint
- High temperature gaskets
- HVAC duct insulation
- Incandescent light fixture backing
- Joint compound/wallboard
- Laboratory hoods/table tops
- Laboratory fume hood
- Mudded pipe elbow insulation
- Nicolet (white) roofing paper
- Packing materials
- Paper fire box in walls
- Pipe insulation/fittings
- Plaster/wall joints
- Poured flooring
- Rolled roofing
- Roofing shingles
- Sink insulation
- Spray-applied insulation
- Stucco
- Sub flooring slip sheet
- Textured paints/coatings
- Vapor barrier
- Vermiculite
- Vinyl floor tile/mastic
- Vinyl sheet flooring/mastic
- Vinyl wall coverings

Note: This is a sample list of products that may contain asbestos. It is intended as a general guide to show which types of materials may contain asbestos and is not all inclusive.

Figure 7.3 Partial list of materials that may contain asbestos.

thereby increasing asbestos levels and endangering people living or working in these spaces.

Asbestos-containing material has become a high-profile public concern after federal legislation known as AHERA (Asbestos Hazard Emergency Response Act) was enacted in 1987. The Environmental Protection Agency and Consumer Product Safety Commission have also banned several asbestos products, and manufacturers have voluntarily limited uses of asbestos. Today, asbestos can still be found in older homes, in pipe and furnace insulation materials, shingles, millboard, textured paints and other coating materials, and floor tiles. Asbestos is considered the most widely recognized environmentally regulated material (ERM) during building evaluations.

Asbestos-containing materials are also found in concealed areas such as wall cavities, below ground level, and other hidden spaces. In many older establishments, asbestos-based insulation was used on heating pipes and on the boiler. An adequate asbestos survey requires the inspector to perform destructive testing (i.e., opening walls, etc.) to inspect areas likely to contain suspect materials.

7.2.1.1.2 Health Hazards

The most dangerous asbestos fibers are too small to be visible. The health danger of asbestos fibers depends mainly upon the quantity of fibers in the atmosphere and the length of exposure. Asbestos is made up of microscopic bundles of fibers that may become airborne when asbestos-containing materials are damaged or disturbed. Impaction and abrasion are typically the chief causes of increased airborne-fiber levels. The type, quantity, and physical condition of the asbestos-based material have a significant bearing on the degree of risk.

The risk of airborne asbestos fibers is generally low when the material is in good condition. However, when the material becomes damaged or if it is located in a high-activity area (family room, workshop, laundry, etc.) the risk increases. Increased levels of exposure to airborne asbestos fibers will cause disease. When these fibers get into the air, they may be inhaled and accumulate in the lung tissue, where they can cause substantial health problems including lung cancer, mesothelioma (a cancer of the chest and abdominal linings), and asbestosis (irreversible lung scarring that can be fatal). Symptoms of these diseases do not show up until many years after exposure began. Studies show that people with asbestos-related diseases were usually exposed to elevated concentrations on the job, although some developed disease from exposure to clothing and equipment brought home from job sites. While the process is slow and years may pass before health problems are evidenced, the results and thus the risks are well established.

7.2.1.1.3 Radon

Radon is a natural odorless, tasteless, radioactive gas that is emitted from the soil as a carcinogenic byproduct of the radioactive decay of radium-226, which is found in uranium ores (although radon itself does not react with other substances). The byproduct can cling to dust particles that, when inhaled, settle in bronchial airways. Generally, radon is drawn into a building environment by the presence of air-pressure differentials. The ground beneath a building is typically under higher pressure than the basement or foundation. Air and gas move from high-pressure areas to low-pressure areas. The gas can enter the building through cracks in walls and floors as well as penetrations associated with plumbing, electrical openings, and sump wells in building spaces coming in close contact with uranium-rich soil. Vent fans and exhaust fans also put a room under negative pressure and increase the draw of soil gas, which can increase the level of radon within a building. Radon exposure becomes a concern when it becomes trapped in buildings and indoor levels of concentrations build up.

Adequate ventilation is necessary to prevent radon from accumulating in buildings to dangerous levels, as this can pose a serious health hazard. Where radon is suspected, a survey should be conducted to measure the concentrations in the air and determine whether any actions will be required to reduce the contamination. Radon levels will vary from region to region, season to season, and building to building. Radon levels are typically at their highest during the coolest part of the day when pressure differentials are at their greatest.

According to the Environmental Protection Agency, radon is the cause of an estimated 14,000 lung-cancer deaths annually. The primary factors affecting radon concentrations in the air are ventilation and the radon source. The most common radon source is the presence of radium-226 in the soil and rock surrounding or adjoining basement walls and cellar floor slabs. Although you cannot see radon, it's not hard to find out if you have a radon problem in your home or office. There are many kinds of inexpensive, state-certified do-it-yourself radon test kits available through the mail or in hardware stores and other retail outlets. When radon decays and is inhaled into the lungs, it releases energy that can damage the DNA in sensitive lung tissue and cause cancer.

When high concentrations of radon in air are detected, it is often an indication of possible radon contamination of the water supply (if a private water supply is present). In this case, a water test for radon is the prudent first step. Should high concentrations of radon be found in the water, then an evaluation of ventilation rates in the structure as well as air-quality tests for radon are highly recommended. Generally speaking, high radon concentrations are more likely to exist where there are large rock masses, such as in mountainous regions. The Environmental Protection Agency (EPA) recommends that buildings should be tested every few years to assess the safety of radon levels.

7.2.1.1.4 Radon Mitigation

Everything being equal, elevated radon levels should not necessarily deter investors from purchasing a property, as the problem can usually be easily be resolved, even in existing buildings, without having to incur great expense. However, lowering high radon levels requires technical knowledge and special skills, which means a trained radon-reduction contractor who understands how to fix radon problems.

With new construction, some builders have starting to incorporate radon-prevention techniques in their designs. Some municipalities also have local building codes requiring prevention systems. The EPA has published several brochures and instructional aids regarding radon-resistant construction. This is perhaps the most cost-effective way to handle a radon problem, as it is easier to build the system into the building rather than to retrofit it later. If your building has a radon system built in, the EPA recommends periodic testing to ensure that the system is working properly and that the radon level in your building has not changed. The development of foundation cracks is one example of how the radon level could change.

EPA studies suggest that elevated radon levels are more likely to exist in energy-efficient buildings than otherwise. The reason is that radon can become hazardous indoors due to the limited outside air to dilute the indoor concentrations afforded by efficient construction. Although energy-efficient construction may save energy bills, it may also increase occupants' exposure to radon and other indoor-air pollutants. It is also recommended that testing for radon should also be conducted when any major renovations are made to the building.

7.2.1.1.5 Health Risks

The principal health hazard associated with exposure to elevated levels of radon is lung cancer. Research suggests that, while swallowing water with high radon levels may also pose risks, these are believed to be much lower than those from breathing air containing radon.

During the decaying process, radon emits alpha, beta, and gamma radiation. However, the real threat is not so much from the gas itself but from the products that it produces when it decays, such as lead, bismuth, and polonium. These products are microscopic particles that readily attach themselves to dust, pollen, smoke, and other airborne particles in a building and are inhaled. Once in the lungs, the particles become trapped, and, as they begin to decay, they expose the sensitive lung tissue to the harmful radiation they emit during the decaying process. The effect is cumulative, and elevated levels of radon exposure cause lung cancer in both smokers and nonsmokers alike. The risk would be greatest for people with diminished lung capacity, asthma sufferers, and smokers.

7.2.1.1.6 Radon Testing Methods

For short-term testing, consultants typically use electret ionization chambers, which generally last about a week. The chamber method works by incorporating a small charged Teflon® plate screwed into the bottom section of a small plastic chamber. When the radon gas enters the chamber, it begins to decay and creates charged ions that deplete the charge on the Teflon® plate. By registering the voltage prior to deployment and then reading the voltage upon recovery, a mathematical formula is used to calculate the radon concentration levels within the building.

For long-term testing, consultants prefer to use either long-term electret chambers (like those used for the short-term measurements) or alpha tracks. The alpha track, when deployed in the building, records the number of alpha particles that scratch the plastic surface inside the detector. The laboratory counts these microscopic indentations and then mathematically calculates the radon levels within the building.

7.2.1.1.7 Lead

For many years now, lead has been recognized as a harmful environmental pollutant to the extent that in late 1991 the Secretary of the Department of Health and Human Services described it as the "number one environmental threat to the health of children in the United States." There are many ways in which humans may be exposed to mineral particles of lead: in ambient air, drinking water, food, contaminated soil, deteriorating paint, and dust. Lead is a heavy metal; it does not break down in the environment and continues to be used in many materials and products to this day. Lead is a natural element; and most lead in use today is inorganic lead, which enters the body when an individual breathes (inhales) or swallows (ingests) lead particles or dust once it has settled. Lead dust or particles cannot penetrate the skin unless it is broken. (Organic lead, however, such as the type used in gasoline, can penetrate the skin.)

7.2.1.1.8 Indoor Lead Levels

Because of its widespread use, lead has for some time been known to be a common contaminant of interior environments. For centuries lead compounds such as white lead and lead chromate have been used as white pigments in commercial paints.

In addition to their pigment properties, these lead compounds were valued because of their durability and weather resistance, which made them viable particularly in exterior paints. In addition to lead's durability properties, it was also added to paint to improve its drying characteristics.

Old lead-based paint is considered the most significant source of lead exposure in the U.S. today. In fact, the majority of homes and buildings built before 1960 contained heavily leaded paint. Even as recently as 1978 there were homes and buildings that used lead paint. This paint may have been used on window frames, walls, the building's exterior, or other surfaces. The improper removal of lead-based paint from surfaces by dry scraping, sanding, or open-flame burning can create harmful exposures to lead. High concentrations of airborne lead particles in a space can also result from lead dust from outdoor sources, including contaminated soil tracked inside; Harmful effects can also result from the use of lead in certain indoor activities such as soldering and stained-glassmaking.

Because of potentially serious health hazards and negative publicity, lead content was gradually reduced until it was eliminated altogether in 1978 (in the U.S.). In commercial buildings, lead was used primarily as a paint preservative.

Lead piping has sometimes been used in older buildings, and, while not legally required to be replaced, it can create a health hazard because frequently the piping deteriorates and leaches into the building's drinking water. In some buildings lead solder has also been used in copper-pipe installation, but in most jurisdictions this procedure has now been banned due to water contamination resulting from the deterioration of the solder. In any case, any evaluation for lead contamination requires that the water content be analyzed by a laboratory and action then taken to mitigate the hazards if lead content is found to be in excess of regulated limits. The potential for water contamination can often be removed by chemical treatment of the water. Where this cannot be accomplished, the piping may have to be replaced.

Unfortunately, lead is still allowed in paint for bridge construction and machinery and thus remains a significant source of exposure. Its continued use is mainly due to its ability to resist corrosion and its ability to expand and contract with the metal surface of a structure without cracking. But even if its use were banned today, there would still be potential exposure to workers and surrounding communities for many years to come because of the many metal structures such as bridges that have been coated with it.

7.2.1.1.9 Health Risks

Lead is a highly toxic substance that affects a variety of target organs and systems within the body, including the brain and the central nervous, renal, reproductive, and cardiovascular systems. High levels of lead exposure can cause convulsions, coma, and even death. However, the nervous system appears to be the main target organ system for lead exposure.

One of the best methods to assess human exposure to lead is to measure the blood lead level, because it is a sensitive indicator of exposure that has been correlated with a number of health endpoints. It also gives an immediate estimate of the level of a person's recent exposure to lead. The negative aspect of measuring blood lead level is that, while it will tell you how much lead is in the bloodstream, it will not tell you what is stored in soft tissues or bones. Additionally, the test will not spell out your body burden of lead or any damage if any that has occurred.

Effects of lead poisoning depend largely on dose exposure. Contact with lead-contaminated dust is the primary method by which most children are exposed to harmful levels of lead. It enters a child's body mainly through ingestion. Lead-contaminated dust is often hard to see but can get into the body and create substantial health risks. Pregnant women, infants, and children are more vulnerable to lead exposure because lead is more easily absorbed into growing bodies, and the tissues of small children are more sensitive to the damaging effects of lead. Ingestion of lead has been proven to cause delays in children's physical and mental development as well as negatively impacting their developing nervous systems, causing lower IQ levels, increased behavioral problems, and learning disabilities.

In adults, high lead levels have many adverse effects, including kidney damage, digestive problems, high blood pressure, headaches, diminishing memory and concentration, mood changes, nerve disorders, sleep disturbances, and muscle or joint pain. A single very high exposure to lead can also result in lead poisoning. Adults' bones and teeth contain about 95 percent of the body burden (total amount of lead stored in the body). Likewise, lead can seriously impact the ability of both men and women to bear healthy children.

7.2.1.1.10 Testing for Lead Paint

Inspectors employ various methods and procedures to assist them in identifying the presence of lead paint. In the field, the most widely applied method is the use of an x-ray fluorescent lead-in-paint analyzer (XRF). The XRF analyzer is normally held up to the surface being tested for several seconds. The analyzer then emits radiation, which is absorbed and then fluoresces (is emitted) back to the analyzer. The XRF unit breaks down the signals to determine if lead is present and if so in what concentration. An XRF analyzer can normally read through up to 20 layers of paint, but these analyzers are expensive and should only be used by trained professionals.

7.2.1.2 Contaminants Generated by Combustion

Some examples of combustion byproducts include fine particulate matter, carbon monoxide (CO), and nitrogen oxides. Tobacco smoke is another source.

Combustion (burning) byproducts are essentially gases and tiny particles that are created by the incomplete burning of fuels. These fuels (such as natural gas, propane, kerosene, fuel oil, coal, coke, charcoal, wood, gasoline, and materials such as tobacco, candles, and incense), when burned, produce a wide variety of air contaminants. If fuels and materials used in the combustion process were free of contaminants and combustion were complete, emissions would be limited to carbon dioxide (CO_2), water vapor (H_2O), and high-temperature reaction products formed from atmospheric nitrogen (NOx) and oxygen (O_2). Sources of combustion-generated pollutants in indoor environments are many and include wood heaters and wood stoves, furnaces, gas ranges, fireplaces, and car exhaust (in an attached garage).

There are several combustion-generated contaminants including CO_2, H_2O, carbon monoxide (CO), nitrogen oxides (NOx) such as nitric oxide (NO) and nitrogen dioxide (NO_2), respirable particles (RSP), aldehydes such as formaldehyde (HCHO) and acetaldehyde, as well as a number of volatile organic compounds (VOCs); fuels and materials containing sulfur will produce sulfur dioxide (SO_2). Particulate-phase emissions may include tar and nicotine from tobacco, creosote from wood, inorganic carbon, and polycyclic aromatic hydrocarbons (PAHs).

Carbon dioxide (CO_2) is a colorless, odorless, heavy, incombustible gas that is found in the atmosphere and formed during respiration. It is typically obtained from the burning of, gasoline, oil, kerosene, natural gas, wood, coal, and coke. It is also obtained from carbohydrates by fermentation, by the reaction of acid with limestone or other carbonates, and naturally from springs. CO_2 is absorbed from the air by plants in a process called photosynthesis. Although carbon dioxide is not normally a safety problem, a high CO_2 level can indicate poor ventilation, which in turn can lead to a buildup of particles and more harmful gases such as carbon monoxide that can negatively impact people's health and safety. CO_2 is used extensively in industry as dry ice or carbon dioxide snow, in carbonated beverages, fire extinguishers, etc.

Carbon Monoxide (CO) is an odorless, colorless, lighter than air, nonirritating gas that interferes with the delivery of oxygen throughout the body. CO is the leading cause of poisoning deaths in the United States; it occurs when there is incomplete combustion of carbon-containing material such as coal, wood, natural gas, kerosene, gasoline, charcoal, fuel oil, fabrics, and plastics.

At low concentrations levels, healthy people may experience fatigue and shortness of breath during exertion. Flushed skin, tightness across the forehead, and slightly impaired motor skills may also occur. People with heart disease may encounter chest pain. The first and most obvious symptom is usually a headache with throbbing temples. Infants, children, pregnant women, the elderly, and people with heart or respiratory problems are most susceptible to carbon-monoxide poisoning. The fact that CO cannot be seen, smelled, or tasted makes it especially dangerous, because you are not aware when you are being poisoned. Moreover, CO poisoning is frequently misdiagnosed by both victims and doctors.

Mild to moderate CO poisoning may cause flulike symptoms or gastroenteritis, particularly in children, including nausea, lethargy, and malaise. As the CO level or exposure time increases, symptoms become more severe and additional ones appear: irritability, chest pain, fatigue, confusion, dizziness, and impaired vision and coordination. Higher levels cause fainting upon exertion, marked confusion, and collapse. At very high concentrations, we may witness coma and convulsion as well as permanent damage to the brain, central nervous system, or heart and finally death.

The four primary sources of CO in the environment are:

- Automobile exhaust from attached garages combined with inadequate ventilation is responsible for two-thirds of all accidental CO deaths. Lethal levels of the gas can accumulate in as little as 10 minutes in a closed garage. Certain occupations are exposed regularly to elevated levels of CO. These include toll-booth attendants, professional drivers, highway workers, traffic officers, and tunnel workers. Likewise, certain indoor events, such as tractor pulls and car and truck exhibitions, if not adequately ventilated, can expose spectators and participants to elevated CO levels.

- Faulty heating equipment accounts for almost one-third of accidental CO fatalities. These fatalities can be caused by home or office heating systems (e.g., leaking chimneys and furnaces or back-drafting from furnaces, gas water heaters, wood stoves, fireplaces; and gas stoves) and also by improperly vented or unvented kerosene and gas appliances, propane space heaters, charcoal grills or hibachis, and Sterno-type fuels. Dangerous amounts of carbon monoxide can be released when there is inadequate fresh air or a flame is not sufficiently hot to completely burn a fuel.
- Fires have been found to raise CO levels in the blood of unprotected persons to 150 times normal in a single minute; CO poisoning is the most frequent cause of immediate death associated with fire. Environmental tobacco smoke can also cause elevated CO levels in both smokers and nonsmokers who are exposed to the smoke.
- Methylene chloride is a solvent used in some paints and varnish strippers that is readily absorbed by the body and changes to CO. Using products that contain methylene chloride for more than a few hours can raise CO levels in the blood 7 to 25 times normal. It is particularly dangerous for persons with preexisting cardiac conditions who use these products in unventilated spaces, as they risk heart attack and death.

An electronic device known as a carbon-monoxide alarm is the only reliable method currently used to test for the presence of carbon monoxide. Most fire departments, gas companies, and some specialized contractors have sophisticated equipment that can measure and record carbon-monoxide levels. In the home, detectors should be placed in areas where the family spends most of its time, such as the family room, bedroom, or kitchen. Detectors should be placed far enough away from obvious and predictable sources of CO, such as a gas stove, to avoid false alarms.

CO poisoning can be prevented by following a number of simple steps:

- Keep gas appliances properly adjusted and have them checked periodically for proper operation and venting.
- Open flues when fireplaces are being used and ensure that flues, chimneys, and vents are clear of debris and working properly.
- Have CO monitors installed at home and in the workplace and ensure that they are properly maintained.
- Refrain from using unvented space heaters, gas stoves, charcoal grills, or Sterno-type fuels as sources of heat. Wood stoves when used should be properly sized and certified to meet EPA emission standards.
- When working around CO sources such as propane-powered forklifts and space heaters, ensure that adequate ventilation is in place. Whenever exposure is unavoidable, CO monitoring badges should be worn.
- The car's exhaust system should be checked regularly and properly maintained at all times. Cars or other gasoline-powered engines should not be run inside a garage, even with the doors open.
- Do not use paint strippers that contain methylene chloride. If the use of solvents containing this substance is unavoidable, make sure the area is properly ventilated.

- A trained professional should be brought in at least once a year to inspect, clean, and tune up the central-heating system (furnaces, flues, and chimneys) and repair any leaks promptly.

Nitrogen dioxide (NO_2) is a colorless, odorless gas that irritates the mucous membranes in the eye, nose, and throat and causes shortness of breath when exposed to high concentrations. Nitrogen dioxide is also a major concern as an air pollutant because it contributes to the formation of photochemical smog, which can have significant impacts on human health. Documented evidence indicates that high concentrations or continued exposure to low levels of nitrogen dioxide increases the likelihood of respiratory problems. Because nitrogen dioxide is relatively nonsoluble in tissue fluids, it enters the lungs, where it may expose lower airways and alveolar tissue. Nitrogen dioxide inflames the lining of the lungs and can reduce immunity to lung infections. There is also documented evidence from animal studies that repeated exposures to elevated nitrogen-dioxide levels may lead or contribute to the development of lung diseases such as emphysema. People at particular risk from exposure to nitrogen dioxide include children with asthma and older people with heart disease or other respiratory diseases.

Exposure can cause problems such as wheezing, coughing, colds, flu, and bronchitis. Increased levels of nitrogen dioxide can also have significant impacts on people with asthma because it can cause more frequent and more intense attacks.

There are numerous combustion-generated contaminants including several mucous-membrane and upper-respiratory-system irritants. Aldehydes such as HCHO are the most common, although in some cases acrolein, RSP, and SO_2 may also be included. Aldehydes cause irritation to the eyes, nose, throat, and sinuses and are discussed in greater detail in the following section. Respirable particles vary in composition, and their primary effect is irritation of the upper-respiratory passages and bronchi. Because of its solubility in tissue fluids, SO_2 can cause bronchial irritation.

7.2.1.3 Organic Contaminants

In today's environment, natural and synthetic organic chemicals comprise many different types and can be found virtually everywhere, including soil, ground water, surface water, plants, and our bodies. Modern industrialized societies have developed such a massive array of organic pollutants that it is becoming increasingly difficult to generalize in a meaningful way as to sources, uses, or impacts. These contaminants find their way into the natural environment through accidental leakage or spills, such as leaking underground storage tanks, or through planned spraying of pesticides to agricultural land and urban areas.

The main organic compounds include volatile organic compounds (VOCs), very volatile organic compounds (VVOCs), semivolatile organic compounds (SVOCs), and solid organic compounds (POMs). Solid organic compounds may comprise components of airborne or surface dusts. Organic compounds often pose serious indoor contamination problems; they include the aldehydes, VOCs, and SVOCs (which include a large number of volatile as well as less volatile compounds, and pesticides and biocides, which are largely SVOCs).

Organic compounds that are known to be contaminants of indoor environments include a large variety of aliphatic hydrocarbons, aromatic hydrocarbons, oxygenated hydrocarbons (such as aldehydes, ketones, alcohols, ethers, esters, and acids), and halogenated hydrocarbons (primarily those containing chlorine and fluorine).

Volatile-organic-compound concentrations levels are generally higher in indoor environments than in outdoor air. Studies suggest that indoor air may contain several hundred different VOCs. Moreover, VOCs can be released from products while being used and to a lesser degree while in storage. Fortunately, the amounts of VOCs emitted tend to decrease as the product ages and dries out.

There has been a steady increase in the number of identified VOCs in recent years. They are characterized by a wide range of physical and chemical attributes – the most important of which are their water solubility and whether they are neutral, basic, or acidic. VOCs are released into the indoor environment by many sources, including building materials, furnishings, paints, solvents, air fresheners, aerosol sprays, adhesives, fabrics, consumer products, building cleaning and maintenance materials, pest-control and disinfection products, humans themselves, office equipment, tobacco smoking, plastics, lubricants, refrigerants, fuels, solvents, pesticides, and many others.

Among the health hazards many VOCs pose is that they are potent narcotics that cause depression in the central nervous system; others can cause eye, nose, and throat irritation; headaches; loss of coordination; nausea; and damage to the liver, kidneys, and central nervous system (Figure 7.4). A number of these chemicals are suspected or known to cause cancer in humans.

Formaldehyde (HCHO) is a colorless, pungent-smelling gas, and one of the more common VOCs found indoors. It is an important chemical used widely by industry to manufacture building materials and household products. It is also a byproduct of combustion and certain other natural processes and thus may be present in substantial concentrations both indoors and outdoors.

Formaldehyde is the most common of the aldehydes and is considered by many as possibly the single most critical indoor pollutant because of its common occurrence and

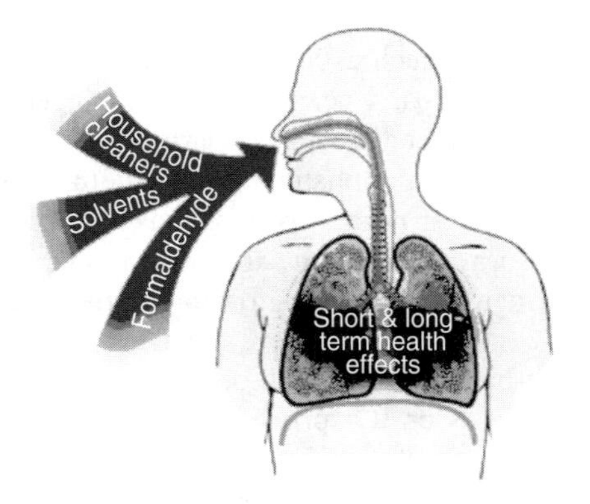

Figure 7.4 Diagram showing inhalation of volatile organic compounds.
Source: Air Advice Inc.

its strong toxicity. Formaldehyde is a colorless, gaseous substance with an unpleasant smell. On condensing it forms a liquid with a high vapor pressure. Due to its high reactivity, it rapidly polymerizes with itself to form paraformaldehyde. Because of this reaction, liquid HCHO needs to be kept at a low temperature or mixed with a stabilizer (such as methanol) to prevent or minimize polymerization.

Formaldehyde, by itself or in combination with other chemicals, serves a number of purposes in manufactured products. For example, it is used to add permanent-press qualities to clothing and draperies, as a component of glues and adhesives, and as a preservative in some paints and coating products. It is also used in a variety of deodorizing commercial products, such as lavatory and carpet preparations.

Formaldehyde is commercially available as both paraformaldehyde and as formalin, an aqueous solution that typically contains 37 to 38 percent HCHO by weight and 6 to 15 percent methanol. HCHO is also used in many different chemical processes, such as the production of urea and phenol-formaldehyde resin.

Urea-formaldehyde (UF) copolymeric resins are present in many building materials such as wood adhesives, which are used in the manufacture of pressed-wood products including particle board, medium-density fiber board (MDF), and plywood and in finish coatings (acid-cured), textile treatments, and urea-formaldehyde foam insulation (UFFI).

People are often unaware that formaldehyde is given off by materials other than UFFI. Certain types of pressed-wood products (composition board, MDF, paneling, etc.), carpeting, and other materials can be formaldehyde sources. Many of these products use a urea-formaldehyde-based resin as an adhesive. Some of these materials will continue to give off formaldehyde much longer than UFFI. Like the majority of VOCs, formaldehyde levels will decrease substantially with time and/or with increased ventilation rates.

For some people formaldehyde can be a respiratory irritant, and continuous exposure to it can be dangerous. More specifically, chronic, low-level, continuous, and even intermittent exposure can cause chemical hypersensitivity and provide an accelerating factor in the development of chronic bronchitis and pulmonary emphysema. HCHO also has the ability to cause irritation, inflammatory-type symptoms, and symptoms of the central nervous system such as headache, sleeplessness, and fatigue. Elevated HCHO exposures (above 0.1 parts per million) can trigger asthma attacks, nausea, diarrhea, unnatural thirst, and menstrual irregularities.

Acetaldehyde is a two-carbon aliphatic aldehyde with a pungent, fruity odor. It is used in a number of industrial processes and is a major constituent of automobile exhaust fumes. It is also a predominant aldehyde found in tobacco smoke. Compared to HCHO, it is a relatively mild irritant of the eye and upper respiratory system.

Acrolein is a three-carbon aldehyde with one double bond. It is highly volatile gas with an unpleasant choking odor. It is primarily used in the production of a variety of compounds and products and is released into the environment as a combustion oxidation product from oils and fats containing glycerol, wood, tobacco, and automobile/diesel fuels. At relatively low exposure concentrations it is a potent eye irritant.

Glutaraldehyde is a five-carbon dialdehyde. It is found in liquid form and has a sharp, fruity odor. Its main application is as an active ingredient in disinfectant formulations widely used by the medical and dental professions. The main health effects associated with glutaraldehyde exposures include irritation of the nose and throat, nausea and headache, and pulmonary symptoms such as chest tightening and asthma.

Polychlorinated biphenyls (PCBs) are oils used primarily as coolants in electrical transformers. Although the production and sale of PCBs were banned by the Environmental Protection Agency (EPA) in 1979, a large number of PCB-filled transformers remain in use. It has also been estimated that some 2,000,000 mineral-oil transformers still contain some percentage of PCBs. PCBs may also be found in light ballasts and elevator hydraulic fluids. PCBs are a suspected carcinogen, but properly sealed or contained PCBs do not pose a hazard. However, PCBs do become a hazard when they catch fire, creating carcinogenic byproducts that can contaminate air, water, and the finishes and contents of a building. Leaking PCBs can also contaminate building materials and soil. PCB evaluations typically focus on identifying the presence of or potential for PCB leakage, measuring of PCB concentration levels, and determining the presence of combustible materials adjacent to PCB-containing equipment.

Hydrocarbons are a class of organic chemical compound consisting only of the elements hydrogen (H) and carbon (C). They are cardinal to our modern way of life and being one of the Earth's most important energy resources. Hydrocarbons are the principal constituents of petroleum and natural gas and are also derived from coal. The bulk of the world's hydrocarbons are used for fuels and lubricants, as well as for electrical-power generation and heating. The chemical, petrochemical, plastic, and rubber industries are also dependent upon hydrocarbons as raw materials. Hydrocarbons are colourless, flammable, toxic liquids.

Symptoms associated with exposure to aliphatic hydrocarbons may include watery eyes, nausea, vomiting, dizziness, weakness, and central-nervous-system effects such as depression, convulsions, and, in extreme cases, coma. Other symptoms may include pulmonary and gastrointestinal irritation, pulmonary edema, bronchial pneumonia, anorexia, anemia, nervousness, pain in the limbs, and numbness. Noticeable odors similar to gasoline and oil all suggest that some hydrocarbon contamination may be present. Likewise, leaking subsurface tanks at fuel stations and other facilities have created significant health and safety problems by contaminating the soil around the buildings. Benzene is found in most hydrocarbons and is considered to be one of the more serious contaminants, known to cause leuaemia. Air-quality tests may be necessary as well as tests for contaminants in the soil around the foundation.

Pesticides are chemical poisons, designed to control, destroy, or repel plants and animals such as insects (insecticides), weeds (herbicides), rodents (rodenticides), and mold or fungus (fungicides). They include active ingredients (those intended to kill the target) and inert ingredients, which are often not "inert" at all. Pesticides are generally toxic and can be absorbed through the skin, swallowed, or inhaled; as such are unique contaminants of indoor environments.

Studies show that approximately 16 million Americans are sensitive to pesticides; their immune systems have been damaged as a result of prior pesticide exposure.

In addition, pesticides have been linked to a wide range of serious and often fatal conditions: cancer, leukemia, miscarriage, genetic damage, decreased fertility, liver damage, thyroid disorders, diabetes, neuropathy, still-births, decreased sperm counts, asthma, and other autoimmune disorders (lupus, etc.).

Federal and state governments carefully regulate pesticides to ensure that they do not pose unreasonable risks to human health or the environment. There are currently more than 1055 active ingredients registered as pesticides, which are formulated into thousands of different products that are available in the marketplace, including some of the most widely used over the past 60 years. These include aldrin, chlordane, DDT, dieldrin, endrin, heptachlor, mirex, toxaphene, and lindane (hexachlorocyclohexane, HCH). Many pesticides (most notably chlordane, used for termite treatment) are serious hazards. It is hoped that ecological methods of pest control will in the future replace the overdependence on chemicals that now threatens our ecosystem.

7.2.1.4 Biological Contaminants

Biological pollutants arise from various sources such as microbiological contamination – e.g., fungi, bacteria, viruses, mites, pollens – and the remains and dropping of pests such as cockroaches. Such pollutants of biological origin can significantly impact indoor-air quality and cause infectious disease through airborne transmission. Of particular concern are those biological contaminants that cause immunological sensitization manifested as chronic allergic rhinitis, asthma, and hypersensitivity pneumonitis.

Moisture in buildings is one of the major contributors to poor indoor-air quality, unhealthy buildings, and mold growth; by controlling the relative humidity level, the growth of some sources of biologicals can be minimized. Standing water, water-damaged materials, rainwater leaks, or wet surfaces also serve as a breeding ground for molds, mildews, bacteria, and insects. Likewise, damp or wet areas such as cooling coils, humidifiers, condensate pans, or unvented bathrooms can serve as suitable breeding grounds. Contaminated central HVAC systems can then distribute these contaminants through the building.

Proactive preventive and remedial measures include rainwater-tight detail design, preventing uncontrolled air movement, reducing indoor-air moisture content, reducing water-vapor diffusion into walls and roofs, selecting building materials that have appropriate water-transmission characteristics, and maintaining proper workmanship quality control.

A proven method for deterring rainwater intrusion into walls is the rain-screen approach, which incorporates cladding, air cavities, drainage planes, and airtight support walls to offer multiple moisture-shedding pathways. The concept of the rain screen is simple; it separates the plane in a wall where rainwater is shed and where air infiltration is halted. In terms of construction, this means that there is an outer plane that sheds rainwater but allows air to freely circulate and an inner plane that is relatively airtight.

7.2.1.4.1 Mold and Mildew

Mold and mildew are types of musty-smelling fungi that thrive in moist environments. Their function in nature is primarily to break down and decompose organic materials such as leaves, wood, and plants. They grow, penetrate, and infect the air we breathe. There is no normal way to eliminate all mold and mold spores in the indoor environment, and thus the key to controlling indoor mold growth is to control moisture content. Exposure to fungi in indoor-air settings has emerged as a significant health problem of great concern in both residential environments as well as in workplace settings. There are three major health hazards associated with fungi: infection, allergies, and toxins.

Fungi are primitive plants that lack chlorophyll and therefore feed on organic matter that they digest externally and absorb. True fungi include yeast, mold, mildew, rust, smut, and mushrooms. Given sufficient moisture they can grow on almost any material, even inorganic. When mold spores land on a damp spot indoors, they can grow and digest whatever they are growing on. There are molds that can grow on wood, paper, carpet, fabric, and foods, but they usually grow best in dark, moist habitats, especially if organic matter is available.

There are thousands of species of molds, including pathogens, saprotrophs, aquatic species, and thermophiles. Molds are part of the natural environment and are present everywhere in nature; their presence is only visible to the unaided eye when mold colonies grow (Figure 7.5).

Molds produce small spores to reproduce; they float through indoor and outdoor air continually. These spores may contain a single nucleus or be multinucleate. Mold spores can be asexual or sexual; many species can produce both types. Some species can remain airborne indefinitely, and many are able to survive extremes of temperature (some molds can begin growing at temperatures as low as 2 degrees C) as well as extremes of pressure. When conditions do not foster or enable growth, most molds have the ability to remain alive in a dormant state within a wide range of temperatures before they die. Different mold species vary enormously in their tolerance to temperature and humidity extremes.

For mold to grow or establish itself, it needs four vital elements: viable spores, a nutrient source (organic matter such as wood products, carpet, and drywall), moisture, and warmth. The mere presence of humid air, however, is not necessarily conducive to fostering mold growth except where it has a relative humidity (RH) level at or above 80 percent and is in contact with a surface. Mold spores are carried by air currents and can reach all surfaces and cavities of buildings. When the surfaces and/or cavities are warm and contain the right nutrients and amounts of moisture, the mold spores will grow into colonies and gradually destroy what they grow on.

The key to controlling mold growth is to control indoor moisture and the temperatures of all surfaces, including interstitial surfaces within walls. Mold generally needs a temperature range between 40 and 100 degrees F to grow, and maintaining relative-humidity levels between 30 and 60 percent will help control many of these known biological contaminants. Humidity control prevents the indoor growth of mold, mildew, viruses, and dust mites. Winter humidification and summer dehumidification controls/modules can supplement central HVAC systems when climate excesses require additional conditioning measures. Likewise, by removing any of the four essential growth elements the growth process will be inhibited or nonexistent.

Typical Molds Found in Damp Buildings

Fungal Species	Substrate	Possible Metabolites	Potential Health Effects
Alternaria alternata	moist window-sills, walls	allergens	asthma, allergy
Aspergillus versicolor	damp wood, wallpaper glue	mycotoxins, VOCs	unknown
Aspergillus fumigatus	house dust, potting soil	allergens	asthma, rhinitis, hypersensitivity pneumonitis
		many mycotoxins	toxic pneumonitis infection
Cladosporium herbarum	moist window-sills, wood	allergens	asthma, allergy
Penicillium chrysogenum	Damp wallpaper, behind paint	mycotoxins	unknown
		VOCs	unknown
Penicillium expansum	Damp wallpaper	mycotoxins	nephrotoxicity?
Stachybotrys chartarum (atra)	Heavily wetted carpet, gypsum board	mycotoxins	dermatitis, mucosal irritation, immunosuppression

Figure 7.5 Table showing partial list of typical types of molds found in damp buildings.

The first step in a mold-remediation project includes determining the root cause of the mold growth. The next step is to evaluate the order of magnitude of the growth; this is usually done through visual examination. Since old mold growth may not always be visible, investigators may need to use instruments such as moisture meters, thermal-imaging equipment, or borescope cameras to identify moisture in building materials or "hidden" mold growth within wall cavities, HVAC ducts, etc. Mold assessments and inspections should always include HVAC systems and their air-handler units, drain pans, coils, and ductwork in their surveys. In addition, depending on the age of the building, the inspector should take samples of the building materials, such as ceiling tiles, drywall joint compounds, and sheet flooring for the presence of asbestos. All these organisms may contribute to poor indoor-air quality and can cause serious health problems.

Figure 7.6 (a) Example of mold on a ceiling, often found in damp buildings (*Source*: mold-kill.com). (b) Example of an extreme case of toxic mold growth in the process of being treated (*Source*: Applied Forensic Engineering, LLC).

Fungi in indoor environments include microscopic yeasts and molds, called microfungi, whereas plaster and wood-rotting fungi are referred to as macrofungi because they produce spores that are visible to the naked eye. Some molds produce toxic liquid or gaseous compounds, known as mycotoxins, in addition to infectious airborne mold spores, which often cause serious health problems to residents and workers. Toxicity can arise from inhalation or skin contact with toxigenic molds. Some of these toxic molds can only produce the dangerous mycotoxins under specific growing conditions. Mycotoxins are harmful or lethal to humans and animals when exposed to high concentrations. Toxic molds and fungi are a significant source of airborne VOCs that create IAQ problems (Figure 7.6).

7.2.1.4.2 Bacteria and Viruses

Tens of millions of people around the world suffer daily from viral infections of varying degrees of severity at immense cost to the economy, including the costs for medical treatment, lost income due to inability to work, and decreased productivity. In fact, viruses have been identified as the most common cause of infectious diseases acquired within indoor environments, particularly those causing respiratory and gastrointestinal infection. The most common viruses causing respiratory infections include influenza viruses, rhinoviruses, corona viruses, respiratory syncytial viruses (RSVs), and parainfluenza viruses (PIVs); viruses responsible for gastrointestinal infections include rotavirus, astrovirus, and Norwalk-like viruses (NLVs). Some of these infections, like the common cold, are very widely spread but are not severe, while influenza-like infections are relatively more severe.

Bacteria and viruses are minute in size and readily become airborne, remaining suspended in air for hours. Airborne bacteria and viruses in interior spaces are a cause of considerable concern due to their ability to transmit infectious diseases. Bacteria, viruses, and other bioaerosols that are common in both the home and the workplace may increasingly contribute to sick-building syndrome (SBS) if humidity levels are either too low or too high, depending on how their growth and our respiratory system are affected.

There are many pathways to infection spread, and among the most significant, from an epidemiological point of view is airborne transport. Microorganisms can become airborne when droplets are given off during speech, coughing, sneezing, vomiting, or atomization of feces during sewage removal. Q fever is another emerging infectious disease among U.S. soldiers serving in Iraq. Fever, pneumonia, and/or hepatitis are the most common signs of acute infection with Q fever.

Liquid and solid airborne particles (aerosols) in indoor air originate from many indoor and outdoor sources. These particles may differ in size, shape, chemical composition, and biological composition. Particle size signifies the most important characteristic affecting particle fate during transport, and it is also significant in affecting biological properties. Bacterial aerosols have also been found to be a means to transmit several major diseases, as shown in Figure 7.7.

Professor Lidia Morawska of Queensland University of Technology, Australia, says that the degree of hazard created by biological contaminants including viruses in indoor environments is controlled by a number of factors:

- The type of virus and its potential health effects
- Mode of exit from the body
- Concentration levels
- Size distribution of aerosol containing the virus
- Physical characteristics of the environment (temperature, humidity, oxygenation, UV light, suspension medium etc.)
- Air-circulation pattern
- Operation of heating, ventilation, and air-conditioning system

The physical characteristics of the indoor environment as well as the design and operation of building ventilation systems are of paramount importance. Ducts, coils, and recesses of ventilation systems often provide fertile breeding grounds for viruses and bacteria that have been proven to cause a wide range of ailments from influenza to tuberculosis. Likewise, a number of viral diseases may be transmitted in aerosols

Disease	Causal Organism
Tuberculosis	*Mycobacterium tuberculosis*
Pneumonia	*Mycoplasma pneumoniae*
Diphtheria	*Corynebacterium diphtheriae*
Anthrax	*Bacillus anthracis*
Legionnaires' disease	*Legionella pneumophila*
Meningococcal meningitis	*Neisseria meningitides*
Respiratory infections	*Pseudomonas aeruginosa*
Wound infections	*Staphylococcus aureus*

Figure 7.7 Major infectious diseases associated with bacterial aerosols.

derived from infected individuals. A number of infectious viral diseases and associated causal viruses transmitted through air are shown below in Figure 7.8.

7.2.1.4.3 Mites

Mites are microscopic bugs that thrive on the constant supply of shed human skin cells (commonly called dander) that accumulate on carpeting, drapes, furniture coverings, and bedding (Figure 7.9). The proteins in the combination of feces and skin shedding are what cause allergic reactions in humans. Dust mites are perhaps the most common cause of perennial allergic rhinitis. Dust mites are the source of one of the most powerful biological allergens and flourish in damp, warm environments. Depending on the person and level of exposure, estimates given suggest that dust mites may be a factor in 50 to

Disease	Causal Organism
Influenza	*Orthomyxovirus*
Cold	*Coronavirus*
Measles	*Paramyxovirus*
Rubella	*Togavirus*
Chickenpox	*Herpes virus*
Respiratory infection	*Adenovirus*

Figure 7.8 Some of the major infectious diseases associated with viral aerosols.

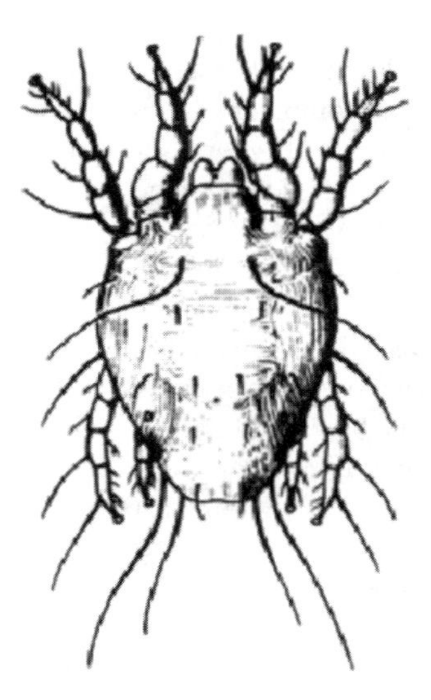

Figure 7.9 Drawing of a dust mite.
Source: EHSO.

80 percent of asthmatics, as well as in numerous cases of eczema, hay fever, and other allergic ailments.

It is estimated that up to 15 percent of people are allergic to dust mites, which due to their very small size (250 to 300 microns in length) and translucent bodies are not visible to the naked eye. To be able to give an accurate identification, one needs at least a 10x magnification. Dust mites have eight hairy legs, no eyes or antennae, a mouth group in front of the body (resembling a head), and a tough, translucent shell. Dust mites, like their insect cousins, have multiple developmental stages. They commence with an egg and develop into larva, several nymph stages, and finally the adult.

Biological contaminants such as dust mites prefer warm, moist surroundings like the inside of a mattress particularly when someone is lying on it, but they may also accumulate in draperies, carpet, and other areas where dust collects. The favorite food of mites appears to be dander (both human and animal skin). Humans generally shed about 0.2 ounce of dander (dead skin) a week.

Dust-mite populations are usually highest in humid regions and lowest in areas of high altitude and/or dry climates. Control measures are needed to reduce concentrations of dust-borne allergens in the living/working environment by controlling both allergen production and the dust that serves to transport it.

7.2.1.4.4 Rodent, Insect, and Animal Allergens

According to the Illinois Department of Public Health, a typical large city in the United States annually receives more than 10,000 complaints about rodent problems and performs tens of thousands of rodent-control inspections and baiting services. Effective measures should therefore be taken to prevent rodents, insects, and pests from entering the home or office. Cockroaches, rats, termites, and other pests have plagued commercial facilities for far longer than computer viruses. Increasingly, research has confirmed and pinpointed pest infestation as the trigger or cause of a host of diseases, and according to the National Pest Management Association pests can cause serious threats to human health, including such diseases as rabies, salmonellosis, dysentery, and staph. But in addition to these serious health concerns they also detract from a facility's appearance and value.

Today, communities of rats exist within and beneath cities, traveling unnoticed from building to building along sewers and utility lines. Each rat colony has its own territory, which can span an entire city block and harbor more than 100 rats. As they explore their territories, rats and mice discover new food sources and escape routes. A rat's territory or "home range" is generally within a 50- to 150-foot radius of the nest, while mice usually have a much smaller range, living within a 10- to 30-foot radius of the nest. In places where all their needs (food, water, shelter) are met, rodents have smaller territories (Figure 7.10).

Rodents are known to carry disease and fleas and to leave waste. Wild and domestic rodents reportedly harbor and spread as many as 200 human pathogens. Diseases include the deadly hantavirus and arena virus. Hantavirus is usually contracted by

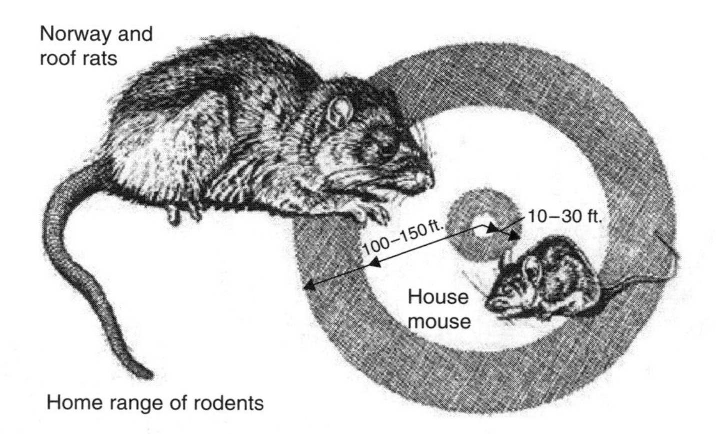

Figure 7.10 Drawing showing typical home range of rodents.
Source: Illinois Department of Public Health.

inhaling airborne particles from rodent droppings, urine, or saliva left by infected rodents or through direct contact with infected rodents.

Today, more than 900,000 species of insects have been identified, and additional species are being identified every day. Some of these insects are known sources of inhalant allergens that may cause chronic allergic rhinitis and/or asthma. They include cockroaches, crickets, beetles, moths, locusts, midges, termites, and flies. Insect body parts are especially potent allergens for some people. Cockroach allergens are also potent and are commonly implicated as contributors to sick-building syndrome in urban housing and facilities with poor sanitation. Most of the allergens from cockroaches come from the insect's discarded skins. As the skins disintegrate over time, they become airborne and are inhaled. This type of allergen can be resolved by eliminating the cockroach population.

Cockroaches are known carriers of serious diseases, such as salmonella, dysentery, gastroenteritis, and other stomach organisms. They adulterate food and spread pathogenic organisms with their feces and defensive secretions. Cockroaches have been reported to spread at least 33 kinds of bacteria, six kinds of parasitic worms, and at least seven other kinds of human pathogens. They can pick up germs on their bodies as they crawl through decaying matter and then carry these onto food surfaces (Figure 7.11). Cockroaches molt regularly throughout their life cycle. The discarded skin becomes airborne and can cause severe asthmatic reactions, particularly to children, the elderly, and people with bronchial ailments.

There are in excess of 20 varieties of ants invading homes and offices throughout the United States, particularly during the warm months of the year (Figure 7.12). Worldwide, there are more than 12,000 species, but of these only a limited number actually cause problems. The one trait all ants share is that they are unsightly and contaminate food. Ants range in color from red to black.

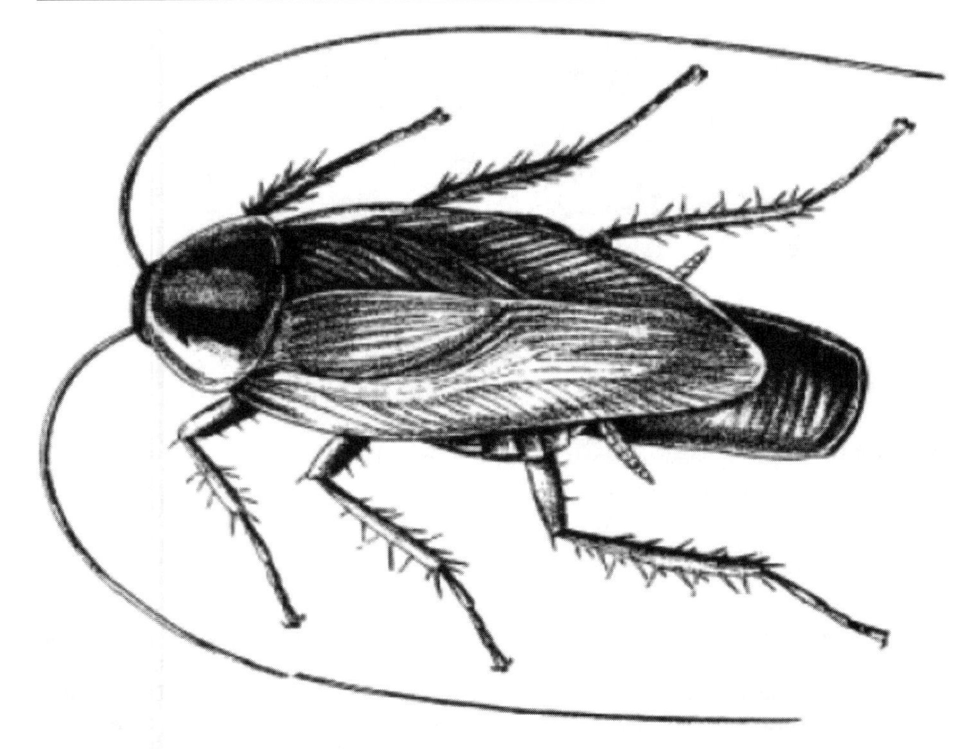

Figure 7.11 Drawing of an American cockroach.

Destructive ant species include fire and carpenter ants. Others types include the pharaoh, honey, house, Argentine, and thief ant. Fire ants are vicious, unrelenting predators and have a powerful, painful sting. More than 32 deaths in the U.S. have been attributed to severe allergic reactions to fire-ant stings.

Termites can pose a major threat to structures, which is why it is important to address any termite infestation as soon as possible. A qualified termite-control company or inspector should look for the many tell-tale signs termites usually provide, such as small holes in wood, straw-shaped mud tubes, crumbling drywall, termite insect wings, and sagging doors or floors (Figure 7.13).

Allergens are produced by many mammalian and avian species; they can be inhaled by humans and cause immunological sensitization as well as symptoms of chronic allergic rhinitis and asthma. These allergens are normally associated with dander, hair, saliva, and urine of dogs, cats, rodents, and birds. Likewise, pollens, ragweed, and a variety of other allergens find their way indoors from the outside. Ragweed is known to cause what is commonly referred to as "hay fever," or what allergist/immunologists refer to as allergic rhinitis. In the United States seasonal allergic rhinitis (hay fever), which is caused by breathing in allergens such as pollen, affects more than 35 million people. Sufferers of allergic rhinitis exposure often experience sneezing, runny noses,

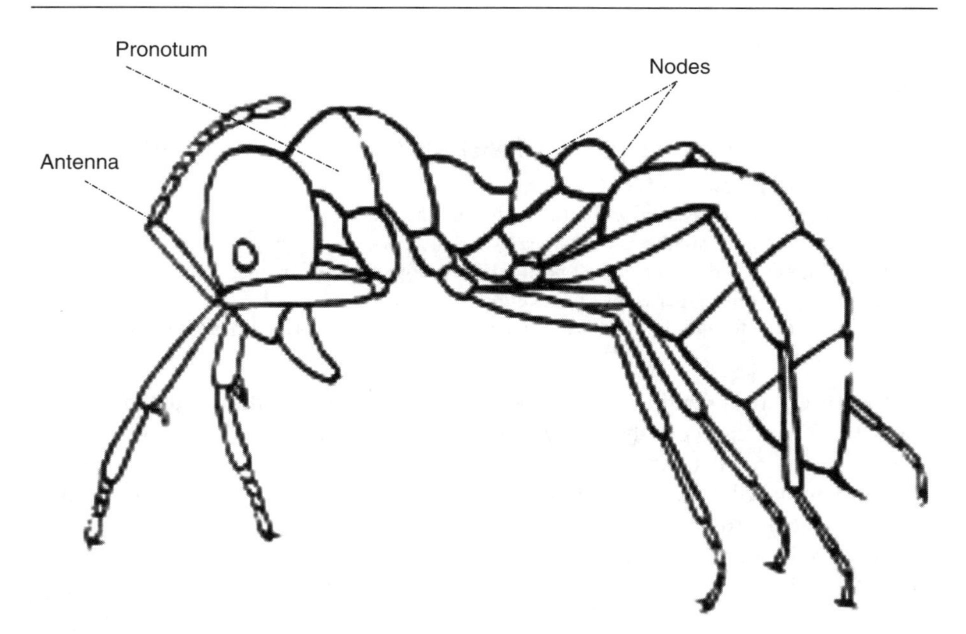

Figure 7.12 Illustration identifying primary features of an ant.

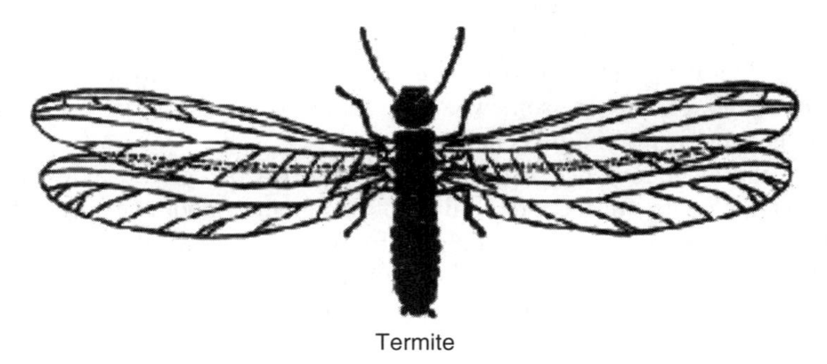

Figure 7.13 Illustration of a termite.

and swollen, itchy, watery eyes. These allergy symptoms can have a major impact on a person's quality of life, including his or her ability to function well at school or work.

7.2.1.5 Reducing Exposure

Resolving air-quality problems in buildings requires first the consideration of all the various contaminants, causes, and concentrations to better be able to address the problems and find appropriate solutions. The location where a person is exposed may also play a pivotal role. While the average worker is exposed approximately 40 hours per week, many individuals are in their home 24/7. Exposure times, therefore, as well as concentration levels become part of the equation.

Pollutant-source removal or modification is the best approach whenever sources are identified and control is possible. These may include:

- Routine maintenance of HVAC systems
- Applying smoking restrictions in the home and the office
- Venting contaminant-source emissions to the outdoors
- Proper storage and use of paints, pesticides, and other pollutant sources in well-ventilated areas and use during periods of nonoccupancy
- Allowing time for building materials in new or remodeled areas to off-gas pollutants before occupancy

Most mechanical ventilation systems in large buildings are designed and operated not only to heat and cool the air, but also to draw in and circulate outdoor air, and one cost-effective method to reduce indoor pollutant levels is to increase ventilation rates and air distribution. At a minimum, HVAC systems should be designed to meet ventilation standards in local building codes. In practice, however, many systems are not operated or adequately maintained to ensure that these design ventilation rates are in place. Often IAQ can be improved by operating the HVAC system to at least its design standard and to ASHRAE Standard 62-2001 if possible. When confronted with strong pollutant sources, local exhaust ventilation may be required to exhaust contaminated air directly from the building. The use of local exhaust ventilation is particularly advised to remove pollutants that accumulate in specific areas such as restrooms, copy rooms, and printing facilities. Air cleaners can also be a useful adjunct to source control and ventilation, although they are somewhat limited in their application. Air cleaners are discussed in Section 7.3.6 of this chapter.

Because the average person spends so much of his or her life indoors (roughly 90 percent), the quality of the indoor air is of paramount importance. In fact, indoor-air pollution is currently ranked among the top four environmental risks in America by the EPA. This may explain why for many forward-thinking real-estate property managers it is becoming a standard of doing business to have their buildings routinely inspected as part of a proactive IAQ monitoring program. Some advantages include:

- Proactive programs build value into the property by having a professional record of indoor environmental conditions that are formally documented over time.
- Having a proactive IAQ program in place can be used as an effective marketing tool to attract quality tenants in a very competitive real-estate market.
- Regular monitoring of ventilation rates, building air flows, and filtration efficiencies can dramatically improve IAQ, thereby creating a more comfortable and productive work environment.
- A proactive program reduces the likelihood of labeling a "healthy" building as causing SBS. Problems or potential problems can be quickly identified and resolved at minimal expense.
- Proactive IAQ monitoring shields owners and facility managers against liability. It shows that the owner has applied due diligence to ensure that his building remains healthy. Additionally, owners with large building portfolios can benefit from these routine inspections as a means to demonstrate that a cohesive, organized, and coordinated IAQ policy is in place.

- A proactive IAQ program enhances management/tenant relations by demonstrating a genuine concern for the tenants and their employees. Less time is needed to investigate and resolve tenant complaints.
- The proactive IAQ report provides critical third-party documentation certifying effective maintenance standards and supporting engineering or operation budgets.

Following are some steps to reduce asbestos exposure:

- Leave undamaged asbestos material alone if it is not likely to be disturbed.
- Trained and qualified contractors should be used for control measures that may disturb asbestos and for cleanup.
- Proper procedures should be followed when replacing wood-stove door gaskets that may contain asbestos.

Following are some steps to reduce radon exposure:

- Test home and office for radon.
- If radon level is found to be 4 picocuries per liter (pCi/L) or higher, it needs to be fixed.
- Radon levels less than 4 pCi/L still pose a risk and in many cases may be reduced.
- For additional information on radon, contact your state radon office or call 800-SOS-RADON.

Following are some steps to reduce lead exposure:

- Leave lead-based paint undisturbed if it is in good condition.
- Do not sand or burn off paint that may contain lead.
- Do not remove lead paint yourself.
- Avoid bringing lead dust into the home or workplace.
- If your work or hobby involves lead, change clothes and use doormats prior to entering your home or workplace.
- Keep areas where children play as dust-free and clean as possible.
- Eat a balanced diet, rich in calcium and iron.

Following are some steps to reduce respirable-particles exposure:

- Vent all furnaces to the outdoors and keep doors to remaining spaces open when using unvented space heaters.
- Choose properly sized wood stoves, certified to meet EPA emission standards, and ensure that doors on all wood stoves fit tightly.
- Use trained professional to inspect, clean, and tune up central heating systems (furnace, flues, and chimneys) annually and promptly repair any leaks.
- Change filters on central heating and cooling systems and air cleaners according to manufacturer's directions.

Following are some steps to reduce formaldehyde exposure:

- Exterior-grade pressed-wood products should be used because they are lower-emitting (because they contain phenol resins, not urea resins).
- Use air conditioning and dehumidifiers to maintain a moderate temperature and reduce humidity levels.
- Increase ventilation, particularly after bringing new sources of formaldehyde into the home or workspace.

Pesticides are products that contain hazardous chemicals designed to kill or repel household pests (insecticides, termiticides, and disinfectants). Products are also used on lawns and gardens that drift or are tracked inside the property. Evidence of pests is typically found in different forms such as droppings (especially from cockroaches and rodents), frass (from wood borers), gnawing, tracks, and grease marks (from rodents), damage (such as powderpost-beetle exit holes), and shed insect skins. The presence of feeding debris or frass is one indication of infestation. Pests are also to be found behind baseboards, under furniture, behind moldings, in floor cracks, behind radiators, and in air ducts. Many pesticides contain harmful VOCs that contribute to poor indoor-air and environmental quality (IAQ/IEQ). Following are some steps to reduce pest and pesticide exposure:

- Pesticides should be applied in strict accordance with manufacturer's instructions.
- Pesticides should be used in recommended quantities only.
- Check around door jambs for evidence of cockroaches and spiderwebs.
- Window sills should be examined regularly, as many pests fly or crawl towards light.
- Increase ventilation when using pesticides indoors. Take plants or pets outdoors when applying pesticides to them.
- Check for presence of moisture, as this can attract moisture-loving pests such as carpenter ants, termites, or mold.
- Use nonchemical methods of pest control where possible.
- If you use a pest-control company, select it carefully.
- Do not store unneeded pesticides inside the home; dispose of unwanted containers safely.
- Store clothes with moth repellents in separately ventilated areas if possible.
- Keep indoor spaces clean, dry, and well ventilated to avoid pest and odor problems.

In addition, evidence of damaged screens, doors, and walls, which could allow pest entry, should be fixed. Sanitation problems should be recorded and rectified. Presence of heavy landscaping near the foundation and plants such as ivy growing on walls need to be controlled or avoided, as they increase the risk of attracting outdoor pests inside. Moisture problems around the foundation, gutters, or air-conditioning units should be monitored and rectified, as they can attract moisture-related pests to a building. Bright exterior lights attract insects to the building's exterior; they may then find their way indoors.

Following are some steps to reduce biological-contaminant exposure:

- Install and use exhaust fans in kitchens and bathrooms and have them vented to the outdoors; clothes dryers should also be vented to the outdoors. One benefit to using kitchen and bathroom exhaust fans is that they can also reduce levels of organic pollutants that vaporize from hot water used in showers and dishwashers.
- Ventilate any attic or crawl spaces to prevent moisture buildup and keep humidity levels in these areas below 50 percent to avoid water condensation on building materials.
- If using cool-mist or ultrasonic humidifiers, clean appliances according to manufacturer's instructions and refill with fresh water daily. These humidifiers are susceptible to becoming breeding grounds for biological contaminants and can potentially cause diseases such as hypersensitivity pneumonitis and humidifier fever. Clean evaporation trays in air conditioners, dehumidifiers, and refrigerators regularly.

- Water-damaged carpets and building materials can harbor mold and bacteria and therefore should be thoroughly cleaned and dried (within 24 hours if possible) or removed and replaced.
- Keep the interior clean. Dust mites, pollens, animal dander, and other allergy-causing agents can be reduced, although not eliminated, through regular cleaning. Allergic individuals should also leave the space while it is vacuumed, as vacuuming can actually increase airborne levels of mite allergens and other biological contaminants. Using central vacuum systems that are vented to the outdoors or vacuums with high-efficiency filters may also be helpful.
- Biological pollutants in basements can be minimized by regularly cleaning and disinfecting the basement floor drain. Operate a dehumidifier in the basement if needed to keep relative-humidity levels between 30 and 50 percent.

Below are a number of relevant standards, codes, and guidelines:

- ASHRAE Standard 62-1999: Ventilation for Acceptable Indoor Air Quality
- ASHRAE 129-1997: Measuring Air-Change Effectiveness
- South Coast Rule #1168, South Coast Air Quality Management District
- Regulation 8, Rule 51, the Bay Area Air Quality Management District
- Canadian Environmental Choice/Ecologo
- Best Sustainable Indoor Air Quality Practices in Commercial Buildings
- Guidelines for Reducing Occupant Exposure to Volatile Organic Compounds (VOCs) from Office Building Construction Materials
- Carpet and Rug Institute Green Label Indoor Air Quality Test Program

An investigation procedure into IAQ may be characterized as a cycle of information gathering, hypothesis formation, and hypothesis testing. It typically begins with a walkthrough inspection of the problem area to gather information relating to the four basic factors that influence indoor-air quality:

- A building's occupants
- A building's HVAC system
- Possible pollutant pathways
- Possible sources of contamination

Typical IAQ investigations include documentation of readily obtainable information regarding the building's history and of any complaints. It would also include identifying known HVAC zones and complaint areas, notifying occupants of the upcoming investigation, and, identifying key individuals who may be needed for information and access. The walkthrough itself entails visual inspection of critical building areas and consultation with occupants and staff. However, if insufficient information is obtained from the walkthrough to formulate a hypothesis or if initial tests fail to reveal the source of the problem, the investigator needs to move on and collect additional information to allow formulation of additional hypotheses. The process of formulating hypotheses, testing them, and evaluating them continues until the problem is resolved.

It is very likely that in the coming years we will witness national and state regulations stipulating that architects, designers, facility managers, and property owners meet mandated indoor-air quality standards. The EPA, OSHA, ASHRAE, ASTM, and other organizations are currently in ongoing discussions concerning national indoor-air quality standards.

7.2.2 Thermal Comfort

Thermal comfort is a state of wellbeing that almost defies definition. It involves temperature, humidity, and air movement among other factors. Perhaps the most frequent complaint facility managers hear from building occupants is that their office space is too cold. That would appear to be an easy enough problem to resolve if it wasn't for the fact that the second most common complaint is that it's too hot. Studies show that people from different cultures generally have different comfort zones; even people belonging to the same family may feel comfortable under different conditions, and keeping everyone comfortable at the same time is an elusive goal at best. Regarding levels of thermal satisfaction, the Center for the Built Environment states, "Current comfort standards specify a 'comfort zone,' representing the optimal range and combinations of thermal factors (air temperature, radiant temperature, air velocity, humidity) and personal factors (clothing and activity level) with which at least 80 percent of the building occupants are expected to express satisfaction." This is the goal outlined by the American Society of Heating, Refrigerating and Air-Conditioning Engineers, Inc. (ASHRAE) in the industry's gold standard of comfort, Standard 55, Thermal Environmental Conditions for Human Occupancy. ASHRAE Standard 55 also specifies which thermal conditions are deemed likely to be comfortable to occupants.

We have shown that employee health and productivity are greatly influenced by the quality of the indoor environment. Poor air quality and lighting levels, off-gassing of chemicals from building materials, and the growth of molds and bacteria can all adversely affect building occupants. Sustainable design supports the wellbeing of building occupants and their desire to achieve optimum comfort by reducing indoor-air pollution. This can be achieved by applying a number of strategies such as the selection of materials with low off-gassing potential, providing access to daylight and views, appropriate ventilation strategies, and controlling lighting, humidity, and temperature levels.

Based on the above, finding the right temperature to satisfy everyone in a space is probably impossible. However, when temperature extremes – too cold or too hot – become the norm indoors, all building occupants suffer. In spaces that are either very hot or very cold, individuals must expend physiological energy to cope with the surroundings – energy that could be better utilized to focus on work and learning, particularly since research has shown that people simply don't perform as well and attendance declines in very hot and very cold workplaces.

7.2.3 Noise Pollution

Noise pollution is considered a form of energy pollution in which distracting, irritating, or damaging sounds are freely audible. Noise and vibration from sources including HVAC systems, vacuums, pumps, and helicopters can often trigger severe symptoms, including seizures, in susceptible individuals.

In the United States, regulation of noise pollution was stripped from the Environmental Protection Agency and passed on to the individual states in the early 1980s. Although two noise-control bills passed by the EPA remain in effect, the EPA can no longer form relevant legislation. Needless to say, a noisy workplace is not conducive to getting work done. What is not so apparent is that constant noise can lead to voice disorders for paraprofessionals in the office, where many employees spend time on the telephone or routinely use their voices at work,. An increasing number of teachers and paraprofessionals are seeking medical care because they are chronically hoarse.

The voice is one of the most important instruments for professionals. What's more, people living or working in noisy environments secrete increased stress hormones, and stress will cause significant distraction from the work and learning at hand. Studies have documented higher stress levels among children and staff whose schools are located on busy streets or near major airports. A number of studies have also shown that office workers consider noise pollution to be a major irritant.

Humans, whether tenants or building occupants, have a basic right to live in an environment relatively free from the intrusion of noise pollution. Unfortunately, this is not always possible in an industrialized/urbanized society that relies heavily on equipment that generates objectionable noise. Although good engineering design can mitigate noise-pollution levels to some extent, it is often not to acceptable levels, particularly if a significant number of individual sources combine to create a cumulative impact.

The City of Berkeley's Planning and Development Department states that to understand noise, one must first have a clear understanding of the nature of sound. It defines sound as pressure variations in air or water that can be perceived by human hearing; the objectionable nature of sound could be caused by its pitch or its loudness. In addition to the concepts of pitch and loudness, there are several methods to measure noise. The most common is the use of a unit of measurement called a decibel (dB). On the decibel scale, zero represents the lowest sound level that a healthy, unimpaired human ear can detect. Sound levels in decibels are calculated on a logarithmic basis. Thus, an increase of 10 decibels represents a tenfold increase in acoustic energy, and a 20-decibel increase is 100 times more intense (10×10), etc. The human ear likewise responds logarithmically, and each 10-decibel increase in sound level is perceived as approximately a doubling of loudness.

Sound is of great value; it warns us of potential danger and gives us the advantage of speech and the ability to express joy or sorrow. But sometimes sound can also prove to be undesirable. Often sound may interfere with and disrupt useful activities. Sometimes, too, sounds such as certain types of music (e.g., pop or opera), may become noise at certain times (e.g., after midnight), in certain places (e.g., a museum), or to certain people (e.g., the elderly). It is therefore a value judgment as to when sound becomes unwanted noise, which is why it is difficult to offer a clear definition of "good" or "bad" noise levels in any attempt to generalize the potential impact of noise on people.

Some sources confirm that elevated noise in the workplace or home can "cause hearing impairment, hypertension, ischemic heart disease, annoyance, sleep disturbance, and decreased school performance. Changes in the immune system and birth defects have been attributed to noise exposure, but evidence is limited." Hearing loss is potentially one of the disabilities that can occur from chronic exposure to excessive noise, but it may also occur in certain circumstances such as after an explosion. Natural hearing loss associated with aging may also be accelerated from chronic exposure to loud noise. In many developed nations the cumulative impact of noise is capable of impairing the hearing of a large fraction of the population over the course of a lifetime. Noise exposure has also been known to induce dilated pupils, elevated blood pressure, tinnitus, hypertension, vasoconstriction, and other cardiovascular impacts.

The Occupational Safety and Health Administration (OSHA) has a noise-exposure standard that is set at just below the noise threshold where hearing loss may occur from long-term exposure. The impact of noise on physical stress reactions can be readily observed when people are exposed to noise levels of 85 dB or higher. The safe maximum level is set at 90 dB averaged over eight hours. If the noise is above 90 dB, the safe exposure dose becomes correspondingly shorter. Adverse stress-type reaction to excessive noise can be broken down into two stages. The first stage is where noise is above 65 dB, making it difficult to have a normal conversation without raising one's voice. The second is the link between noise and socioeconomic conditions, which may further lead to undesirable stress-related behavior, increase workplace accident rates, or in many cases stimulate aggression and other antisocial behaviors.

Most people accept the premise that, all else being equal, it is preferable to live in a house that is quiet than in a noisy one. This implies that there is an economic penalty associated with noise exposure. However, noise is not the only factor that can influence this decision. People living along heavily traveled roads may experience greater problems with traffic safety, air pollution, exhaust odor, crime, or loss of privacy. Cumulatively, these factors can significantly depress property values. Commercial uses may be mixed in with residential uses, which may further reduce the desirability of a property. Upon considering all of these factors together, it becomes difficult to isolate the level of economic impact directly attributable to noise alone. New purchasers and renters may be unaware of how intrusive the noise can be so that the undesirability level of living in a noisy environment may increase over time. Noise levels therefore may not significantly impact property values, especially when you take into consideration all of the other variables, bearing in mind that there may be a significant negative reaction to the noise levels encountered in the future.

The prevailing sources of artificial noise pollution in today's urban communities that are outside the control of affected individuals include:

- Transportation: cars, trucks, buses, trains near railroad tracks, and aircraft near airports
- Routine activities of daily life
- Construction activity
- Industrial-plant equipment noise

The main difference between transportation and nontransportation noise sources is that a municipality can generally impose controls over the level and duration of noise at the property line of any nontransportation source of noise. Cities can only adopt noise-exposure standards for noise emanating from trucks, trains, or planes and prohibit certain land uses in areas prone to excessive noise for an intended use. Cities also play a role in enforcing state vehicle code requirements regarding muffler operation and may set speed limits or weight restrictions on certain streets. However, a city's actions are typically proactive with regard to nontransportation sources and reactive for sources outside the city's control.

Noise abatement and reduction of excessive noise exposure can be accomplished using three basic approaches:

- Reduce the noise level at the source.
- Increase the distance between the source and the receiver.
- Place an appropriate obstruction between the noise source and the receiver.

A noise wall is sometimes the only practical solution, since vehicular noise is exempt from local control and relocation of sensitive land uses away from freeways or major roads is not practical. Yet noise walls have both positive and negative aspects. On the positive side, they can reduce the noise exposure to affected persons or other sensitive uses by effectively blocking the line of sight between source and receiver. A properly sited wall can reduce noise levels by almost 10 dB, which for most people translates to being about one-half as loud as before. Unfortunately, the social, economic, and aesthetic costs of noise walls are high. While noise walls would screen the traffic, it may also block beautiful views of trees, parks, and water and may also give drivers a claustrophobic feeling of being surrounded by massive walls.

The construction cost of a noise wall is not cheap, averaging between $100 and $200 per foot. This essentially means that one mile of wall would cost between $500,000 and $1,000,000. More importantly, many people have expressed great disappointment after completion of a sound wall because, while the noise problem was reduced, it did not disappear as had been their expectation. Caltrans, for example, has a number of noise-abatement programs in place that focus on employing walls or berms to reduce noise intrusion from state and/or federal highways. Likewise, Caltrans will generally support design features that minimize local objections, providing their design standards are met. Those standards include the following:

- Walls must reduce noise levels by at least 5 dB.
- Walls must be capable of blocking truck exhaust stacks that are located at 11.5 feet above pavement levels.
- Walls constructed within 15 feet of the outside of the nearest travel lane must be built upon safety-shaped concrete barriers.

Concrete and masonry are the preferred wall materials. The effectiveness of a material in stopping sound transmission is called the transmission loss (TL).

7.2.4 Daylighting and Daylight Factor (DF)

The sun has been our main source of light and heat for millions of years, and through evolution humans have become to totally depend on it for health and survival. The world and in particular the sustainable-design movement are now returning to nature because of an increasing concern with global warming, carbon emissions, and sustainable design

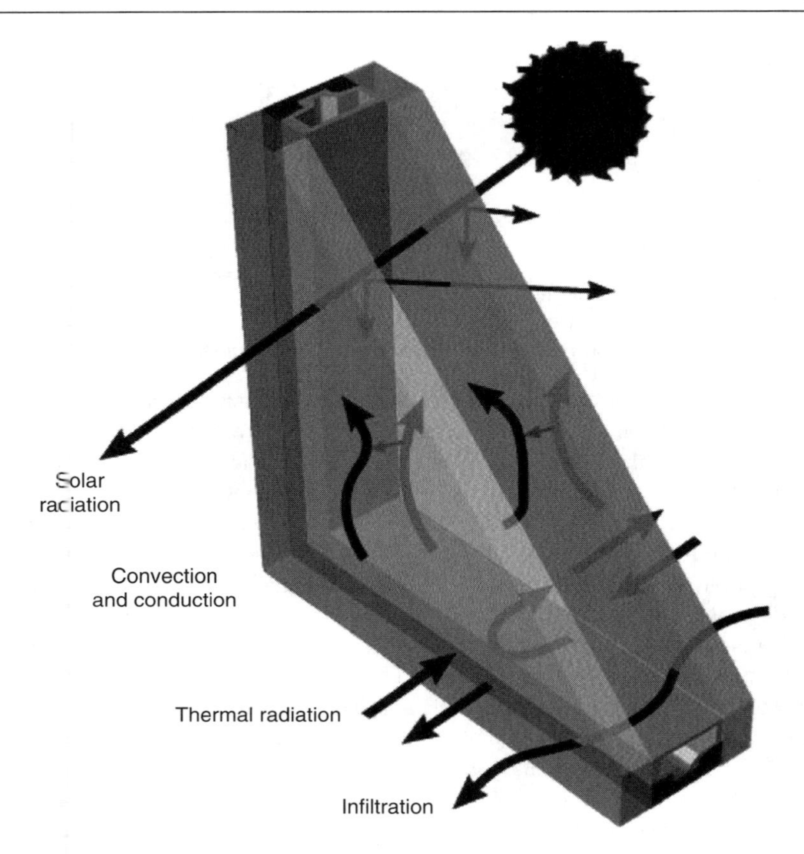

Figure 7.14 The three major types of energy flows that occur through windows: nonsolar heat losses and gains in the form of conduction, convection, and radiation; solar heat gains in the form of radiation; and air flow, both intentional as ventilation and unintentional as infiltration. *Source*: Department of Energy.

and have started to take positive steps to increase use of managed admission of natural light in both residential and nonresidential buildings. Daylighting has come to play a pivotal role in programs such as LEED™ and now has increased recognition in California's Title 24 energy code.

Craig DiLouie of the Lighting Controls Association defines daylighting as "the use of daylight as a primary source of illumination to support human activity in a space." Daylight is simply visible radiation that is generated by the sun and that can reach us in one of three different forms: direct sunlight, skylight (i.e., sunlight that has been scattered in the atmosphere), and sunlight or skylight that has been reflected off the ground. Of the three, direct sunlight is the most powerful source and has the greatest impact on our lives; it not only provides visible light but also provides ultraviolet and infrared (heat) radiation (Figure 7.14).

Area	Footcandles
Building surrounds	1
Parking area	5
Exterior entrance	5
Exterior shipping area	20
Exterior loading platforms	20
Office corridors & stairways	20
Elevators & escalators	20
Reception rooms	30
Reading or writing areas	70
General office work areas	100
Accounting/Bookkeeping areas	150
Detailed drafting areas	200

Recommended Illumination Levels

Figure 7.15 Table showing recommended illumination levels.

Evaluation of daylight quality in a room traditionally consisted of a manual average daylight-factor calculation or a computerized version of the manual method. But in light of recent technological advances, daylight design is rapidly moving forward and is now able to provide the kind of information that would accommodates all of the requirements of the daylight consultant, the architect, and the end user. The ideal package should integrate natural and electrical lighting calculations and also take into account an evaluation of the thermal impact on window design. Figure 7.15 shows a table of recommended illumination levels for various locations and functions.

The term "average daylight factor" is sometimes construed to be the average factor on all surfaces, whereas the output from most computer-based calculations reflects average daylight factors derived from a series of points on the working plane.

The daylight factor (DF) is a very common measure for expressing the daylight availability in a room. Chris Croly, a building-services engineering associate with BDP Dublin and Martin Lupton, however, states that: "The calculation of daylight factors using traditional methods becomes particularly difficult when trying to assess the effects of transfer glazing, external overhangs, or light shelves. Modern radiosity or ray-tracing calculations are now readily available and are easy to use but still generally offer results in the form of daylight factors or lux levels corresponding to a particular static external condition." They define daylight factor as "the ratio of the internal illuminance to that on a horizontal external surface located in an area with an unobstructed view of a hemisphere of the sky." The DF thus describes the ratio of outside luminance over inside luminance, expressed in a percent. The higher the DF, the more natural light is available in the room. The impact of direct sunlight on both illuminances must be considered separately and is not included. The DF can be expressed as:

$$\mathrm{DF} = 100 * E_{\mathrm{in}}/E_{\mathrm{ext}}$$

where E_{in} represents the inside illuminance at a fixed point and E_{ext} represents the outside horizontal illuminance under an overcast (CIE) or uniform sky.

7.2.4.1 Daylighting Strategies

Documented research continues to demonstrate that effective daylighting saves energy while increasing the quality of the visual environment and reduces operating costs while improving occupant satisfaction. Thus, while daylight can reduce the amount of electric light needed to adequately illuminate a workspace and therefore reduce potential energy costs, allowing too much light or solar radiation into a space can have a negative effect, resulting in heat gain and offsetting any savings achieved by reducing lighting loads. Some architectural/design firms known for their sustainable design, such as HOK and Gensler, design the majority of their buildings to be internally load-dominated, meaning that the buildings need to be cooled for most of the year. It's important to fine-tune the glazing system to harvest as much daylight as possible without creating the negative effects of too much heat gain.

Strategies for improved daylighting include the use of miniature optical light shelves, light-directing louvers, light-directing glazing, clerestories, roof monitors and skylights, light tubes, and heliostats. It is important to appreciate that whatever tools are applied in the daylighting design process, to be successful they will involve the integration of several key disciplines including architectural, mechanical, electrical, and lighting. As with sustainable design in general, these team members need to be brought into the design process early to ensure that daylighting concepts and strategies are satisfactorily implemented throughout the project (Figure 7.16).

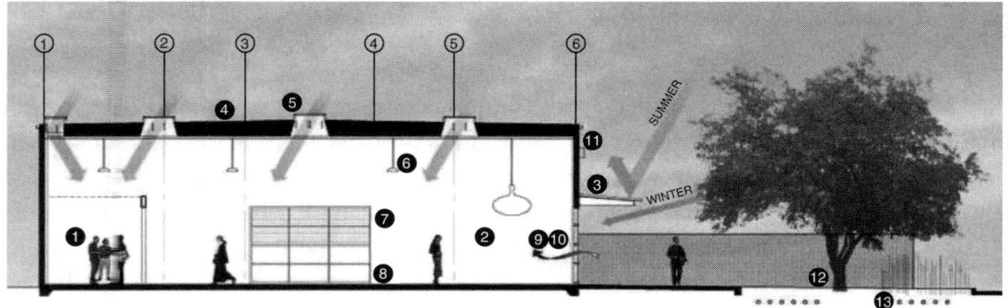

1. Private office
2. Open office
3. Sunshade with building integrated photovoltaics
4. Roof with building integrated photovoltaics
5. Skylight
6. Energy efficient & occupancy sensor controlled light fixtures
7. Electrochromic glass
8. Radiant heat floor
9. Natural ventilation
10. High performance glass
11. Reduction of outdoor light pollution
12. Water efficient landscaping
13. Ground source heat pump

Figure 7.16 Schematic drawing illustrating an integrated approach to the design of a building's various systems in the new IDeAs headquarters.
Source: Integrated Design Associates, Inc.

The application of innovative, advanced daylighting strategies and systems can significantly improve the quality of light in an indoor environment as well improving energy efficiency by minimizing lighting, heating, and cooling loads, thereby reducing a building's electricity consumption. By providing a direct link to the outdoors, daylighting helps create a visually stimulating and productive environment for building occupants, at the same time reducing as much as one-third of total building energy costs.

When light hits any surface, part of it is reflected back. This reflection is normally diffused (non-directional) and is dependent on the object's reflection. The reflection of the outside ground is usually in the order of 0.2, or 20 percent. This signifies that in addition to the direct sunlight and skylight components, there also exists an indirect component, which can make a significant contribution to the lighting inside a building, especially since the light reflected off the ground will hit the ceiling, which is usually very bright.

7.2.4.2 Daylighting and Visual Comfort

The challenge that designers often face with natural light is the fluctuations in light levels, colors, and direction of the light source. This led architects and engineers to make some unwise design decisions, which in the 60s and 70s culminated in hermetically sealed office blocks that were fully air-conditioned and artificially lit. This in turn led to a sharp increase in complaints and symptoms attributed to building-related illnesses (BRIs) and sick-building syndrome. Gregg Ander, FAIA, says: "In large measure, the art and science of proper daylighting design is not so much how to provide enough daylight to an occupied space but how to do so without any undesirable side effects."

The designer should adopt practical design strategies for sustainable daylighting design that will go a long way to achieving visual comfort by applying three primary approaches:

- Environmental: Using the natural forces that impact design, resource, and energy conservation.
- Architectonic: What has made daylighting design so difficult until now is the lack of specific design tools. Today most large architectural practices have a diverse team of consultants and design tools that enable them to undertake complex daylighting analysis, whereas the typical school or small office does not have this capability or the budget for it.
- Human factors: The impact on people and their experience; visual comfort. Designers need to achieve the best lighting levels possible while avoiding glare and high-contrast ratios. These can usually be avoided by not allowing direct sunlight to enter a workspace – e.g., through the use of shading devices.

These basic lighting approaches reflect the strategies of sustainability and thus support the larger ecological goal.

A study conducted by the Heschong Mahone Group (HMG), a California architectural consulting firm, concluded that students who received their lessons in classrooms with more natural light scored as much as 25 percent higher on standardized tests than students in the same school district whose classrooms had less natural light. This appears to confirm what many educators have suspected – i.e., that children's capacity to learn is greater under natural illumination from skylights or windows than from artificial lighting. The logical explanation given is that daylighting enhances learning by boosting the eyesight, mood, and/or health of students and their teachers.

Another investigation by HMG looked at the relationship of natural light to retail sales. The study analyzed the sales at 108 stores that were part of a large retail chain. The stores were all one story and virtually identical in layout except that two-thirds of the stores had skylights while the others did not. The study specifically focused on skylighting as a means to isolate daylight as an illumination source and avoid all of the other qualities associated with daylighting from windows. When they compared the sales figures for the various stores, they discovered a statistically compelling connection between skylighting and retail-sales performance and found that stores with skylight systems had increased sales by 40 percent – even though the design and operation of all the store sites was remarkably uniform except for the presence of skylights.

Skylights were found to have a major positive impact on the overall operation of the chain and were positively and significantly correlated to higher sales. The study showed that all other things being equal, an average nonskylit store in the chain would likely increase its sales by an average of about 40 percent just by adding skylights. HMG professes that this was determined with 99 percent statistical certainty.

Many architects are now specifying high-performance glass with spectrally selective coatings that only allow visible light to pass through the glass while keeping out the infrared wavelength. This eliminates most of the infrared and ultraviolet radiation while allowing the majority of the visible-light spectrum through the glass. But even with high-performance glass, much of the light can be converted to heat. Glass with high visible-light transmittance still allows light energy into a building, and when this light energy hits a solid surface, it is absorbed and reradiated into the space as heat.

The combination of daylighting and efficient electric-lighting strategies can provide substantial energy savings. The building's planning module can often give indications as to how best to organize the lighting. In any case, the lighting system must correlate to the various systems in place, including structural, curtain wall, ceiling, and furniture. Likewise, initial lighting costs may rise when designing for sustainability and implementing energy-efficient strategies. These energy-saving designs may require items such as dimmable ballasts, photocells, and occupancy sensors, all of which are not typically covered in most initial project budgets (Figure 7.17). However, this could easily be compensated for by reducing building-system and load costs and even possibly reallocating some of this money to the lighting budget.

One of the main attributes of daylight is that it enhances the psychological value of space. Likewise, the introduction of daylight into a building reduces the need for electric lighting during the day while helping to provide a connection between indoor spaces and the outdoors for the building occupants. However, natural light is not without its negatives. These include glare, overheating, variability, and privacy issues. The designer needs to find ways to increase the positive aspects of using natural light in buildings while at the same time minimizing the negative.

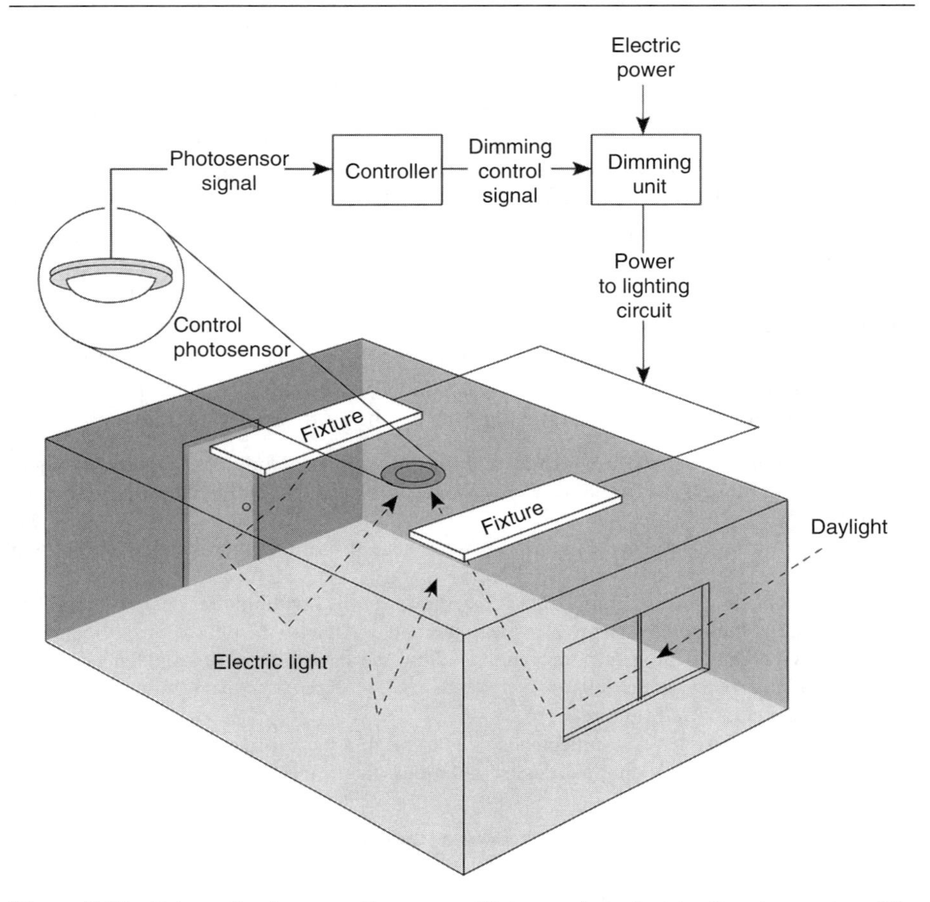

Figure 7.17 Schematic diagram of a room utilizing a photoelectric dimming system. The ceiling-mounted photosensor reads both electric light and daylight within the space and adjusts the electric lighting as required to maintain the design level of total lighting.
Source: Ernest Orlando, Lawrence Berkeley Laboratory.

Addressing glare means keeping sunlight out of the field of view of building occupants while protecting them from disturbing reflections. Addressing overheating means adding appropriate exterior shading, filtering incoming solar radiation, or even using passive control means such as thermal mass. Furthermore, addressing the variability and privacy issues requires creative ways to block or alter light patterns and compensate with other light sources.

It is only recently that daylighting strategies have started to be considered at an early stage of a building design. This is because tools that could predict the performance of advanced daylighting strategies were not available to the designer. For best results, daylight design strategies should be in place at the outset of the design process and the daylighting consultant should be involved very early in the process. The data output from the daylighting studies could be extremely useful for fine-tuning and finalizing the building's orientation, massing, space planning, and interior finishes.

Innovative daylighting systems are designed to redirect sunlight or skylight to areas where it is most needed while avoiding glare. These systems use optical devices that initiate reflection, refraction, and/or use the total internal reflection of sunlight and skylight. And with today's advancing technology daylighting systems can be programmed to actively track the sun's movement or passively control the direction of sunlight and skylight.

The financial and competitive pressures of powerful market forces are driving some owners and design teams to seek architectural solutions such as the utilization of highly glazed, transparent façades. While these trends may offer clear potential benefits, such approaches also expose owners to real risks and costs as well. The general interest in potential benefits from these design solutions can be summarized as follows:

- Most building owners desire daylight and find concepts and buildings that employ highly transparent façades preferable to the dark-tinted or reflective buildings of the 1970s and 1980s.
- Building owners are generally aware of the potential health and productivity benefits of daylight.
- The evident shift toward highly glazed façades is coupled with interior designs that reflect the desire of building owners to provide views and daylight to their employees. Open plans and low-height partition furniture layouts allow the daylit zones to be extended from a conventional 10 to 15 feet (3.0 to 4.6 m) depth to a 20- (6.1 m) or even 30-foot (9.1 m) depth from the window wall (Figure 7.18).
- The increased use of low-reflectance, higher-brightness flat-screen LCD monitors has allowed architects to employ design solutions that involve increasing daylight and

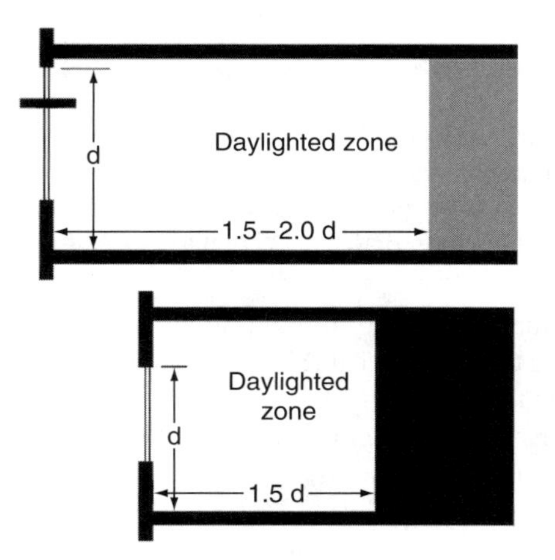

Figure 7.18 A rule of thumb for daylight penetration with typical depth and ceiling height is 1.5 times head height for standard windows and 1.5 to 2.0 times head height with light shelf for south-facing windows under direct sunlight.
Source: Ernest Orlando, Lawrence Berkeley Laboratory.

luminance levels within buildings. But to take full advantage of natural daylight and avoid potential dark zones, it is critical that the lighting designer plans the lighting circuits and switching schemes in relation to the building's fenestration system.

At the same time there are potential risks associated with highly glazed façades:

- Adequate tools are not always available to reliably predict the thermal and optical performance of components and systems and to assess environmental quality.
- Increased sun penetration and excessive brightness levels that exceed good practice may heighten visual discomfort.
- Elevated cost of automated shading systems; possible need to purchase lighting controls utilizing dimming ballasts and difficulty in commissioning systems after installation.
- High cost and technical difficulty of reliably integrating dimmable lighting and shading controls with each other and with building automation systems to ensure effective operation over time.
- Buildings utilizing transparent glazing generally use greater cooling loads and cooling energy, which has the potential for thermal discomfort (Figure 7.19).
- Uncertainty of occupant behavior with the use of automated, distributed controls in open landscaped office space and the potential for conflict between different needs and preferences.

To cash in on the potential benefits and minimize possible risks, there is a growing recognition that at least in workspaces (as distinct from corridors, lobbies, etc.), large glazed spaces require much better sun and glare control. Appropriate solutions must be delivered by systems that can rapidly respond to exterior climate and interior needs.

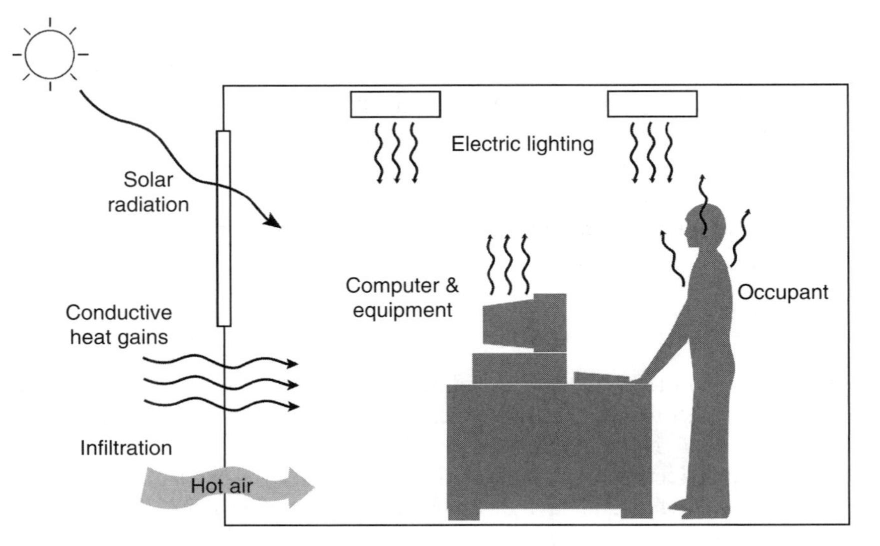

Figure 7.19 Various sources of cooling loads.
Source: Ernest Orlando, Lawrence Berkeley Laboratory.

One of the challenges facing manufacturers is how to provide such needed functionality at lower cost and risk to owners. Due to their various advantages and disadvantages, lighting consultants usually recommend the use of switching for spaces with nonstationary tasks such as corridors and continuous dimming for spaces where users perform stationary tasks, such as offices. It has been shown that daylight harvesting using continuous dimming equipment automatically controlled by a photosensor can generate 30 to 40 percent savings in lighting energy consumption, thereby significantly reducing operating costs for the owner.

7.2.4.3 Shades and Shade Controls

To achieve the greatest benefit of harvesting daylight, it is essential to implement a shading strategy tailored to the building. In hot climates, exterior shading devices have been found to work well to both reduce heat gain and diffuse natural light prior to entering the workspace (Figure 7.19). Examples of such devices include light shelves, overhangs, vertical louvers, horizontal louvers, and dynamic tracking or reflecting systems. Thus, for example, exterior shading of the glass can eliminate up to 80 percent of the solar heat gain. Shades and shade control strategies are based upon the perception that occupants of commercial buildings typically prefer natural light to electric light. Shade-system goals would normally include:

- Maximizing use of natural light within a glare-free environment.
- Avoid direct solar radiation on occupants through interception of sunlight penetration.
- Facilitate occupant connectivity with the outdoors through increased glazing and external views.
- Provide manual-override capability for occupants.

The overall determination is to ensure that the shades are operational as much of the time as is possible without causing thermal or visual discomfort (Figure 7.20). Thermal comfort is maintained by solar tracking and the design of the external sun screens. Visual comfort is affirmed by managing the luminance on the window wall. The specifications for the manual-override system are based upon postoccupancy surveys of office-building occupants with automated shade systems. The prime recorded complaint in these studies was an occupant's inability to control the operation of a shade device when required.

7.2.5 Views

Recent studies show that windows providing daylight and ample views can dramatically affect building occupants' mental alertness, productivity, and psychological wellbeing. David Hobstetter, a principal in Kaplan-McLaughlin-Diaz, a San Francisco-based architectural practice, reaffirms this, saying, "Dozens of research studies have confirmed the benefits of natural daylight and views of greenspace in improving a person's productivity, reducing absenteeism, and improving health and wellbeing."

In many countries around the world views, whether high-rise or otherwise, are normally considered mere perks, but recent research suggests that the view from a

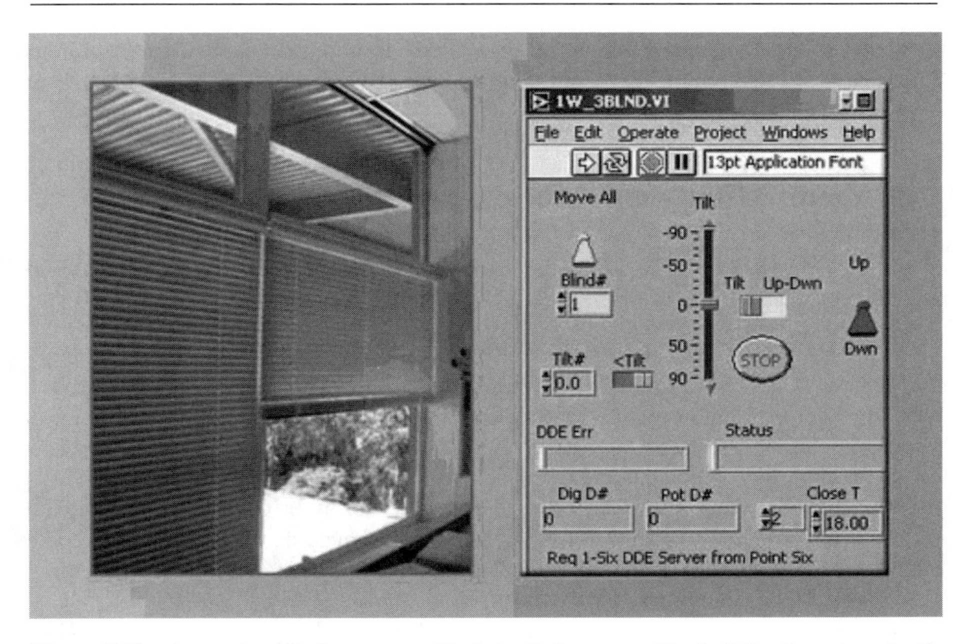

Figure 7.20 A venetian-blind system at a Berkeley Laboratory office building is equipped with a "virtual instrument" panel for IBECS control of blind settings.
Source: HPCBS.

window may be even more important than the daylight it admits. The California Energy Commission's 2003 study of workers in the Sacramento Municipal Utility District's call center found that better views were consistently associated with better performance: "Workers with good views were found to process calls 7 percent to 12 percent faster than colleagues without views. Workers with better views also reported better health conditions and feelings of wellbeing, while their counterparts reported higher fatigue." Another revealing study noted that computer programmers with views spent 15 percent more time on their primary task, while workers without views spent 15 percent more time talking on the phone or to one another.

Though some educators are of the opinion that views out of windows may be unnecessarily distracting to students, the CEC's 2003 study of the Fresno school district found that a varied view out of a window that included vegetation or human activity and objects in the far distance supported better learning. Such findings confirm results of earlier research, such as a 1984 hospital study that concluded that postoperative patients with a view of vegetation took far fewer painkillers and experienced faster recovery times than patients looking at concrete walls.

Most building occupants relish contact with the outside world, even if only through a windowpane, and landscapes not surprisingly are preferred to cityscapes. Researchers have concluded that views of nature improve attention spans after extended mental activity has drained a person's ability to concentrate. Among the main building types that can most benefit from the application of daylighting are educational buildings such as schools, administrative buildings such as offices, storage facilities such as warehouses, and maintenance facilities.

7.3 Ventilation and Filtration

Throughout history buildings, whether a Babylonian palace, an Egyptian temple, or a Roman castle, were ventilated naturally using either *badgeer* (also known as *malqafs*) – wind shafts or towers – or some other innovative method (Figure 7.21), since mechanical systems did not exist at the time. Andy Walker of the National Renewable Energy Laboratory says, "Wind towers, often topped with fabric sails that direct wind into the building, are a common feature in historic Arabic architecture." The incoming air is often routed past a fountain to achieve evaporative cooling as well as ventilation. At night the process is reversed, and the wind tower acts as a chimney to vent room air." It is not surprising that with today's increased awareness of the cost and environmental impacts of energy use, natural ventilation has once again come to the fore and become an increasingly attractive method for reducing energy use and cost and for providing acceptable indoor environmental quality.

Natural ventilation systems utilize the natural forces of wind and buoyancy – i.e., pressure differences – to move fresh air through buildings. These pressure differences can be a result of wind, temperature differences, or differences in humidity. The amount and type of ventilation achieved will depend to a large extent on the size and placement of openings in the building. Inadequate ventilation is one of the main culprits of indoor pollutant levels by not bringing in enough outdoor air to dilute emissions from indoor sources and by not carrying indoor-air pollutants out of the home or workplace.

7.3.1 Ventilation and Ductwork

Ventilation is vital for the health and comfort of building occupants. It is specifically needed to reduce and remove pollutants emitted from various internal and external sources. Good design combined with optimum airtightness is prerequisite to ensuring healthy air quality, occupant comfort, and energy efficiency. Sufficient air supply and movement can be tested and analyzed to determine the efficiency of an HVAC system. Regular maintenance of ductwork is pivotal to achieving both better indoor environment and system stability. Ductwork can be evaluated, cleaned, and sealed to prevent air-flow and quality issues. All ductwork should be analyzed by a professional trained and certified by the National Air Duct Cleaning Association.

7.3.2 Air Filtration

It is unfortunate that to date no federal standards have yet been adopted for air-filter performance. Air-cleaning filters are designed to remove pollutants from indoor air, so by properly filtering your facility's air of harmful particles, you can improve the indoor

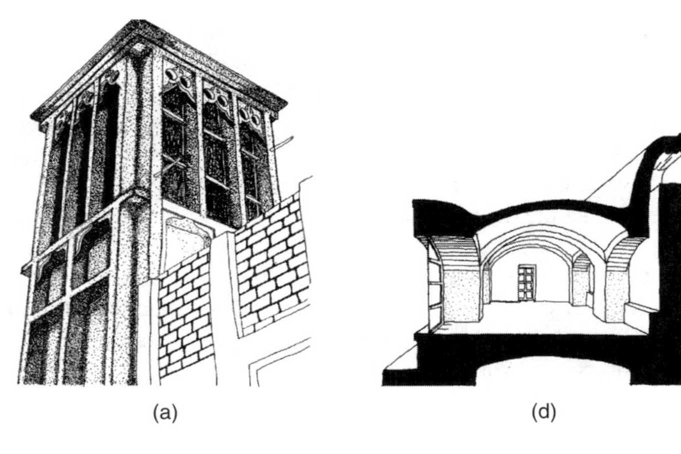

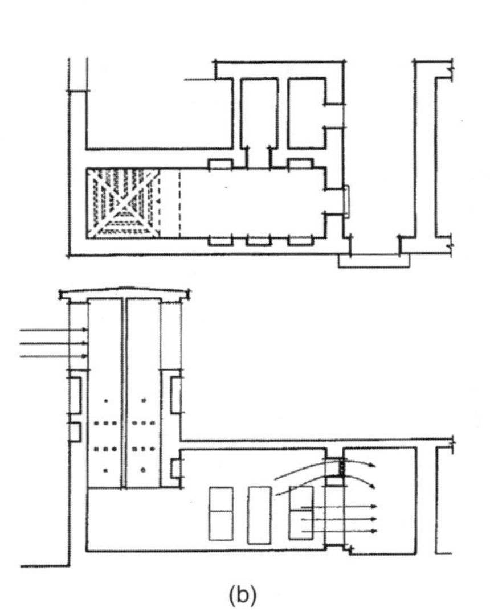

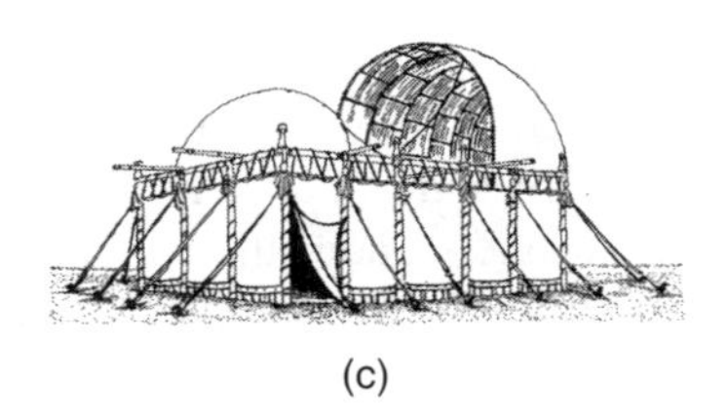

Different types of traditional wind catches.

Figure 7.21 Drawings depicting various types of wind catchers (*badgeer/malqaf*) used in traditional and ancient architecture. (a) Multidirectional traditional Dubai wind catcher. (b) Plan and section of Dubai wind catcher. (c) Ancient Assyrian wind catcher. (d) Section through traditional wind scoop. (e) Traditional Pakistani wind catchers.

environment and breathe cleaner air. Proper filtration removes dirt, dust, and debris from the air you breathe. It also reduces pollen and other allergens, which can cause asthmatic attacks and allergic reactions. But while air-cleaning devices may help to control the levels of airborne allergens, particles, and in some cases gaseous pollutants in a facility, they may not decrease adverse health effects from indoor-air pollutants.

There are several kinds of air-cleaning devices to choose from – some are designed to be installed inside the ductwork of a facility's central heating, ventilating, and air-conditioning (HVAC) system to clean the air in the whole facility. Other types include portable room air cleaners, which are designed to be used to clean the air in a single room or specific areas and are not intended for complete facility filtration. There are various types of air filters currently on the market, such as: mechanical filters, electronic filters, hybrid filters, gas-phase, and ozone generators.

7.3.3 Air Purification

Some sources, such as building materials, furnishings, and household products including air fresheners, release pollutants more or less continuously. Other sources, related to activities carried out in the home or workplace, release pollutants intermittently. These include smoking, unvented or malfunctioning stoves, furnaces, space heaters, solvents in cleaning and hobby activities, paint strippers in redecorating activities, and cleaning products and pesticides. High pollutant concentrations can remain in the air for long periods after some of these activities have ended if action is not taken.

While air filtration removes particulate, air purification is required to remove what a filter does not, such as odors and gases. Chemicals in paints, carpets, and other building materials (i.e., VOCs) are harmful to building occupants and should be removed through air purification. There is also an increasing concern regarding the presence of biological infectious agents, and air purification on a regular basis is one way to address these potential problems.

7.3.4 Amount of Ventilation

Outdoor air normally enters and leaves a building through various means, particularly infiltration, natural ventilation, and mechanical ventilation. Outdoor air can infiltrate a building through openings, joints, and cracks in walls, floors, and ceilings and around windows and doors. Natural ventilation involves air moving through opened windows and doors. Air movement associated with infiltration and natural ventilation is a consequence of air-temperature differences between the indoor and outdoor air and of wind. When insufficient outdoor air enters a home, pollutants can accumulate to a degree that they can pose health and comfort problems.

In the event that natural ventilation is insufficient to achieve good air quality, there are a number of mechanical ventilation devices, from outdoor-vented fans that will intermittently remove air from rooms such as bathrooms and kitchens to air-handling systems that utilize fans and duct systems to continuously remove indoor air and distribute filtered and conditioned outdoor air to strategic points throughout the building.

The rate at which outdoor air replaces indoor air is known as the air-exchange rate. Insufficient air infiltration, natural ventilation, or mechanical ventilation means that the air-exchange rate is low and can result in rising pollutant levels.

Sometimes residents or occupants are in a position to take appropriate action to improve the indoor-air quality of a space by removing the source, altering an activity, unblocking an air-supply vent, or opening a window to temporarily increase the ventilation; in other cases, however, the building owner or manager is the only person in a position to remedy the problem. Building management should be prevailed upon to follow guidance in EPA's IAQ Building Education and Assessment Model (I-BEAM). I-BEAM expands and updates EPA's existing Building Air Quality guidance and is considered to be a complete state-of-the-art program for managing IAQ in commercial buildings. Building management should also be encouraged to follow guidance in EPA and NIOSH's Building Air Quality: A Guide for Building Owners and Facility Managers. The BAQ guidance is available as a downloadable PDF file.

7.3.5 Ventilation Improvements

As previously stated, another approach to lowering the concentrations of indoor-air pollutants is to increase the amount of outdoor air coming indoors. Many heating and cooling systems, including forced-air heating systems, do not mechanically bring fresh air into the house. This can often be addressed by opening windows and doors or running a window air conditioner with the vent control open to increase the outdoor ventilation rate. In residences, local bathroom or kitchen fans that exhaust outdoors can be used to remove contaminants directly from the room where the fan is located while increasing the outdoor-air ventilation rate.

Good ventilation is especially important when undertaking short-term activities that can generate high levels of pollutants, such as painting, paint stripping, or heating with kerosene heaters. Such activities should preferably be executed outdoors whenever possible.

Advanced designs of new homes have recently come on the market that feature mechanical systems that bring outdoor air into the home as well as energy-efficient heat-recovery ventilators (also known as air-to-air heat exchangers).

The following design recommendations can help achieve better ventilation in buildings:

- Naturally ventilated buildings should preferably be narrow, as wide buildings pose greater difficulty in distributing fresh air to all areas using natural ventilation.
- Occupants should be able to operate window openings.
- Use of mechanical cooling is advised in hot, humid climates.
- Determine whether an open- or closed-building ventilation approach potentially offers the best results. A closed-building approach will work better in hot, dry climates, where there is a large diurnal temperature range from day to night. An open-building approach is more effective in warm and humid areas, where the temperature difference between day and night is relatively small.
- Consideration should be given to the use of fan-assisted cooling strategies. Andy Walker says that "Ceiling and whole-building fans can provide up to 9 degrees F effective

temperature drop at one-tenth the electrical energy consumption of mechanical air-conditioning systems."

- When possible, provide ventilation to the attic space, as this greatly reduces heat transfer to conditioned rooms below. Ventilated attics have been found to be approximately 30 degrees F cooler than unventilated attics.
- Maximize wind-induced ventilation by siting buildings so that summer wind obstructions are minimal.

7.3.6 Air Cleaners

There are a variety of types and sizes of air cleaners presently on the market, ranging from relatively inexpensive tabletop models to larger, more sophisticated and expensive systems. Some air cleaners are highly effective at particle removal, while others, including most tabletop models, are much less so. It should be noted that air cleaners are generally not designed to remove gaseous pollutants.

An air cleaner's effectiveness is expressed as a percentage efficiency rate. It depends on how well it collects pollutants from indoor air and how much air it draws through the cleaning or filtering element. The latter is expressed in cubic feet per minute. Even an efficient collector that has a low air-circulation rate will not be effective; neither will an air cleaner with a high air-circulation rate but a less efficient collector. The long-term performance of any air cleaner relies largely on maintaining it in accordance with the manufacturer's directions.

Another critical factor in determining the effectiveness of an air cleaner is the level and strength of the pollutant source. Tabletop air cleaners, in particular, may not be capable of adequately reducing amounts of pollutants from strong nearby sources. Persons who are sensitive to particular pollutant types may find that air cleaners are useful mostly when used in conjunction with collaborative efforts to remove the source.

The EPA does not currently recommend using air cleaners to reduce levels of radon and its decay products. The effectiveness of these devices is questionable because they only partially remove the radon decay products and do not diminish the amount of radon entering the home. EPA is planning to undertake additional research on whether air cleaners are or could become a viable means of reducing the health risk from radon.

For most indoor-air-quality problems, source control is the most effective solution. The EPA issued "Ozone Generators that are Sold as Air Cleaners" to provide accurate information regarding the use of ozone-generating devices in indoor-occupied spaces. This information is based on the most credible scientific evidence currently available. The document explains what air-duct cleaning is, offers guidance to help consumers decide whether to have the service performed, and provides useful information on choosing a duct cleaner, determining if duct cleaning was done correctly, and how to prevent contamination of air ducts.

7.3.7 Ventilation Systems

Mechanical ventilation systems in large commercial buildings are normally designed and operated to heat and cool the air as well as to draw in and circulate outdoor air.

On the other hand, ventilation systems themselves can be a source of indoor pollution and contribute to indoor-air problems if they are poorly designed, operated, or maintained. They can sometimes spread harmful biological contaminants that have steadily multiplied in cooling towers, humidifiers, dehumidifiers, air conditioners, or inside surfaces of ventilation ducts. For example, problems arise when, in an effort to save energy, ventilation systems are incorrectly programmed and bring in inadequate amounts of outdoor air. Other examples of inadequate ventilation occur when the air supply and return vents within a space are blocked or placed in a manner that prevents the outdoor air from reaching the breathing zone of building occupants. Improper location of outdoor-air intake vents can also bring in contaminated air, particularly from automobile and truck exhaust, fumes from dumpsters, boiler emissions, or air vented from bathrooms and kitchens.

For mechanically ventilated spaces designers should refer to ASHRAE Standard 55-1992, Addenda 1995, "Thermal Environmental Conditions for Human Occupancy." For naturally ventilated spaces, refer to California High Performance Schools (CHPS) Best Practices Manual Appendix C, "A Field Based Thermal Comfort Standard for Naturally Ventilated Buildings."

7.4 Building Materials and Finishes Emittance Levels

Several studies over the years have investigated the impact of pollution emitted by building materials on indoor-air quality and related it to ventilation requirements. However, there has been a lack of systematic experiments in which building materials are initially ranked according to their pollution strength and the impact on the indoor-air quality of using these materials in real rooms analyzed. Such studies would allow us to quantify the extent to which using low-polluting building materials would reduce the energy needed for ventilation of buildings without compromising indoor-air quality. One of the primary objectives of an ongoing research project is to quantify this energy-saving potential based on the effects on the perceived air quality.

7.4.1 Unhealthy Building Materials

Healthy Building Network (HBN) identifies the primary toxic building materials that have unacceptably high VOC emittance levels.

HBN singled out polyvinyl chloride, or vinyl, for elimination because of its uniquely wide and potent range of chemical emissions throughout its life cycle. It is virtually the only material that requires phthalate plasticizers, frequently includes heavy metals, and emits large amounts of VOCs. In addition, during manufacture it produces a large quantity of highly toxic chemicals including dioxins (the most potent carcinogens measured), vinyl chloride, ethylene dichloride, and PCBs, among others. Moreover, when burned at the end of its useful life, whether in an incinerator or landfill fire, it releases hydrochloric acid and more dioxins. It is therefore prudent if not imperative to avoid products made with PVC.

Volatile organic compounds (VOCs) essentially consist of thousands of different chemicals, such as formaldehyde and benzene, that evaporate readily into the air. Depending on the level of exposure, they can cause dizziness; headaches; eye, nose, and throat irritation; asthma; and in some case, cancer; they induce longer-term damage to the liver, kidney, and nervous system,\ and stimulate higher sensitivity to other chemicals.

When dealing with wet products such as paints, adhesives, and other coatings, ensure that the products contain no or low VOC. Look for the Green Seal when using certified paints or paints with less than 20 g/l VOCs. For adhesives and coatings make sure they are SCAQMD (South Coast Air Quality Management District)-compliant.

For flooring and carpet, wall covering, ceiling tiles, and furniture, it is advisable to use only CA 01350-compliant products. A number of programs currently use the CA 01350 testing protocol to measure the actual levels of individual VOCs emitted from the material and compare it to allowable levels set by the state of California. These include CHPS, CRI's Green Label Plus, SCS's Indoor Advantage, RFCI's FloorScore, and GreenGuard's Schools and Children. Try to avoid flooring that requires waxing and stripping, processes that will release more VOCs than the original material.

Ensure that all composite wood products and insulation have no added formaldehyde. The CA 01350 program has set limits on formaldehyde emissions; however, for these products there are options that completely exclude the presence of formaldehyde.

DEHP and other phthalates have attracted considerable adverse publicity for their use in PVC medical products and in toys, and concerns have been raised about their impact on the development of young children. Phthalates are, however, also used widely in flexible PVC building materials and have been linked to bronchial irritation and asthma. It is important therefore to avoid using products with phthalates (including PVC).

Even though heavy metals are known to be hazardous to your health, they continue to be used for stabilizers or other additives in building materials. Lead, mercury, and organotins are all known potent neurotoxins that are particularly damaging to the brains of fetuses and growing children. Cadmium is a carcinogen and can cause a variety of kidney, lung, and other damage. Look for products that do not contain heavy metals.

The use of **halogenated** flame retardants (HFRs) in many fabrics, foams, and plastics is known to have saved many lives over the years. However, these retardants, including PBDEs and other brominated flame retardants (BFRs), have been found to disrupt thyroid and estrogen hormones, which can cause developmental effects such as permanent changes to the brain and to the reproductive system. Being persistent and bioaccumulative, they are rapidly accumulating to dangerous levels in humans and have now become the subject of an increasing number of bans and phaseouts. Avoid all products that use halogenated flame retardants.

Numerous treatments for fabric and some building materials have been based on perfluorocarbons (PFCs) that – like HFRs – are characteristically highly persistent and bioaccumulative and hence are concentrating at alarming levels in humans. PFOA is a major component of treatment products such as Scotchguard, Stainmaster, Teflon, and Gore-Tex and has been linked to a range of developmental and other adverse health effects. Avoid all products that are treated with a PFC-based material.

Here is a checklist for healthy materials:

- Avoid PVC (polyvinyl chloride, vinyl)
- Low or no VOC (volatile organic compounds)
- CA 01350-compliant
- Ensure no added formaldehyde
- Ensure no phthalates or heavy metals (lead, mercury, cadmium, organotins) are used
- Ensure no HFRs (PBDEs, BFRs, and other halogenated flame retardants) are used
- Ensure no PFCs (perfluorocarbons) are present

7.4.2 Resources for Locating Healthy Building Materials

Because of the great diversity of materials used in the construction and manufacturing industries, we are unable to produce a single building-materials list or certification that covers all of the relevant health and environmental issues. For example, we find that of the various programs listed below that certify products to meet the CA 01350 VOC-emissions standards, none also screen for phthalates and flame retardants, and you will find many PVC/vinyl products on those lists. Furthermore, some 01350 VOC programs are managed by trade associations. It is prudent to always ask to see the actual lab certification of 01350 VOC emissions when first screening a material and ask about PVC/vinyl, heavy metals, HFRs, and PFCs.

HBN PVC Alternatives Database is a CSI-prepared listing of PVC-free alternatives for a wide range of different building materials. Website: www.healthybuilding.net/pvc/alternatives.html.

Paints and coatings that meet the GreenSeal VOC content standards are listed in the Green Seal Certified Products list; these materials do not contain certain excluded chemicals and meet typical LEED™ performance requirements.

EcoLogo is an environmental standard and certification mark. EcoLogo was founded in 1988 by the Government of Canada and is now recognized internationally. The EcoLogo Program provides customers – public, corporate, and consumer – with assurance that the products and services bearing the logo meet stringent standards of environmental leadership. The EcoLogo Program is a Type I ecolabel as defined by the International Organization for Standardization (ISO) and is one of two such programs in North America that have been successfully audited by the Global EcoLabeling Network (GEN) as meeting ISO 14024 standards for ecolabeling. Among the products EcoLogo certifies are carpet, paint, and adhesives.

CHPS (Collaborative for High Performance Schools) maintains a table listing products that have been certified by the manufacturer and an independent laboratory to meet the CHPS Low-Emitting Materials criteria, Section 01350, for use in typical classrooms. Certified materials include adhesives, sealants, concrete sealers, acoustical ceilings, wall panels, wood flooring, composite wood boards, resilient flooring, and carpet. Note: This list also includes paint listings, but CA 01350 is not yet a replacement for low VOC screening. The website is: www.chps.net/manual/lem_table.htm.

The GreenLabel Plus designation means that the Carpet and Rug Institute (a trade association) assures customers that approved carpet products meet stringent requirements for low chemical emissions and furthermore certifies that these carpets and adhesives meet CA 01350 VOC requirements. Thus any architect, interior designer, government specifier, or facility administrator who is committed to using green-building products can assume that the Green Label Plus logo signifies that the carpet product has been tested and certified by an independent laboratory and meets stringent criteria for low emissions.

FloorScore Scientific Certification Systems certifies for the Resilient Floor Covering Institute (a trade association) that resilient flooring meets CA 01350 VOC requirements. Website is: www.scscertified.com/iaq/floorscore_1.html.

Air Quality Sciences certifies for GreenGuard that furniture and indoor finishes meet the lower of CA 01350 VOC or 1/100 of TLV. The GreenGuard Environmental Institute (GEI) has formulated performance-based standards to define goods with low chemical and particle emissions for use indoors; these primarily include building materials, interior furnishings, furniture, cleaning and maintenance products, electronic equipment, and personal-care products. The standard establishes certification procedures including test methods, allowable emissions levels, product-sample collection and handling, testing type and frequency, and program-application processes and acceptance. GEI now certifies products across multiple industries.

No listings are yet screened for BFRs, HFRs, or PFCs. The flame retardants are added to plastics, particularly fabrics and foams. PBDEs are the most widely used. All halogen-based flame retardants, however, are likely to be problematic.

7.5 Best Practices for IEQ

Indoor environmental quality is a critical component of sustainable buildings. As we have already seen, numerous documented studies have confirmed the effect of the indoor environment on the health and productivity of building occupants. Ventilation, thermal comfort, air quality, and access to daylight and views are all important factors that play a significant role in determining indoor environmental quality.

The Architectural and Transportation Barriers Compliance Board (Access Board), which is an independent federal agency devoted to accessibility for people with disabilities, contracted with the National Institute of Building Sciences (NIBS) to establish an Indoor Environmental Quality Project as a first step in implementing an action plan. NIBS issued a project report on indoor environmental quality in July 2005 that revealed that a growing number of people in the United States suffer a range of debilitating physical reactions caused by exposures to everyday materials and chemicals found in building products, floor coverings, cleaning products, fragrances, and others products. This condition is known as multiple chemical sensitivity (MCS). The range and severity of these reactions are varied. In addition, the Access Board received numerous complaints from other people who reported adverse reactions from exposures to electrical devices and frequencies, a condition referred to as electromagnetic sensitivity (EMS).

In response to these concerns, the Access Board sponsored a study on ways to tackle the problem of indoor environmental quality for persons with MCS and EMS as well as for the general population. In conducting this study for the Board, NIBS brought together a number of interested parties to explore the relevant issues and to develop an appropriate action plan. While the focus of the project was on commercial and public buildings, many of the issues addressed and recommendations offered can also be applied to residential settings. The report includes, among other things, recommendations on improving indoor environmental quality that address building products, materials, ventilation, and maintenance.

A steering committee was established for the project, which included representation from MCS and EMS organizations, experts on indoor environmental quality, and representatives from the building industry. Committee members examined various methods and strategies for collecting and disseminating information, focusing on specific areas, increasing awareness of relevant issues, encouraging extended project participation, identifying potential partners for further study and outreach, and developing practical recommendations for best practices.

Below are some of the steps and best practices that can be applied to ensure good IEQ:

- Conduct a facility-wide IEQ survey/inspection, noting odors, unsanitary conditions, visible mold growth, staining, presence of moisture in inappropriate places, poorly maintained filters, personal air cleaners, hazardous chemicals, uneven temperatures, and blocked vents.
- Determine operating schedule and design parameters for the HVAC system and ensure adequate fresh air is provided to prevent the development of indoor-air-quality problems and to contribute to the comfort and wellbeing of building occupants. Maintain complete and up-to-date ventilation-system records.
- Ensure that appropriate preventive maintenance (PM) is performed on HVAC system including but not limited to outside-air intakes, inside of air-handling unit, distribution dampers, air filters, heating and cooling coils, fan motor and belts, air-distribution ducts and VAV boxes, air humidification and controls, and cooling towers.
- Manage and review processes with potentially significant pollution sources such as renovation and remodeling, painting, shipping and receiving, pest control, and smoking. Ensure adequate controls are instituted on all renovation and construction projects and evaluate control impacts on IEQ.
- Control environmental tobacco smoke by prohibiting smoking within buildings or near building entrances. Designate outdoor smoking areas at least 25 feet from openings serving occupied spaces and air intakes.
- Control moisture inside buildings to inhibit mold growth, particularly in basements. Dehumidify when necessary and respond promptly to floods, leaks, and spills. Use of porous materials in basements should be monitored and restricted whenever possible.
- When mold growth is evidenced, immediate action should be taken to remediate.

- Choose low-emitting materials with minimal or no volatile organic compounds. This particularly applies to paints, sealants, adhesives, carpet and flooring, furniture, composite wood products, and insulation.
- Monitor carbon dioxide (CO_2), and install carbon-dioxide and air-flow sensors in order to provide occupants with adequate fresh air when required.
- To maintain occupants' thermal comfort, include adjustable features such as thermostats or operable windows.
- Window size, location, and glass type should be selected to provide adequate daylight levels in each space.
- Window sizes and positions in walls should be designed to take advantage of outward views and have high visible transmittance rates (greater than 50 percent) to ensure maximum outward visibility.
- Incorporate design strategies that maximize daylight and views for building occupants' visual comfort.
- Educate cleaning staff regarding use of appropriate methods and products, cleaning schedules, material storage and use, and trash disposal.
- A process for complaint procedures should be established and IEQ complaints promptly responded to.
- Discuss with occupants how they can participate in maintaining acceptable indoor environmental quality.
- Permanent entryway systems such as grilles or grates should be installed to prevent occupant-borne contaminants from entering the building.
- A construction IAQ management plan should be in place so that during construction materials are protected from moisture damage and particulates controlled through the use of air filters.

There are a number of suggested IEQ-related recommendations that tenants should follow to ensure that a healthy indoor environment is maintained for all building occupants:

- The use of air handlers during construction must be accompanied by the use of filtration media with a minimum efficiency reporting value (MERV) of 8 at each return grille as determined by ASHRAE 52.2-1999.
- Replace all filtration media immediately prior to occupancy, and conduct when possible a minimum two-week flush-out with new filtration media and 100 percent outside air after construction is completed and prior to occupancy of the affected space.
- Contractors should notify property manager 48 hours prior to commencement of any work that may cause objectionable noise or odors.
- Protect stored on-site materials and installed absorptive materials from moisture damage.
- All applied adhesives must meet or exceed the limits of the South Coast Air Quality Management District Rule #1168. Sealants used as fillers must meet or exceed Bay Area Air Quality Management District Reg. 8, Rule 51.
- Ensure that all paints and coatings meet or exceed the VOC and chemical component limits of GreenSeal requirements.
- Ensure that carpet systems meet or exceed the Carpet and Rug Institute Green Label Indoor Air Quality Test Program.
- Composite wood and agrifiber products should not contain any added urea-formaldehyde resins.

- Contractors should provide protection and barricades where needed to ensure personnel safety and should comply with OSHA at a minimum.

In conclusion, it should be remembered that the air quality in your building is one of the most important factors in maintaining employee productivity and health. Towards this end, IEQ monitoring will help minimize tenant complaints of building-related illnesses (BRI) and sick-building syndrome (SBS). A cohesive, proactive IEQ monitoring program is a powerful tool that can be used to achieve this goal.

It is also likely that in the coming years we will witness national and state regulations stipulating that designers, facility managers, and property owners meet specified indoor-air-quality standards. Several national and international organizations including the EPA, OSHA, ASHRAE, ASTM, USGBC, and others are currently in discussions concerning formulating new standards and updating and improving existing national indoor-air-quality standards.

8 Water Efficiency and Sanitary Waste

8.1 General Issues

The current popularity of sustainability in the building industry has helped make taking care of natural resources become part of our everyday culture. One of the more prominent issues facing us today is water conservation. Recent estimates place the amount of fresh water – i.e., water needed for drinking, industry, and sanitation – at about 2.5 percent of the world's total water supply. Roughly one-third of this is readily accessible to humans via lakes, streams, and rivers. Demand for fresh water continues to rise, and, if current trends continue, experts project that demand will double within the next three decades. Since 1950 the U.S. population has increased by almost 90 percent. In that same time span, public demand for water increased by 209 percent. Americans now use an average of 100 gallons of water per person each day. This increased demand has put tremendous stress on water supplies and distribution systems, threatening both human health and the environment.

South Nevada Water Authority has a Water-Efficient Technologies program that offers financial incentives for capital expenditures when businesses retrofit existing equipment with more water-efficient technologies. The EPA has launched WaterSense, a water-oriented counterpart to the ENERGY STAR® program that promotes water efficiency and aims to boost the market for water-efficient products, programs, and practices.

Local codes do not always keep pace with some of the emerging technologies that are not yet code-compliant but are available in the marketplace. These include gray-water systems, rain-water collection systems, high-efficiency irrigation systems, recirculating-shower systems, recirculation of hot water, insulation of hot-water piping, demand-type tankless water heaters, water softeners, and drinking-water treatment systems, all of which are being implemented through EPA WaterSense.

According to EPA, toilets account for approximately 30 percent of the water used in residences, and Americans annually waste 900 billion gallons by using old, inefficient toilets. By replacing an older toilet with a WaterSense-labeled model, a family of four could reduce total indoor water use by about 16 percent and, depending on local water and sewer costs, save more than $90 annually. It is estimated that just by replacing one old toilet in every home with a WaterSense-labeled high-efficiency toilet, the total water savings could supply nearly 10 million U.S. households with water for a year.

Moreover, water conservation translates into energy conservation and savings. If just one in every 10 homes in the United States installed WaterSense-labeled faucets or aerators in their bathrooms, in aggregate this could result in a saving of about 6 billion gallons of water and more than $50 million in energy costs to supply, heat, and treat

DOI: 10.1016/B978-1-85617-691-0.00008-4

that water. The EPA also estimates that if the average home were retrofitted with water-efficient fixtures, there would be a savings of 30,000 gallons of water per year. If only one out of every 10 homes in the U.S. upgraded to water-efficient fixtures (including ENERGY STAR® -labeled clothes washers), the resultant savings could reach more than 300 billion gallons and nearly $2 billion annually. This could have a significant positive economic impact on small plumbing contractors and small businesses throughout various sectors.

Due to increased demand and focus on water efficiency, the emerging water- and energy-conservation market has great potential to revitalize not only the plumbing industry but also traditional construction and small businesses across the country at a time when most business owners are suffering due to tough economic times and a lethargic construction industry.

8.2 Waste-Water Strategy: Water Reuse/Recycling

The Department of Energy (DOE) estimates that commercial buildings consume 88 percent of the potable water in the United States. This offers facility managers a unique opportunity to make a huge impact on overall water consumption. Benchmarking a facility's water use and implementing measures to improve overall efficiency will go a long way to achieving this goal.

Likewise, in spite of the limited emphasis by LEED™ on water efficiency, water-efficient design should be one of the main goals of any project, particularly since our nation's growing population is placing considerable stress on available water supplies. While the U.S. population has nearly doubled in the last five or six decades, public demand for water has more than tripled! This increased demand is putting additional stress on water supplies and distribution systems, and depleting reservoirs and ground water can put water supplies, human health, and the environment at serious risk. According to the EPA, lower water levels can contribute to higher concentrations of natural or human pollutants. Using water more efficiently helps maintain supplies at safe levels, protecting human health and the environment.

The Environmental Protection Agency estimates that an American family of four uses about 400 gallons of water per day. Roughly 30 percent of this is used outdoors for various purposes including landscaping, cleaning sidewalks and driveways, washing cars, and maintaining swimming pools.

Nationally, landscape irrigation counts for almost one-third of all residential water use. That amounts to more than seven billion gallons per day. Water efficiency is one of the principal categories of the LEED® Rating System, and the number of WE credits available depends on the type of certification sought. The main WE topics to know for LEED™ certification are:

- Water-efficient landscaping: reduce use by 50 percent, no potable use or no irrigation
- Innovative wastewater technologies
- Water-use reduction: by 20 or 30 percent

8.2.1 Water Efficient Landscaping

The purpose of water-efficient landscaping is to reduce (by at least 50 percent) or eliminate the amount of potable water and natural surfaces used for landscape irrigation. However, when landscape-design strategies alone are unable to reach a project's irrigation-efficiency goals, attempts should be made to meet efficiency demands through optimization of the irrigation-system design. For example, use of high-efficiency drip, micro-, and subsurface systems can reduce the amount of water required to irrigate a given landscape. The USGBC reports that drip systems alone can reduce water use by 30 to 50 percent. Climate-based controls, such as moisture sensors with rain shutoffs and weather-based evapotranspiration controllers, can further reduce demands by allowing naturally occurring rainfall to meet a portion of irrigation needs. For LEED™ certification, one point is awarded for a 50 percent reduction in water consumption for irrigation from a calculated midsummer baseline case, and a total of two points for a 100-percent water reduction.

To assist in greening the supply, it is necessary to tap alternate water sources. LEED™ recognizes two alternate water sources: rain-water collection and waste-water recovery. Rain-water collection involves collecting and holding on-site rainfall in cisterns, underground tanks, or ponds. This water can then be used during dry periods by the irrigation system. Waste-water recovery can be achieved either on site or at the municipal scale. On-site systems capture gray water (which does not contain human or food-processing waste) from the building and apply it to irrigation. Reductions can be attributed to any combination of the following approaches:

- Use a high-efficiency micro-irrigation system, such as drip, micro-misters, and subsurface irrigation systems.
- Replace potable (drinking) water with captured rain water, recycled waste water (gray water), or treated water.
- Use water treated and conveyed by a public agency that is specifically planned for non-potable purposes.
- Plant species and install landscaping that does not require permanent irrigation systems.
- Apply xeriscape principles to all new development whenever possible.

When a landscaping design incorporates rain-water collection or waste-water recovery in particular, it is essential to assemble a team of experts and establish project roles at an early stage in the process. Rain-water-collection and waste-water-treatment systems stretch over multiple project disciplines, making it particularly important to clearly articulate responsibilities. Having an experienced landscape architect on board is pivotal for a water-efficient landscape and irrigation-system design. To take advantage of the available LEED™ points for water-efficient landscaping credits, early planning is strongly advised.

Many of the LEED™ credits deal with gray water and black water. Gray water is typically considered to be untreated waste water that has not come into contact with toilet waste, such as water from showers, sinks (other than the kitchen), bathtubs, wash basins, and clothes washers. Gray-water use includes indoor and outdoor reuse. When used outdoors, gray water is usually filtered and then used for watering landscapes. Indoor gray-water use, on the other hand, is recycled and used mainly for flushing toilets. Gray water has other applications, including construction activities, concrete mixing, and cooling for power plants. Black water lacks a specific definition that is accepted nationwide but is generally considered to constitute toilet, urinal, and kitchen-sink water (in most jurisdictions). However, depending on the jurisdiction, implementing gray-water systems that reuse waste water from showers and sinks for purposes such as flushing toilets or irrigation may encounter code-compliance restrictions.

Recycling water is commonly thought of in terms of two different water-usage strategies: reclaimed water and gray water, and it is important to distinguish between these systems (although some incorrectly use them interchangeably).

Simply put, reclaimed water is waste-water effluent/sewage that has been treated according to high standards at municipal treatment facilities and that meets the reclaimed-water effluent criteria. Its treatment takes place off-site, and it is delivered to a facility. Reclaimed water is most commonly used for nonpotable purposes, such as landscaping, agriculture, dust control, soil compaction, and processes such as concrete production and cooling for power plants.

The use of reclaimed water is increasing in popularity, especially in states such as California, where openness to innovative, environmentally friendly concepts prevails, especially in the face of a very real and critical water crisis. Orange County, California, for example, started delivering purified waste water last year, providing one of the first "toilet-to-tap" systems in the United States.

Gray water, on the other hand, is the product of domestic water use in showers, washing machines, and sinks. It does not normally include waste water from kitchen sinks, photo-lab sinks, dishwashers, or laundry water from soiled diapers. These sources are typically considered to be black-water producers because they contain serious contaminants and therefore cannot be reused. Gray-water use is a point-of-source strategy – i.e., gray water collected from a building will be reused in the same building.

8.2.2 Innovative Wastewater Technologies

The intent of the Innovative Wastewater Technologies credit is to reduce waste-water generation and potable-water demand and increase the recharge of local aquifers. You can reduce potable-water demand by using water-conserving fixtures and reusing nondrinking water for flushing or, if treated on-site to tertiary standards, on site. Tertiary treatment is the final stage of treatment before water can be discharged back into the environment. If tertiary treatment is used, the water must be treated by biological systems, constructed wetlands, or a high-efficiency filtration system.

A water-efficient technologies program is now in place that offers financial incentives to commercial and multifamily property owners who install water-efficient devices and implement new, water-saving technologies. Examples of effective approaches include:

- Ultra-high-efficiency toilets and efficient retrofits
- Efficient shower heads and retrofits
- Waterless and high-efficiency urinals

- Other ultra-low water-consumption products
- Converting from grass to an artificial surface for sports fields
- Retrofitting standard cooling towers with qualifying high-efficiency drift-elimination technologies

Strategies for meeting one of water-efficiency compliance requirements, that of reducing potable-water use for sewage conveyance, falls into two categories that can be implemented either independently or in concert. As mentioned above, the use of ultra high-efficiency plumbing fixtures can reduce the water required for sewage conveyance in excess of the 50 percent requirement. To use a typical example, composting toilets (not normally used in commercial facilities) and waterless urinals use no water. These two technologies alone can eliminate a facility's use of potable water for sewage conveyance, qualifying both for this credit's point plus potentially a LEED™ Innovation in Design point for exemplary performance. Should the selected plumbing fixtures alone prove to be inadequate to reach the 50-percent reduction threshold or if ultra-high-efficiency plumbing fixtures are not selected, the water necessary for toilet and urinal flushing can be reduced by a minimum of 50 percent or eliminated entirely by applying rain-water collection or waste-water treatment strategies.

The Southface Eco Office in Atlanta provides an example of how this credit can be achieved (Figure 8.1). The facility, targeting LEED™ Platinum certification, completely eliminates the use of potable water for sewage conveyance by using a variety of complementary strategies. Foam-flush composting toilets and waterless urinals are used in the staff restrooms; composting toilets require only six ounces of water per use, which dramatically reduces the volume of water required for sewage conveyance. Water requirements in the public restrooms are reduced through the employment of a combination of dual-flush toilets, ultra-high-efficiency toilets, and waterless urinals. The remaining reduced volume of water required for sewage conveyance is supplied by rain water collected from a roof-mounted solar array and stored in a rooftop cistern and an in-ground storage tank.

A knowledgeable team and early involvement of local code officials are critical components for the successful design and implementation of nonpotable water-supply systems. Furthermore, dual plumbing lines for nonpotable water supply within the building are fairly easy to plan for during the design phase but much more difficult to retrofit after construction is complete.

8.2.3 Water-Use Reduction

The intent of the Water Use Reduction credit, according to LEED™, is to "maximize water efficiency within buildings to reduce the burden on municipal water supply and waste-water systems." One point is awarded for reducing water use by 20 percent, and this is increased to two points for reducing water use by 30 percent. The fixtures governed by this credit include water closets, urinals, lavatory faucets, showers, and kitchen sinks. Water-using fixtures and equipment, such as dishwashers, clothes washers and mechanical equipment (nonregulated uses), which are not addressed by this credit, may qualify for the LEED™ Innovation in Design point.

Figure 8.1 The new Eco Office in Atlanta, Georgia, is a 10,000-square-foot facility seeking LEED™ Platinum rating. It is designed as a model for environmentally responsible commercial construction that is achievable utilizing existing off-the-shelf materials and technology. The new Eco Office provides a showcase of state-of-the-art energy, water, and waste-reducing features. *Source*: PolySteel U.S. Development.

Employing proven, cost-effective technologies can facilitate achieving the 30 percent reduction necessary to earn both points for this credit. The use of low-flow lavatory faucets with automatic controls (0.5 gallons per minute, 12 seconds per use) is normally sufficient to achieve a 20 percent reduction in water use, qualifying for one point. An additional 14 percent reduction can be achieved by the use of waterless urinals, which, when combined with low-flow faucets, should exceed the 30 percent reduction threshold, thereby earning both points.

Here, too, the first step in the optimization process, reducing demand, does not apply. It's not possible to design away occupants' needs to use the restroom, wash their hands, or take a shower. Strategies for water-use reduction, therefore, fall into the same two categories identified for Innovative Wastewater Technologies – either meeting demand efficiently or fulfilling the demand in alternate, more environmentally appropriate means. The two credits complement each other, and water savings related to the Innovative Waste Water Technology credit will also contribute to the Water Use Reduction credit.

John Starr, AIA, and Jim Nicolow, AIA, of Lord, Aeck, and Sargent, Architects, state that among the LEED™ Water Efficiency credits, "Water Use Reduction can often be achieved without the early planning and design integration required by the other two credits. Most alternative plumbing fixtures use conventional plumbing supply and waste lines, allowing these fixtures to be substituted for less-efficient standard fixtures at any point in the design process and even well into the construction process."

Recycling is a term usually reserved for waste such as aluminum cans, glass bottles, and newspapers. Water can also be recycled, and, indeed, through the natural water cycle, the earth has recycled and reused water for millions of years. Water recycling, though, generally refers to using technology to speed up these natural processes.

As local municipalities and individual facilities continue to struggle to meet water needs in the face of dwindling supplies, a variety of reclaimed-water and gray-water system approaches have emerged.

Reclaimed and gray-water systems range in size and complexity. Toward the high end are the multibuilding installations that draw waste water from municipal sources, followed by the middle tier, which includes buildings that have installed storage tanks capable of collecting thousands of gallons of water from rain water, sinks, and steam condensate, that is then treated and funneled to water reuse sources. There are also the simpler, more affordable undercounter systems that provide on-the-spot treatment of water that flows down sink drains, which is then pumped directly into toilet tanks.

The more complex systems should be built into new construction rather than retrofitted later, whereas on-the-spot collection systems can be implemented at any time. It is important when specifying sustainable systems and technologies to remain within budget as a matter of setting goals and performing research up front to determine the additional value and payoff.

The amount of gray water produced in a particular building depends largely on the facility type. A typical office building, for example, may not yield as much gray water as a college dorm or multiuse retail and condominium building. The benefits are all about economies of scale and deriving value from the system, no matter how large or how small it may be.

Consider the amount of potable water that can be saved in a typical four-person household. On average, each person uses 80 to 100 gallons of water per day, with toilet flushing being the largest contributor to this use. The combined use of kitchen and bathroom sinks is only 15 percent of the water that comes into a home, which is significant, considering that 100 percent of the water that comes into the home has been treated and made potable for drinking.

With the largest single source of fresh water in the home capable of using gray water instead of potable water, the household is able to make real gains by reusing water that is perfectly suitable for toilet flushing. For household and small commercial facilities, the most appropriate solution may be to use a gray-water system that incorporates a reservoir, which is installed under the sink and attached to the toilet. These gray-water systems are designed so that the toilet draws first from the collected water in the reservoir. However, the system remains connected to the fresh-water pipes so that, should flushing deplete the amount of water stored in the reservoir, the toilet can then secondarily draw from outside water. Because toilets are the largest consumers of water in households, this type of system is able to save up to 5000 gallons per year.

Differing gray-water policies and regulations among states have considerable impact on the extent to which facilities and homeowners can deploy gray-water systems. Arizona, for example, has gray-water guidelines to educate residents on methods to build simple, efficient, and safe gray-water irrigation systems. For those who follow these guidelines, their system falls under a general permit and automatically becomes "legal," which means that the residents don't have to apply or pay for any permits or inspections. Likewise, California has a gray-water policy but one that is very restrictive, which usually makes it difficult and unaffordable to install a permitted system. Many states have no gray-water policy and don't issue permits at all, while others issue experimental permits for systems on a case-by-case basis.

The recycling of water by whichever means provides substantial benefits, including reduction of stress on potable-water resources, reduction of nutrient loading to waterways, reducing strain on failing septic tanks or treatment plants, and using less energy and chemicals. All of these benefits result in significant savings in water, energy, and money.

It is scarcely believable that as little as 5 to 10 years ago, purchasing environmentally friendly building components that met LEED™ compliance standards could have added more than 10 percent to total building costs, whereas today plumbers, engineers, and other specifiers are now discovering that they can adopt higher sustainability standards without necessarily spending extra. And if they do have to spend extra, the payoff more than compensates when you factor in long-term operating costs, including water and waste-water utility bills plus the energy it takes to heat water for faucets and showerheads, etc.

According to Flex Your Power, California's energy-efficiency marketing and outreach campaign, utilities account for about 30 percent of an office building's expenses. A 30 percent reduction in energy consumption can lower operating costs by $25,000 a year for every 50,000 square feet of office space.

The public is showing greater awareness and taking greater notice of how companies and facilities expend water and energy, and both users and communities are holding building owners accountable for their use of precious local resources. Engineers need to stay abreast and monitor water- and energy-efficiency options in restrooms and elsewhere in their facilities to minimize operating costs and help ensure that buildings meet LEED™ standards.

8.2.4 Construction Waste Management

Commercial construction typically generates between 2 and 2.5 pounds of solid waste per square foot – the majority of which is recyclable. Salvaging and recycling construction and demolition (C&D) waste can substantially reduce demand for virgin resources and the associated environmental impacts. Additionally, effective construction waste management, including appropriate handling of nonrecyclables, can reduce contamination from and extend the life of existing landfills. Whenever feasible, therefore, reducing initial waste generation is environmentally preferable to reuse or recycling.

A construction waste-management plan should recognize project waste as an integral part of overall materials management. The premise is that waste management is a part of materials management, and the recognition that one project's wastes are materials available for another project facilitates efficient and effective waste management. Moreover, waste-management requirements should be included as a topic of discussion during both the preconstruction phase and at ongoing regular job meetings to ensure that contractors and appropriate subcontractors are fully aware of the implications of these requirements on their work prior to and throughout construction.

Waste management should be coordinated with or made part of a standard quality-assurance program, and waste-management requirements should be addressed regularly throughout the project. All topical applications of processed clean wood waste and ground gypsum board as a soil amendment must be implemented in accordance with local and state regulations. When possible, adherence to the plan would be facilitated by tying completion of recycling documentation to one of the payments for each trade contractor.

8.3 Water Fixtures and Conservation Strategies

There are thousands of plumbing fixtures and fittings in today's mainstream market that can help save water, energy, and money. These include but are not limited to aerators, metering and electronic faucets, and prerinse spray valves. When selecting energy-efficient equipment, it is vital to select quality products that meet conservation requirements without compromising performance. The product should deliver the consistent flow required while maintaining the water and energy savings the industry demands. With restroom fixtures accounting for most of a typical commercial building's water consumption, the best opportunities for increasing efficiency can be found there. The good news is that there is increased public awareness, while higher-efficiency plumbing fixtures are more widely available.

Perhaps the best way to increase water efficiency in buildings is through plumbing-fixture replacement and implementation of new technologies. Replacing older, high-flow water closets and flush valves with models that meet current UPC and IPC requirements is important. While current codes require the lower flow rate for new fixtures, existing buildings often have older, high-flow flush valves. Despite the tremendous water savings available by updating the fixtures, facility managers often avoid the upgrade due to concerns about clogging. Solid-waste removal must be 350 grams or greater. Fixtures pass or fail based on whether they can completely clear all test media in a single flush in at least four of five attempts. Toilets that pass qualify for the EPA WaterSense label. It should be noted that when the Energy Policy Act of 1992 was first enacted, many facility managers at the time experienced clogging problems. Those problems have long since been addressed (Figure 8.2).

Comparison of Plumbing-Fixture Water-Flow Rates			
Plumbing Fixture	**Before 1992**	**EPA 1992**	**Current Plumbing Codes**
Toilets	4 to 7 gpf	1.6 gpf	1.6 gpf
Urinals	3.5 to 5 gpf	1.0 gpf	1.0 gpf
Faucets*	5 to 7 gpm	2.5 gpm	0.5 gpm
Showerheads*	4.5 to 8 gpm	2.5 gpm	2.5 gpm

*At 80 psi flowing water pressure.

Figure 8.2 Table illustrating a comparison of plumbing-fixture water-flow rates.
Source: Domestic Water Conservation Technologies, Federal Energy Management Program, U.S. Department of Energy, Office of Energy Efficiency and Renewable Energy, National Renewable Energy Laboratory, October 2002.

Selecting water-efficient fixtures will not only reduce sewer and water bills, but efficient water use reduces the need for expensive water-supply and waste-water treatment facilities and helps maintain healthy aquatic and riparian environments. Moreover, it reduces the energy needed to pump, treat, and heat water. Water is employed in a product's manufacture, during a product's use, and in cleaning, which means that water efficiency and pollution prevention can occur during several product life-cycle stages.

Mark Sanders, product manager for Sloan Valve Company's AQUS Greywater System, says, "Gray-water and reclaimed-water strategies make good use of water resources, especially when implemented in conjunction with efficient plumbing systems."

The maximum volume of water that may be discharged by the toilet when field adjustment of the tank trim is set at its maximum water-use setting must not exceed the following amounts:

- For single-flush fixtures: 1.68 gpf
- For dual-flush fixtures: 1.40 gpf in reduced-flush mode and 2.00 gpf in full-flush mode

The maximum volume of water discharged, using both original equipment tank trim and after market closure seals, should be tested according to the protocol detailed on the WaterSense web-site. There are basically two approaches to measuring water volume: gallons per flush for toilets and urinals or gallons per minute for flow-type fixtures such as lavatories, sinks, and showers. Metered faucets with controlled flow rates for preset time periods are measured in gallons per cubic yard.

For LEED™ purposes, baseline calculations should be computed by determining the number and gender of the users. As a default, LEED™ lets you assume that females use toilets three times per day and males once per day in addition to using the urinal two times per day. Both males and females will use the bathroom faucets three times each day and the kitchen sink once for 15 seconds each.

8.3.1 Toilets

By using water more efficiently, we can help preserve water supplies for future generations, save money, and protect the environment.

8.3.1.1 High-Efficiency Toilets

The National Energy Policy Act was signed into law in 1994, requiring that toilets sold in the United States use no more than 1.6 gallons (6 liters) per flush. This mandate to conserve has nudged manufacturers to produce a new generation of high-efficiency toilets (HETs) that use technologies such as pressure-assist, gravity-flush, and dual-flush to remove waste using as little water as possible. Of these new technologies, the dual-flush method has the advantage of intuitive flushing; the operator can decide electively that less water is required and so use one gallon (3 liters) or less per flush instead of the 1.6 gallon maximum.

Two types of toilet fixtures dominate today's marketplace: ultra-low-flush toilets (ULFTs), also known as low-flow or ultra-low-flow, and high-efficiency toilets (HETs). ULFTs are defined by a flush volume in the range between 1.28 and 1.6 gpf. The HET is defined as a fixture that flushes at 20 percent below the 1.6 gpf maximum or less, equating to a maximum of 1.28 gpf. Dual-flush fixtures are included in the HET category.

This 20 percent reduction threshold serves as a metric for water authorities and municipalities designing more aggressive toilet-replacement programs and in some cases establishing an additional performance tier for financial incentives such as rebate and voucher programs. It is also a part of the water-efficiency element of many green-building programs that exist throughout the United States. Unfortunately, this standard currently applies only to tank-type toilets. Flushometer valve toilets have not been studied in the same way as tank types, and testing needs to be performed on the flushometer valve with the various bowls on the market so that the pair can then be rated.

Although toilets purchased for new construction and retrofits are required to meet the new standards, millions of older inefficient toilets remain in use. As water and sewer costs keep rising, low-flow toilets are becoming increasingly attractive to the American consumer, while local and state governments use rebates and tax incentives to encourage households to convert to these new technologies.

The advantages of low-flow toilets in conserving water and thus reducing the demand on local water-treatment facilities are obvious. According to the EPA, the elimination of inefficient toilets would save the nation about two billion gallons of water a day. With a growing population, an antiquated water-treatment infrastructure, and the potential threat of global warming contributing to uncertain weather, water conservation will remain a major concern to the public.

8.3.1.2 Dual-Flush Toilets

Dual-flush toilets can help make bathrooms more environmentally friendly. They handle solid and liquid waste differently from standard American toilets, giving the user a choice of flushes. The interactive toilet design helps conserve water and has become popular especially in countries where water is in short supply and in areas where water-supply and -treatment facilities are older or overtaxed. The Environmental Protection Agency (EPA) estimates that by the year 2013 36 states may experience water shortages as a result of increased water usage and inefficient water management from aging regional infrastructures. Using less water to flush liquid waste, while logical, may face cultural biases in the United States that make accepting such an innovative approach to personal waste removal harder.

The simple process of handling bodily waste is apparently a delicate topic, indeed so much so that culture can be as much a factor in effecting change as necessity. For this reason bodily functions are kept under wraps, and any changes in our approach to handling them may sometimes create culture shock and resistance. Interest in low-flow and dual-flush toilets is on the rise in the United States, partly due to increased government regulation, the rising cost of water, and the introduction of incentives in many states for making changes in the way we use the commode.

The way water is used to remove waste from the bowl impacts the amount of water needed to get the job done. Standard toilets use siphoning action, which basically employs a siphoning tube to discharge waste. A high volume of water enters the toilet bowl when the toilet is flushed, fills the siphon tube, and pulls the waste and water down the drain. When air enters the tube, the siphoning action stops. Dual-flush toilets employ a larger trapway (a hole at the bottom of the bowl) and a wash-down flushing design that pushes waste down the drain. Because no siphoning action is involved, the system requires less water per flush, and the larger trapway diameter facilitates the exit of waste from the bowl. Combined with the savings from using only half-flushes for liquid waste, a dual-flush toilet can save up to 68 percent more water than a conventional low-flow toilet.

Use of a larger-diameter trapway is the main reason a dual-flush toilet doesn't clog as often as a conventional toilet, while requiring less water to flush efficiently and saving more water than a low-flow toilet when flushing liquid waste. Among the downsides to buying dual-flush units is that they are a little more expensive than comparable low-flow toilet designs. In addition, there is the problem of aesthetics. Dual-flush toilets typically retain only a small amount of water in the bowl, and flushing doesn't always remove all the waste. Even in full flush mode, some occasional streaking will occur. With a dual-flush toilet, you'll probably need to use the toilet brush more often to keep the bowl clean.

8.3.1.3 Composting Toilets

Although not widely used at present, public health professionals have started to appreciate composting toilets as an environmentally sound method for dealing with human wastes. By their very nature they require little or no water to function effectively and are therefore particularly suitable (although not exclusively) for use in locations where mains water and sewerage connections are unavailable or in locations where water-consumption needs to be minimized to the greatest extent possible (Figure 8.3).

One estimate shows that the average American uses 74 gallons (280 liters) of water per day, one-third of which splashes down a flushing toilet. While an older toilet may swallow up to 7 gallons (26.5 liters) per flush and federal law stipulates only 1.6 gallon

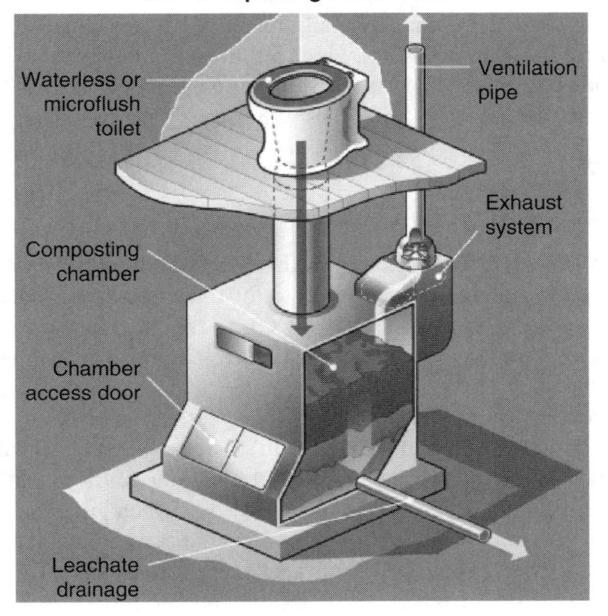

Figure 8.3 Drawing showing how a composting toilet works. Composting toilets use the natural processes of decomposition and evaporation to recycle human waste. The waste that enters the toilets is over 90 percent water, which is evaporated and carried back to the atmosphere through the vent system. The small amount of remaining solid material is converted to useful fertilizing soil by natural decomposition.
Source: HowStuffWorks, Inc.

(6.1 liter) for low-flow models in new homes, a composting toilet can save more than 6600 gallons (24,984 liters) of water per person a year.

Not using water to flush a toilet also cuts out all the energy expended down the line, from the septic system to the treatment plant. That could be beneficial to our waterways. An example of this is the Chesapeake Bay, which receives approximately 1.5 billion gallons (5.7 billion liters) of waste-water flow every day from some 500 sewage-treatment plants. This has caused much of the water to become a dead zone, incapable of keeping animals and plants alive.

For a self-contained composting toilet to work properly, appropriate ventilation is required to keep the smell out of your bathroom while providing enough oxygen for the compost to break down. Some toilets achieve this by employing fans and a heater powered by electricity. (Some models do not require electricity.) The composter also has to be kept at a minimum temperature of 65 degrees F (18.3 degrees C), so, for those living off the grid, a heater could potentially require more electricity than is used in the rest of the house. The heater doesn't have to run all the time, however, and one model may only operate at a maximum level of 540 watts for about 6 hours a day. Because the self-contained models are relatively small, the power use for fans is fairly minimal. It may range from about 80 to 150 watts, which is the same amount of power used by a light bulb. n possible alternative would be to use solar panels to power the fans and heater.

8.3.2 Urinals

8.3.2.1 High-Efficiency Urinals

High-efficiency urinals (HEU) are urinals that use 0.5 gallon per flush (gpf) or less – at least one-half the amount of water used to flush the average urinal. The California Urban Water Conservation Council (CUWCC), in cooperation with water authorities and local agencies, defined these urinals as fixtures that have an average flush volume lower than the mandated 1.0 gallon per flush; some models are zero water-consumption urinals. Based on data from studies of actual usage, these urinals save 20,000 gallons of water per year with an estimated 20-year life. High-efficiency urinals are making a significant difference in water usage, water bills, and our environment.

In addition to saving water and sewer cost, zero-water urinals are a significant improvement over traditional urinals in both maintenance and hygiene. These fixtures use a special trap with lightweight biodegradable oil that lets urine and water pass through but prevents odor from escaping into the restroom (Figure 8.4). There are no valves to fail and no flooding.

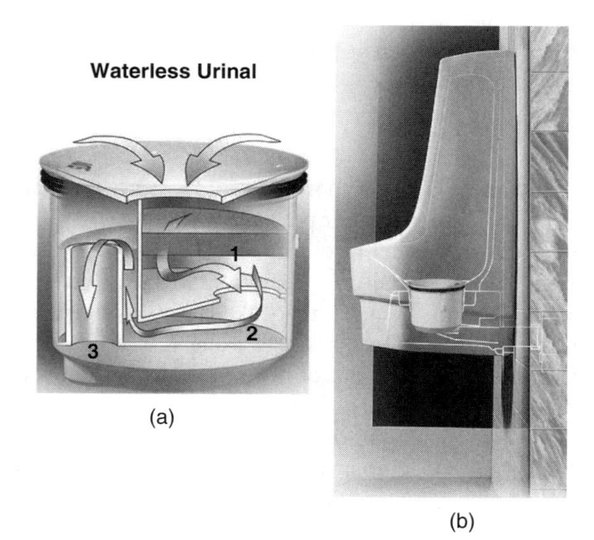

Figure 8.4 A waterless-urinal diagram showing how it functions. The cartridge acts as a funnel directing flow through the liquid sealant, preventing any odors from escaping. The cartridge then collects uric sediment. The remaining liquid, which is noncorrosive and free of hard water, is allowed to pass freely down the drainage pipe (Courtesy: Sloan Valve Company).

8.3.2.2 Water-Free Urinals

As their name implies, water-free urinals do not use water and therefore provide additional savings of water as well as sewage and water-supply line costs. A biodegradable sealant liquid is contained within the cartridge. A barrier is formed between the drain and the open air by the liquid sealant, thereby eliminating odors. Installation is easy whether in new or retrofit applications. The initial cost of a water-free urinal is often less than conventional no-touch fixtures, lowering your initial investment. The urinal can be used to accumulate water-efficient LEED™ credits, including innovation points.

Zero-water and high-efficiency urinals are part of the next generation of water-efficient plumbing products and contribute to U.S. Green Building Council LEED™ credits for water-use reduction.

Ultra-low-water urinals utilize only 1 pint (0.125 gallon) of water to flush. These systems combine the vitreous-china fixture with either a manual or sensor-operated flush valve. They provide effective, low-maintenance flushing in public restrooms while reducing water consumption by up to 88 percent.

8.3.3 Faucets

The EPA includes specific requirements for faucet flow rates. For example, residential lavatory faucets must be regulated by an aerator to 2.2 gallons per minute (gpm) or less, kitchen faucets to 2.5 gpm or less. Commercial faucet requirements vary according to fixture type: handle-operated models are regulated by an aerator to 0.5 gpm, while self-closing and sensor-operated models are limited to less than 0.25 gallon per cycle.

Again we find that the technology available greatly exceeds EPA regulations. While kitchen faucets may require about a 2.5 gpm flow rate in order to fill a pot in a timely fashion, studies have shown that residential lavatory faucets would be satisfactory for the user even when reduced to a 0.5 gpm flow rate. Conservation-minded specifiers have started to recommend aerators that deliver that flow rate.

8.3.3.1 Electronic Faucets

The electronic faucet is an easy way to save energy, and, although it is more costly than a traditional faucet, it will pay for itself in water and energy savings in a short period of time. The electronic faucet has a sensor feature that prevents the faucet from being left on and from excess dripping. According to ENERGYSTAR.gov, "Hot water leaking at a rate of one drip per second from a single faucet can waste up to 1661 gallons of water over the course of a year." An electronic faucet should come equipped with several standard features, including a choice of electric plug-in (AC) and battery (DC) power options.

8.3.3.2 Metering Faucets

Metering faucets and aerators are less expensive than electronic faucets yet can deliver similar energy-saving results. Metered faucets, which are common in commercial washrooms, are generally mechanically operated fixtures that deliver water (at no more than .25 gal/cycle) and then self-close. The manual push feature prevents faucets from being left on after use and prevents unnecessary waste while scrubbing hands. The typical metering faucet's cycle time can be adjusted to deliver the desired amount of water per minute. Many of these devices are designed to allow the user to adjust the temperature before operation. However, in the majority of commercial washrooms we find that sensors are becoming the standard. Engineers appear to have reached the limit of water efficiency for sensor models: 0.08 gallon per cycle. However, it is not user demand or engineering limitations that have determined this to be the limit; it is due to the fact that other environmental considerations come into play. For example, at less than 0.08 gal/cycle, battery disposal (or power use) takes a greater toll on the environment than is represented by the water that might be saved.

8.3.3.3 Aerators

Aerators are one of the most common faucet accessories; they add air into the water stream to increase the feeling of flow. Aerators are capable of controlling the flow to less than 1.5 gpm and provide a simple and inexpensive low-flow solution. They come in a variety of models to provide the exact flow that complies with local plumbing codes.

8.3.4 Shower Heads

A 10-minute shower can use between 25 and 50 gallons of water because a typical high-flow shower head uses between 6 to 10 gpm. Flow-optimized showerheads have a flow of about 1.75 gpm and use 30 percent less water; they can contribute towards maximizing LEED™ points. The flow-optimized single- and three-function shower heads outperform the standard 2.5 gpm flow rate without sacrificing performance.

The flow rate of 2.5 gallons per minute is both the EPA requirement and the LEED™ baseline. Attempts to reduce the flow rate still further are mostly met with very unhappy users. Some users even remove the flow restrictors from their fixtures, producing rates of 4 to 6 gpm, which is clearly not green by any standard. Likewise, flow rates below 2.5 gpm risk failure of certain types of thermostatic mixing valves, leading to scalding. Before specifying valves and shower heads, it may be prudent to consult the manufacturer of the valve; the information may help alleviate this problem altogether.

8.3.5 Baseline Water-Consumption Calculations

To achieve the LEED™ credit, one must first determine the baseline model for water usage in the building. The primary factors in determining this calculation are the types of fixtures in the building, the number of occupants, and the flow or flush rate for the specified fixtures. When evaluating a building's water-use efficiency, the USGBC offers a helpful method that allows one to benchmark annual water use and compare that use to current standards.

The first step is to establish water use based on past annual-use records or on estimates of building occupancy. This is followed by estimating a theoretical water-use baseline based on the types of fixtures in the building and the number of building occupants. To determine the number of occupants in the building, the number of full-time-equivalent (FTE) building occupants must be known (acquired from the LEED™ administrator). The FTE will typically be broken down 50/50 for men and women except in cases where the type of building is primarily suited to one gender – for example, a gym for women. In cases that do not adhere to a strict 50/50 split for male and female occupants, an explanation of the design-case ration is recommended. This can be included in the narrative section of the LEED™ online template for this credit. The FTE should include the transient building occupants (visitors) the building is designed for in addition to the primary occupants. For projects that will have both FTE and transient occupants, separate calculations are required for each type of occupancy.

In Figure 8.5a we have an example used by the USGBC to illustrate the calculation process. It represents potable-water calculations for sewage conveyance for a two-story office building with a capacity of 300 occupants. The calculations are based on a typical eight-hour workday and a 50/50 male/female ratio. Male occupants are assumed to use water closets once and urinals twice in a typical day (default), and females are assumed to use water closets three times (default). The reduction amount is the difference between the design case and the baseline case.

In Figure 8.5b we show the baseline case being used in line with the Energy Policy Act of 1992 fixture-flow rates. When undertaking these calculations, the number of

Fixture Type	Daily Uses	Flow Rate (GPF)	Occupants	Sewage Generation (gal)
Low-Flow Water Closet (Male)	0	1.1	150	0
Low-Flow Water Closet (Female)	3	1.1	150	495
Composting Toilet (Male)	1	0.0	150	0
Composting Toilet (Female)	0	0.0	150	0
Waterless Urinal (Male)	2	0.0	150	0
			Total Daily Volume (gal)	495
			Annual Work Days	260
			Annual Volume (gal)	128,700
			Rain-water or Gray-water Volume (gal)	(36,000)
		Total Annual Volume (gal)		**92,700**

Figure 8.5a Design Case.
Source: USGBC.

Fixture Type	Daily Uses	Flow Rate (GPF)	Occupants	Sewage Generation (gal)
Water Closet (Male)	1	1.6	150	240
Water Closet (Female)	3	1.6	150	720
Urinal (Male)	2	1.0	150	300
			Total Daily Volume (gal)	1,260
			Annual Work Days	260
		Total Annual Volume (gal)		**327,600**

Figure 8.5b Baseline Case. The USGBC requires that the baseline case must use the flow rates and flush volumes established by EPAct 1992.
Source: USGBC.

occupants, number of workdays, and frequency data should remain the same. Furthermore, gray-water or rain-water harvesting volumes should not be included. The baseline case here estimates the amount of potable water per year used for sewage conveyance to be 327,600 gallons. This means that a reduction of 72 percent has been achieved in potable-water volumes used for sewage conveyance. Using this strategy can earn one point in LEED™'s rating system.

Of note, the baseline calculation is based on the assumption that 100 percent of the building's indoor-plumbing fixtures comply with the requirement of the 2006 Uniform Plumbing Code or the 2006 International Plumbing Code fixture and fitting performance requirements. Once the baseline has been established for the building, the actual use can be compared and measures can be implemented to reduce water use and increase overall water efficiency. Although this baseline methodology is specific to LEED™, it can nevertheless be used in buildings that are not seeking LEED™ certification.

For calculation purposes in LEED™ projects, the precise number of fixtures is not important unless there are multiple types of the same fixture specified throughout the building. For example, if there are public restrooms with different water closets on the second floor, their use is to be accounted for as a percentage of the FTE in the LEED™ credit template calculations.

When the Energy Policy Act's fixture and flow rates are applied to FTE building occupants, the baseline quantity use can be established. FTE calculations for the project are used consistently throughout the baseline- and design-case calculations to determine the estimated use by the building occupants.

8.4 Retention Ponds, Bioswales, and Other Systems

Storm-water runoff is generally generated when precipitation from rain and snowmelt events flows over land or impervious surfaces and is unable to percolate into the ground. As the runoff flows over the land or impervious surfaces (paved streets, parking lots, and building rooftops), it accumulates debris, chemicals, sediment, or other pollutants that could adversely affect water quality if the runoff is discharged untreated. The most appropriate method to control storm-water discharges is the application of best management practices (BMPs). Because storm-water discharges are normally considered point sources, they will require coverage under an NPDES permit.

Utilizing rain-water collection systems such as cisterns, underground tanks, and ponds can substantially reduce or eliminate the amount of potable water used for irrigation. Rain water can be collected from roofs, plazas, and paved areas and then filtered by a combination of graded screens and paper filters prior to its use in irrigation.

A retention pond is basically a body of water that is used to collect storm-water runoff for the purpose of controlling its release. Retention ponds have no outlets, streams, creek ditches, etc. The water collects and then is released through atmospheric phenomena such as evaporation or infiltration.

Retention ponds differ from detention ponds in that a detention pond has an outlet such as a pipe to discharge the water to a stream. A detention pond is defined as a body of water that is used to collect storm-water runoff for the purpose of controlling its release. The pipe that a detention pond contains is sized to control the release rate of the runoff. Neighborhood ponds serve several purposes, but none of those purposes includes swimming or wading.

Ponds must be of sufficient depth (at least 8 to 10 feet) to prevent stagnation and algae growth and to handle the amount of storm-water runoff that is expected to enter.

Most ponds typically have a "safety ledge" at the edge to keep those who unintentionally enter the pond from getting into deep water immediately. This safety ledge is generally no wider than 10 feet and leads directly to much deeper water. The slope off the safety ledge varies greatly, as does the depth of water it leads to.

One main problem that is encountered with ponds is the buildup of bacteria such as *E. coli*. Because of the limited water flow and the tendency of wildlife including geese to gather around ponds, they can become breeding grounds for dangerous bacteria. With proper design and maintenance, ponds can be very attractive, but they may require more planning and more land. Without maintenance these ponds can turn into a major liability for the owner.

A decade ago most detention ponds were nothing more than ugly holes in the ground hidden as far from view as possible. Today, most developers are attempting to incorporate their detention ponds as amenities, whether they have a permanent pool, walking trails, picnic areas, or playgrounds. Ponds are today less a "waste of land" and more of a beneficial use of land.

Local requirements for rain-water harvest and waste-water treatment will vary greatly from location to location, which is why early involvement of and input from local code officials are important. The owner should assemble an experienced team, including architect, landscape architect, civil and plumbing engineers and rain-water system designer, early in the design process if realistic, cost-effective efficiency goals are to be achieved.

Often developers will try to do away with retention ponds and replace them with pervious concrete pavement, which is perhaps more expensive than typical concrete pavement but can be partially or fully offset by reducing or eliminating the need for drainage systems, retention ponds, and their associated maintenance costs. In addition to the cost savings, elimination of retention ponds can also help meet the goal of reducing site disturbance found in LEED™ and therefore help earn additional LEED™ points.

The practicality of whether to incorporate detention ponds or not will be dependent largely on site development, not large-scale land development. Costs of building an amenity-type detention pond vs. a traditional detention pond should be studied, bearing in mind that the same storage volume will have to be provided. Then, after adding the cost for pumps, controls, additional storage, and thousands of linear feet of pipe, the decision needs to be made whether all that extra cost outweighs the cost of losing, say, 12 to 15 percent of your land to traditional detention.

In the past, this water was conveniently forced into the city storm drains or into retention ponds, thus becoming someone else's problem. Water from rainstorms and snowmelt needs to be carefully managed in order to conserve water in time of need, to better clean water before it starts its journey back to local aquifers, and to lessen the burden of excessive water runoff on municipal drainage systems.

A system of interlocking, porous pavers resting on a multilayer bed of crushed stones and gravel of different sizes can be used for the parking area. This will allow water to diffuse through the surface of the parking lot, slowing the rush of water into the ground and permitting the surrounding landscaping to absorb the water while being diverted toward bioswales surrounding the property. Bioswales are gently sloped areas of the property designed to collect silt and other rain-water runoff while slowing the speed with which water collects (Figure 8.6). The swales are shaped so that water is diverted in a manner so as not to encourage erosion of the ground and soil.

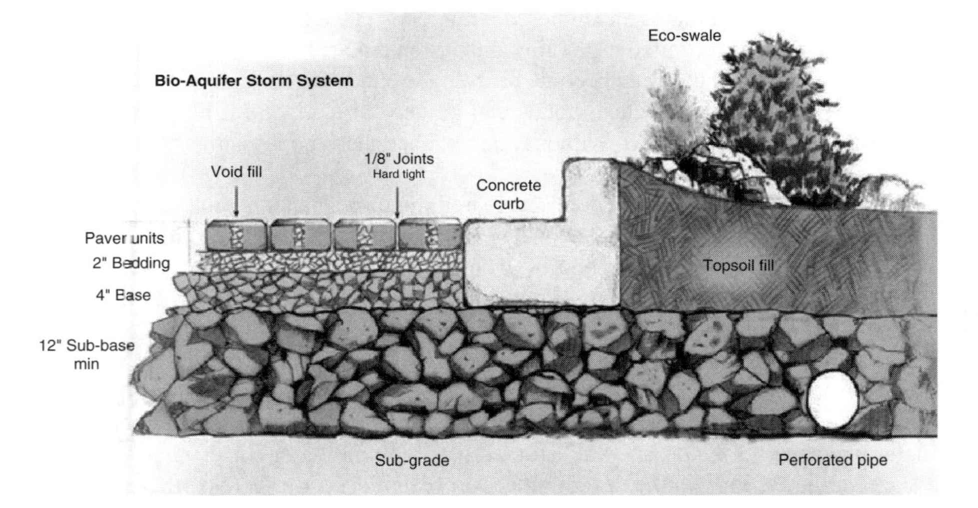

Figure 8.6 Drawing of a typical bioswale. It consists of gently sloped areas of the property designed to collect silt and other rain-water runoff and slow down the speed with which water collects. The swales are shaped so that water is diverted but not so sharply as to encourage erosion of the ground and soil.
Source: Other World Computing.

Native vegetation can be planted in the bioswale to facilitate water absorption, and lengthy root systems will prevent soil erosion while needing minimum maintenance. Native plants are hearty plants that can manage well during periods of dry, hot weather, yet manage to make use of and manage the flow of water from unexpected storms. The ability to combine nature with a well-planned surface system can provide an extremely efficient source of water management and filtering in addition to an attractive design.

9 Impact of Energy and Atmosphere

9.1 Introduction

With the threat of global economic collapse, it has become critically important to revise our attitudes toward sustainability and green building and to make our buildings cost-effective and healthy places to live and work. This can now be achieved through the use of integrated design processes that enable us to create high-performance buildings wherein all systems and components work together to produce overall functionality and environmental performance while meeting the needs of owner and tenant. Moreover, through integrated design we can now create net-zero-energy buildings (NZEB): buildings that, on an annual basis, draw from outside sources equal or less energy than is produced on-site from renewable energy sources.

This chapter will deal with the many aspects that impact the design and construction of high-performance, intelligent buildings that are both healthy and cost-effective. First, however, is an overview of the new LEED™ Rating System as it applies to the Energy and Atmosphere category, including some of the changes and new requirements for acquiring LEED™ credits. It is important to note that many of the LEED™ exam questions tend to focus on energy credits, especially strategies to optimize energy performance. It would therefore be especially prudent to pay particular attention to this category. For the latest updates relating to the LEED™ 2009 tests and certification requirements, visit the GBCI and USGBC websites, www.gbci.org and www.usgbc.org.

The newly adopted LEED™ 2009 requirements have changed significantly from their predecessors, with an increased emphasis on sustainable sites, water efficiency, and energy and atmosphere. In terms of possible credits and points, Energy and Atmosphere must be considered the most important of the seven categories in the new LEED™ 2009 Rating System. For certification purposes, Energy and Atmosphere can earn up to 35 points out of 100 + 10 (Figure 9.1). It should be stressed, however, that no individual product or system in itself can be LEED™-certified; it can only help contribute to the completion of LEED™ credits.

The significance of the dramatic changes to the LEED™ 2009 scoring system cannot be overstated, particularly in how the system relates to energy modeling. Energy and Atmosphere Prerequisite 2 (minimum energy performance) and Credit 1 (optimize energy performance) have changed significantly. Thus, the threshold for the prerequisite has changed from 14 to 10 percent and the points awarded in the performance credit have increased from a 1 to 10-point scale to a possible 1 to 19-point scale, awarding

DOI: 10.1016/B978-1-85617-691-0.00009-6

Credit	NC	CI	EB	C&S	Schools
Sustainable Sites	26	21	26	28	24
Water Efficiency	10	11	14	10	11
Energy and Atmosphere	35	37	35	37	33
Materials/Resources	14	14	10	13	13
Indoor Environmental Quality	15	17	15	12	19
Innovation in Design	6	6	6	6	6
Regionalization	4	4	4	4	4
Total	**110**	**110**	**110**	**110**	**110 Points**

Figure 9.1 Table showing point allocations for different categories in LEED™ 2009 Rating System.

9 extra points for the same percentage improvement over the baseline building. But what is perhaps even more interesting is that the baseline itself has changed. LEED™ project teams are mandated to use the ASHRAE standard referenced in the applicable Reference Guide and are permitted to use addenda within the most recent Supplement to that Standard. The new LEED™ 2009 is largely governed by the 2007 update of ASHRAE 90.1 (as opposed to the previous version, ASHRAE 90.1 2004). The main modifications relating to LEED™ requirements include mandatory compliance with Appendix G of ASHRAE 90.1 2007.

For example, ASHRAE has made changes to the baseline HVAC system types themselves. In the 2004 version, the cutoff in baseline criteria between commercial systems 3 to 4 (single-zone rooftop equipment) and 5 to 6 (VAV rooftop equipment) was three floors or less and larger than 75,000 square feet. The new standard substantially reduces that limit to 25,000 square feet. Whereas a 60,000-square-foot office building that used VAV HVAC could pick up 10 to 15 percent in fan-energy savings alone using the old system, it will now be modeled against a similar baseline HVAC system.

Over the years building-construction values in the United States have become stricter. For example, in climate zone 3A, minimum compliance for roof insulation has increased from R-15 to R-20, and wall insulation has increased from R-13 to R-16.8. Although glass compliance has remained unchanged, different U-values have been introduced based on the type of glass, with the former assembly values of U-0.57 and SHGC-0.25 remaining consistent. Thus, using highly efficient glass remains an appropriate method for earning percentage points against the baseline. The significance of this change will be dramatically increased in building types where skin loads represent a large percentage of the peak HVAC load (i.e., office buildings) but less significant in spaces where persons and ventilation loads dominate the sizing of the HVAC equipment (e.g., in assembly areas and schools).

Most state and local governments adopt commercial energy codes to establish minimum energy-efficiency standards for the design and construction of buildings, and in the United States the majority of energy codes are based on ASHRAE 90.1 or IECC. It should be noted that several other organizations have also produced standards for energy-efficient buildings, but ASHRAE and the U.S. Green Building Council

(USGBC) are perhaps the best known. ASHRAE Standard 90.1 2007, Energy Standard for Buildings Except Low-Rise Residential Buildings, was established by the USGBC as the commercial-building reference standard for the new rating program, LEED™ 2009, that launched on April 27, 2009. The latest ASHRAE version is the last in a long succession to the original ASHRAE Standard 90 1975 standard, which is becoming increasingly more stringent.

In terms of qualifying for LEED™ certification, HVAC and lighting control systems earn very few points on their own, perhaps three points or so. However, with the addition of the necessary sensors and building controls, the number of achievable points grows considerably – to as many as 25 points and even more with the use of fully integrated building systems. This is because with integrated systems the building can earn multiple LEED™ points as well as cost savings by taking such measures as having a zone's occupancy sensor control both lighting and HVAC.

A comprehensive building-operation plan needs to be developed that addresses the heating, cooling, humidity-control, lighting, and safety systems. Additionally, there is a need to develop a building automation-control plan as well.

Energy-efficiency measures (ECMs) are recommended to ensure that the building will have the highest percentage of energy savings below the baseline building for the lowest upfront capital costs. Many recommended measures may involve minor upfront capital costs.

For Commercial Interiors, the EA Credit 3, Measurement and Verification, points can be earned in one of two ways. For projects less than 75 percent of the total building area, either by installing submetering equipment to measure and record energy use within the tenant space (2 points) or by negotiating a lease whereby the tenant pays the energy costs, which are not included in the base rent (3 points). For projects that constitute 75 percent or more of the total building area, continuous metering equipment must be installed for one of several end uses such as lighting systems and controls or boiler efficiencies (5 points).

Other factors to pay particular attention to in preparing for the LEED™ exam in the Energy and Atmosphere category are discussed below.

9.1.1 LEED™ EA Prerequisite 1: Fundamental Commissioning of Building Systems

The commissioning plan involves verification that the facility's energy-related systems are all installed, calibrated, and performing according to the owner's project requirements (OPR) and basis of design (BOD). The building is to comply with the mandatory and prescriptive requirements of ASHRAE 90.1 2007 in order to establish the minimum level of energy efficiency for the building type. The plans and data produced as a result of the building commissioning will lay the groundwork for later energy-efficiency savings. This prerequisite is discussed in detail in Chapter 5. This is an extremely important prerequisite and should be completely understood. Several questions almost always turn up on the tests.

9.1.2 LEED™ EA Prerequisite 2: Minimum Energy Performance

The intent of this prerequisite is to establish the minimum level of energy efficiency for the project. The important aspect to remember here is to comply with both the mandatory and prescriptive provisions of ASHRAE 90.1 2007 or state code, whichever is more stringent.

9.1.3 LEED™ EA Prerequisite 3: Fundamental Refrigerant Management

The intent is to reduce ozone depletion. This can be achieved by zero use of CFC-based refrigerants in new HVACR systems. For existing construction a comprehensive CFC phaseout conversion prior to project completion is required if you are reusing existing HVAC equipment, as per the Montreal Protocol (1995).

9.1.4 LEED™ EA Credit 1: Optimize Energy Performance

The intent is to increase levels of energy performance in comparison to prerequisite standards. Option 1 is a whole-building energy simulation using an approved energy-modeling program (1 to 19 points for NC and Schools and 3 to 21 points for SC). Option 2 is to use a prescriptive compliance path: Comply with ASHRAE's Advanced Energy Design Guide (1 point) appropriate to the project scope; the facility must be 20,000 square feet or less and be an office or retail occupancy. Option 3 is to use a prescriptive compliance path: Comply with the Advanced Buildings Core Performance" Guide (1–3 points). The facility must be less than 100,000 square feet.

For project teams pursuing Option 1, new construction must exceed ASHRAE 90.1 2007 Appendix G baseline performance rating by 50 percent (previously 45.5 percent for NC), and existing buildings must exceed by 46 percent (previously 38.5 percent for NC) to be considered under the Innovation in Design category.

9.1.5 LEED™ EA Credit 2: On-site Renewable Energy

The intent is to encourage increase of renewable-energy self-supply and reduce impacts associated with fossil-fuel energy use. For the minimum renewable energy percentage for each point threshold see the reference guide.

For NC and Schools, projects can earn credit for exemplary performance by showing that on-site renewable energy accounts for at least 15 percent of annual building energy cost. For the CS category, on-site renewable energy must account for at least 5 percent of the annual building energy cost to earn an exemplary performance credit.

9.1.6 LEED™ EA Credit 3: Enhanced Commissioning

This credit is discussed in Chapter 5. The basic intent is to begin commissioning early in the design process and implement additional activities after systems performance verification.

For NC, CS and Schools, projects that conduct comprehensive envelope commissioning may be considered for an innovative credit. These projects will need to demonstrate the standards and protocols by which the envelope was commissioned.

9.1.7 LEED™ EA Credit 4: Enhanced Refrigerant Management

The intent is to reduce ozone depletion while complying with the Montreal Protocol. Option 1 is not to use refrigerants. Option 2 is to select refrigerants and HVACR that minimize or eliminate the emission of compounds that contribute to ozone depletion and global warming. You must meet or exceed requirements set by the maximum threshold for the combined contributions to ozone-depletion and global-warming potential.

9.1.8 LEED™ EA Credit 5: Measurement and Verification

The intent is to provide for ongoing measurement and accountability of building energy consumption. Option 1 is to develop and implement a measurement and verification (M&V) plan consistent with Option D: Calibrated Simulation as specified in the International Performance Measurement and Verification Protocol (IPMVP), Volume III: Concepts and Options for Determining Energy Savings in New Construction, April 2003. The M&V period must last not less than one year of postconstruction occupancy. Option 2 is to develop and implement a M&V plan consistent with Option B: IPMVP, Volume III: Concepts and Options for Determining Energy Savings in New Construction, April 2003. The M&V period must last not less than one year of postconstruction occupancy.

9.1.9 LEED™ EA Credit 6: Green Power

The intent is to encourage and develop the use of grid-source, renewable-energy technologies. Option 1 is to use the annual electricity consumption results of EA Credit 1: Optimization Energy Performance to determine baseline electricity use. Option 2 is to determine the baseline electricity consumption by using the DOE Commercial Buildings Energy Consumption Survey database. Renewable-energy certificates (RECs) provide the renewable attributes associated with green power. RECs can be provided at a much more competitive cost than from local utilities. A facility is not required to switch its current utility in order to procure off-site renewable energy for this credit.

For NC, CS, and Schools, projects that purchase 100 percent of their electricity from renewable sources may be considered for an innovation in design credit.

Air filters are discussed in Section 9.3.6. It suffices to say that a building's HVAC air-filtration system provides measureable and discernable ways to improve indoor-air quality (IAQ) and energy efficiency, two main tenets of the LEED™ program. In addition, proper filtration systems can contribute to the completion of LEED™ credits and prerequisites.

Finally, fire-protection systems should never be compromised, as they serve the purpose of life safety. However, just like other building systems, they should be designed, sourced, installed, and maintained in a manner that is environmental friendly and reduces their impacts on the environment, as discussed in Section 9.6.

9.2 Intelligent Energy-Management Systems

Building automation is becoming one of the landmarks of today's society; it basically consists of a programmed, computerized, "intelligent" network of electronic devices that monitor and control the mechanical and lighting systems in a building. In fact, more and more buildings are incorporating central communications systems, and the computer-integrated building has not only become a reality but an integral part of mainstream America. The intent is to create an intelligent sustainable building and reduce energy and maintenance costs. In addition, increasing consumer demand for clean, renewable energy and the deregulation of the utilities industry have spurred growth in green power – solar, wind, geothermal, steam, biomass, and small-scale hydroelectric sources of power. Small commercial solar-power plants are beginning to emerge and have started to serve some energy markets.

9.2.1 Building Automation and Intelligent Buildings

There have been various definitions of intelligent and sustainable design. Regarding sustainability, the *ASHRAE* GreenGuide defines it as "providing for the needs of the present without detracting from the ability to fulfill the needs of the future." An intelligent building, on the other hand, can be said to be one that provides a productive and cost-effective environment through optimization of its basic elements: structure, systems, services, and management and the interrelationships among them. Thomas Hartman, P.E., a building automation expert, believes that there are three cardinal elements of an intelligent building:

1. Occupants: An intelligent building is one that provides easy access; keeps people comfortable, environmentally satisfied, and secure; and provides services to encourage productivity.
2. Structure and systems: An intelligent building is one that at a bare minimum significantly reduces environmental disruption, degradation, or depletion while ensuring a long-term useful functional capacity.
3. Advanced technologies: An intelligent building is one that because of its climate and/or use is challenged to meet the needs listed above and succeeds in meeting those challenges through the use of appropriate advanced technologies.

Most engineers today understand an intelligent building to be one that incorporates computer programs to coordinate many building subsystems and to regulate interior temperature, HVAC, and power. The goal is usually to reduce the operating cost of the building while maintaining the desired environment for the occupants (Figure 9.2). Many people fail to realize that it is really the role of advanced technologies to dramatically improve the comfort, environment, and performance of a building's occupants while minimizing the external environmental impact of its structure and systems. The key phrase here is "comfort of its occupants." In the final analysis, intelligent buildings

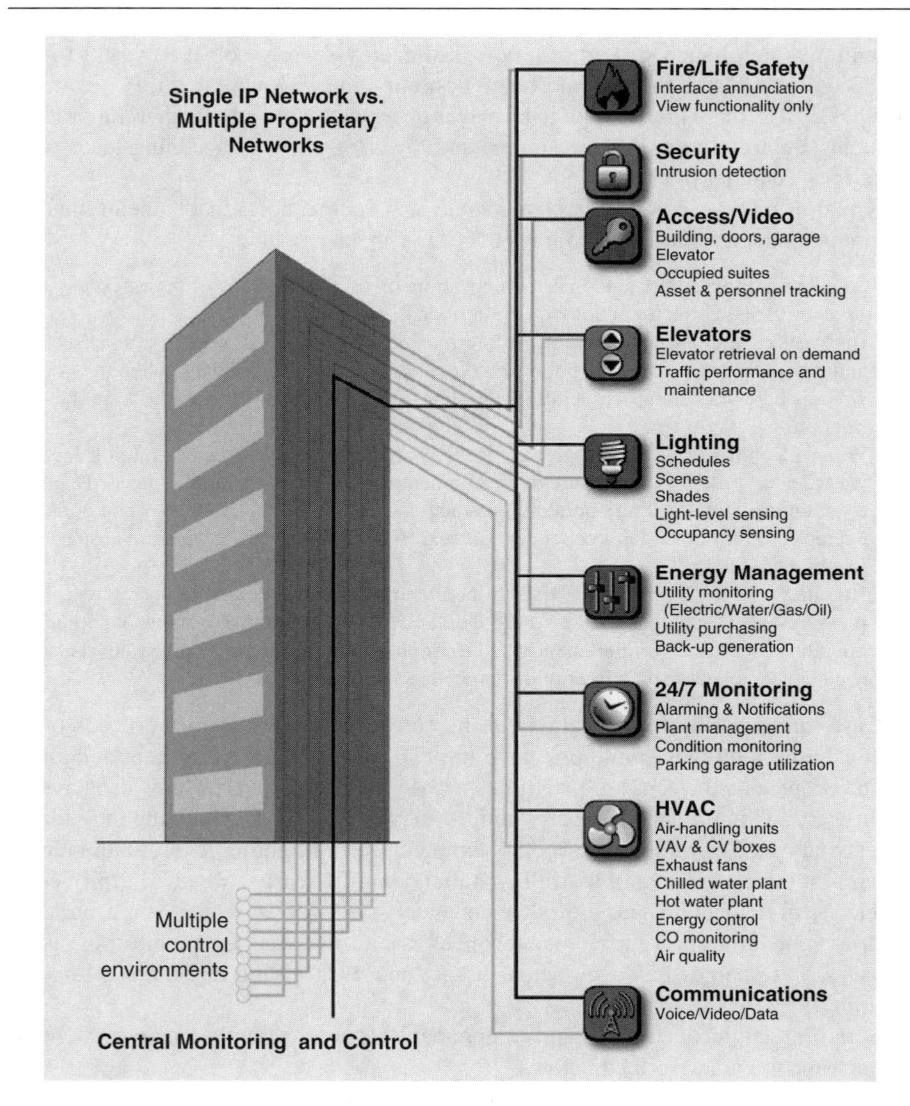

Figure 9.2 An intelligent building is one that can merge building management requirements with IT systems to achieve optimized system performance as well as simplifying general facility operations. In this illustration a single IP network is compared with a multiple proprietary network.
Source: BPG Properties Ltd.

help property owners and developers as well as tenants to achieve their objectives in the areas of comfort, cost, safety, long-term flexibility, and marketability.

There are a number of commercial off-the-shelf building-automation systems (BAS) now on the market, and the majority of facility and building managers recognize the potential value of such system as a powerful energy-saving tool; if it wasn't for the initial costs involved, there would be no hesitation in employing them. For example, one basic BAS that is readily available saves energy by widening temperature ranges, reducing lighting in unoccupied spaces, and lowering costs by shedding loads when electricity is higher priced.

Kristin Kamm, a senior research associate at E Source, notes that some of the most common strategies that BAS employs to cut energy use include:

- Scheduling turns equipment on or off depending on time of day, day of the week, day type, or other variables such as outdoor air conditions.
- Lockouts ensure that equipment doesn't turn on unless it's necessary. For example, a chiller and its associated pumps can be locked out according to calendar date, when the outdoor air falls below a certain temperature, or when building cooling requirements are below a minimum.
- When equipment operates at greater capacity than necessary to meet building loads, it wastes energy. A BAS can ensure that equipment operates at the minimum needed capacity by automatically resetting operating parameters to match current weather conditions. For example, as the outdoor air temperature decreases, the chilled water temperature can be reset to a higher value.
- Building operators who use a BAS to monitor information such as temperatures, flows, pressures, and actuator positions may use that data to determine whether equipment is operating incorrectly or inefficiently and to troubleshoot problems. Some systems also the use the data to automatically provide maintenance bulletins.

Customized building automation can be complex depending on the needs of the client. Some intelligent buildings have the capability to detect and report faults in the mechanical and electrical systems, especially critical systems. Many also have the ability to track individual occupants and to adapt building systems to the individual's wants and needs (e.g., setting a room's temperature and lighting levels automatically when a homeowner enters), as well as anticipating forecasted weather, utility costs, or electrical demand. There are other nonenergy uses for automation in a building, such as scheduling preventive maintenance, monitoring security, monitoring rent or consumables charges based on actual usage, and even giving directions within the building.

Some of the typical elements and components that are frequently employed in building automation are described below.

Today's controllers allow users to take advantage of networking and gain real-time access to information from multiple resource segments in a building's network. Controllers come in a wide range of sizes and capabilities to control common devices. Usually the primary and secondary buses are chosen based on what the controllers provide.

There are different types of occupancy sensors, such as infrared, ultrasonic, and dual tech, which are designed to meet a wide range of applications. Occupancy is usually based on time of day, but override is possible through different means. Some buildings can sense occupancy in their internal spaces by an override switch or sensor. Sensors can be either ceiling- or wall-mounted, depending on the type and application (Figure 9.3).

Figure 9.3 A multicriteria intelligent sensor, Acclimate™, incorporates both thermal and photoelectric technologies that interact to maximize detection. Acclimate™ is a fire sensor with an on-board microprocessor with advanced software that makes adjustments to reduce false alarms.
Source: System Sensor.

With a building-automation system lighting can be turned on and off based on time of day or occupancy sensors and timers. One typical example is to turn the lights in a space on for a half hour after the last motion was sensed. A photocell placed outside a building can sense darkness and the time of day and modulate lights in outer offices and the parking lot.

With most air handlers less temperature change is required because they typically mix the return and outside air. This in turn can save money by using less chilled or heated water – not all AHUs use chilled/hot-water circuits. Some external air needs to be introduced to keep the building's air quality healthy. The supply fan (and return if applicable) is started and stopped based on either time of day, temperature, building pressure, or a combination of all three.

A constant-volume air-handling unit (CAV) is a less efficient type of air handler because the fans do not have variable-speed controls. Instead, CAVs open and close dampers and water-supply valves to maintain temperatures in the building's spaces. They heat or cool the spaces by opening or closing chilled or hot-water valves that feed their internal heat exchangers. Generally one CAV serves several spaces, but larger buildings may incorporate many CAVs.

A variable-volume air-handling unit (VAV) is a more efficient unit than the CAV. VAVs supply pressurized air to VAV boxes, usually one box per room or area. A VAV air handler can change the pressure to the VAV boxes by changing the speed of a fan or blower with a variable frequency drive or (less efficiently) by moving inlet guide vanes to a fixed-speed fan. The amount of air is determined by the needs of the spaces served by the VAV boxes.

VAV hybrid systems are a variation in between VAV and CAV systems. In this system the interior zones operate as in a VAV system, but the outer zones differ in that the heating is supplied by a heating fan in a central location, usually with a heating coil fed by the building boiler. The heated air is ducted to the exterior dual duct mixing boxes and dampers controlled by the zone thermostat, which calls for either cooled or heated air as needed.

The function of a central plant is to supply the air-handling units with water. It may supply a chilled-water system, hot-water system and condenser-water system, as well as transformers and an auxiliary power unit for emergency power. If well managed, these systems can often help each other. For example, some plants generate electric power at periods with peak demand, using a gas turbine, and then use the turbine's hot exhaust to heat water or power an absorptive chiller.

A chilled-water system is typically used to cool a building's air and equipment. Chilled-water systems usually incorporate chiller(s) and pumps. Analog temperature sensors are used to measure the chilled-water supply and return lines. The chiller(s) are sequenced on and off to ensure that the water supply is chilled.

In a condenser water system cool condenser water is supplied to the chillers through the use of cooling tower(s) and pumps. In order to ensure that the condenser water supply to the chillers is constant, speed drives are commonly employed on the cooling-tower fans to control temperature. Proper cooling-tower temperature assures the proper refrigerant head pressure in the chiller. The cooling-tower set point used depends upon the refrigerant. Analog temperature sensors measure the condenser water supply and return lines.

The hot-water system supplies heat to the building's air-handling units or VAV boxes. The hot-water system will have a boiler(s) and pumps. Analog temperature sensors are placed in the hot-water supply and return lines. Some type of mixing valve is typically incorporated to control the heating water loop temperature. To maintain supply, the boiler(s) and pumps are sequenced on and off.

Most building-automation systems today incorporate some alarm and security capabilities. If an alarm is detected, it can be programmed to notify someone. Notification can be implemented via a computer, pager, cellular phone, or audible alarm. Security systems can also be interlocked to a building-automation system. If occupancy sensors are present, they can also be used as burglar alarms. This is discussed in greater detail in Section 9.6 of this chapter.

There are a large number of propriety protocols and industry standards on the market, including: ASHRAE, BACnet, DALI, DSI, Dynet, ENERGY STAR®, KNX, LonTalk, and ZigBee. Details of these systems are outside the scope of this book.

9.3 Active Mechanical Systems: Zoning and Control Systems

9.3.1 General

The majority of people living today in urban American cities now take for granted that the buildings they live and work in will have appropriate mechanical heating,

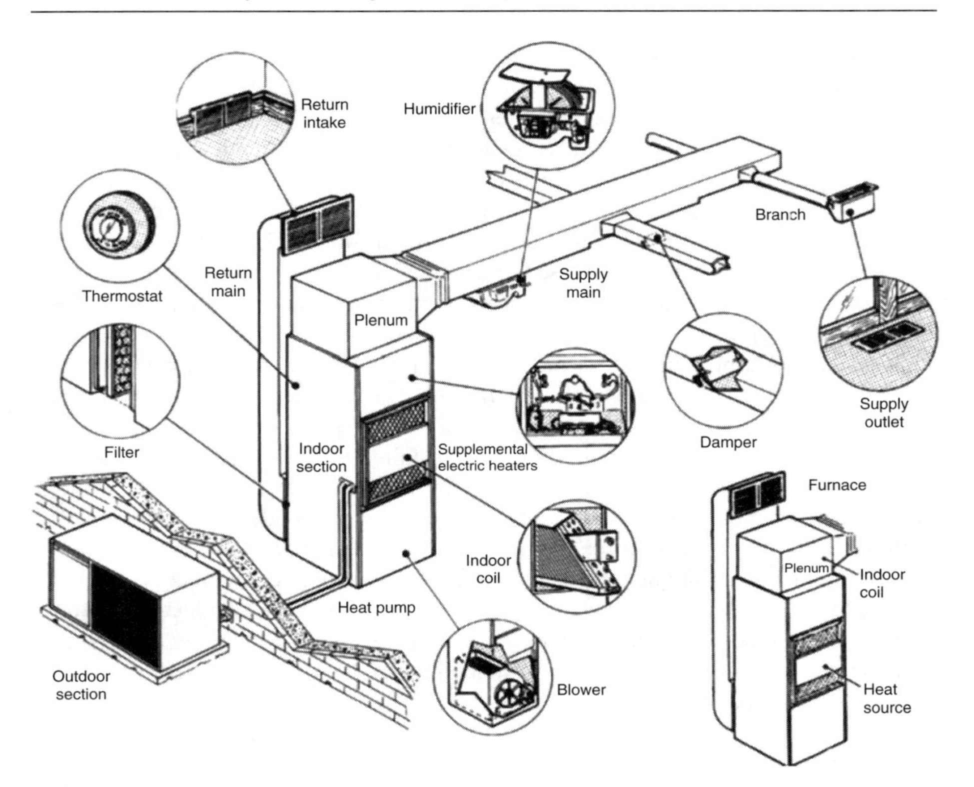

Figure 9.4 Diagram illustrating the basics of HVAC systems.
Source: Southface Energy Institute.

ventilation, and air-conditioning (HVAC) systems in place. It is understood that these systems are designed to provide air at comfortable temperature and humidity levels, free of harmful concentrations of air pollutants (Figure 9.4). The continuous development of air-conditioning systems has brought about fundamental changes in the way we design projects because it has allowed investors to build larger, higher, and more efficient buildings than was previously possible. But even as buildings today are being designed with increasingly sophisticated energy management and control systems (EMCS) for monitoring and controlling the conditions of a building's interior space, we nevertheless frequently discover that a building's heating, ventilating, and air-conditioning (HVAC) equipment routinely fails to satisfy the performance expectations of its designers and owners, what is more surprising is that such failures often go unnoticed for extended periods.

New technologies and recent developments in computers and electronics equipment have made it possible to create HVAC systems that are smarter, smaller, and more efficient. These advances have reshaped how the systems are installed, how they are maintained, and how they operate. Other important recent developments in HVAC equipment design include the introduction of variable air volume (VAV). This involves a technique for controlling the capacity of a heating, ventilating, and/or air-conditioning (HVAC) system. This means that with these systems persons who have conditioned air circulating in, on, or around them can control the temperature in their own particular personal space. For example, if two individuals are on the same system and one seeks to increase the temperature, the system can heat that person's space and cool the other. Another advantage is that VAV can change the volume of air delivered to the space and also damper off a space that is not used or occupied, thereby increasing efficiency. The fan capacity control, especially with modern electronic variable-speed drives, reduces the energy consumed by fans, which can be a substantial part of the total cooling energy requirements of a building. Dehumidification with VAV systems is also greater than it is with constant volume systems, which modulate the discharge air temperature to attain part of the load cooling capacity.

Some estimates suggest that buildings in the United States annually consume about 42 percent of America's energy and 68 percent of its electricity. Of this, HVAC systems consume a significant percentage. Energy sources that provide power to an HVAC system are usually gas, solid fuels, oil, or electricity, and the conducting medium is usually water, steam, or gas. The heating and cooling source equipment is comprised of components that use the energy source to heat or cool the conducting medium. The heating and cooling units (such as air conditioners and air-handling units) are the components of the system that are instrumental in modifying the air temperatures in the interior spaces.

Scientists and others have known for many years that physical comfort is critical to work effectiveness, satisfaction, and physical and mental wellbeing. As discussed in Chapter 7 on indoor environmental quality, we know that uncomfortable conditions in the workplace, such as noise, inadequate lighting, uncomfortable temperature, high humidity, ergonomics and other physiological stressors, invariably restrict the ability of people to function to their full capacity, leading in many cases to lower job satisfaction and increases in building-related illness (BRI) symptoms. And since humans generally spend most of their time indoors, health, wellbeing, and comfort in buildings are crucial issues.

9.3.2 Choosing Refrigerants

A refrigerant is a chemical compound that is used as the heat carrier; changes from gas to liquid and then back to gas in the refrigeration cycle. Refrigerants are used primarily in refrigerators/freezers, air conditioners, and fire-suppression systems. Mike Opitz, Certification Manager, LEED™ for Existing Buildings (USGBC), states: "Chemical refrigerants are the heart of a large majority of building HVAC and refrigeration equipment. These manufactured fluids provide enormous benefits to society but in recent decades have been found to have harmful consequences when released to the atmosphere. All refrigerants in common use until the 1990s caused significant damage to the protective ozone layer in the earth's upper atmosphere, and most also enhanced the greenhouse effect, leading to accelerated global warming." For example, in the

EA Credit 4 (Enhanced Refrigerant Management) for NC, Schools, and CS, points can be earned by either not using refrigerants (Option 1) or by selecting environmentally friendly refrigerants and heating, ventilating, air conditioning, and refrigeration (HVACR) that minimize or eliminate the emission of compounds that contribute to ozone depletion and global warming (Option 2).

To address these and other relevant issues, on September 21, 2007, parties to the Montreal Protocol, including the United States, overwhelmingly agreed to accelerate the phaseout of hydrochlorofluorocarbons (HCFCs) to protect the ozone and combat climate change with adjustments beginning in 2010 to production and consumption allowances for developed and developing countries. This refrigerant phaseout of CFCs (production of CFCs ceased in 1995) and HCFCs will have a significant impact on proposed real-estate purchases that still utilize this equipment. This means that owners and administrators need to take the long view when making decisions about their capital investments.

The 2007 Montreal Protocol definitely energized the green-building movement and equipment manufacturers to undergo changes in the types of refrigerants used in certain equipment because of general environmental concerns and to seek more suitable environmental alternatives. It is no longer a question of whether facilities managers will upgrade their HVAC and other equipment but when and in what way.

The Environmental Protection Agency (EPA), in accordance with the Montreal Protocol, is now obligated to phase out hydrochlorofluorocarbon (HCFC) refrigerants used in heat-pump and air-conditioning systems because of their impact on ozone depletion. Chlorofluorocarbon (CFC) refrigerant manufacture has been banned in the United States since 1995. To date, the main alternatives are hydrofluorocarbons (HFCs) and HFC blends, although there are several potential non-HFC alternatives as well. DuPont has produced a complete family of easy-to-use, non-ozone-depleting HFC retrofit refrigerants for CFC and HCFC equipment. But while HFCs may be suitable as short- to medium-term replacements, they may not be suitable for long-term use due to their high global-warming potential (GWP).

In some categories of the LEED™ Rating System, Fundamental Refrigerant Management is included as a prerequisite, the intent being to reduce ozone depletion.

The LEED™ Requirements are to have zero use of CFC-based refrigerants in new HVACR systems. When reusing existing base building HVAC, a comprehensive CFC phaseout conversion must be conducted prior to project completion. Various categories of the U.S. Green Building Council LEED™ programs award one credit point for using non-ozone-depleting HFC refrigerants.

9.3.3 Types of HVAC Systems

While there are a wide variety of HVAC systems in use in today's real estate, no system is right for every application. In order to service the various needs, there are a number of different types of systems available (e.g., single-zone/multiple-zone, constant-volume/variable-air-volume). The most common classification of HVAC systems is by the carrying mediums used to heat or cool a building. The two main transfer mediums for this purpose are air and water. On smaller projects, electricity is often used for heating, although some systems now use a combination of transfer media. HVAC systems range in complexity from stand-alone units that serve individual rooms or zones to large, centrally controlled systems serving multiple zones in a building.

Heating systems are used to regulate temperature in commercial and public offices, residences, and other facilities. Heating systems can be either central or local. The most commonly used setup is the central heating system, where the heating is concentrated in a single central location, from where it is then circulated for various heating processes and applications.

Electric heating is a device that transforms electrical energy into heat. Every electric heater contains an electric resistor, which acts as its heating element. The practice of using electricity for heating is becoming increasingly popular in both residences and public buildings. Electric heating generally costs more than energy obtained from combustion of a fuel, but its convenience, cleanliness, and reduced space needs have often justified its use. The heat can be provided from electric coils or strips used in varying patterns, such as convectors in or on the walls, under windows, or as baseboard radiation in part or all of a room. By the incorporation of a heat pump the overall cost of electric heating can be substantially reduced.

Electric baseboard heaters are a fairly common heat source, heating the room by using a process called electric resistance. These types of heaters are zonal heaters controlled by thermostats located within each room. Inside baseboard heaters are electric cables, and it is these that warm the air that passes through. Electric baseboard heaters are typically installed along the lower part of outside walls to provide perimeter heating. Room air heated by the resistance element rises and is replaced by cooler room air, establishing a continuous convective flow of warm air while in operation.

The vast majority of modern commercial buildings, including office buildings, high-rise residential buildings, hotels, and shopping malls, are today provided with some form of central heat. There are many different types of central heating systems on the market, most of which consist of a central boiler (which is actually a heat generator because the water is not "boiled"; it peaks at 82 to 90 degrees C) or furnace to heat water, pipes to distribute the heated water, and heat exchangers or radiators to conduct this heat to the air. However, there is no such thing as a standard central-heating system, and each project requires the system to be tailored to meet its own requirements. With advanced controls a correctly programmed central-heating system when optimized will be able to constantly monitor and automatically adjust the system on its own.

Central heating differs from local heating in that the heat generation occurs in one place, such as a furnace room in a house or a mechanical room in a large building. The most common method to generate heat involves the combustion of fossil fuel in a furnace or boiler. The resultant heat then gets distributed: typically by forced-air through ductwork, by water circulating through pipes, or by steam fed through pipes. Increasingly, many buildings utilize solar-powered heat sources, in which case the distribution system normally uses water circulation.

Most modern systems have a pump to circulate the water and ensure an equal supply of heat to all the radiators. The heated water is often fed through another heat exchanger inside a storage cylinder to provide hot running water. Forced-air systems send air through ductwork, which can be reused for air conditioning, and the air can be filtered or put through air cleaners. The heating elements (radiators or vents) are ideally located in the coldest part of the room, typically next to the windows. A central-heating system provides warmth to the whole interior of a building (or portion of a building) from one point to multiple rooms.

A furnace is a heating-system component designed to heat air for distribution to various building spaces. Furnaces can be used for residential and small commercial heating systems. Furnaces use natural gas, fuel oil, and electricity for the heat source as well as on-site energy collection and heat transfer. Natural-gas furnaces are available in condensing and noncondensing models. The cooling can be packaged within the system, or a cooling coil can be added. When direct expansion systems with coils are used, the condenser can be part of the package or remote. The efficiency of new furnaces is measured by the annual fuel-utilization efficiency (AFUE), a measure of seasonal performance. Today's furnaces are designed to be between 78 and 96 percent AFUE. Traditional "power combustion" furnaces are 80 to 82 percent AFUE. Above 90 percent AFUE, a furnace is "condensing," which generally means that it recaptures some of the heat wasted in traditional systems by condensing escaping water vapor.

Radiant heating is increasing in popularity because it is clean, quiet, efficient, dependable, and invisible is provided in part by radiation in all forms of direct heating, but the term is usually applied to systems in which floors, walls, or ceilings are used as the radiating units. Steam or hot-water pipes are placed in the walls or floors during the construction process, and radiant-heating systems circulate warm water through continuous loops of tubing. The tubing system transfers the heat into the floor and upward into virtually any surface, including carpeting, hardwood, parquet, quarry and ceramic tile, vinyl flooring, or concrete. If electricity is used for heating, the panels containing heating elements are mounted on a wall, baseboard, or ceiling of the room. Radiant heating provides uniform heat and is both efficient and relatively inexpensive to operate.

There are basically two types of warm-air heating systems: gravity and forced-air. The gravity system operates by air convection and is based on the principle that when air is heated it expands, becomes lighter, and rises. Cooler air is dense and therefore falls. The difference in air temperature creates the convection or motivation for air movement. The return of a gravity system must be unrestricted, and even a filter is considered too restrictive. This is necessary to develop positive convection and better distribution. The furnace consists of a burner compartment (firebox) and a heat exchanger. The heat exchanger is the medium used to transfer heat from the flame to the air, which moves via ducts to the various rooms. Besides being the medium of heat transfer, the heat exchanger keeps the burned fuels separate from the air. Often the furnace is arranged so the warm air passes over a water pan in the furnace for humidification before circulating through the building. As the air is heated, it passes through the ducts to individual grilles or registers (which may be opened or closed to control the temperature) in each room of the upper floors.

As with gravity warm-air heating systems, the heat exchanger is the medium of heat transfer and separates the burned fuel from the air that moves through the building. Forced-circulation systems typically have a fan or blower placed in the furnace casing, which blows air through an evaporator coil that cools the air (Figure 9.5). This cool air is routed throughout the intended space by means of a series of air ducts, thus ensuring the circulation of a large amount of air even under unfavorable conditions. When combined with cooling, humidifying, and dehumidifying units, forced-circulation systems may be effectively used for heating and cooling. The ability to utilize the same equipment to provide air conditioning throughout the year has given added impetus to the use of forced-circulation warm-air systems in residential installations.

Hot-water systems typically have a central boiler, in which water is heated to a temperature of from 140 to 180 degrees F (60 to 83 degrees C) and is then circulated by means of pipes to some type of coil units, such as radiators, located in the various rooms. Circulation of the hot water can be accomplished by pressure and gravity, but

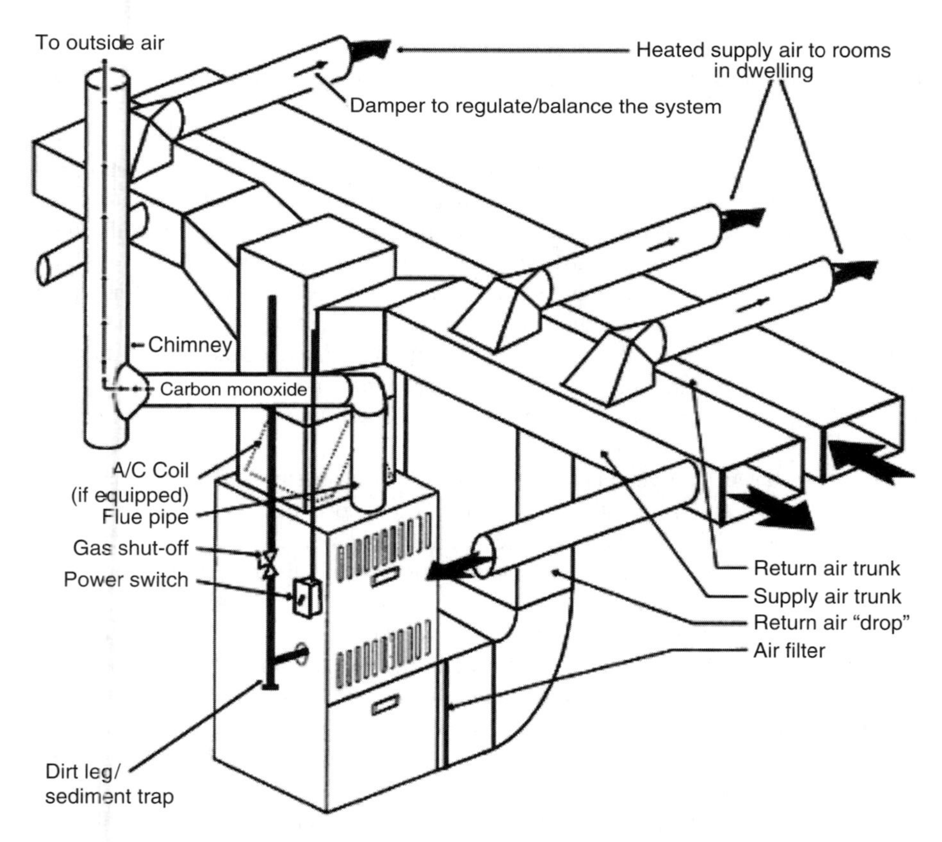

Figure 9.5 Diagram describing how a forced warm-air system operates.
Source: warmairinc.com.

forced circulation using a pump is more efficient because it provides flexibility and control. In the rooms, the emitters give out the heat from their surfaces by radiation and convection. The cooled water is then returned to the boiler. In addition, there are combination systems that use ducts to supply air from the central air-handling unit, and water to heat the air before it is transferred into the conditioned space. Combination boiler heating systems are the most commonly used in central-heating systems. Running on mains pressure, water eliminates both the need for tanks to be placed in the loft and the need for a hot-water cylinder, as the water is instantly heated when needed. Although hot-water circulating systems generally provide convenience and save water, they can also expend large amounts of energy and may not be cost-effective.

In boiler hydronic systems there are different approaches to arrange the piping, depending on the budget at installation and the efficiency level required. Generally speaking, hot-water systems use either a one-pipe or a two-pipe system to circulate the heated water. The one-pipe system uses less pipe than the two-pipe arrangement, which is why it is less expensive to install. However, it is also less efficient, because larger radiators or longer baseboards are required at the end of the loop, as this part of the loop gets less heat. The operation of a one-pipe system is fairly simple: water enters each radiator from the supply side of the main pipe, circulates through the radiator, and flows back into the same pipe.

Most modern layouts use a two-pipe layout, in which radiators are all supplied with hot water at the same temperature from a single supply pipe, and the water from these radiators flows back to the furnace to be reheated through a common return pipe. Although the two-pipe system requires more pipework, it is more efficient and easier to control than the one-pipe system. Another advantage of the two-pipe direct-return and reverse-return loop over the one-pipe series loop is that it can be zoned. Zoning offers additional control over where and when heat is required, which in turn can reduce heating costs. As with many hydronic loop systems, the two-pipe direct return needs balancing valves, and in both systems an expansion tank is required to compensate for variations in the volume of water in the system. Another system that is sometimes used is the sealed hot-water system. This is basically a closed system that does not need water tanks because the hot water is supplied directly from the mains.

Steam heating systems are similar to water heating systems except that steam is used as the heating medium instead of water. Steam is often used to carry heat from a boiler to consumers as heat exchangers, process equipment, etc. Sometimes steam is also used for heating purposes in buildings. Steam heating systems closely resemble their hydronic counterparts except that steam rather than hot water is circulated through the pipes to the radiators and no circulating pumps are required. The steam condenses in the radiators and/or baseboards, giving up its latent heat. Both one-pipe and two-pipe arrangements are employed for circulating the steam and for returning to the boiler the water formed by condensation.

Each heating unit in a one-pipe gravity-flow system has a single pipe connection through which it simultaneously receives steam and releases condensate. All heating units and the end of the supply main are sufficiently above the boiler water line so that condensate is able to flow back to the boiler by gravity. In a two-pipe system, steam supply to the heating units and condensate return from heating units are through separate pipes. Air accumulation in piping and heating units discharges from the system through the open vent on the condensate pump receiver. Piping and heating units must be installed with proper pitch to provide gravity flow of all condensate to the pump receiver. The three main types of steam systems generally employed are: air-vent, vacuum or mechanical-pump, and vapor systems.

Heat pumps are actually air conditioners that run in reverse to bring heat from outdoors into the interior. This works by heating up a piped refrigerant in the outdoor air, then pumping the heat that is generated by the warmed refrigerant inside to warm the indoor air. This type of system works best in moderate climates and becomes less efficient in very cold winter temperatures when electrical heat is needed for auxiliary heating demands. This is because, as the outdoor temperature begins to drop, the heat loss of a space becomes greater, requiring the heat pump to operate for longer stretches of time to be able to maintain a constant indoor temperature. As with furnaces, heat pumps are usually controlled by thermostats. Figure 9.6 illustrates a typical residential application of a water pump system.

A recently introduced heat-pump variant, called a reverse-cycle chiller, heats or cools an insulated tank of water and then distributes the heating or cooling either through fans and ducts or radiant floor systems. The need for auxiliary electric heating coils and defrosting cycles to prevent icing of the refrigerant is eliminated, thus making these systems more suitable for cold climates. Newer models also now offer solar-powered hot-water heating for the unit. These systems still require an exterior condenser unit similar to traditional HVAC systems.

A geothermal heat pump (GHP) is a relatively new technology that is gaining wide acceptance for both residential and commercial buildings. Studies show that

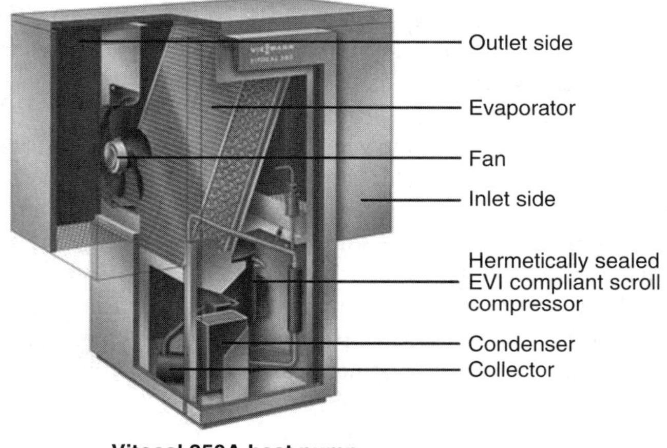

Figure 9.6 The Vitocal 350A-G air/water heat pump; heating output is 10.6 to 18.5 kW. *Source*: Viessmann Werke.

approximately 70 percent of the energy used in a geothermal heat-pump system is renewable energy from the ground. This system utilizes the relatively constant temperature of the ground or water several feet below the earth's surface as the source of heating and cooling. The earth's constant temperature is what makes geothermal heat pumps one of the most efficient, comfortable, and quiet heating and cooling technologies available today. Geothermal heat pumps are appropriate for retrofit or new facilities where both heating and cooling are desired, and business owners around the United States are now installing geothermal heat pumps to heat and cool their buildings. These systems can also be located indoors, because there is no need to exchange heat with the outdoor air. Although this technology may be more expensive to install than traditional HVAC systems, geothermal systems will greatly reduce gas or electric bills through reduced energy, operation, and maintenance costs, thus allowing for relatively short payback periods (Figure 9.7). Conventional ductwork is generally used to distribute heated or cooled air from the geothermal heat pump throughout the building.

When selecting ground-source heat pumps, models should be chosen that qualify for the ENERGYSTAR® label or that meet the recommended levels of coefficient of performance (COP) and energy-efficiency ratio (EER). Efficiency is measured by the amount of heat a system can produce or remove using a given amount of electricity. A common measurement of this performance is the seasonal energy-efficiency ratio (SEER). The

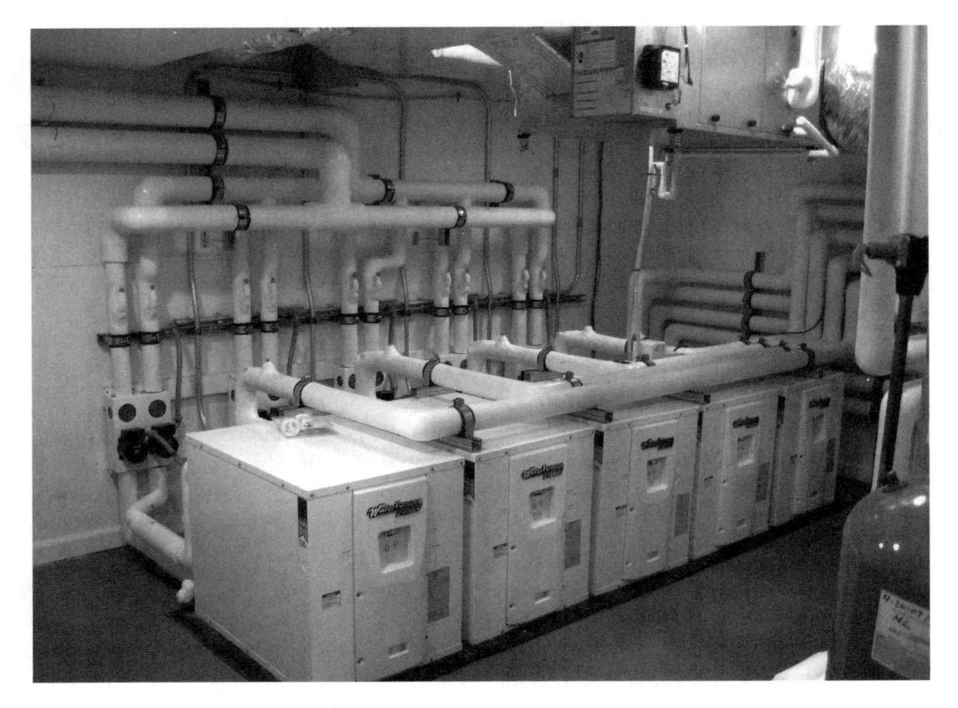

Figure 9.7 A residential geothermal heat-pump system.
Source: Climate Heating and Cooling, Inc.

Federal Appliance Standards, which took effect on January 23, 2006, will require that central air conditioners be a minimum of 13 SEER. Most manufacturers now offer SEER 10, 11, 12, and 13 models, and some offer SEER 14. This translates into five separate efficiency options, with model numbers usually keyed to the SEER numbers so they are easy to recognize. The most efficient models, however, generally involve dual compressor systems and increased heat-exchange area, and thus cost significantly more.

Heat-pump performance continues to improve with the ongoing development of new technologies and innovations. The introduction of two-speed compressors allows heat pumps to operate close to the heating or cooling capacity that is needed at any particular moment. This saves large amounts of electrical energy and reduces compressor wear. Some heat pumps are equipped with variable-speed or dual-speed motors on their indoor fans, outdoor fans, or both. The variable-speed controls for these fans attempt to keep the air moving at a comfortable velocity, minimizing cool drafts and maximizing electrical savings. Another advance in heat-pump technology is a device called a scroll compressor, which compresses the air or refrigerant by forcing it into increasingly smaller areas. This device uses two interleaved scrolls to pump, compress, or pressurize fluids such as liquids and gases.

A vapor-compression refrigeration unit uses a circulating liquid refrigerant as the medium; it absorbs and removes heat from the space to be cooled and subsequently rejects that heat elsewhere. It is the most common refrigeration system used today for air conditioning of large public buildings, private residences, hotels, hospitals, theaters, and restaurants. These systems have four essential components: a gas compressor, an evaporator, a condenser (heat transfer), and an expansion valve (also called a valve).

Solar thermal collectors are considered to be one of the renewable-energy technologies with the best economics. A solar collector specifically intended to collect heat – that is, to absorb sunlight to provide heat – may be used to heat air or water for building heating purposes. A solar collector operates on a very simple basis. The radiation from the sun heats a liquid, which goes to a hot-water tank. The liquid heats the water and flows back to the solar collector. Water-heating collectors may replace or supplement a boiler in a water-based heating system. Air-heating collectors may replace or supplement a furnace. Solar collectors have an estimated life time of 25 to 30 years or more and require very little maintenance – only control of antifreeze and pressure in the system.

There are various types of solar systems, which can be either active or passive. These terms refer to whether the systems rely on pumps or only thermodynamics to circulate water. As solar energy in an active solar system is typically collected at a location remote from the spaces requiring heat, solar collectors are normally associated with central systems. Solar water-heating collectors may also provide heated water that can be used for space cooling in conjunction with an absorption refrigeration system. Since the sun provides free energy, a saving of up to 70 percent can be made from the energy that would otherwise be used for heating the water. Besides the economical reward there is also a significant environmental advantage. By using solar collectors for heating water an average family can save up to one ton of CO_2 per year. Figure 9.8 illustrates an example of an indirect active solar system.

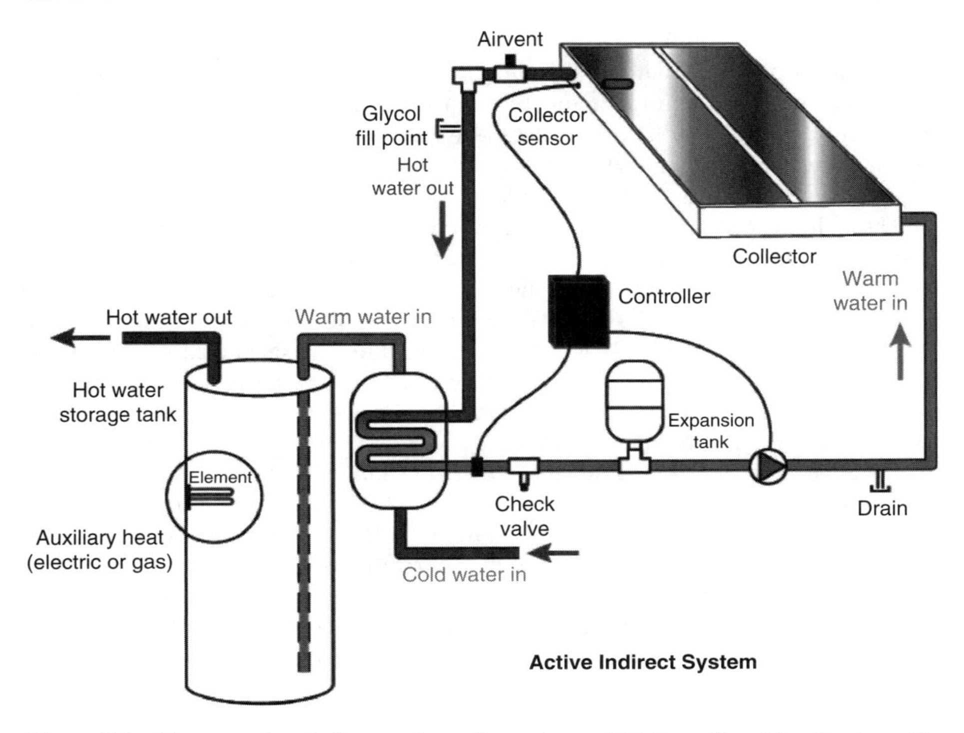

Figure 9.8 Diagram of an indirect active solar system, which is preferred in climates with extended periods of below-freezing temperatures.
Source: Southface Energy Institute.

According to ASHRAE, an air-conditioning system is a system that accomplishes four specific objectives simultaneously: to control air temperature; to control air humidity; to control air circulation; and to control air quality. The fact is that air-conditioning systems are typically designed to provide heating, cooling, ventilation, and humidity control for a building or facility. It is much easier to incorporate air-conditioning systems into new offices and public buildings under construction than to retrofit existing buildings because of the bulky air ducts that are required.

There are many types of air-conditioning systems; some use direct expansion coils for cooling, such as window units, package units, split-system air conditioners, packaged terminal air conditioners like those used in hotels, and mini-split ductless air conditioners. Other types of air conditioners utilized chilled water; these are typically used in large commercial buildings. But whichever type of air conditioner is used, the coils in the system are required to be brought to a temperature colder than the air.

When operating properly, air conditioners use the direct expansion coils or chilled-water coils to remove the heat from the air as air is blown across them. The evaporator coil in an air-conditioning system is responsible for absorbing heat. The evaporator and condenser both are comprised of tubing, usually copper, surrounded by aluminum fins.

As air (or water in a chiller) passes over the evaporator coils, a heat-exchange process takes place between the air and the refrigerant. The refrigerant absorbs the heat and evaporates in the indoor evaporator coils, drawing the heat out of the air and cooling the facility. Finally, the hot refrigerant gas is pumped (through the compressor) into the outdoor condenser unit, where it returns to liquid form. There are several different types of systems that are currently available.

Through-the-wall or window electric air-conditioning units are often used in single zone applications that do not have central air conditioning installed, such as small buildings, studio apartments, and trailers. They are also used in retrofit situations in conjunction with an existing system. They are small ductless units with casings extending through the wall, generally noisy, and designed to cool small areas, though some larger units may be able to cool larger spaces. The primary advantage of a wall unit over a window unit is that it does not occupy window space. When a window unit is installed, the window becomes unusable.

Removal of the cover of an unplugged window unit will show that it is comprised of a compressor, an expansion valve, a hot coil (on the outside), a chilled coil (on the inside), two fans, and a control unit (Figure 9.9). The wall unit works by removing hot air from the room into the unit, and the hot air that enters the unit is brought over the air-conditioning condenser and cooled. The cooled air is then pushed back into the room. Many of the newer units incorporate significant innovations such as electronic touchpad controls, energy-saver settings, and digital temperature readouts. Some units also incorporate timers that allow you to set the AC to cycle on and off at certain times of the day.

9.3.3.1 Central Air Conditioning Systems

Air conditioning is one of those amenities that is frequently taken for granted, and often, particularly in relatively warm climates, we find it has become more the rule than the exception. In addition to cooling, A/C systems dehumidify and filter air, making it more comfortable and cleaner.

Many central HVAC systems today tend to serve one or more thermal zones, with their main components located outside the zone or zones being served – usually at a convenient central location in, on, or near the building. Central air-conditioning systems are widely installed in offices, public buildings, theaters, stores, restaurants, and other building types. Although centralized air-conditioning systems provide fully controlled heating, cooling, and ventilation, they need to be installed during construction. In recent years, these systems have increasingly become automated by computer technology for purposes of energy conservation. In older buildings, indoor spaces may be equipped with a refrigerating unit, blowers, air ducts, and a plenum chamber in which air from the interior of the building is mixed with air from the exterior. Such installations are used for cooling and dehumidifying during the summer months, and the regular heating system is used during the winter. Figure 9.10 shows the general components of a small central air-conditioning system that can be used in small commercial buildings or residences.

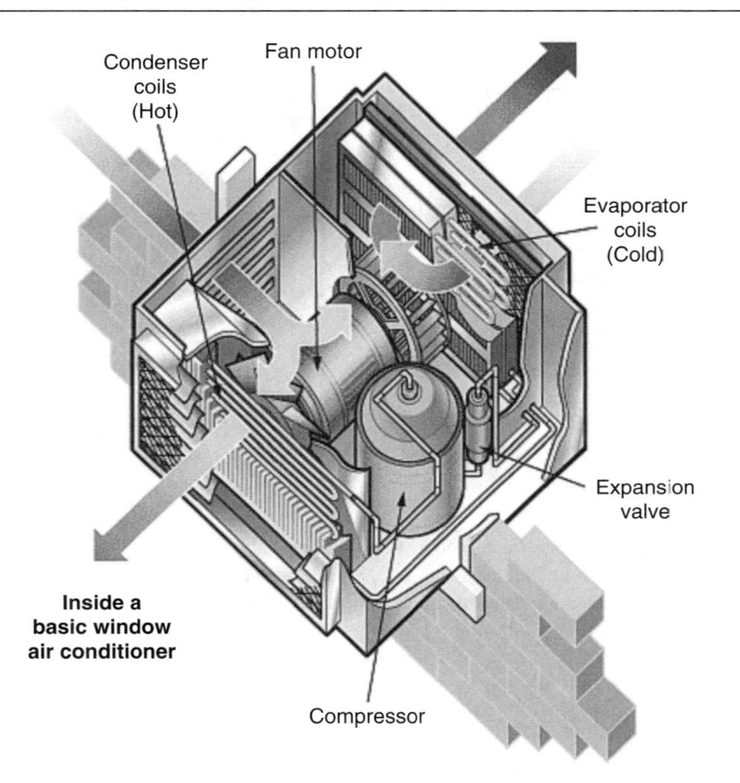

Figure 9.9 Components of a basic window A/C unit: a compressor, an expansion valve, a hot coil (on the outside), a chilled coil (on the inside), two fans, and a control unit.
Source: HowStuffWorks.com.

Central air-conditioning systems consist of three main components: the outdoor unit (condenser and compressor), the indoor unit (blower coil or evaporator), and the indoor thermostat (Figure 9.9). The success of a central air system is dependent upon these three systems appropriately functioning together. Likewise, the design of an air-conditioning system depends, among other things, on the type of structure in which the system is to be placed, the amount of space to be cooled, the function of that space, and the number of occupants using it. For example, a room or building with large windows exposed to the sun or an indoor office space with many heat-producing lights and fixtures requires a system with a larger cooling capacity than a space with minimal windows in which cool fluorescent lighting is used. A space in which the occupants are allowed to smoke will require greater air circulation than a space of equal capacity in which there is no smoking. Air-conditioned buildings often have sealed windows, because open windows would disrupt the attempts of the HVAC system to maintain constant indoor-air conditions.

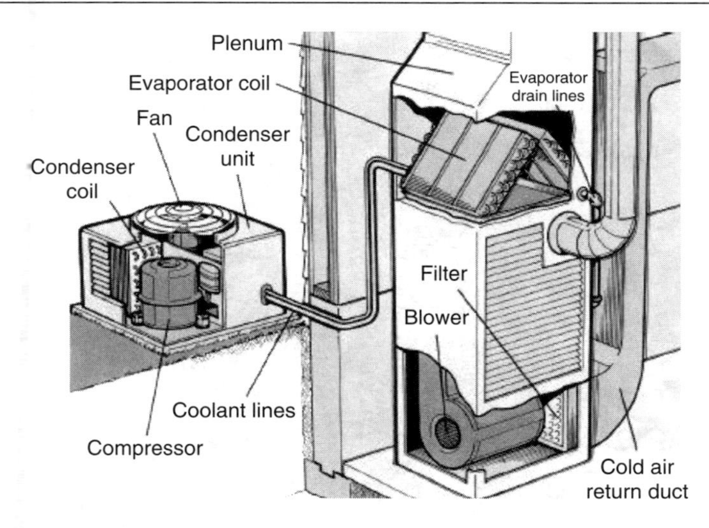

Figure 9.10 Domestic central air conditioners are made up of two basic components: the condenser unit, located outside the house on a concrete slab, and the evaporator coil, above the furnace. These components in turn also are comprised of several elements.
Source: HowStuffWorks.com.

9.3.3.2 Split Systems

A split A/C system uses at least two pieces of hardware and generally implies that the condenser and compressor are located in an outdoor cabinet and the refrigerant metering device and the evaporator stored in an indoor cabinet. With many split-system air conditioners, the indoor cabinet also houses a furnace or a part of the heat pump. The cabinet or main supply duct of this furnace or heat pump also houses the air conditioner's evaporator coil. For reverse-cycle applications, the heat exchangers can swap roles, with the heat exchanger exposed to outside air becoming the evaporator and the inside heat exchanger becoming the condenser (Figure 9.11). Split systems may have a variety of configurations. The four basic components of the vapor-compression refrigeration cycle – compressor, condenser, refrigerant metering device, and evaporator – can be grouped in several ways. The grouping of components is based on practical considerations, such as available space, ease of installation, reducing noise, etc. Split systems are generally more expensive to purchase but potentially less expensive to install, and a ductless system is practical for homes that don't already have ductwork. The big advantage of a ductless mini-split system is that you can adjust temperature levels for individual rooms or areas. However, split systems require more maintenance than single-box central air conditioners.

The condensing unit can sometimes be quite massive, and it is normally located on the roof in the case of department stores, businesses, malls, warehouses, etc. An alternative is to have many smaller units on the roof, each attached inside to a small air handler that cools a specific zone within the building. The split-system approach may

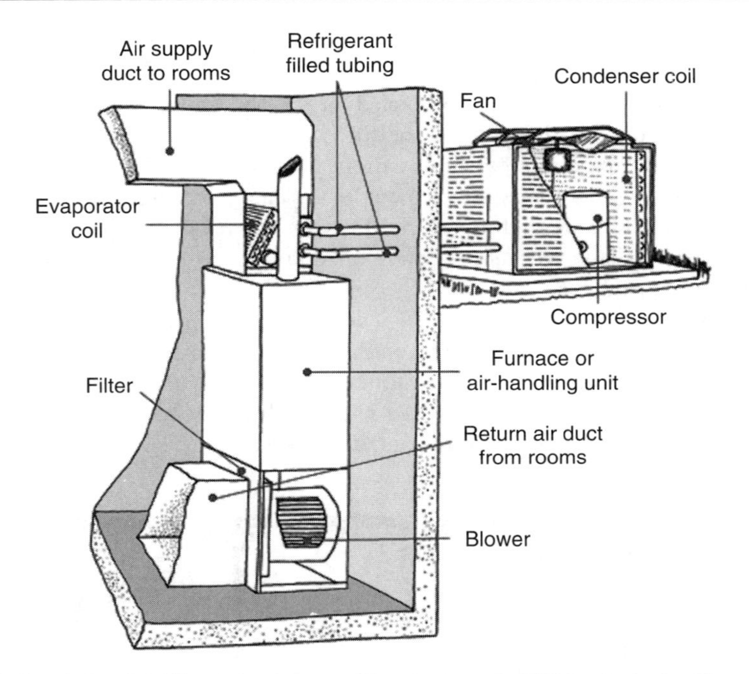

Figure 9.11 A drawing illustrating how a split system works. With a typical split system, the condenser and the compressor are located in an outdoor unit; the evaporator is mounted in the air-handling unit, which is often a forced-air furnace. With a package system, all of the components are located in a single outdoor unit, which may be located on the ground or on the roof of the house.
Source: HomeTips.com.

not always be suitable for larger buildings, particularly multistory buildings, because problems start to appear: the distance between the condenser and the air handler may exceed pipe distance limitations or the amount of ductwork and the length of ducts may cease to be viable.

9.3.3.3 Packaged Systems

Packaged air-conditioner models are used for medium-sized halls and multiple rooms on the same floor. They are usually used for applications where air conditioning of more than five tons is required. The main difference between a package and a split-system unit is that the latter uses indoor and outdoor components to provide a complete comfort system, whereas the former requires no external coils, air handlers, or heating units. Packaged units commonly use electricity to cool and gas to heat. All their components are typically located in a single outdoor unit located either on the ground or roof. A packaged rooftop unit is a self-contained air-handling unit (AUH), typically used in low-rise buildings and mounted directly onto roof curbs, discharging conditioned air into the building's air-duct distribution system. AUHs come in many capacities, from units of just over one ton to systems of several hundred tons that contain multiple compressors and are designed for single- or multiple-zone applications.

All-air systems transfer cooled or heated air from a central plant via ducting, distributing air through a series of grilles or diffusers to the room or rooms being served. They represent the majority of systems currently in operation. The overall energy used to cool buildings with all-air systems includes the energy necessary to power the fans that transport cool air through the ducts. Because the fans are usually placed in the airstream, fan movement heats the conditioned air (Figure 9.12), thus adding to the thermal-cooling peak load.

Robert McDowall, author of *Fundamentals of HVAC Systems*, says: "When the ventilation is provided through natural ventilation, by opening windows or other means, there is no need to duct ventilation air to the zones from a central plant. This allows all processes other than ventilation to be provided by local equipment supplied with hot and chilled water from a central plant. These systems are grouped under the term

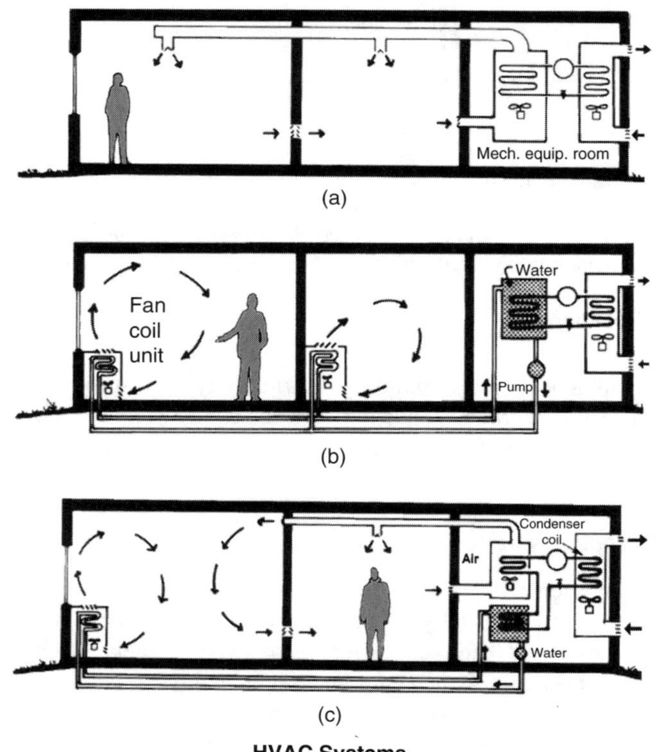

Figure 9.12 Diagram illustrating three basic HVAC systems. (a) All-air system (b) All-water system. (c) Air-water system.
Source: N. Lechner.

all-water systems. Most all-water systems are heating systems." McDowall also notes that "Both the air-and-water and all-water systems rely on a central supply of hot water for heating and chilled water for cooling."

The conditioning effect in these systems is distributed from a central plant to conditioned spaces via heated or cooled water. Water is an effective heat-transfer medium, thus distribution pipes are often of relatively small in volume (compared to air ducts). On the other hand, water cannot be directly dumped into a space through a diffuser but requires a more sophisticated delivery device. All-water heating-only systems employ a variety of delivery devices, such as baseboard radiators, convectors, unit heaters, and radiant floors. All-water cooling-only systems are rare, with valance units being the most common delivery device for such systems. When full air conditioning is contemplated, the most appropriate delivery device may be the fan-coil unit. All-water systems are generally the most expensive to install and own and are classed as the least energy-efficient in terms of transfer of energy (Figure 9.12).

Air-water systems are central HVAC systems that distribute the conditioning effect by means of heated or chilled water and heated or cooled air (Figure 9.12).

9.3.4 HVAC System Requirements

Building spaces such as cavities between walls can support platforms for air handlers, and plenums constructed with materials other than sealed sheet metal, duct board, or flexible duct must not be used for conveying conditioned air including return air and supply air. Care should be taken to ensure that ducts installed in cavities and support platforms are not compressed in a manner that would cause reductions in the cross-sectional area of the ducts. Connections between metal ducts and the inner core of flexible ducts must be mechanically fastened, and openings must be sealed with mastic, tape, or other duct closure systems that meet the codes and standards of local jurisdictions.

National building codes generally stipulate that access be provided to certain components of mechanical and electrical systems. This is usually required for maintenance and repair and includes such elements as valves, fire dampers, heating coils, mechanical equipment, and electrical junction boxes. Commercial construction usually takes advantage of ceiling plenums to run horizontal ducts, while vertical ducts are contained within their own chases. Depending on the type of structure and depth of the plenum, large ducts may occupy much of this depth, leaving little if any space for recessed light fixtures. Where the plenum is used as a return air space, most local and national building codes prohibit the use of combustible materials such as wood or exposed wire within the space in commercial building projects.

Access flooring (typically in computer-room applications), which consists of a false floor of individual panels raised by pedestals above the structural floor, is sometimes used in commercial construction. This is designed to provide sufficient space to run electrical and communication wiring as well as HVAC ductwork. Sometimes small pipes are designed to run within a wall system, whereas larger pipes may need deeper walls or even chase walls to accommodate them. Fan systems that exhaust air from the building to the outside must be provided with backdraft or automatic dampers. Gravity ventilating systems must have an automatic or readily accessible, manually operated damper in all openings to the exterior, except combustion inlet and outlet air openings and elevator-shaft vents.

9.3.5 Common HVAC Deficiencies

HVAC deficiencies are usually maintenance-related. When maintenance of the equipment is deferred or performed by unqualified personnel, the system will increasingly experience problems. When properly maintained, a building's HVAC system can enjoy a substantial life span. HVAC deficiencies fall into two main categories: issues that are fairly simple to address, such as filter or belt replacement, and complex issues requiring the attention of specialized personnel, such as pump or boiler replacement. Another deficiency often encountered is inadequacy of the system for the size of the facility. Most designers of mechanical systems now utilize computer software to determine heating and cooling loads, but a general rule of thumb in an assessment is to compare the actual tonnage of the unit to the standard design tonnage using the formulas below:

$$\text{BTU of unit} \div 12,000 = \text{actual tonnage of unit}$$
$$\text{Square footage of the building} + 350 = \text{design tonnage}$$

Deficiencies and other issues to be checked for conformance when conducting HVAC evaluations include:

- Evidence of abnormal component vibrations or excessive noise
- Evidence of unsafe equipment conditions, including instability or absence of safety equipment (guards, grills, or signage)
- Identify the location of the thermostats
- Does the building have an exhaust system, and are the toilets vented independently or mixed with the common-area venting system?
- Insufficient air movement to reach all parts of the room or space being cooled or, alternatively, the presence of drafts in the room or space being cooled
- Return air at the return registers is not at least 10 to 15 degrees F warmer than the supply air
- Evidence of leaking caused by inadequate seals
- Evidence of fan alignment deficiencies, deterioration, corrosion, or scaling
- Is there a fresh air makeup system in the building?

9.3.6 HVAC Components and Systems

The principal function of an HVAC system is to provide building occupants with healthy and comfortable interior thermal conditions. HVAC systems generally include a number of active mechanical and electrical systems to provide thermal control in buildings. Control of the thermal environment is one of the key objectives of virtually all occupied buildings. Numerous systems and components are used in combination to provide fresh air as well as temperature and humidity control in both residential and commercial properties (Figure 9.13).

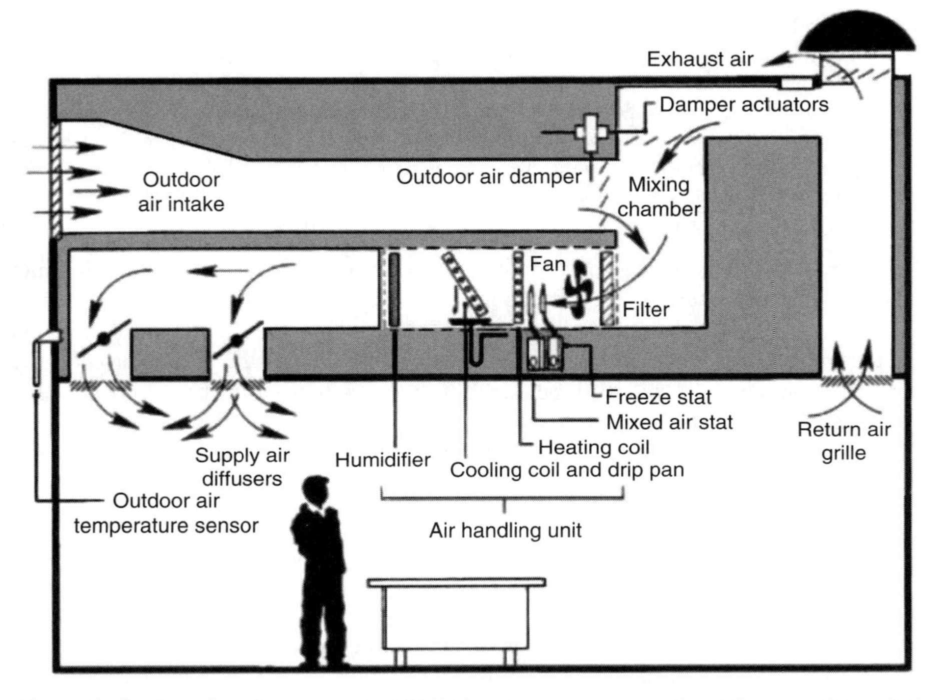

Figure 9.13 Drawing showing typical HVAC system components that deliver conditioned air to a building or space to maintain thermal comfort and indoor-air quality.
Source: Terry Brennan, Camroden Associates.

HVAC system components may best be grouped into three functional categories: source components, distribution components, and delivery components:

- Source components generally provide or remove heat or moisture. There are four basic types of heat sources employed in buildings: on-site combustion (fuel such as natural gas or coal); electric resistance (converting electricity to heat); solar collector on roof or furnace, and heat pump in furnace. Choosing a heat source for a given building depends on several factors such as source availability, fuel costs, required system capacity, and equipment costs.
- Distribution components are used to convey a heating or cooling medium from a source location to portions of a building that require conditioning. Central systems produce a heating and/or cooling effect in a single location, which is then transmitted to the various spaces in a building that require conditioning. Three transmission media are commonly used in central systems: air, water, and steam. Hot air can be used as a heating medium, cold air as a cooling medium. Hot water and steam can be used as heating media, while cold water is a common cooling medium. When a central system is used, it will always require distribution components to convey the heating or cooling effect from the source to the conditioned locations.
- Delivery components basically serve as an interface between the distribution system and occupied spaces. The heating or cooling effect produced at a source and distributed by

a central system to all the spaces within a building has to be properly delivered to each space to promote comfort and wellbeing. In air-based systems, the dumping of heated or cooled air into each space does not provide the control over air distribution required of an air-conditioning system. Likewise, with water-based systems, the heated or cooled media (water or steam) cannot just be dumped into a space. Some means of transferring the conditioning effect from the media to the space is required. Devices designed to provide the interface between occupied building spaces and distribution components are collectively termed delivery devices.

It should be noted that compact systems that only serve a single zone of a building frequently incorporate all three functions in a single piece of equipment, whereas systems that are intended to condition multiple spaces in a building (central systems) usually employ distinctly different equipment elements for each function. Furthermore, for each commercial property type, some systems will perform better than others. However, from a lender's perspective, performance is judged by how well the needs of tenants and owners are met regarding comfort, operating costs, aesthetics, reliability, and flexibility.

9.3.6.1 Ductwork

The primary objective of duct design is to provide an efficient distribution network of conditioned air to the various spaces within a building. To achieve this, ducts must be designed to facilitate air flow and minimize friction, turbulence, and heat loss and gain. Optimal air-distribution systems have correctly sized ducts with minimal runs, smooth interior surfaces, and minimum direction and size changes. Ducts that are badly designed and installed can result in poor air distribution, poor indoor-air quality (IAQ), occupant discomfort, additional heat losses or gains, increased noise levels, and increased energy consumption. Duct-system design and construction requirements can be impacted by the design of the building envelope.

A duct system's overall performance can also be impacted by the materials used. Fiberglass insulation products are currently used in the majority of duct systems installed in the United States and serve as key components of well-designed, well-operated, and well-maintained HVAC systems that provide both thermal and acoustical benefits for the life of the building. Other materials commonly used for low-pressure duct construction include sheet metal (galvanized steel), black carbon steel, aluminum, stainless steel, fiberglass-reinforced plastic, polyvinyl steel, concrete, and copper.

It is strongly advised to put in place a preventive maintenance inspection program that will help identify system breaches that are typically more prevalent at duct intersections and flexible connections. Supply and return air may both utilize ductwork that may be located in the ceiling cavity or below the floor slab, depending on the configuration of the system. In single-duct systems, both cool and hot air utilize the same duct, and in double-duct systems, separate ducts are used for cooling and heating. It should be noted that air flowing through a duct system will encounter friction losses through contact with the duct walls and in passing through the various devices such as dampers, diffusers, filters, and coils. To overcome these friction losses, a fan is utilized; it provides the energy input required to overcome friction losses and to circulate air through the system. Depending on the size of the ductwork, a central HVAC system may require the use of several fans for air supply and return and for exhaust air.

Where ducts penetrate a firewall, the use of fire dampers will be required. Modern fire dampers contain a fusible link that melts and separates when a particular temperature is reached, causing it to slam shut in the event of a fire. Another major factor that impacts the fabrication and design of duct systems is that of acoustics; unless the ducting system is properly designed and constructed, it can act as a large speaker tube transmitting unwanted noise throughout the building. Splitters and turning vanes are often used to reduce the noise that is generated within the ducts, while at the same time mitigating friction losses by reducing turbulence within the ductwork.

Grilles, registers, and diffusers: These are employed in conjunction with ductwork and assist in controlling the return, collection, and supply of conditioned air in HVAC systems. A grille is basically a decorative cover for return air inlets; it does not have an attached damper, and in most cases it has no moving parts. However, a grille can be used for both supply air and return air. The same is not true for a register or diffuser. Grilles are also used to block sightlines to prevent persons from seeing directly into return-air openings.

A diffuser is an air-flow device designed primarily to discharge supply air into a space, mix the supply air within the room air, and minimize unwelcome drafts. The location of supply-air diffusers and return-air grilles should be integrated with other ceiling elements, including lights, sprinkler heads, smoke detectors, speakers, and the like, so that the ceiling is aesthetically pleasing and well planned. To assist in this integration, these components are usually connected to the main ductwork with flexible ducting to allow some adjustability in their placement.

Registers are adjustable grille-like devices that cover the opening of a duct in a heating or cooling system, providing an outlet for heated or cooled air to be released into a room. They are similar to diffusers except that they are designed and used for floor or side-wall air-supply applications or sometimes as return-air inlets. Supply-air registers are essentially grilles equipped with double-deflection adjustable vanes at the face and a damper behind the face for balance and to control the direction of flow and/or flow rate. A diffuser performs the task of diffusing the gas. The grilles, registers, and diffusers introduce and blend the fresh air with the air of another location. When fitted together, the grilles take in fresh air, the registers blow the air out, and the diffusers scatter the air in the required space.

Thermostats are devices whose main function is to control the operation of HVAC systems, turning on heating or cooling as required. Boilers and heating systems use thermostats to prevent overheating and also control the temperature of the circulating water. Thermostats are fitted to hot-water cylinders, boilers, and radiators and in rooms. In addition to providing the basic function of maintaining comfortable indoor temperatures, modern thermostats are capable of being programmed to automatically raise or lower the temperature of a facility according to predefined schedules and allow input of weekday and weekend schedules. More sophisticated thermostats will also allow humidity control and outdoor air ventilation and even signal when the system filters need changing. Some modern thermostats can also include a communications link and demand management features that can be used to reduce air-conditioning-system energy use during periods of peak electrical demand or high electricity costs. Thermostat location should be coordinated with light switches, dimmers, and other visible control devices. They are typically placed 48 to 60 inches above the floor and away from exterior walls and heat sources.

Zoning and high-efficiency equipment can substantially increase the overall energy performance of a home or office while keeping rising energy costs to a manageable level. Zoning systems can automatically direct the flow of the conditioned air to those zones needing it and at the same time automatically switch over and provide the opposite mode to other zones, eliminating the need for constant balance and outlet adjustments based upon continuously changing indoor conditions. But zoning represents just one of several actions that are designed to improve HVAC performance and that gives building occupants personal climate control throughout their work environment.

HVAC systems can have single-zone or multizone capabilities. In a single-zone system, the entire building is considered one area, whereas in a multizone system, the building is divided into various zones, allowing specific control of each area. In fact, many of today's newest HVAC systems are designed to incorporate individually controlled temperature zones to improve occupant comfort and provide the ability to manage the heating or cooling of individual rooms or spaces. Additionally, they allow us to adjust individual room temperatures for individual preferences and to close off air flow to areas that are rarely used. A zoned HVAC system is typically provided with a series of dampers. The use of zone dampers can save on the installation cost of multiple-unit systems. But whether you are using one HVAC unit with zone dampers or multiple HVAC units, zoning can save on utility and maintenance costs. More importantly perhaps, the design of the duct system for today's zoning is an important factor in a comfortable and efficient zoning system.

9.3.6.2 Boilers

These are heating-system components designed to generate steam or hot water for distribution to various building spaces (Figure 9.14). As water cannot be used to directly heat a space, boilers are only used in central systems where hot water is circulated to delivery devices (such as baseboard radiators, unit heaters, convectors, or air-handling units). Once the delivery device is heated with the hot water, the water is returned to the boiler to be reheated, and the water circulation loop continues. Generally speaking, hot-water boilers are more efficient than steam boilers for several reasons. For example, there is less heat loss throughout the hot-water piping and the shell of the boiler, because a hot-water boiler operates at a lower temperature than a steam boiler. This means that there is less heat loss throughout the entire boiler and piping system. Also, because a hot-water boiler operates at a lower temperature, it requires less fuel or energy to convert into heat.

An on-site solar-energy collection system may serve in lieu of a boiler. Heat-transfer systems (heat pumps) likewise may serve as a substitute for a boiler. Constructed of

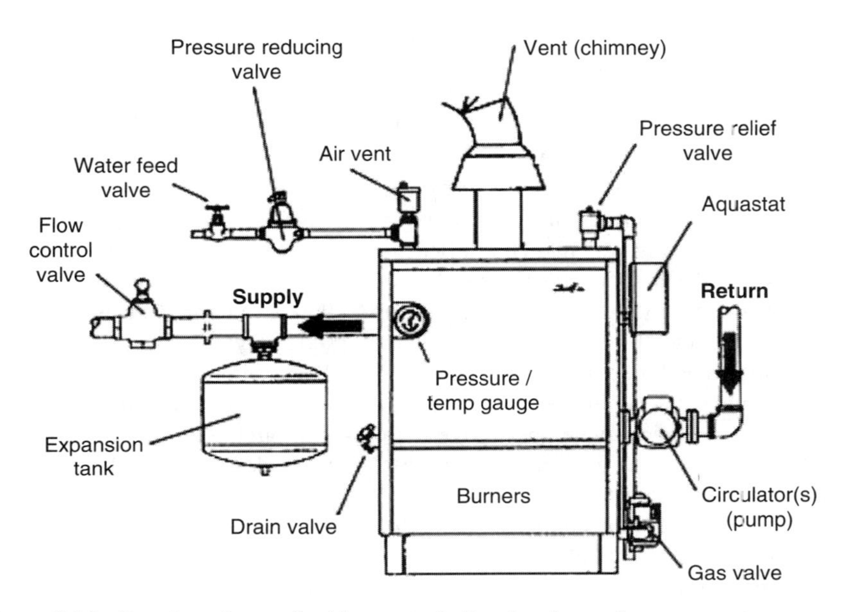

Figure 9.14 Drawing of a gas-fired hot-water boiler showing main components. *Source*: Home-Cost.com.

cast iron, steel, and occasionally copper, boilers can be fired by various fuel sources including, natural or propane gas, electricity, coal, oil, steam, hot water, and wood.

Chilled-water systems are cooling systems that remove heat from one element (water) and move it into another element (ambient air or water). They are a key component of air-conditioning systems for large buildings, although they typically use more energy than any other piece of equipment. It is similar to an air-conditioning system in that it is compressor-based, but a chiller cools liquid while an air-conditioning system cools air. Other components are a reservoir, re-circulating pump, evaporator, condenser, and temperature controller. Chillers vary in terms of condenser cooling method, cooling specifications, and process-pump specifications. The cooling fluid used is usually a mix of ethylene glycol and water.

Chillers can be air-cooled, water-cooled, or evaporatively cooled. Water-cooled chillers incorporate the use of cooling towers, which improve the chillers' thermodynamic effectiveness as compared to air-cooled chillers. This is due to heat rejection at or near the air's wet-bulb temperature rather than the higher – sometimes much higher – dry-bulb temperature. Evaporatively cooled chillers offer efficiencies better than air-cooled but lower than water-cooled. Air-cooled chillers are usually located outside and consist of condenser coils cooled by fan-driven air. Water-cooled chillers are typically located inside a building, and the heat from these chillers is carried by recirculating water to outdoor cooling towers. Evaporatively cooled chillers are basically water-cooled chillers in a box. These packaged units cool the air by humidifying it and then evaporating the moisture.

Chilled-water systems are mainly employed in modern commercial and industrial cooling applications, although there are some residential and light commercial systems in use. One of the reasons behind the popularity of chilled-water systems is because they use water as a refrigerant. Water is much less expensive than refrigerant, which makes the systems cost-effective especially in commercial HVAC air-conditioning applications. Thus, instead of running refrigerant lines over a large area of the building, water pipes are run throughout the building to evaporator coils in air handlers for HVAC air-conditioning systems. The chilled water is pumped through these pipes from a chiller, where the evaporator coil absorbs heat and returns it to the chiller to reject. Maintaining them well and operating them smartly can yield significant energy savings.

There are two main types of chillers commonly used today: the compression chiller and the absorption chiller. Absorption chillers use a heat source such as natural gas or district steam to create a refrigeration cycle that does not use mechanical compression. During the compression cycle, the refrigerant passes through four major components within the chiller: the evaporator, the compressor, the condenser, and a flow-metering device such as an expansion valve. The evaporator is the low-temperature (cooling) side of the system, and the condenser is the high-temperature (heat-rejection) side of the system. Compression chillers, depending on the size and load, use different types of compressors for the compression process. Mechanical compression chillers, for example, are classified by compressor type: reciprocating, rotary screw, centrifugal, and frictionless centrifugal.

Important factors that impact the choice of a water-cooled chiller or an air-cooled unit include whether a cooling tower is available or not. The water-chiller option is often preferred to the air-cooled unit because it costs less, has a higher cooling capacity per horsepower, and consumes less energy per horsepower. Compared to water, air is a poor conductor of heat, making the air-cooled chiller much larger and less efficient, which is why it is less frequently used unless it is not possible to construct a water cooling tower.

9.3.6.3 Cooling Towers

The main function of cooling towers is to remove heat from the water discharged from the condenser so that the water can be discharged to the environment or recirculated and reused. Cooling towers are used in conjunction only with water-cooled chillers and vary in size from small rooftop units to very large hyperboloid structures. Cooling towers are also characterized by the means by which air is moved. Mechanical draft cooling towers are the most widely used in buildings; they rely on power-driven fans to draw or force the air through the tower. They are normally located outside the building.

Mechanical draft towers can be of two types that are common to the HVAC industry: induced-draft and forced-draft. Induced-draft towers have a large propeller fan at the top of the tower (discharge end) to draw air upward through the tower while warm condenser water spills down. This type requires much smaller fan motors for the same capacity than forced-draft towers (Figures 9.15). Forced-draft towers utilize a fan at the bottom or side of the structure. Air is forced through the water spill area and discharged

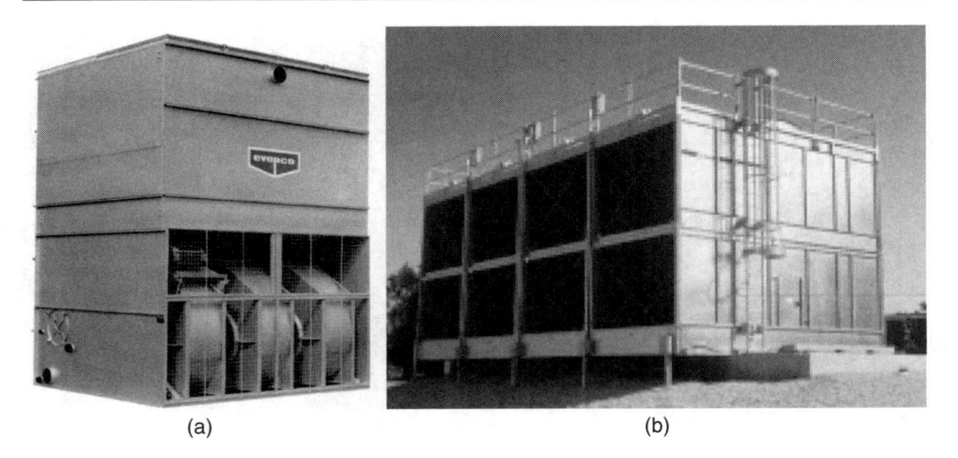

(a) (b)

Figure 9.15 (a) Forced-draft towers with fans on the air inlet to push air either counterflow or crossflow to the movement of the water. (b) Induced-draft towers have a large propeller fan at the top of the tower to draw air counterflow to the water. Induced-draft towers are considered to be less susceptible to recirculation, which can result in reduced performance.
Source: McQuay International.

out the top of the structure. After the water has been cooled in the cooling tower, it is pumped to a heat exchanger or condenser in the refrigeration unit, where it picks up heat again and is returned to the tower.

9.3.6.4 Condensers

A condenser is one of the critical components used in air-conditioning systems to cool and condense the refrigerant gas that becomes hot during the evaporation stage of the cooling process. There are two common condenser types: air-cooled and cooling tower. Achieving the cooling process is accomplished through the use of air, water, or both.

9.3.6.5 Air Filters

The primary function of air filters is to remove particles from the air. They are a critical component of the air-conditioning system; without them systems would become dirty and the interior environment would be filled with pollutants and become unhealthy (Figure 9.16). Although no individual product or system in itself can be LEED™-certified, proper employment of air-filtration systems provides tangible ways to improve indoor-air quality (IAQ) and energy efficiency and can contribute to the completion of LEED™ prerequisites and credits. This is why the right filter medium strategy is important to help buildings become "greener" and meet LEED™ and other green-building rating-system criteria.

The LEED™ 2009 Reference Guide (depending on the LEED™ certification targeted) should be consulted for more detailed information relating to achievable credits. For example for IEQ Credit 3.1 (Construction IAQ Management Plan: During

Figure 9.16 A photograph showing a clogged filter that has been removed from an A/C unit. It is critically important to change filters periodically to minimize pollution and improve IAQ.

Construction): "If permanently installed air handlers are used during construction, filtration media with a minimum efficiency reporting value (MERV) of 8 shall be used at each return air grille, as determined by ASHRAE 52.2-1999." Upon completion of construction and prior to occupancy (after all interior finishes are completed), new MERV 13 filters should be installed, followed by a two-week building flush-out achieved by supplying a total air volume of 14,000 cubic feet of outdoor air per square foot of floor area. For NC IEQ Credit 5: Indoor Chemical and Pollutant Source Control, the reference guide states: "In mechanically ventilated buildings, provide regularly occupied areas of the building with new air-filtration media prior to occupancy that provides a minimum efficiency reporting value (MERV) of 13 or better. Filtration should be applied to process both return and outside air that is to be delivered as supply air."

An understanding of ASHRAE 52.1 and 52.2 is pivotal in identifying what to look for when selecting the right filter to meet IAQ and energy-efficiency requirements to help meet green-building standards. On this point, Dave Matela of Kimberly-Clark Filtration Products says: "One of the biggest determining factors is filtration efficiency, which defines how well the filter will remove contaminants from air passing through the HVAC system. Initial and sustained efficiency are the primary performance indicators for HVAC filters. Initial efficiency refers to the filter's efficiency out of the box or immediately after installation. Sustained efficiency refers to efficiency levels maintained throughout the service life of the filter. Some filters have lower initial efficiency and do not achieve high efficiency until a 'dirt cake' builds up on the filter. Other filters offer both high initial as well as sustained efficiency, meaning they achieve an ideal performance level early and maintain that level."

There are many types of air-conditioning filters; the most common are: conventional fiberglass disposable filters (1 and 2 inches thick), pleated fiberglass disposable filters (1 and 2 inches thick), electrostatic filters, electronic filters, and carbon filters. Most air-conditioning filters are sized 1.5 to 2 square feet for each ton of capacity for a home or commercial property. Applying the MERV rating is a good way to help evaluate the effectiveness of a filter. (MERV was developed by the American Society of Heating, Refrigeration and Air Conditioner Engineers.) MERV values vary from 1 to 16; the higher the MERV value, the more efficient the filter will be in trapping airborne particles. Another consideration is air flow through the HVAC system. Leaving a dirty air filter in place or using a filter that is too restrictive may result in low air flow and possibly cause the system to malfunction. Of note, there are various types of filters with a MERV 13 rating, each having different design requirements and pressure drops. It would be prudent therefore for building owners and designers to consult with a certified air-filter specialist (CAFS) to obtain the best information on the optimum filters and prefilters to obtain LEED™ certification. Filters should be selected for their ability to protect both the HVAC system components and general indoor-air quality.

9.4 Electrical Power and Lighting Systems

9.4.1 General

Electrical systems, which provide a facility with accessible energy for heating, cooling, lighting, equipment (telecommunication devices, personal computers, networks, copiers, printers, etc.), and appliance operation, have witnessed dramatic developments in the last few decades. They are the fastest-growing energy load within a building. More than ever, facilities today need electrical systems to provide power with which most of the vital building systems operate. These systems control the energy required in the building and distribute it to the location utilizing it. Most frequently, distribution line voltage carried at utility poles is delivered at 2400 to 4160 volts. Transformers step down this voltage to predefined levels for use within buildings. In an electric-power distribution grid, the most common form of electric service is through the use of overhead wires known as a service drop, which is an electrical line running from a utility pole to a customer's building or other premises. It is the point where electric utilities provide power to their customers.

In residential installations in North America and countries that use the same system a service drop consists of two 120-volt lines and a neutral line. When these lines are insulated and twisted together, they are referred to as a triplex cable. In order for these lines to enter a customer's premises, they must usually first pass through an electric meter and the main service panel, which will usually contain a "main" fuse or circuit breaker. This circuit breaker controls all of the electrical current entering the building at once, and a number of smaller fuses/breakers protect individual branch circuits. There is always a main shutoff switch to turn off all power; when circuit breakers are used, this is provided by the main circuit breaker. The neutral line from the pole is connected to an earth ground near the service panel; often a conductive rod driven into the earth.

In residential applications the service drop provides the building with two separate 120-volt lines of opposite phase, so 240 volts can be obtained by connecting a circuit between the two 120-volt conductors, while 120-volt circuits are connected between either of the two 120-volt lines and the neutral line; 240-volt circuits are used for high-power devices and major appliances, such as air conditioners, clothes dryers, ovens, and boilers, while 120 volt circuits are used for lighting and ordinary small appliances. It should be noted that these are "nominal" numbers, meaning that the actual voltage may vary.

In Europe and many other countries, a three-phase 416Y/230-volts system is used. The service drop consists of three 240-volt wires, or phases, and a neutral wire that is grounded. Each phase wire provides 240 volts to loads connected between it and the neutral. Each of the phase wires carries 50 Hz alternating current, which is 120 degrees out of phase with the other two. The higher voltages, combined with the economical three-phase transmission scheme, allow a service drop to be longer than in the North American system, and a single drop can service several customers.

For commercial and industrial service drops, which are usually much larger and more complex, a three-phase system is used. In the United States, common services consist of 120Y/208 (three 120 V circuits 120 degrees out of phase, with 208 V line to line), 240 V three-phase, and 480 V three-phase; 575 V three-phase is common in Canada, and 380 V to 415 V or 690 V three-phase is found in many other countries. Generally, higher voltages are used for heavy industrial loads and lower voltages for commercial applications.

The difference between commercial and residential electrical installations can be quite significant, particularly with large installations. While the electrical needs of a commercial building can be simple, consisting of a few lights for some small structures, they are often quite complex, with transformers and heavy industrial equipment. When electrical- or lighting-system deficiencies become evident and need attention, they are usually measurable and include power surges, tripped circuit breakers, noisy ballasts, and other more obvious conditions such as inoperative electrical receptacles or lighting fixtures that are frequently discovered or observed during a review of the system. As illustrated in Figures 9.17 and 9.18, there are a number of typical deficiencies found in both the electrical and the lighting systems.

In many commercial buildings, the major load placed on a given electrical system comes from the lighting requirements; thus the distribution and management of electrical and lighting loads must always be monitored on a regular basis. Lighting management should also be periodically checked because space uses change and users relocate within the building. It is also highly advisable for the lighting system to be integrated with the electrical system in the facility. Lighting systems are designed to ensure adequate visibility for both the interior and exterior of a facility and are comprised of an energy source and distribution elements normally consisting of wiring and light-emitting equipment.

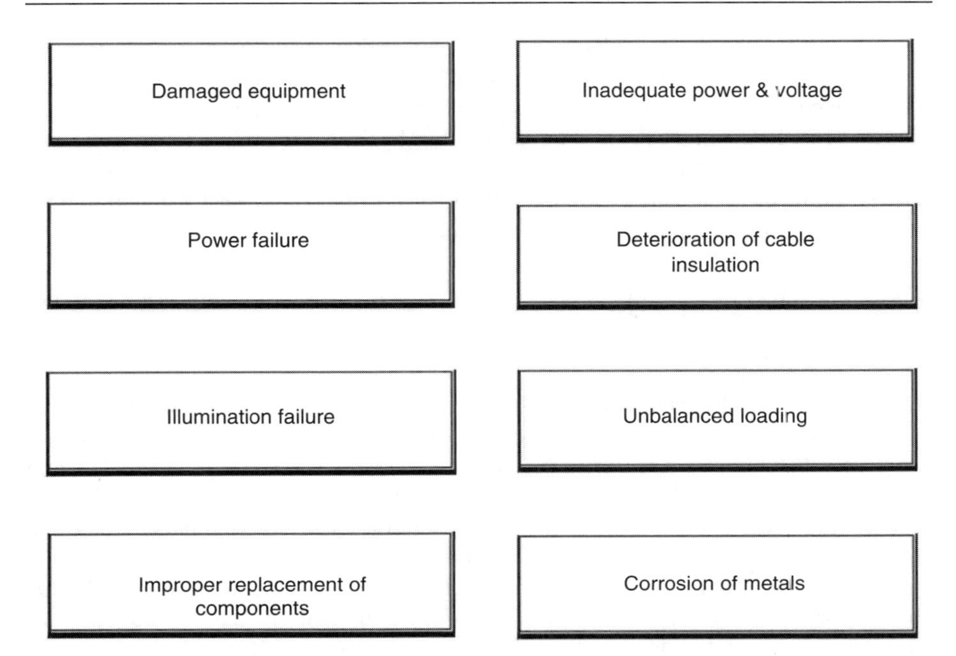

Figure 9.17 Diagram showing typical deficiencies found in electrical systems.

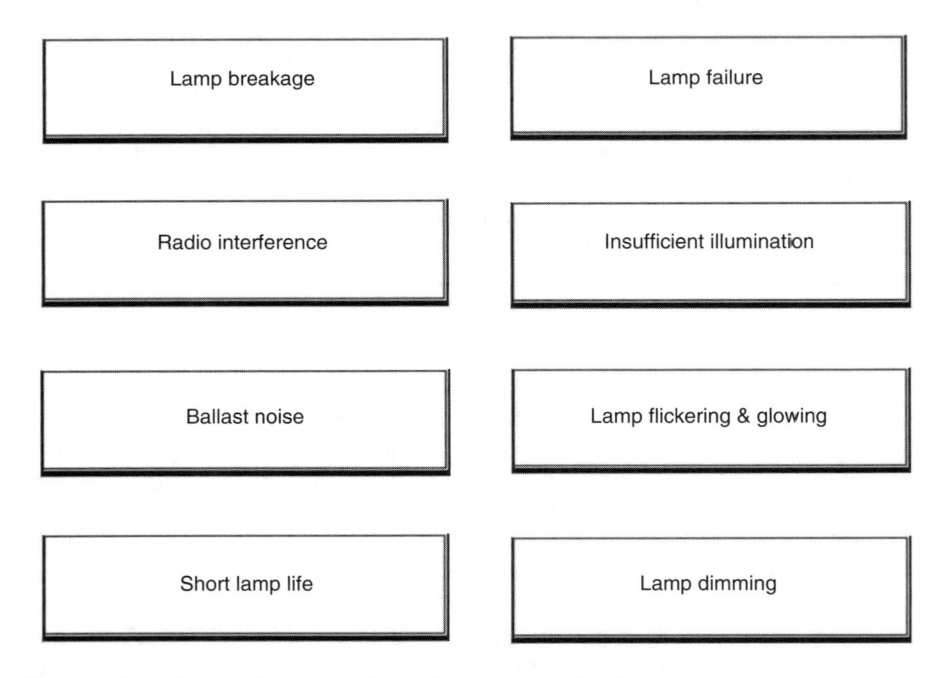

Figure 9.18 Diagram showing typical deficiencies found in lighting systems.

There are several different electrical codes enforced in various jurisdictions throughout the U.S. Some of the larger cities, such as New York and Los Angeles, have created and adopted their own electrical codes. The National Electrical Code (NEC) and the National Fire Protection Code (NFPC), published by the National Fire Protection Association (NFPA), cover almost all electrical-system components. The NEC is commonly adopted in whole or in part by municipalities. Inspection of the electrical and lighting system should include a determination of general compliance with these codes.

9.4.2 Basic Definitions

9.4.2.1 Amperage

This is a unit of electrical current. The amperage (A or amp) provided by an electrical service is the flow rate of electrical current that is available. An appliance will typically have an amp rating; if only a wattage is quoted, use the formula: amperage = wattage/voltage. Practically speaking, the voltage level provided by an electrical service, combined with the ampacity rating of the service panel, determine the electrical load or capacity. Branch-circuit wire sizes and fusing or circuit breakers typically set the limit on the total electrical load or the number of electrical devices that can run at once on a given circuit. Thus, for example, if you have a 100 A current-flow rate in place, you may be able to run approximately ten 10-amp electric heaters simultaneously. If you have only 60 A available, you won't be able to run more than six such heaters without risking overheating the wiring, tripping a circuit breaker, blowing a fuse, or causing a fire.

The service ampacity and voltage must be known to be able to determine the amount of electrical service a facility receives. The safe and proper service amperage available at a property is set by the smallest of: the service conductors, the main disconnect fuse or switch, or the rated capacity of the electric panel itself. The main fuse/circuit breaker (CB) is the only component that actively limits amperage at a property by shutting off loads drawing more than the main fuse rating. The main breakers or fuses are allowed to have lower overcurrent protection than the capability of the service equipment (panel) and conductors (entrance cables).

9.4.2.2 Volt

A volt (V) can be defined as the potential difference across a conductor when a current of one ampere dissipates one watt of power. Practically speaking, a volt is a measure of the strength of an electrical source at a given current or amperage level. If we bring 100 A into a building at 240 V, we have twice as much power available as if we bring in 100 A at 120 V. However, if we exceed the current rating of a wire, it will get hot, risking a fire. This is why fuse devices are employed to limit the current flow on electrical conductors to a safe level and thereby prevent overheating and potential fires.

As mentioned earlier, a 240 V circuit is a nominal rating that implies that the actual voltage level will vary. In many countries the actual voltage level varies around the nominal delivered voltage rating, and in fact, depending on the quality of electrical power delivered on a particular service, voltage will also vary continuously around its actual rating. Most electrical-power systems are prone to slight variations in voltage due to demand or other factors. Generally, this difference is inconsequential, as most appliances are built to tolerate current a certain percentage above or below the rated voltage. However, severe variations in current can damage electrical equipment. It is always advisable to install a voltage stabilizer where sensitive electronic equipment is used.

9.4.2.3 Watt

In electricity, a watt is equal to current (in amperes) multiplied by voltage (in volts). The formula: watts = volts × amps basically describes this relationship. In buildings the unit of electricity consumption measure is the watt-hour, which is usually measured in thousands, called kilowatt-hours (kwh). In larger buildings, not only is the total consumption rate measured but the peak demand as well.

9.4.3 Components of the Electrical System

The service connection is a device that provides a connection between the power-company service and the facility and also measures the amount of electricity a facility uses. From here a meter either feeds a disconnect switch or a main breaker or fuse panel. The connection can be located either overhead or underground. The service connection should be checked for type (i.e., voltage, amperage), general condition, and whether the total power adequately serves the facility's requirements. The equipment should be clean and free from debris or overgrown planting.

The function of switching equipment is to control the power supply in the facility and all the services arriving on the site (service drop). Switchgear is used to interrupt or reestablish the flow of electricity in a circuit. It is generally used in combination with metering, a disconnect switch, and protective and regulating equipment to protect and control motors, generators, transformers, and transmission and distribution lines. The main service switch is the system disconnect for the entire electrical service. To avoid excessive voltage drop and flicker, the distance from the transformer to the meter should not exceed 150 feet.

Switchgears are typically concentrated at points where electrical systems make significant changes in power, current, or routing, such as electrical-supply substations and control centers. Switchgear assemblies range in size from smaller, ground-mounted units to large walk-in installations and can be classified as outdoor or indoor units. Commercial and industrial assemblies are usually indoors, while utilities and cogeneration facilities are more likely to have outdoor gear. Manufactured for a variety of functions and power levels, all switchgear conforms to standards set by the Institute of Electrical and Electronic Engineers (IEEE), the American National Standards Institute (ANSI), or the National Electrical Manufacturers Association (NEMA).

A switchboard is comprised of one or more panels with various switches and indicators that are used to route electricity and operate circuits. The main switchboard controls and protects the main feeder lines of the system. Switchgear and switchboards should be readily accessible, in good condition, and have protective panels and doors. Moreover, they should be checked for evidence of overloading or burn marks. Switchboard covers should not normally be removed.

Measurement of electric consumption in a building can be accomplished in one of two ways. In residential applications, only the total electric consumption is measured. In larger facilities, both the total consumption rate and the peak demand are measured. This is because large peaks require the utility company to build more power-generating capacity to meet the demand. Commercial services of up to 200 amps single-phase may have service panels similar to those found in residences. Larger services may require stand-alone switchboards with one or more meters.

9.4.3.1 Panelboards

Electrical panelboards and their cabinets house an assembly of circuit breakers where power generation can be monitored and the power generated can be distributed. Panelboards are designed to control and protect the branch circuits in addition to providing a central distribution point for the circuits for a building, a floor, or part of a floor. Each breaker serves a single circuit, and the overload protection is based on the size and current-carrying capacity of the wiring in that circuit. A building may have a number of panelboards and a main panel, with a disconnect switch for the entire building.

Evidence of a tag (normally paper) or embossed rating on fuse pull-outs on the panel itself often includes the amperage rating of the panel. This information is usually present in newer panels on a panel side or on the panel cover. Actual dimensions of an electric panel are not a reliable determinant of ampacity. For example, many larger panels can be fitted with a variety of bus-bar and main switch assemblies of varying ampacity.

Aluminum wiring was used in some residences mainly between 1965 and early 1973, but this application has been largely discontinued because aluminum-wired connections have been found to have a very high probability of overheating compared with copper-wired connections and were therefore a potential fire hazard. A large number of connection burnouts have occurred in aluminum-wired homes, and according to the U.S. Consumer Product Safety Commission, many fires have resulted, some involving injury and death. The main problem with aluminum wiring is a phenomenon known as cold creep. When aluminum wiring warms up, it expands, and when it cools down, it contracts. However, unlike copper, when aluminum goes through a number of warm/cool cycles, it begins to lose some of its tightness. To add to the problem, aluminum oxidizes, or corrodes when in contact with certain types of metal, so the resistance of the connection increases. This causes it to heat up and corrode/oxidize still more until eventually the wire may become very hot and melt the insulation or fixture it is attached to and possibly cause a fire without ever tripping the circuit breaker.

Aluminum-wire "alloys" were introduced in the early1970s, but they did not address most of the connection-failure problems. Aluminum wiring is still permitted and used for certain applications, including residential service-entrance wiring and single-purpose higher amperage circuits such as 240 V air-conditioning or electric-range circuits. Although the fire risk from single-purpose circuits is much less than for branch circuits, field reports indicate that these connections remain a potential fire hazard.

A simple method of identifying aluminum wiring is to examine the wire sheathing for the word "aluminum." If you can't find it embossed in the wire sheathing, then look for silver-colored wire instead of the copper-colored wire used in modern wiring. Without opening any electrical panels or other devices, it is possible to look for printed or embossed letters on the plastic wire jacket where wiring is visible at the electric panel. Some aluminum wire has a specific brand name, such as Kaiser, Alcan, Aluminum, or AL/2 plainly marked on the plastic wire jacket. Some white-colored plastic wire jackets are inked in red; others have embossed letters without ink and are hard to read. Shining a light along the wire may make it easier to identify. Of note, the fact that no aluminum wiring was evident in the panel does not necessarily mean that none is present. Aluminum may have been used for parts of circuits or for some but not all circuits in the building.

9.4.3.2 Service Outlets and Receptacles

Service outlets include convenience receptacles, motors, lights, and appliances. Receptacles are commonly known as outlets or sometimes erroneously as wall plugs. (A plug is what actually goes into the outlet.) It is preferable for outlets to be three-prong, where the third prong is grounded. For large spaces or areas, all of the outlets should not be on the same circuit so that when a fuse or circuit breaker trips due to an overload, the space will not be plunged into complete darkness.

The grounding of a service to earth is a safety precaution and is necessary mainly to protect against lightning strikes or other high-voltage line strikes. Earth grounding in a commercial building might be to a grounding rod inside a switchboard, to a steel cold-water pipe in the plumbing system, or to the steel frame of a building. Other methods of grounding are also used depending on the equipment or system to be grounded. Ground wires are typically covered with green insulation or sometimes may be without cover.

9.4.3.3 Motors

Motors are devices that convert any form of energy into mechanical energy, especially an internal-combustion engine or an arrangement of coils and magnets that converts electric current into mechanical power. There are basically four types of motors in general use.

The DC motor is a rotating electric machine designed to operate from a direct voltage source. It is used for small-scale applications and for elevators, where continuous and smooth acceleration to a high speed is important.

Stepper/switched reluctance (SR) motors are brushless, synchronous electric motors that can divide a full rotation into a large number of steps. The motor's position can be controlled precisely without any feedback mechanism. Stepper motors are basically similar to SR motors – in fact, the latter are very large stepping motors with a reduced pole count and generally closed-loop commutated. The main advantage of stepper motors is that they can achieve accurate position control without the requirement for position feedback. Stepper motors operate differently from normal DC motors, which rotate when voltage is applied to their terminals.

Induction motors can be either three-phase AC or single-phase AC. The three-phase AC induction motor is a rotating electric machine designed to operate from a three-phase source of alternating voltage and usually applies to larger motors. These motors are characterized by extreme reliability and remain constant in rpm unless heavily overloaded. The single-phase motor is a rotating machine that has both main and auxiliary windings and a squirrel-cage rotor. The fourth type of motor in general use is the universal motor, which is a rotating electric machine similar to a DC motor but designed to operate either from direct current or single-phase alternating current; it varies in speed based on the load. The universal motor is usually found in mixers, hand drills, and similar appliances. Motors should always be protected against overload by thermal relays, which shut off the power when any part of the motor or housing overheats.

9.4.3.4 Switches and Controls

These are devices that direct the flow of power service to the electrical equipment. Safety switches are installed in locations where service cut-off is available in case of emergencies. These include toggle switches, dials, and levers.

9.4.3.5 Emergency Power and Emergency Lighting

Large facilities often have standby power to ensure continued electrical service when a shutdown of the standard power service takes place. Emergency power is required for life-support systems, fire- and life-safety circuits, elevators, exits, and emergency lighting. Facilities that require full operation during emergencies or disasters, such as hospitals and shelters, always have backup power. Computer facilities, to ensure continued storage and survival of the data, also commonly have emergency power. For major equipment, a diesel-engine generator with automatic starting and transfer switches is often provided for emergency power (Figure 9.19), while for lighting, battery units are installed. The typical AC power frequency in the United States is 60 cycles per second or 60 Hertz, whereas in Europe 50 Hertz is the standard.

Having an emergency lighting system in place is necessary in the event of a power failure or other emergency; emergency lighting in a facility enables the occupants to exit safely. Emergency lighting can consist of individual battery units placed in all corridors and areas that may require sufficient lighting for exiting and in interior and some exterior exitways. These batteries are continuously recharged while power is on and take over when power is lost. Alternatively, the lighting can be powered by a central battery unit. Fluorescent lamps will require some method of power conversion, as batteries are typically 12-volt.

9.4.3.6 Transformers

Transformers are devices that convert an alternating-current (AC) circuit of a certain voltage to a higher or lower value without change of frequency by electromagnetic

Figure 9.19 Photo of an Amtrak generator for emergency backup power inside a building.

induction. A step-up transformer receives a low voltage and converts it into a higher voltage, and a step-down transformer does just the reverse. Transformers are often used to step up voltage in order to transmit power over long distances without excessive losses and subsequently to step down voltage to more usable levels. While a transformer changes the voltage in a circuit, it has practically no effect on the total power in the circuit.

There are two distinct types of transformers, wet or dry. There are also subcategories of each main type. For wet or liquid-filled transformers, the cooling medium can be conventional mineral oil. Some wet-type transformers use less flammable liquids, such as high-fire-point hydrocarbons and silicones. Wet transformers are typically more efficient than dry types and usually have a longer life expectancy. There are some drawbacks, however. For example, fire prevention is more important with liquid-type units because the liquid cooling medium used may catch fire (although dry-type transformers are also susceptible) or even explode. Wet-type transformers typically contain a type of fire-resistive fluid or mineral oil such as PCBs, and, depending on the application, they may require a containment trough for protection against possible leaks of the fluid, which is why they are predominantly used outdoors.

For lower-voltage indoor-installed distribution transformers of 600 V and below, the dry-type transformer is preferred even though it has minimal requirements for insulation and avenues for ventilation of heat generated by voltage changes. Dry-type transformers come in enclosures that have louvers or are sealed. Location of transformers should be carefully considered, and there should be clear access surrounding exterior transformers and adequate ventilation and access for interior transformers, which should

Figure 9.20 Illustration of a transformer located outside a facility and protected by bollards.

be inside a fireproof vault. On-site transformers in parking lots may require bollards or other protection (Figure 9.20). Transformers should be analyzed for PCBs and their registration number noted. In addition, transformers tend to make a certain amount of noise (hum), and it is desirable to address this.

9.4.4 Lighting Systems

The main objectives of a good lighting system are to provide the visibility required based on the task to be performed while achieving the desired economic results. Another primary objective must be to minimize energy usage while achieving the visibility, quality, and aesthetic values desired by users. The quality and quantity of lighting affects the ambience, security, and function of a facility as well as the performance of its employees. Divergent artificial sources produce different kinds of light and vary significantly in their efficiency, which is the calculated lumen output per watt input. Regrettably, U.S. lighting design does not readily translate overseas – not when different regions have their own voltage, product standards, construction methods, and conceptions about what lighting is meant to achieve.

9.4.4.1 Interior Lighting

It goes without saying that without a light source we cannot see and without surfaces to reflect light there is nothing to see. To understand this relationship between a source of light, the surfaces that reflect light, and how we see light, we need to have a common comprehensive lighting language. This is particularly important because lighting typically accounts for a significant percentage of the annual commercial and residential electric bills. Advances in lighting technology can make significant reductions in the amount of money that is spent on lighting a facility.

Today we find many types of interior-lighting systems that enable us to make full use of a facility around the clock. The most common categories are:

- Fluorescent lamps: This type of fixture has long been preferable to incandescent lighting in terms of energy efficiency. Fluorescent lighting is far more efficient and has an average life of 10 to 20 times longer. Fluorescent lamps last up to 20,000 hours of use and use roughly one-third as much electricity as incandescent bulbs with comparable output. Compact fluorescent lamps (CFLs) are similar in operation to standard fluorescent lamps but are manufactured to produce colors similar to incandescent lamps. New developments in fluorescent technology, including the high-efficacy T-5, T-8, and T-10 lamps, have pushed the energy-efficiency envelope further. Recently, attention has also been paid to the mercury content of fluorescents and the consequences of mercury releases into the environment. As with all resource use and pollution issues, reduction is the best way to limit the problem. Even with low-mercury lamps, however, recycling of old lamps remains a high priority. A fluorescent fixture typically consists of the lamp and associated ballast, which controls the voltage and the current to the lamp. Replacing standard incandescent light bulbs with CFLs will reportedly slash electrical consumption in homes and offices where incandescent lighting is widely used. By reducing the amount of electricity used, corresponding emissions of associated carbon dioxide, sulfur dioxide, and nitrous oxide are reduced. CFL technology continues to develop and evolve and is now capable of replacing most of the light fixtures that were originally designed for incandescent light bulbs.
- Incandescent lamps: Incandescent lamps have relatively short lives (typically 1000 to 2000 hours of use) and are the least efficient of common light sources. In fact, only about 15 percent of the energy they use comes out as light; the rest becomes heat. Nevertheless, they remain popular because they produce a pleasant color that is similar to natural sunlight, and they are the least expensive to buy. Incandescent lamps come in various shapes and sizes with different characteristics. Environmental issues include lamp efficacy (lumens per watt), luminaire efficiency, controllability of the light source, potential for PV power, and control of light pollution.
- Tungsten halogen lamps are a type of incandescent lamp that has become increasingly popular in recent years. They produce a whiter, more intense light than standard incandescent lamps and are typically used for decorative, display, or accent lighting. They are about twice as efficient as regular incandescent lamps and last two to four times longer.
- High-intensity discharge (HID): This category of high-output light source consists of a lamp within a lamp that runs at a very high voltage. There are basically four types of HID lamps: high-pressure sodium (HPS), mercury vapor, metal halide gas, and low-pressure sodium. As with fluorescent lights, HID lights require a ballast for proper operation. The efficiency of HID sources varies widely from mercury vapor, with a low efficiency – almost as low as incandescent – to low-pressure sodium, which is an extremely efficient light source. Color rendering also varies widely from the bluish cast of mercury vapor lamps to the distinctly yellow light of low-pressure sodium.
- Fiberoptics: This is an up and coming technology, providing an alternative that is superior to conventional interior and exterior lighting systems. The technology is based on the use of

hair-thin transparent fibers to transmit light or infrared signals. The fibers are flexible and consist of a core of optically transparent glass or plastic, surrounded by a glass or plastic cladding that reflects the light signals back into the core. Light signals can be modulated to carry almost any other sort of signal, including sounds, electrical signals, and computer data, and a single fiber can carry hundreds of such signals simultaneously, literally at the speed of light. A typical fiberoptic lighting system can be broken down into two basic components: a light source, which generates the light, and the fiber optics, which delivers the light. Although fiberoptic lighting offers unique flexibility compared to conventional lighting, it does have its limitations. Areas of high ambient light should be avoided, as they tend to "wash out" the color. However, often fiberoptics can be installed in areas not accessible to conventional lighting. Good ventilation is very necessary for all illuminators. Light-colored reflective surfaces are preferable for end-light or side-light applications. Dark surfaces absorb light and should only be used to provide contrast. Typical applications include cove lighting, walkway lighting, and entertainment illumination (Figure 9.21).

Interior lighting should meet minimum illumination levels (Figure 9.22). It is important to determine the amount of light required for the activity that will take place in a space. Typically, the light needed for visibility and perception increases as the size of details decreases, as contrast between details and their backgrounds is reduced, and as

Figure 9.21 Photo showing glass block walls with fiberoptic lighting.
Source: Lumenyte International Corp.

Area	Footcandles
Building Surrounds	1
Parking Area	5
Exterior Entrance	5
Exterior Shipping Area	20
Exterior Loading Platforms	20
Office Corridors and Stairways	20
Elevators and Escalators	20
Reception Rooms	30
Reading or Writing Areas	70
General Office Work Areas	100
Accounting/Bookkeeping Areas	150
Detailed Drafting Areas	200

Figure 9.22 Table of recommended illumination levels for various functions.

task reflectance is reduced. However, interior lighting must not exceed allowed power limits. Interior lighting includes all permanently installed general and task lighting shown on the plans but does not include specialized lighting for medical, dental, or research purposes and display lighting for exhibits in galleries, monuments, and museums.

The useful life of a lighting installation becomes progressively less efficient during its operation due to dirt accumulation on the surface and aging of the equipment. The rate of reduction is influenced by the equipment choice and the environmental and operating conditions. In lighting-scheme design we must take account of this deficiency by the use of a maintenance factor and plan suitable maintenance schedules to limit the decay.

9.4.4.2 Exterior Lighting Systems

Outdoor lighting is common around buildings, and there are many innovative, energy-efficient lighting solutions for outdoor applications. Exterior lighting should be carefully designed and sufficient thought given to its placement, intensity, timing, duration, and color. It should meet the requirements of the Illuminating Engineering Society of North America (IESNA). Outdoor lighting is used to illuminate statues, signs, flags, or other objects mounted on a pole, pedestal, or platform; spotlighting or floodlighting used for architectural or landscape purposes must use full cutoff or directionally shielded lighting fixtures that are aimed and controlled so that the directed light is substantially confined to the object intended to be illuminated. Facility evaluations are often required to identify inadequate exterior-lighting conditions. Full cutoff lighting fixtures are required for all outdoor walkway, parking lot, canopy, and building/wall-mounted lighting and all lighting fixtures located within those portions of open-sided parking structures that are above ground. (An open-sided parking structure is a parking structure that contains exterior walls that are not fully enclosed between the floor and ceiling.) The most common incandescent outdoor lighting options are metal halide, and high-pressure sodium.

To control light pollution, full-cutoff luminaires should be specified. It also makes very good sense to use whenever possible environmentally friendly, commercial outdoor lighting systems. For example, ENERGY STAR®\lights consume only about 20 percent of the energy consumed by traditional lighting products. This provides substantial savings in money spent, energy consumed, and reduction of greenhouse gas emissions.

To meet code requirements, automatic controls are typically required for all exterior lights. The control may be a directional photocell, an astronomical time switch, or a building-automation system with astronomical time-switch capabilities. The control should automatically turn off exterior lighting when daylight is available. Lights in parking garages, tunnels, and other large covered areas that are required to be on during daylight hours are exempt from this requirement. Incandescent and high-intensity discharge are the most common types for exterior lighting. Illumination levels should be adequate and in good condition.

9.4.4.3 Emergency Lighting

NFPA 101 2006 stipulates that emergency illumination (when required) must be provided for a minimum period of 1.5 hours to compensate for the possible failure of normal lighting. NFPA also requires emergency lighting to be arranged to provide initial illumination of not less than an average of one footcandle and a minimum at any point of 0.1 footcandle measured along the path of egress at floor level. In all cases an emergency-lighting system must be designed to provide illumination automatically in the event of any interruption of normal lighting (NFPA 101 2006 7.9.2.3). Emergency lighting and LED signs typically use relatively small amounts of energy and have a long life expectancy. Although LED fixtures may cost more than incandescent fixtures, reduced energy costs and labor savings will often quickly make up the difference.

9.4.5 Harmonics Distortion

Loads connected to electricity-supply systems may be broadly categorized as either linear or nonlinear. There was a time when almost all electrical loads were linear – those that weren't made up such a small portion of the total that they had little effect on electrical-system operation. That all changed, however, with the arrival of the solid-state electronic revolution. Today, we are immersed in an environment rich in nonlinear loads, such as UPS equipment, inverters, induction motors, computers, variable-speed drives, and electronic fluorescent-lighting ballasts. Operation of these devices represents a double-edged sword. While they provide greater efficiency, they can also cause serious consequences to power-distribution systems – in the form of harmonic distortion. In reality, total harmonic distortion is hardly perceptible to the human ear, and even though the voltage distortion caused by the increasing penetration of nonlinear loads is often accommodated without serious consequences, power quality is compromised in other cases unless steps are taken to address this phenomenon.

While the majority of loads connected to the electricity-supply system draw power that is a linear (or near linear) function of the voltage and current supplied to it, these linear loads do not normally cause disturbance to other users of the supply system. Some types of loads, however, cause a distortion of the supply voltage/current waveform due to their nonlinear impedance. Harmonic distortion can surface in electric-supply systems through the presence of nonlinear loads of sufficient size and quantity. The severity of problems depends upon the local and regional supply characteristics, the size of the loads, the quantity of these loads, and how the loads interact with one another. Utility companies are clearly concerned about emerging problems caused by increasing concentrations of nonlinear loads resulting from the growing proliferation of electronic equipment, particularly computers and their AC-to-DC power-supply converters and electronic controllers.

Reducing harmonic voltage and current distortion from non-linear distribution loads such as adjustable frequency drives (AFDs) can be achieved through the use of several basic approaches. However, in the presence of excessive harmonic distortion, it is highly recommended to bring in a specialized consultant to correct the issue. Some of the methods used by harmonic specialists to reduce harmonic distortion may include the use of a DC choke, line reactors, 12-pulse converters, 12-pulse distribution, harmonic trap filters, broadband filters, and active filters. Whatever approach is applied to achieve reduction, it must meet the guidelines of the Institute of Electrical and Electronics Engineers (IEEE).

9.5 Solar-Energy Systems

There are two types of solar energy systems: active and passive.

9.5.1 Active Solar-Energy Systems

Solar Photovoltaics: Photovoltaic (PV) materials convert sunlight into useful, clean electricity. By adding PV to your home or office, you can generate renewable energy, reduce your own environmental impact, enjoy protection from rising utility costs, and reduce greenhouse-gas emissions. Electricity is only one of many uses for solar energy. The sun, of course, is essential to your garden, and it can heat water very cost-effectively, but the most fundamental use of solar energy is in overall building design. Good design uses solar radiation to passively and/or actively heat your building and to help keep it cool. Solar energy is also increasingly being used for street lighting (Figure 9.23).

Building integrated photovoltaic systems (BIPV) offer additional design options, allowing electricity to be generated by windows, shades and awnings, roofing shingles, and PV-laminated metal roofing, for example. BIPV options can be used in retrofits or new construction.

Solar energy is a renewable resource that is environmentally friendly. Unlike fossil fuels, solar energy is available in abundance and is free and immune to rising energy prices. Solar energy can be used in many ways – to provide heat, lighting, mechanical power, and electricity. It helps minimize the impact of pollution from energy generation, which is considered to be the single largest contributor to global warming. Renewable

Figure 9.23 Example of solar-powered LED street lighting with automatic onoff; it lasts for four to five nights after full charge.
Source: Hankey Asia Ltd.

energy could clean the air, stave off global warming, and help eliminate our nation's dependence on fossil fuels from overseas. The recent upsurge in consumer demand for clean, renewable energy and the deregulation of the utilities industry have spurred growth in green power – solar, wind, geothermal, steam, biomass, and small-scale hydroelectric. Small commercial solar-power plants are emerging and starting to serve some of this energy demand.

For decades solar technologies in the United States have used the sun's energy and light to provide heat, light, hot water, electricity, and even cooling for homes, businesses, and industry. The types of renewable technologies available for a particular facility depend largely on the application and what sort of energy is required, as well as a building's design and access to the renewable energy source. Building facilities can use renewable energy for space heating, water heating, air conditioning, lighting, and refrigeration. Commercial facilities include assembly and meeting spaces, educational facilities, food sales, food service, healthcare, lodging, stores and service businesses, offices, and warehouses.

In the LEED™ rating system, the on-site renewable-energy credits are not always easily achieved, particularly in urban locations. Essentially, you need to generate 2.5 to 7 percent of the building's electricity from wind, water, or solar energy, which, due to the many site constraints in a city environment, leaves only solar energy to focus on.

9.5.1.1 Solar Electric-System Basics

Solar electric systems, also known as photovoltaic (PV) systems, convert sunlight into electricity. When interconnected solar cells convert sunlight directly into electricity, they form a solar panel or module, and several modules connected together electrically form an array. Most people picture a solar electric system as simply the solar array, but a complete system consists of several other components. The working of a solar collector is very simple (Figure 9.24). The energy in sunlight takes the form of electromagnetic radiation from the infrared (long) to the ultraviolet (short) wavelengths. This radiation from the sun heats a liquid, which goes to a hot-water tank. The liquid heats the water and flows back to the solar collector. The solar energy that strikes the earth's surface at any particular time largely depends on weather conditions as well as location and orientation of the surface, but overall it averages approximately 1000 watts per 10 square feet (equivalent to 1 square meter) under clear skies with the surface directly perpendicular to the sun's rays. This solar thermal heat is able to provide hot water for an entire family during the summer. The collector size needed per person is just over 16 square feet (1.5 m^2). An average family of four people, therefore, needs a collector of about 65 square feet (6 m^2).

Understanding the basic components of PV systems and how they function is not particularly difficult. The most common component equipment generally used in on-grid and off-grid solar-electric systems is discussed below (of note, systems vary and not all equipment is necessary for every system type).

The technical term for solar panels that create electricity is photovoltaic (PV). Photovoltaic material, most commonly utilizing highly purified silicon, converts sunlight

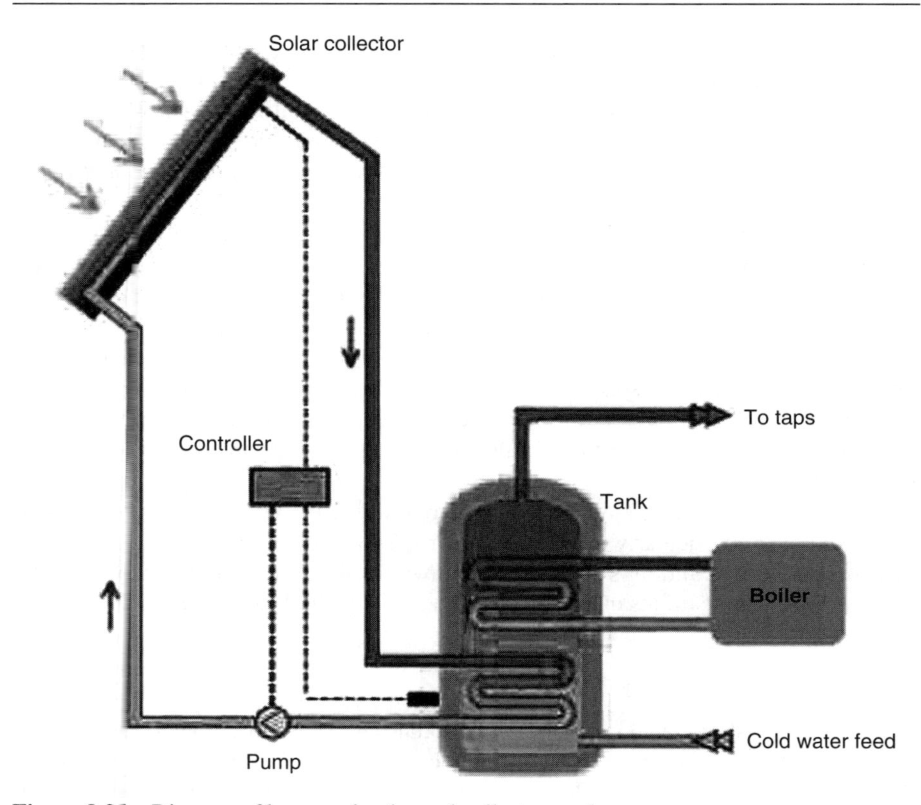

Figure 9.24 Diagram of how a solar thermal collector works.
Source: Nicolaj Stenkjaer – Nordic Folkecenter for Renewable Energy.

directly into electricity. When sunlight strikes the material, electrons are dislodged, creating an electrical current that can be captured and harnessed. The photovoltaic materials can consist of several individual solar cells or a single thin layer, which make up a larger solar panel. Panels are usually mounted on either a stationary rack or a tracking rack that follows the movement of the sun (Figure 9.25). Life expectancy of a typical system is 40 to 50 years. Panels are generally warranted for 20 to 25 years. As they have no moving parts solar electric panels operate silently.

Over recent years photovoltaics have been making significant inroads as supplementary power for utility customers already served by the electric grid. In fact, grid-connected solar systems now comprise a larger market share than off-grid applications. However, compared to most conventional fuel options, photovoltaics remain a very small percentage of the energy makeup both within the U.S. and globally. Still, with increasing concerns of global warming, more and more individuals, companies, and communities are choosing PV for a variety of reasons, including environmental, economic, emergency backup, and fuel and risk diversification. The economics of a photovoltaic system for your home or business take into account not just the

Figure 9.25 Photo taken by Airman First Class Nadine Barclay of the Nellis Solar Power Plant, which is the largest photovoltaic power plant in North America. The 70,000 solar panels sit on 140 acres of unused land on the Nellis Air Force Base, Nevada, forming part of a solar photovoltaic array that will generate in excess of 25 million kilowatt-hours of electricity annually and supply more than 25 percent of the power used at the base.
Source: Wikimedia Commons.

solar resource but rather a combination of the solar resource, electricity prices, and local/national incentives.

Siting is a critical aspect of solar design. For example, panels like full sun, facing within 30 degrees of south and tilting within 30 degrees of the site's latitude. A one-kilowatt system requires about 80 square feet of solar electric panels. Stationary racks can be roof or pole-mounted. Tracking racks are pole-mounted.

An inverter converts the direct-current (DC) electricity produced by the PV modules into usable 120-volt alternating-current (AC) electricity, which is the most common type for powering lights, appliances, and other needs. Grid-tied inverters are utilized to synchronize the electricity they produce with the grid's utility-grade AC electricity, allowing the system to feed solar-made electricity to the utility grid. Inverters are typically warranted for 5 to 10 years.

Mounting racks provide a secure platform on which to anchor the PV panels, ensuring that they are fixed in place and correctly oriented. Panels can be mounted on a rooftop, on top of a steel pole set in concrete, or at ground level. A photovoltaic array is the complete power-generating unit, comprised of one or more solar PV modules (solar panels) that convert sunlight into clean solar electricity. The solar modules need to be mounted facing the sun and avoiding shade for best results. Solar panels generate DC power, which can be converted to AC power with an inverter.

Selecting the correct size and type of wire will enhance the performance and reliability of the PV system. The size of the wire must be sufficiently large to carry the maximum current expected without undue voltage losses.

Batteries are used to store solar-produced electricity for evening or emergency backup power. Batteries may be required in locations that have limited access to power lines, as in some remote or rural areas. If batteries are part of the system, a charge controller may be required to protect the batteries from being overcharged or drawn down too low. Depending on the current and voltages for certain applications, the batteries are wired in series and/or parallel.

The main function of a charge controller is to protect the battery bank from overcharging. This is done by monitoring the bank; when it is fully charged, the controller interrupts the flow of electricity from the PV panels. Modern charge controllers usually incorporate maximum power point tracking (MPPT), which optimizes the PV array's output, thereby increasing the energy it produces.

System meters are used to measure and display several different aspects of a solar-electric system's performance and status, tracking how full the battery bank is; how much electricity the solar panels are producing or have produced; and how much electricity is in use.

The DC disconnect is used to safely interrupt the flow of electricity from the PV array. It is an essential component when system maintenance or troubleshooting is required. The disconnect enclosure houses an electrical switch rated for use in DC circuits and, if required, may also integrate either circuit breakers or fuses.

Disconnect switches are required in battery-based systems to allow the power from a solar electric system to be turned off for safety purposes during maintenance or emergencies. It also protects the inverter-to-battery wiring against electrical fires. A disconnect typically consists of a large, DC-rated breaker mounted in a sheet-metal enclosure.

The AC breaker panel is the point at which all of a property's electrical wiring meets with the provider of the electricity, whether that is the grid or a solar-electric system. The AC breaker panel typically consists of a wall-mounted panel or box that is normally installed in a utility room, basement, garage, or on the exterior of the building. It contains a number of labeled circuit breakers that route electricity to the various spaces throughout a structure. These breakers allow electricity to be disconnected for servicing and also protect the building's wiring against electrical fires.

Homes and businesses with a grid-tied solar-electric system will often have AC electricity both coming from and going to the electric utility grid. A bidirectional kilowatt-hour meter is able to simultaneously keep track of how much electricity flows in each direction, which tells you how much electricity is being used, and how much the solar-electric system is producing. Intertied-capable meters are typically provided by the utility company free of charge.

Off-grid solar-electric systems can be sized to provide electricity during cloudy periods when the sun doesn't shine. But sizing a system to cover a worst-case scenario, such as several cloudy weeks during the winter, can result in an unduly large system that will rarely be used to its full capacity. Backup engine generators can be fueled with biodiesel, petroleum diesel, gasoline, or propane, depending on the design. These generators produce AC electricity that a battery charger (either standalone or incorporated into an inverter) converts to DC energy, which is stored in batteries.

Solar electric systems are attracting increasing attention because they are environmentally friendly and do not generate emissions of greenhouse gasses or other pollutants, thereby reducing global climate impacts. Solar panels reflect visible demonstration of concern for the environment, community education, and proactive forward thinking.

9.5.1.2 Types of Solar Energy Systems

The three most widely used types of solar-electric systems are grid-tied, grid-intertied with battery backup, and off-grid (stand-alone). Each has distinct applications and component needs.

A grid-tied system (alternating current), also known as on-grid, or grid-intertied, does not require storage equipment (i.e., batteries) because it generates solar electricity and routes it to the electric utility grid, offsetting a home's or business' electrical consumption and in some instances even turning the electric meter backwards. Living with a grid-connected solar-electric system doesn't really differ from living with grid power except that some or all of the electricity used comes from the sun. The crucial issue relative to the photovoltaic panel (PV) systems is the technical aspect of tying into the electricity grid.

Applications of this type require the use of grid-tied inverters that meet the requirements of the utility. They must not emit "noise," which can interfere with the reception of equipment (e.g., televisions), switch off in the case of a grid failure, and retain acceptable levels of harmonic distortion (i.e., quality of voltage and current output waveforms). This type of system tends to be an optimum configuration from an economic viewpoint because all the electricity is utilized by the owner during the day and any surplus is exported to the grid. Meanwhile, the cost of storage to meet nighttime needs is avoided, because the owner simply draws on the grid in the usual way. With access to the grid, the system does not need to be sized to meet peak loads. This arrangement is termed net metering or net billing. The specific terms of net-metering laws and regulations vary from state to state and utility to utility, which is why the local electricity provider or state regulatory agency should be consulted for guidelines.

A standalone grid-tied system with battery backup (alternating current) is the same as the grid-tied system except that battery storage (battery bank or generator backup) is added to enable power to be generated even when the electricity grid fails. Incorporating batteries into the system requires more components, is more expensive, and lowers the system's overall efficiency. But for homeowners and businesses that regularly experience utility outages or have critical electrical loads, having a backup energy source is invaluable. The additional cost to the customer can be quantified against the value of knowing that the power supply will not be interrupted.

A standalone off-grid system without energy storage (direct current) consists of a PV system whose output is dependent upon the intensity of the sun. In this system, the electricity generated is used immediately, and therefore, the application must be capable of work on both direct current (DC) and variable power output. Standalone off-grid electric systems are most common in remote locations where there is no utility grid service. These systems operate independently from the grid to provide all the electricity required by a household or small business. The choice to live off-grid may be because of the prohibitive cost of bringing utility lines to remote locations, the appeal of an independent lifestyle, or the general reliability a solar-electric system provides. However, those who choose to live off-grid often need to make adjustments to when and how they use electricity to allow them to live and work within the limitations of the system's capabilities.

To meet the greatest power needs in an off-grid location, the PV system may need to be configured with a small diesel generator. This increases the capability of the PV system, as it no longer has to be sized to cope with the worst sunlight conditions available during the year. The diesel generator can also provide the backup power, but its use is minimized during the rest of the year to keep fuel and maintenance costs to a minimum.

For any module with a defined peak power, the actual amount of electricity in kilowatt-hours (kWh) that it generates will depend primarily on the amount of sunlight it receives. The electrical power output of a PV module is the current that it generates (dependent on its surface area) multiplied by the voltage at which it operates. The larger the module or the solar array – the number of modules connected together – the more power is generated. A linear current booster can be added to convert excess voltage into amperage to keep a pump running in low light conditions. An LCB can boost pump output by 40 percent or more. For safety reasons, PV arrays are normally earthed.

Each kilowatt of unshaded stationary solar electric panels generates about 1200 kilowatt-hours of electricity per year. A one-kilowatt, dual-axis tracking system will generate about 1600 kilowatt-hours per year. Power is generated during peak daylight hours. Solar power exhibits a very good peak coincidence with commercial-building electrical loads. Dual-axis tracking systems, in which the panels follow the sun, will require periodic maintenance.

9.5.2 Passive Solar-Energy Systems

The sun's energy arrives on earth in the primary form of heat and light. Passive solar heating and cooling represent an important strategy for displacing traditional energy sources in buildings and are effective methods of heating and cooling through utilization of sunlight. To be successful, building designs must carefully balance their energy requirements with the building's site and window orientation. The term "passive" indicates that no additional mechanical equipment is used other than the normal building elements. Solar gains are generally introduced through windows, and minimum use is made of pumps or fans to distribute heat or effect cooling. Passive cooling minimizes the effects of solar radiation through shading or generating air flows with convection ventilation.

Correct building orientation, thermal mass, and insulation are specified in conjunction with careful placement of windows and shading. The thermal mass absorbs heat during the day and radiates it back into the space at night. To do this, passive-solar techniques make use of building elements such as walls, windows, floors, and roofs in addition to exterior building elements and landscaping to control heat generated by solar radiation. Solar heating designs collect and store thermal energy from direct sunlight. The effect is a quiet, comfortable, energy-efficient space with stable year-round temperatures.

Another solar concept is daylighting design, which optimizes the use of natural daylight and contributes greatly to energy efficiency. The quantity and quality of light around us help determine how well we see, work, and play. Light impacts our health, safety, comfort, morale, and productivity. Whether at home or in the office, it is possible to save energy and still maintain good light quantity and quality.

The benefits of using passive-solar techniques include simplicity, price, and the design elegance of fulfilling one's needs with materials at hand. Some of the advantages of passive-solar designs include:

- At little or no cost, passive solar design can easily be designed into new construction and can in some cases be retrofitted into existing buildings.
- It pays dividends over the life of the building through reduced or eliminated heating and cooling costs.
- Indoor air quality is improved through elimination of forced-air systems.
- Retrofitting is rarely as effective as initially designing for this method.
- Sites with good southern exposure are most suitable.

LEED™ offers credits in its Indoor Environmental Quality section, Daylight and Views. The intent of the credits appear to be to reduce electric lighting, increase productivity, and provide building occupants with a connection between indoor and outdoor spaces by incorporation of daylight and views into regularly occupied spaces.

LEED™ requirements for the credits are to achieve daylight (through computer simulations) in a minimum of 75 or 90 percent of regularly occupied spaces and achieve a daylight illuminance level of a minimum of 25 footcandles and a maximum of 500 footcandles in a clear-sky condition on September 21 at 9.00 AM and 3.00 PM. A combination of side-lighting and/or top-lighting may be used to achieve the total daylighting zone required, which is at least 75 percent of all the regularly occupied spaces. Sunlight redirection and/or glare-control devices may be provided to ensure daylight effectiveness. The provision of daylight redirection and/or glare-control devices to avoid high-contrast situations should be provided to avoid impeding visual tasks. Exceptions for areas where tasks would be hindered by daylight are considered on their merits.

It should be stressed that the USGBC Reference Guide or website should be consulted for the latest updated requirements including possible exemplary performance credits.

9.5.3 Federal Tax Credits for Energy Efficiency

A number of tax credits are available from ENERGY STAR® for energy efficiency.

Tax credits for consumers are available at 30 percent of the cost, up to $1500, in 2009 and 2010 (for existing homes only) for:

- Windows and doors
- Insulation
- Roofs (metal and asphalt)
- HVAC
- Water heaters (nonsolar)
- Biomass stoves

Tax credits are available at 30 percent of the cost, with no upper limit, through 2016 (for existing homes and new construction) for:

- Geothermal heat pumps
- Solar panels
- Solar water heaters
- Small wind-energy systems
- Fuel cells

Tax credits are now available for home improvements:

- Must be placed in service, i.e., when the property is ready and available for use, from January 1, 2009 through December 31, 2010.
- Must be for taxpayer's principal residence, except for geothermal heat pumps, solar water heaters, solar panels, and small wind-energy systems, where second homes and rentals qualify.
- $1500 is the maximum total amount that can be claimed for all products placed in service in 2009 and 2010 for most home improvements, except for geothermal heat pumps, solar water heaters, solar panels, fuel cells, and small wind-energy systems, which are not subject to this cap and are in effect through 2016.
- Must have a manufacturer certification statement to qualify.
- For record keeping, save your receipts and the manufacturer certification statement.
- Improvements made in 2009 will be claimed on your 2009 taxes (filed by April 15, 2010); use IRS Tax Form 5695 (2009 version) – it will be available in late 2009 or early 2010.
- If you are building a new home, you can qualify for the tax credit for geothermal heat pumps, photovoltaics, solar water heaters, small wind-energy systems and fuel cells but not the tax credits for windows, doors, insulation, roofs, HVAC, or nonsolar water heaters.

Fuel-efficient vehicles and energy-efficient appliances and products mitigate the adverse impacts on our environment, as well as providing many other benefits including better gas mileage (and therefore lower gasoline costs), fewer emissions, lower energy bills, increased indoor comfort, and reduced air pollution.

Starting January 1, 2009, there is a new tax credit for plug-in hybrid electric vehicles, starting at $2500 and capped at $7500, for cars and trucks. The credit is based on the capacity of the battery system. The first 250,000 vehicles sold get the full tax credit, then it phases out like the hybrid-vehicle tax credits.

Tax credits are available to buyers of hybrid gasoline-electric, diesel, battery-electric, alternative-fuel, and fuel-cell vehicles. The tax-credit amount is based on a formula determined by vehicle weight, technology, and fuel economy compared to base-year models. These credits are available for vehicles placed in service starting January 1, 2006. For hybrid and diesel vehicles made by each manufacturer, the credit will be phased out over 15 months starting after that manufacturer has sold 60,000 eligible vehicles. For vehicles made by manufacturers that have not reached the end of the phase-out, the credits will end for vehicles placed in service after December 31, 2010. Check the IRS website for updated information.

Homebuilders are eligible for a $2000 tax credit for a new energy-efficient home that achieves 50 percent energy savings for heating and cooling over the 2004 International Energy Conservation Code (IECC) and supplements. At least one-fifth of the energy savings must come from building-envelope improvements. This credit also applies to contractors of manufactured homes conforming to federal manufactured-home construction and safety standards.

There is also a $1000 tax credit for the producer of a new manufactured home achieving 30 percent energy savings for heating and cooling over the 2004 IECC and supplements (at least one-third of the savings must come from building-envelope improvements) or a manufactured home meeting the requirements established by EPA under the ENERGY STAR® program.

Please note that, with the exception of the tax credit for an ENERGY STAR®-qualified manufactured home, these tax credits are not directly linked to ENERGY STAR®. Therefore, a builder of an ENERGY STAR®-qualified home may be eligible for a tax credit but it is not guaranteed.

These tax credits apply to new homes located in the United States whose construction is substantially completed after August 8, 2005 and that are acquired from the eligible contractor for use as a residence from January 1, 2006, through December 31, 2009.

A tax deduction of up to $1.80 per square foot is available to owners or designers of new or existing commercial buildings that save at least 50 percent of the heating and cooling energy of a building that meets ASHRAE Standard 90.1-2001. Partial deductions of up to $0.60 per square foot can be taken for measures affecting any one of three building systems: the building envelope, lighting, or heating and cooling systems. These tax deductions are available for systems placed in service from January 1, 2006, through December 31, 2013.

9.6 Fire-Suppression Systems

9.6.1 General

As more and more high-rise, high-performance buildings are built both nationwide and globally, the planning for fire protection has taken on a real urgency. Fire-suppression design requires an integrated approach in which system designers need to analyze building components as a total package. As with other aspects of sustainable design, to achieve the most beneficial symbiosis between these components, an experienced system designer, such as a fire-protection engineer, should be involved early in the planning and design process and should be an integral part of the project team. Moreover, moving forward, we should start seeking out sustainable, environmentally friendly fire-suppression approaches to reduce the environmental impacts during design and testing and also to help a project earn LEED™ credits.

Fire-protection systems play an important role in overall building design and construction and should never be comprised, because they serve the purpose of life safety. Indeed, it is frequently argued that the life-safety system is the most important system to be evaluated in a facility, particularly when it comes to high-rise structures. Furthermore, like any other building system, green concepts and specifications can be applied to their design, installation, and maintenance in a manner that reduces their harmful impacts on the environment. Moreover, there have been significant advances recently in fire-detection technology and fire-suppression systems in addition to an ongoing development of international and national codes and standards, all of which have made possible the "greening" of facility fire-safety systems and their increasing importance for building owners and property developers.

For maximum efficiency, the various components of modern fire-protection systems should work together to detect, contain, control, and/or extinguish a fire in its early stages – and to help people survive during the fire. And the installation of environmentally friendly fire-protection technology can help earn credits under the U.S. Green Building Council's Leadership in Energy and Environmental Design (LEED™) Green Building Rating System for new or retrofitted buildings.

A facility's type, size, and function will generally determine the complexity of the life-safety system used. In some of the smaller structures, the system may consist of only smoke detectors and fire extinguishers. In other larger, more complex buildings, a complete fire-suppression system such as fire sprinklers is installed throughout the facility. An important aspect in the assessment of any life-safety system includes verification that periodic maintenance, inspection, and testing of the main components of the system are being conducted.

As illustrated in Figure 9.26, there are several types of life-safety systems normally employed to address fire-safety requirements. Each of these gives rise to its own set of issues, which need to be taken into account in facility surveys. The extent of a life-safety-system survey and the expertise required to perform such an evaluation vary greatly from facility to facility.

Fortunately, fire detection and prevention technologies have become increasingly sophisticated, intelligent, and powerful in recent years. Frank Monikowski and Terry Victor of SimplexGrinnell specify some of the advances and emerging technologies that can be found in today's systems:

- Control-mode sprinklers, standard manufactured sprinklers that limit fire spread and stunt high heat release rather than extinguish a fire; they also "prewet" adjacent combustibles.
- Suppression sprinklers operate quickly for high-challenge fires and are expected to extinguish a fire by releasing a high density of water directly to its base.
- Fast-response sprinklers provide quicker response and are now required for all light-hazard installations.

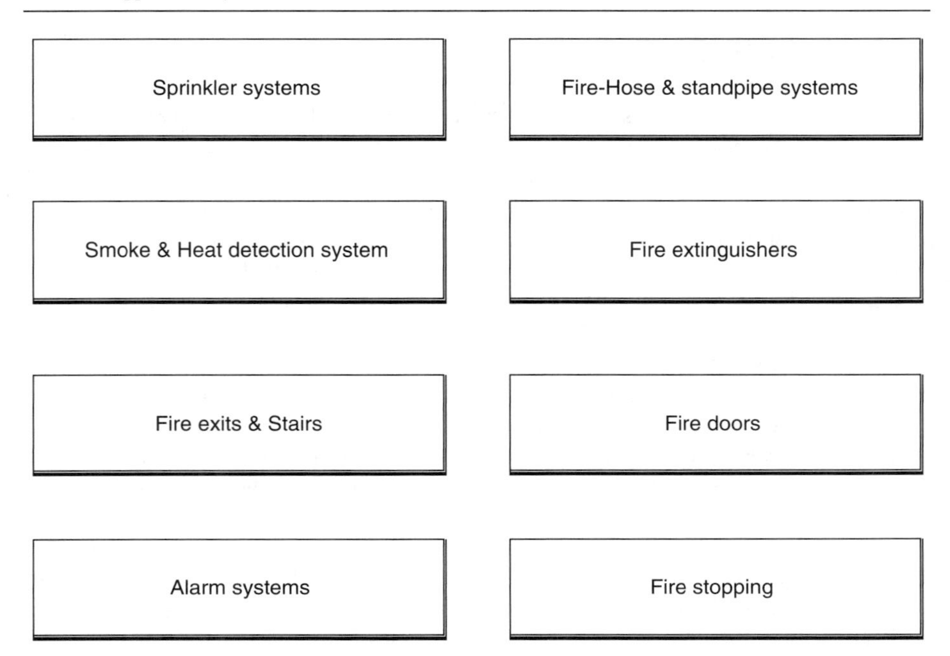

Figure 9.26 Typical fire-suppression-system components.

- Residential sprinklers, designed specifically to increase the survivability of an individual who is in the room where a fire originates.
- Extended-coverage sprinklers, designed to reduce the number of sprinklers needed to protect a given area. They come in quick-response, residential, and standard-response types and are also available for both light- and ordinary-hazard occupancies.
- Special sprinklers, such as early-suppression fast-response (ESFR), designed for high-challenge rack storage and high-pile storage fires. In most cases, these sprinklers can eliminate the expense and resources needed to install in-rack sprinkler heads.
- Low-pressure sprinklers provide needed water coverage in multistory buildings where pressure may be reduced. They bring a number of benefits: reduced pipe size, reduction or elimination of a fire pump, and overall cost savings.
- Low-profile, decorator, and concealed sprinklers, designed to be more aesthetically pleasing.
- Sprinkler system valves that are smaller, lighter, and easier to install and maintain and, therefore, less costly.
- Fluid-delivery-time computer program that simulates water flowing through a dry system in order to accurately predict critical "water-to-fire" delivery time for dry-pipe systems.
- The use of cost-efficient CPVC piping for light-hazard and residential sprinkler systems.
- Advanced coatings on steel pipes, designed to resist or reduce microbiologically influenced corrosion (MIC) and enhance sprinkler-system life.
- Corrosion monitoring devices to alert users of potential problems.
- More efficient coordination in evaluating building sprinkler-system need, including site surveys, accurate measurements, and the use of CAD and hydraulics software to ensure that fire sprinkler-system designs respond to the specific risks and the physical layout of the premises.

The National Fire Protection Association (NFPA) recently issued a new Emergency Evacuation Planning Guide for People with Disabilities. This document provides general information to assist designers in identifying the needs of people with disabilities related to emergency evacuation planning. This guide covers five general categories of disabilities: mobility impairments, visual impairments, hearing impairments, speech impairments, and cognitive impairments. The four elements of evacuation information needed by occupants are: notification, wayfinding, use of way, and assistance.

Fire-suppression systems are only indirectly referenced in LEED™ certification documents. For example, LEED™ for New Construction (LEED™NC) v3 Energy and Atmosphere (EA) Credit 4, Enhanced Refrigerant Management, and LEED™ for Existing Buildings: Operations and Maintenance (LEED™ EBOM) v3 EA Credit 5, Refrigerant Management, have as their intent reducing ozone depletion, supporting compliance with the Montreal Protocol, and minimizing direct contributions to global warming. It appears that credits can be earned with the installation/operation of fire-suppression systems that do not contain ozone-depleting substances such as chlorofluorocarbons (CFCs), hydrochlorofluorocarbons (HCFCs), and halons.

Likewise, LEED™ credits in the Innovation in Design category can also be obtained for fire-suppression systems. The LEED™ Reference Guide in the relevant category should be consulted, but generally, to earn those points, it is necessary to document and substantiate the innovation and design processes used.

9.6.2 Components

9.6.2.1 Sprinkler-System Types

9.6.2.1.1 Automatic Sprinkler Systems

Optimized automatic sprinkler-system designs offer an effective means of addressing environmental impact and sustainability. They are the most common, widely specified, and most effective fire-suppression system in commercial facilities – particularly in occupied spaces. Furthermore, automatic sprinkler systems are now not only required in new high-rise office buildings, but in many American cities it is mandated by code that existing high-rises be retrofitted with automatic sprinkler systems.

There are several types of sprinkler systems that are commonly used in commercial facilities: wet- and dry-pipe, pre-action, deluge, and fire-cycle systems. Of these, wet-pipe and dry-pipe are the most common. In a wet-pipe system, the sprinklers are connected to a water supply, enabling immediate discharge of water at sprinkler heads triggered by the heat of the fire. In a dry-pipe system, the sprinklers are under air pressure, which, when the pressure is eased by the opening of the sprinkler heads, fills the system with water.

With optimized sprinkler-system designs, effective use of the available water source is made, requiring the minimal necessary number of components. They also employ techniques and technologies that are adaptable to future building modifications. A well-designed valve or pump room can maximize the life of the system and facilitate any modifications that may be required in the future. By minimizing variations in piping, construction waste can be reduced and a more efficient installation achieved. In some cases the need for a fire pump may be eliminated altogether, thus reducing water waste.

In the design of automatic sprinkler systems careful attention should be given to proper connections for flow and flow testing. Likewise, flexible connections and armovers may be employed to provide a means for facilitating the relocation of sprinklers with minimal need for additional materials if the system designer incorporates appropriate flow restrictions due to friction losses.

Sometimes sprinklers are not feasible due to special considerations (e.g., water from sprinklers would damage sensitive equipment or inventory), and alternative fire-suppression systems such as gaseous/chemical suppression may be considered. But in the final analysis the type of sprinkler system decided upon depends mainly on a building's function. Of note, the majority of today's fire sprinklers incorporate the latest advances in design and engineering technologies, thereby providing a very high level of life-safety and property protection. The features and benefits now available are making fire sprinkler systems more efficient, reliable, and cost-effective.

As the benefits of sprinkler systems become better understood and more obvious and the cost more affordable, their installation in residential structures is becoming more common. However, these sprinkler systems typically fall under a residential classification and not a commercial one. The main difference between commercial and residential sprinkler systems is that a commercial system is designed to protect the structure and the occupants from a fire, whereas most residential systems are primarily designed to suppress a fire in a manner that allows for the safe escape of the building occupants. While these systems will often also protect the structure from major fire damage, this consideration nevertheless remains of secondary importance.

In residential structures sprinklers are typically omitted from closets, bathrooms, balconies, and attics because a fire in these areas would not normally impact an occupant's escape route. When a system is operating as intended, fire sprinkler systems are highly reliable, but, like any other mechanical system, they require periodic maintenance and inspection in order to sustain proper operation. In the rare event that a sprinkler system does fail to control a fire, the root cause of failure has often been found to be the lack of proper maintenance. Figure 9.27 is an example of a fire-sprinkler control-valve assembly including pressure switches and valve monitors.

9.6.2.1.2 Wet-Pipe Systems

Wet-pipe systems are by far the most widely used and have the highest reliability. Wet systems are typically used in buildings where there is no risk of freezing. The systems are simple, with the only operating component being the automatic sprinkler. A water supply provides pressure to the piping, and all of the piping is filled with water adjacent to the sprinklers. The water is held back by the automatic sprinklers (Figure 9.28) until activated. When one or more of the automatic sprinklers is exposed to sufficient heat, the heat-sensitive element releases, allowing water to flow from that sprinkler. Each sprinkler operates individually. Sprinklers are manufactured to react to a specific range of temperatures, and only sprinklers subjected to a temperature at or

Figure 9.27 Photo of a fire-sprinkler control-valve assembly installation.
Source: Wikipedia.

above their specific temperature rating will operate. Figure 9.29 shows a drawing of a typical wet-pipe sprinkler system.

The principal disadvantage of these systems is that they are not suited for subfreezing environments. Another potential concern is if piping is subject to severe impact damage and could consequently leak – e.g., in warehouses.

9.6.2.1.3 Dry-Pipe Systems

After the wet-pipe system, the dry-pipe system is the most widely used. A dry-pipe sprinkler system is one in which pipes are filled with pressurized air or nitrogen rather

Figure 9.28 Example of ceiling-mounted sprinkler head.
Source: Sujay Fire & Safety Equipment.

than water. This air holds a remote valve, known as a dry-pipe valve, in a closed position. The dry pipe valve is located in a heated space and prevents water from entering the pipe until a fire causes one or more sprinklers to be activated. Once this happens, the air escapes and the dry-pipe valve releases. To prevent the larger water-supply pressure from forcing water into the piping system, the design of the dry-pipe valve intentionally includes a larger valve clapper area exposed to the specified air pressure compared to the water pressure. Water then enters the pipe, flowing through open sprinklers onto the fire. However, regulations (NFPA 13 2007, Sections 7.2 and A7.2) typically stipulate that these systems can only be used in spaces in which the ambient temperature may be cold enough to freeze the water in a wet-pipe system, thus rendering it inoperable. For this reason we often find dry-pipe systems used in refrigerated coolers and in unheated buildings.

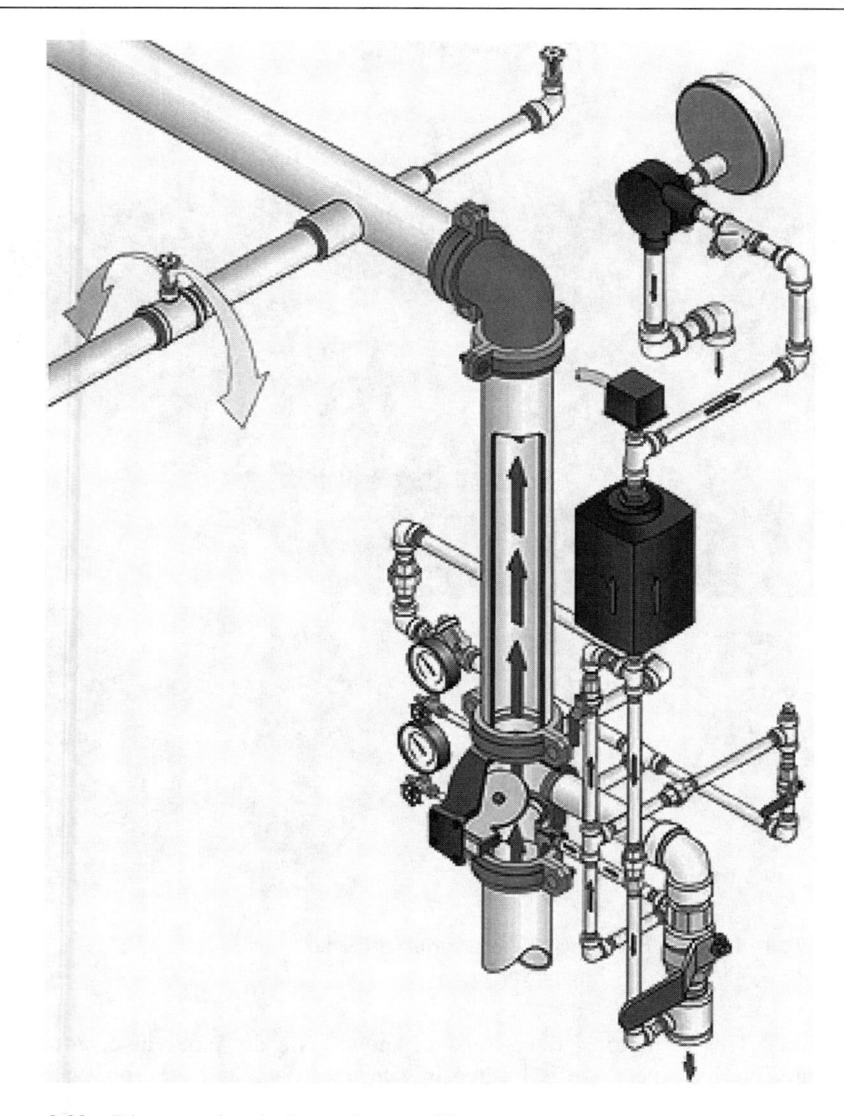

Figure 9.29 Diagram of typical wet-pipe sprinkler system.

The system is activated (becomes operational) when one or more of the automatic sprinklers are exposed to sufficient heat, allowing the maintenance air to vent from that sprinkler. Each sprinkler operates individually. As the air pressure in the piping drops, the pressure differential across the dry-pipe valve changes, allowing water to enter the piping system. Delays can be experienced in dry-pipe systems, since the air pressure must drop before the water can enter the pipes and suppress the fire. Dry-pipe systems are therefore not as effective as wet-pipe systems in fire control during the initial stages of the fire. Dry-pipe valves may employ quick opening devices Other important characteristics of dry-pipe systems include:

- In a dry-pipe system the fire will continue to grow after the activation of a sprinkler. As the water travels the pipework in the sprinkler system, the fire size grows. The dry-pipe system should therefore be designed to be as small as possible.
- Dry-pipe systems are not intended to limit accidental damage from broken or open sprinklers.
- Dry-pipe sprinkler systems cannot be gridded.
- Sprinklers may only be installed in the upright position unless specifically listed dry-pipe sprinklers are utilized or horizontal sidewall sprinklers installed so that water is not trapped.
- The location of the dry-pipe riser is required to be heated to a minimum of 40 degrees F.
- Piping must be pitched back to the riser so that water is not trapped.

9.6.2.1.4 Deluge Systems

Less common are the deluge system and the preaction system. These systems are used in environments that require special sprinkler protection and are activated by fire-detection systems. These systems represent only a small percentage of sprinkler systems in operation.

A deluge system is similar to a preaction system except the sprinkler heads are open and the pipe is not pressurized with air. This means that in a deluge system the heat-sensing operating element is removed during installation so that all sprinklers connected to the water-piping system remain open by the operation of a smoke or heat-detection system. These detection systems are normally installed in the same area as the sprinklers so that, when the detection system is activated, water readily discharges through all of the sprinkler heads in the system.

These systems provide a simultaneous application of water over the entire hazard and are typically used in high-hazard areas where rapid fire spread is a major concern, such as power plants, aircraft hangars, and chemical-storage facilities. Water is not present in the piping until the system operates. Because the sprinkler orifices are open, the piping is at ambient air pressure. To prevent the water-supply pressure from forcing water into the piping, a deluge valve is used in the water-supply connection, which is a mechanically latched nonresetting valve that stays open once tripped.

Because the heat-sensing elements present in the automatic sprinklers have been removed (resulting in open sprinklers), the deluge valve must be opened as signaled by a specialized fire-alarm system. The type of fire-alarm activation device used is based largely on the hazard (e.g., smoke detectors or heat detectors). The activation/initiation device signals the fire-alarm panel, which in turn signals the deluge valve to open. Activation can also be achieved manually, depending on the system goals. Manual activation is usually via an electric or pneumatic fire-alarm pull station, which signals the fire-alarm panel, which in turn signals the deluge valve to open, allowing water to enter the piping system. Water flow effectively takes place from all sprinklers simultaneously.

9.6.2.1.5 Preaction Systems

These sprinkler systems employ the basic concept of a dry-pipe system in that water is not normally contained within the pipes. It differs from the dry-pipe system, however, in that water is held from entering the piping by an electrically operated valve, known as a preaction valve. Valve operation is controlled by independent flame, heat, or smoke detection. A preaction system is generally used in locations where it is not acceptable to have the pipes full of water until there is a fire (e.g., computer suites) or where accidental activation is undesired, such as in museums with rare artworks, manuscripts, or books. Preaction systems are hybrids of wet, dry, and deluge systems, depending on the exact system goal.

There are two subtypes of preaction systems: single interlock and double interlock. The operation of single interlock systems is similar to dry systems except that these systems require that a "preceding" and supervised event (typically the activation of a heat or smoke detector) takes place prior to the "action" of water introduction into the system's piping due to opening of the preaction valve (i.e., a mechanically latched valve). The operation of double interlock systems is similar to a deluge system except that automatic sprinklers are used. Upon detection of the fire by the fire-alarm system, it basically converts from a dry system into a wet system.

Water-mist systems normally force water and pressurized gas together through stainless-steel tubes that are much narrower in diameter than the pipes used in traditional sprinkler systems. The water-mist system produces a fine mist with a large surface area that absorbs heat efficiently through vaporization. Because water-mist systems utilize water as the extinguishing medium, they are totally safe for humans. These systems suppress fire with three main mechanisms:

1. As the water droplets contact the fire, they convert to steam. This process absorbs energy from the surface of the burning material.
2. As the water turns into steam, it expands greatly. This removes heat and lowers the temperature of the fire and the air surrounding it.
3. The water and the steam act to block the radiant heat and prevent the oxygen from reaching the fire (thus starving it of oxygen) so the fire smothers.

These systems can be useful for suppressing fires in gas-turbine enclosures and machinery spaces and are FM (Factory Mutual)-approved for such applications. Water-mist systems are ideally suited for cultural heritage buildings where large amounts of water can potentially cause unacceptable damage to irreplaceable items and in retrofits where space is often limited. Water-mist systems are also often used to protect passenger cruise ships, where the system's excellent performance and low total weight have made them very popular.

9.6.2.1.6 Foam Systems

Foam water-sprinkler systems are essentially a special application system, discharging a mixture of water and low-expansion foam concentrate, resulting in a foam spray from the sprinkler. These systems are generally more economical than a water-only system when evaluated for the same risk and provide for actual extinguishment of the fire and a lower water demand. The conversion assists in the reduction of property loss, loss of life, and in many cases reduction of insurance rates.

Foam concentrates and expanded foams are generally considered to be safe for human exposure but, unless specifically indicated, can adversely impact the environment if allowed to flow freely into watershed areas. The base ingredients of typical foaming agents include nitrates, phosphorous, and organic carbon. It should be noted that the use of halons in fire-suppression systems was phased out in the early 1990s to comply with the Montreal Protocol, because they were determined to cause significant damage to the ozone layer. Moreover, halons have a long life in the atmosphere and a high global-warming potential (GWP).

One of the characteristics of this system is that almost any sprinkler system – wet, dry, deluge, or preaction – can be readily adapted to include the injection of AFFF foam concentrate in order to combat high-risk situations. These systems are typically used with special hazard occupancies associated with high-challenge fires, such as flammable liquids, and airport hangars. Added components to the sprinkler-system riser include bladder tanks to hold the foam concentrate, concentrate control valves to isolate the sprinkler system from the concentrate until activation, and proportioners for mixing the appropriate amount of foam concentrate with the system supply water. The main standard that delineates the minimum requirements for the design, installation, and maintenance of foam-water sprinkler and spray system is NFPA 16: Standard for the Installation of Foam-Water Sprinklers and Foam-Water Spray Systems.

The checklist below is provided by the New York Property Insurance Underwriting Association to help identify general problems that may arise in typical sprinkler systems. This checklist is intended to identify what is required to assure that the sprinkler system is properly maintained:

- Are sprinkler heads free of paint, dust, and grease?
- Are the sprinkler heads obstructed by stored material? There should be no less than 18 inches of clearance at each head. Obstructions will diminish the operation of the head.
- Are the sprinkler pipes used to support lighting or other objects?
- Are there extra sprinkler heads and wrenches located in the control area for maintenance purposes?
- Is the outside screw and yoke valve chained in an open position to avoid disabling of the system?
- Are the sprinkler heads directed properly for their location?
- Is there a sprinkler contractor who supervises and inspects the system as required by NFPA and ISO? Is a service log maintained?
- Are the sprinkler alarms activated to protect your property in the event of accidental discharge or fire?
- Has the occupancy classification of the material in the building changed since its installation so that the sprinkler system is now ineffective?
- Is the heat supply in the premises adequate for the operation of a wet-pipe system?

9.6.2.2 Fire-Hose and Standpipe Systems

Michael O'Brian, president of Code Savvy Consultants, says: "Standpipes are a critical tool that requires preplanning of first-responding apparatus in order to be used effectively. The initial approval process for these systems is critical, and the fire-prevention bureau can assist responding crews by ensuring proper installation and maintenance of these systems. Standpipe systems vary in design, use, and location. These factors vary based on the adopted code and the use, size, and type of building they are installed in. Typically, model codes refer to NFPA 14, Standpipe and Hose Systems, for the design, installation, and maintenance of these systems."

Standpipe systems consist of piping, valves, outlets, and related equipment designed to provide water at specified pressures. They are installed exclusively for fire-department use for the fighting of fires. The system is used in conjunction with sprinklers or hoses and basically consists of a water-pipe riser running vertically through the building, although sometimes a building is provided with only piping for the standpipe system. Standpipe systems can be wet or dry. Dry systems are normally empty and are not connected to a water source. A siamese fitting is located at the bottom end of the pipe, allowing the fire department to pump water into the system. In a wet-type system, the pipe is filled with water and attached to a tank or pump. This type also contains siamese fittings for the fire department's use.

O'Brian says that "Many buildings are required to have an Automatic Class I standpipe system with a design pressure of 100 psi. Based on friction loss, municipal water supply, and pressure loss for the height of the standpipe, a fire pump may need to be designed into the system. Due to the pressure requirements standpipes are limited to a maximum height of 275 feet. Those buildings over 275 feet in height will require the standpipe systems to be split in different pressure zones."

Model fire and building codes stipulate, among other things, the requirements for the installation of standpipe systems. The specific type of system is based on the occupancy classification and building height. Standpipe systems have three common classifications:

- Class I standpipe signifies that it equipped with a 2.5-inch fire-hose connection for fire-department use and those trained in handling heavy fire streams. These connections must match the hose thread utilized by the fire department and are typically found in stairwells of buildings.
- Class II standpipe is one directly connected to a water supply and serves a 1.5-inch fire-hose connection, providing a means for the control or extinguishing of incipient stage fires. They are typically found in cabinets and are intended for trained occupant use. They are spaced according to the hose length. The hose length and connection spacing are intended for all spaces of the building.
- Class III standpipe s is a combined standpipe system (having both Class I and II connections), directly connected to a water supply and for the use of in-house personnel capable of furnishing effective water discharge during the more advanced stages of fire in the interiors of workplaces. Many times these connections will include a 2.5-inch reducer to a 1.5-inch connection.

When a standpipe-system control valve is located within a stairwell, the maximum length of hose should not exceed 100 feet. If the control valve is located in areas other than the stairwell, the length of hose should not exceed 75 feet. Code requires that fire hose on Class II and Class III standpipe systems be equipped with a shutoff-type nozzle.

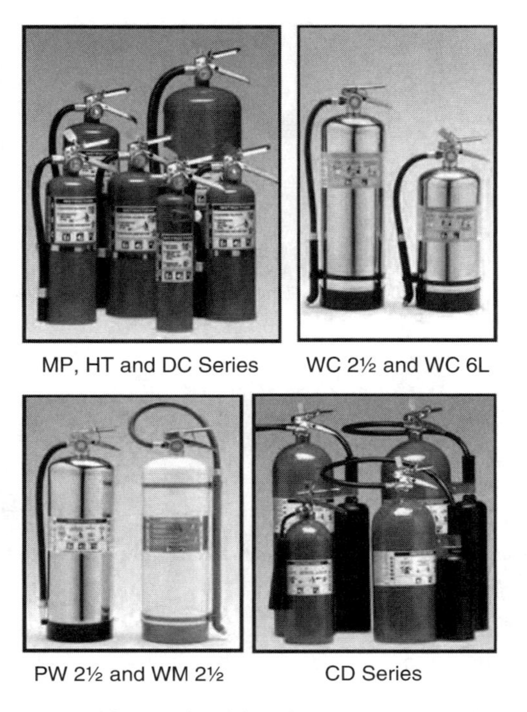

Figure 9.30 Various types of fires extinguishers in common use: MP series multipurpose dry chemical, DC series regular dry chemical, WC series wet chemical, WM series water mist, CD series carbon dioxide, HT series Halotron I.
Source: Larsen's Manufacturing Co.

9.6.3 Hand-Held Fire Extinguishers

There are several different classifications of fire extinguishers, each of which extinguishes specific types of fire (Figure 9.30). Newer fire extinguishers use a picture/labeling system to designate which types of fires they are to be used on, whereas older fire extinguishers are labeled with colored geometrical shapes with letter designations (Figure 9.31).

Class A fire extinguishers are designed to put out fires caused by organic solids and ordinary combustibles such as wood, textiles, paper, some plastics, and rubber. The numerical rating for this class of fire extinguisher refers to the amount of water the fire extinguisher holds and the size of fire it will extinguish. To extinguish a Class A fire, extinguishers utilize either the heat-absorbing effects of water or the coating effects of certain dry chemicals. Class A fire extinguishers should be clearly marked with a triangle containing the letter "A." If in color, the triangle should be green.

Class B fire extinguishersare used to put out fires involving flammable and combustible liquids and gases. They work by starving the fire of oxygen and interrupting

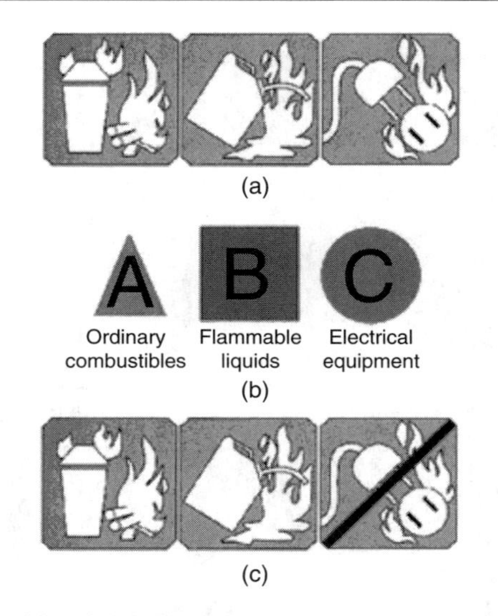

Figure 9.31 New- and old-style labeling systems indicating suitability for use on Class A, B, and C fire extinguishers.

the fire chain by inhibiting the release of combustible vapors. Class B fires are caused by gasoline, oil, and paraffin. The numerical rating for this class of fire extinguisher states the approximate number of square feet of a flammable liquid fire that a non-expert person can expect to extinguish. This includes all hydrocarbon- and alcohol-based liquids and gases that will support combustion. Class B fire extinguishers should be clearly marked with a square containing the letter "B." If in color, the square should be red.

Class C fire extinguishers are most effective for use on fires that involve live electrical equipment where a nonconducting material is involved. This class of fire extinguishers does not have a numerical rating, but the presence of the letter "C" indicates that the extinguishing agent is nonconductive. Class C fire extinguishers should be clearly marked by a circle containing the letter "C." If in color, the circle should be blue.

Class D fire extinguishers are special types designed and approved for specific combustible materials (metals) such as magnesium, titanium, zirconium, potassium, sodium, etc., which require an extinguishing medium that does not react with the burning metal. Class D fire extinguishers should be clearly marked by a five-point painted star containing the letter "D." If in color, the star should be yellow. These extinguishers generally have no rating, nor are they given a multipurpose rating for use on other types of fires.

Class K fire extinguishers are effective for fighting fires involving cooking fats, grease, oils, etc., in commercial cooking environments. These fire extinguishers work on the principle of saponification, which takes place when alkaline mixtures such as potassium acetate, potassium citrate, or potassium carbonate are applied to burning cooking oil or fat. The alkaline mixture, combined with the fatty acid, creates soapy foam on the surface, which holds in the vapors and steam and extinguishes the fire. Class K fire extinguishers should be clearly marked by a hexagon with the letter K. If in color, the hexagon should be black.

If a multipurpose extinguisher is being used, each classification unit should be clearly labeled so that users can identify it quickly in case of an emergency. The approved marking system combines pictographs of both recommended and unacceptable extinguisher types on a single identification label. Many extinguishers available today can be used on different types of fires and will be labeled with more than one designator – e.g., A-B, B-C, or A-B-C. It should also be noted that British standards and classifications differ slightly from American standards.

9.6.3.1 Types of Fire Extinguishers

Dry-chemical extinguishers come in a variety of types and are usually rated for multiple-purpose use (Class A, B, and C fires). They are filled with a foam or powder extinguishing agent and use a compressed, nonflammable gas as a propellant. One advantage a dry chemical extinguisher has over a CO_2 extinguisher is that it leaves a nonflammable substance on the extinguished material, reducing the likelihood of reignition.

Water or APW (air-pressurized water) extinguishers are filled with water and pressurized with oxygen. APW extinguishers should only be used on Class A (ordinary combustible) fires and never on grease fires, electrical fires, or class D fires – the flames will only spread and likely make the fire bigger.

Carbon-dioxide (CO_2) extinguishers contain carbon dioxide, a nonflammable gas, and are highly pressurized. They are most effective on Class B and C (liquid and electrical) fires. Since the gas disperses quickly, these extinguishers are only effective from three to eight feet. The carbon dioxide is stored as a compressed liquid in the extinguisher; as it expands, it cools the surrounding air. The cooling will often cause ice to form around the "horn" where the gas is expelled from the extinguisher. These extinguishers don't work very well on Class A fires because they may not be able to displace enough oxygen to put the fire out, causing it to re-ignite. The advantage of CO_2 extinguishers over dry-chemical extinguishers is that they don't leave a harmful residue and are therefore a good choice for an electrical fire on a computer or other favorite electronic device such as a stereo or TV.

Halon extinguishers contain a gas that interrupts the chemical reaction that takes place when fuels burn. Halon is an odorless, colorless gas that can cause asphyxiation, and these extinguishers have a limited range, usually four to six feet. An advantage of halon is that it is a clean agent because it leaves no corrosive or abrasive residue after release, minimizing cleanup, which makes it more suitable for valuable electrical equipment, computer rooms, telecommunication areas, theaters, etc. However, pressurized fire-suppression-system cylinders can be hazardous, and if not handled properly are capable of violent discharge. Moreover, the cylinder can act as a projectile, potentially causing injury or death. Halon has been banned from new production, except for military use, since January 1, 1994, because its properties contribute to ozone depletion and a long atmospheric lifetime, usually 400 years. However, halon reuse is still permitted in the United States.

NFPA Code 10 addresses all the issues pertaining to portable fire extinguishers and contains clear, widely accepted rules for distribution and placement, maintenance, operation, inspection, testing, and recharging. Portable extinguishers are recognized as a first line of defense against fires and, when maintained and operated properly on a small containable fire, can prevent fires from spreading beyond their point of origin. NFPA Code 10 requires owners of extinguishers to have monthly inspections performed and to maintain records of the inspections.

9.6.3.2 Smoke- and Heat-Detection Systems

Fire-detection systems play a pivotal role in green buildings. Kate Houghton, director of marketing for Kidde Fire Systems, says: "By detecting a fire quickly and accurately (i.e., by not sacrificing speed or causing false alarms) and providing early-warning notification, a fire-detection system can limit the emission of toxic products created by combustion as well as global-warming gases produced by the fire itself. These environmental effects often are overlooked but undoubtedly occur in all fire scenarios. Therefore, reducing the likelihood of a fire is an important part of designing a green building."

A smoke detector or smoke alarm is a device that detects smoke and issues an alarm to alert nearby people of the threat of a potential fire. A household smoke detector is typically mounted in a disk-shaped plastic enclosure about six inches (150mm) in diameter and one inch (25mm) thick, but the shape can vary by manufacturer (Figure 9.32).

Figure 9.32 Drawing of ceiling-mounted smoke detector.
Source: Scott Easton.

Because smoke rises, most detectors are mounted on the ceiling or on a wall near the ceiling. It is imperative that smoke detectors are regularly maintained and checked that they operate properly. This will ensure early warning to allow emergency responses to occur well before a fire causes serious damage. It is not uncommon for modern types of systems to detect smoldering cables or overheating circuit boards. Early detection can save lives and help limit damage and downtime. Laws governing the installation of smoke detectors differ depending on the jurisdiction.

Smoke detectors are typically powered by one or more batteries, but some can be connected directly to a building's wiring. Often the smoke detectors that are directly connected to the main wiring system also have a battery as a power-supply backup in case the facility's wiring goes out. Batteries should be checked and replaced periodically to ensure appropriate protection.

The majority of smoke detectors work either by optical detection or by ionization, and in some cases both detection methods are used to increase sensitivity to smoke. A complete fire-protection system will typically include spot smoke detectors that can signal a fire control panel to deploy a fire-suppression system. Smoke detectors can either operate alone. be interconnected to cause all detectors in an area to sound an alarm if one is triggered, or be integrated into a fire-alarm or security system. Smoke detectors with flashing lights are also available for the deaf or hearing-impaired. Smoke detectors cannot detect carbon monoxide unless they come with integrated carbon-monoxide detectors.

Aspirating smoke detectors (ASD) can detect the early stages of combustion and are 1000 times more sensitive than conventional smoke detectors, giving early warning to building occupants and owners. An aspirating smoke detector generally consists of a central detection unit that sucks air through a network of pipes to detect smoke and in most cases requires a fan unit to draw in a representative sample of air from the protected area through its network of pipes. Aspirating smoke detectors are extremely sensitive and are capable of detecting smoke before it is even visible to the human eye. However, their use is not recommended in unstable environments due to the wide range of particle sizes that are detected.

Optical smoke detectors are light sensors. They include a light source (infrared LED), a lens to collimate the light into a beam like a laser, and a photodiode or other photoelectric sensor at right angles to the beam to detect light. Under normal conditions (i.e., in the absence of smoke) the sensor device detects no light signal and therefore produces no output. The source and the sensor device are arranged so that there is no direct "line of sight" between them. When smoke enters the optical chamber into the path of the light beam, some light is scattered by the smoke particles, and some of the scattered light is detected by the sensor. An increased input of light into the sensor sets off the alarm.

Projected-beam detectors are employed mainly in large interior spaces, such as gymnasiums and auditoriums. A unit on the wall transmits a beam, which is either received by a receiver or reflected back via a mirror. When the beam is less visible to the "eye" of the sensor, it sends an alarm signal to the fire-alarm control panel. Optical smoke detectors are generally quick in detecting slow-burning, smoky fires.

Ionization detectors are sometimes known as an ionization-chamber smoke detectors (ICSDs). They are capable of quickly sensing flaming fires that produce little smoke. They employ a radioactive material to ionize the air in a sensing chamber; the presence of smoke affects the flow of the ions between a pair of electrodes, which triggers the alarm. While over 80 percent of the smoke detectors in American homes are of this type and they are less expensive than optical detectors, they are frequently rejected for projects seeking LEED™ certification for environmental reasons. The majority of residential models are self-contained units that operate on a nine-volt battery, but construction codes in some parts of the country now require installations in new homes to be connected to the house wiring, with a battery backup in case of a power failure.

A heat detector is a device that can detect heat. It can be either electrical or mechanical in operation. Most heat detectors are designed to trigger alarms and notification systems before smoke even becomes a factor. The most common types of heat detectors are thermocouple and electropneumatic, both of which respond to changes in ambient temperature. Typically, if the ambient temperature rises above a predetermined threshold, an alarm signal is triggered.

The main benefit of good detection (beyond triggering the alarm system) is that in many cases there is a chance to extinguish a small, early blaze with a fire extinguisher. Intelligent smoke detectors can differentiate between different alarm thresholds. These systems typically have remote detectors located throughout the facility which are connected to a central alarm station.

9.6.3.3 Fire Doors

Fire doors are doors made of fire-resistant material that can be closed to prevent the spread of fire and are designed to provide extra fire-spread protection for certain areas of a building (Figure 9.33). The fire-rating classification of the wall into which a door is installed dictates the required fire rating of the door. The location of the wall in the building and the prevailing building code establish the wall's fire rating.

Fire doors are commonly installed at staircases from corridors or rooms and cross-corridor partitions in laboratories, plant rooms, workshops, storerooms, machine rooms, service ducts, and kitchens as well as in defined fire compartments. They are also employed in circulation areas that extend the escape route from the stair to a final exit or place of safety, entrances and lobbies; routes leading onto external fire escapes, and corridors that are protected from adjoining accommodation by fire-resisting construction.

According to the National Fire Protection Association (NFPA) doors are rated with respect to the number of hours they can be expected to withstand fire before burning through. There are 20-, 30-, 45-, 60-, and 90-minute rated fire doors as well as two- and four-hour rated fire doors that are certified by an approved laboratory such as Underwriters Laboratories (UL). The certification only applies if all parts of the installation are correctly specified and installed. For example, fitting the wrong kind of glazing may severely reduce the door's fire-resistance rating.

Because fire doors are rated physical fire barriers that protect wall openings from the spread of fire, they are required to provide automatic closing in the event of fire

Interior Fire-Rated Doors and Glass Lites

			Glass lite size allowed		
Class	**Fire rating**	**Location and use**	**Ares**	**Height**	**Width**
A	3 hour	Fire walls separation buildings or various fire areas within a building. 3 to 4 hour walls	None allowed	None allowed	None allowed
B	1½ hours (H.M.) 1 hour (other)	Vertical shafts and enclosures such as stairwells, elevators, and garbage chutes. 2 hour walls	100 inches	33 inches	10–12* inches
B	1 hour	Vertical shafts in low-rise buildings and discharge corridors. 1 to 1½ hour walls	100 inches	33 inches	10–12* inches
C	¾ hour	Exit access corridors and exitway enclosures. 1 hour walls	1296 inches	54 inches	54 inches
N/A	20-minute (⅓ hour)	Exit access corridors and room partitions. 1 hour walls	No limit	No limit	No limit

* Final size depends on code publication used.

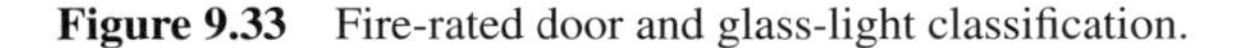

Figure 9.33 Fire-rated door and glass-light classification.

detection. Fire doors should usually be kept closed at all times, although some fire doors are designed to stay open under normal circumstances and to close automatically or manually in the event of a fire. Fire-door release devices are electromechanical devices that enable automatic-closing fire doors to respond to alarm signals from detection devices such as smoke detectors, heat detectors, and central alarm systems. This permits closing the door before high temperatures melt the fusible link. Fusible links should always be used as a backup to the releasing device.

Door assemblies must comply with NFPA 80 and be listed and labeled by UL for the fire ratings indicated, based on testing according to NFPA 252. Assemblies must be factory-welded or come complete with factory-installed mechanical joints and must not require job-site fabrication.

9.6.3.4 Fire Exits and Stairs

OSHA defines an exit route as a continuous and unobstructed path of travel from any point within a workplace to a place of safety. Every building has fire exits, which enable users to exit safely in the event of an emergency. Well-designed emergency exit signs are necessary for emergency exits to be effective. In the United States fire-escape signs often display the word "exit" in large, well-lit, green or red letters. An exit route must be permanent and must be separated by fire-resistant materials. Construction materials used to separate an exit from other parts of the workplace must have a one-hour fire-resistance rating if the exit connects three or fewer stories and a two-hour fire-resistance rating if the exit connects four or more stories.

Unless otherwise stipulated by code, at least two exit routes must be provided in a workplace to permit prompt evacuation of employees and other building occupants during an emergency. The exit routes must be located as far away as practical from each other so that, if one exit route is blocked by fire or smoke, employees can evacuate using the second route. More than two exit routes must be available in a workplace if the number of employees, the size of the building, its occupancy, or the arrangement of the space is such that all employees would not be able to evacuate safely during an emergency. Likewise, a single exit route is permitted if the number of employees, the size of the building, its occupancy, or the arrangement of the space is such that all employees would be able to evacuate safely during an emergency.

Exit routes must be arranged so that employees are not required to travel toward a high-hazard area unless the path of travel is appropriately shielded from the area by suitable partitions or other physical barriers. Exit routes must be free and unobstructed. No materials or equipment may be placed, either permanently or temporarily, along the exit route. The exit access must not go through a room that can be locked, such as a bathroom, nor may it lead into a dead-end corridor. Stairs and ramps are required where the exit route is not substantially level.

OSHA stipulates that each exit discharge must lead directly to the exterior or to a street, walkway, refuge area, public way, or open space with access to the outside. The street, walkway, refuge area, public way, or open space to which an exit discharge leads must be large enough to accommodate the building occupants likely to use the exit route. Exit stairs that continue beyond the level on which the exit discharge is located must be interrupted at that level by doors, partitions, or other effective means that clearly indicate the direction of travel leading to the exit discharge.

Exit doors must not be locked from the inside, and each doorway or passage along an exit access that could be mistaken for an exit (such as a closet) must be marked "not an exit" or some similar designation or be identified by a sign indicating its actual use. Furthermore, exit-route doors must be free of decorations or signs that obscure the visibility of the door, and employees must be able to readily open an exit-route door from the inside at all times without keys, tools, or special knowledge. A device such as a panic bar that locks only from the outside is permitted on exit-discharge doors. Exit-route doors may be locked from the inside only in mental, penal, or correctional facilities and then only if supervisory personnel are continuously on duty and the employer has a plan to remove occupants from the facility during an emergency.

When an outdoor exit route is used, it must have guardrails to protect unenclosed sides where a fall hazard exists. The outdoor exit route must be covered if snow or ice is likely to accumulate along the route, unless it can be demonstrated that any snow or ice accumulation will be removed before it presents a slipping hazard. The outdoor exit route must be reasonably straight and have smooth, solid, substantially level walkways, and it must not have a dead end that is longer than 20 feet (6.2 m).

9.6.3.5 Fire Stopping

The protection of people and property during building fires requires the employment of three essential design elements: alarms to provide early warnings, automatic sprinklers or other suppression systems, and fireproof compartments to contain flames and smoke. These elements work together to give occupants time to escape and firefighters time to arrive. Eliminating any one of the three fire-protection elements – detection, suppression, or compartmentation – would compromise the integrity of the building.

Building regulations in most jurisdictions stipulate that large buildings need to be divided into compartments and that these compartments must be maintained should a fire occur. In order to do this, there are a range of fire-stopping products and methods available offering between 30 and 240 minutes of fire-compartmentation protection for construction movement joints and service penetrations.

Provisions for fire safety may require dividing a building into fire-isolated compartments, which restricts a fire to an area of a building until it can be extinguished. But to be effective, the walls, floor, and ceiling need to contain flames and smoke within the compartment. These components must also provide sufficient insulation to prevent excessive heat from radiating outside the compartment.

A fire compartment can therefore be defined as a space within a building extending over one or several floors that is enclosed by separating members such that the fire spread beyond the compartment is prevented during the relevant fire exposure. Fire compartments are sometimes referred to as fire zones. Compartmentation is critical to preventing fire from spreading into large spaces or into the whole building. It involves the specification of fire-rated walls and floors sealed with fire-stop systems, fire doors, fire dampers, etc.

The division of the building into discrete fire zones offers perhaps the most effective means of limiting fire damage. Compartmentation techniques are designed to contain the fire to within the zone of origin by limiting vertical and horizontal fire spread. Compartmentation also provides at least some protection for the rest of the building and its occupants even if first-aid firefighting measures are used and fail. It also provides protection for inventory and business operations and delays the spread of fire prior to the arrival of the fire brigade.

Determining the required fire resistance for a compartment depends largely upon its intended purpose and on the expected fire. The separating members enclosing the compartment should resist the maximum expected fire or contain the fire until occupants are evacuated. The load-bearing elements in the compartment must always resist the complete fire process or be classified to a certain resistance measured in terms of periods of time, which must be equal or longer than the requirement of the separating members. The most important elements to be upgraded are the doors, floors, and walls; penetrations through floors and walls; and cavity barriers in the roof spaces. Halls and landings should typically be separated from staircases to prevent a fire from traveling vertically up or down the stairwell to the other floors. However, creation of new lobbies can have an unacceptable negative impact on the character of a fine historic interior. To be effective, compartmentation needs to be correctly planned and implemented.

The main function of fire stopping is to stop the spread of fire between floors of a building. Flame-retardant material is installed around floor openings designed to contain conduit and piping. A fire stop is a product that, when properly installed, impedes the passage of fire, smoke, and toxic gases from one side of a fire-rated wall or floor assembly to another. Typical fire-stop products include sealants, sprays, mechanical devices (fire-stop collars), foam blocks, or pillows. These products are installed primarily in two applications: around penetrations that are made in fire-resistive construction for the passage of pipes, cables, or HVAC systems and where two assemblies meet, forming a expansion joint such as the top of a wall, curtain wall (edge of slab), or floor-to-floor joint. Typical opening types include electrical through-penetrations, mechanical through-penetrations, structural through-penetrations, nonpenetrated openings (e.g., openings for future use), reentries of existing fire stops, control or sway joints within fire-resistance rated wall or floor assemblies, junctions between fire-resistance rated wall or floor assemblies, and head-of-wall (HOW) joints, where non-load-bearing wall assemblies meet floor assemblies.

Compliance with all applicable laws and regulations relating to a building is the owner's responsibility including the adopted and enforced fire code within a specific jurisdiction. Fire codes govern the construction, protection, and occupancy details that affect the fire safety of buildings throughout their life span. Numerous fire codes have been adopted throughout the United States – the vast majority of which are similar and based on one of the model codes available today or in the past. One requirement in all of these model codes is that fire-safety features incorporated into a building at the time of its construction must be maintained throughout a building's life. Therefore, any fire-resistance rated construction must be maintained (Figure 9.34).

9.6.3.6 Alarm and Notification Systems

Fire-alarm systems are essential to any facility, particularly in large buildings where there may be visitors or personnel who are unfamiliar with their surroundings. Bruce Johnson, Regional Manager for Fire Service Activities with the International Code Council, says: "Fire-alarm systems and smoke alarms are life-safety systems that save countless lives each year, both civilians and firefighters. The International Residential Code requires interconnected, hard-wired smoke alarms in all new construction (Section R313), and the International Building Code and International Fire Code (Section 907.2) call for manual or automatic fire-alarm systems in most commercial buildings with high life occupancy or other hazards. In addition to new construction, the International Fire Code also has provisions for fire-alarm systems and smoke alarms in existing structures (Section 907.3)." Fire alarms alert building occupants of a fire and emergency public responders (police and fire) through a central station link to initiate an appropriate response.

Mass-notification systems (MNS) are invaluable in the protection of a wide range of facilities and MNS use both audible and visible means to distribute potential lifesaving messages. A MNS is much more than an alarm system. By using the technologies based on fire-alarm codes and standards, fire-system manufacturers are able to produce a robust life-safety and security system.

Fire-alarm control panels (FACP) or fire-alarm control units (FACU) consist of electric panels that function as the controlling components of a fire-alarm system (Figure 9.35). The FACP receives information from environmental sensors designed to

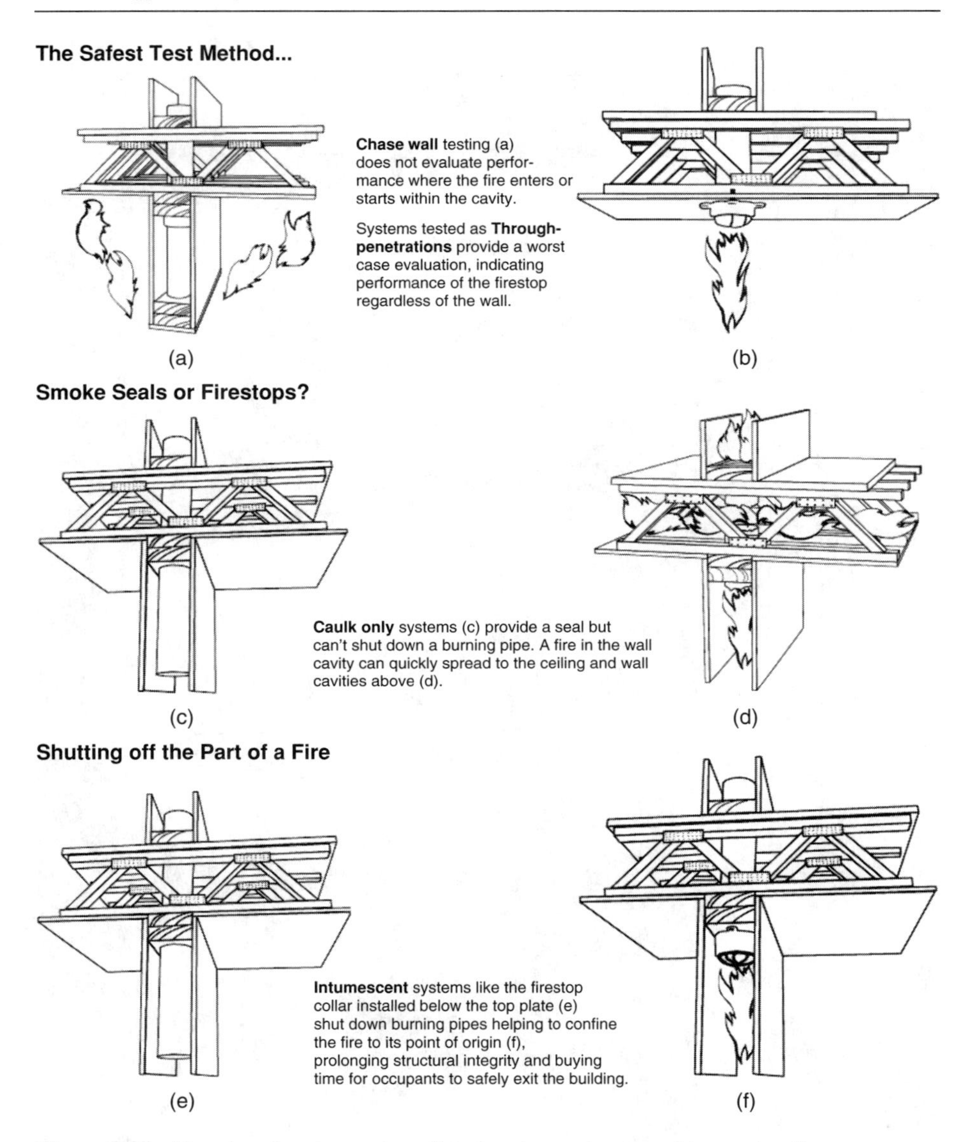

Figure 9.34 Drawing showing various fire-stopping systems used in construction.

detect any changes associated with fire. It also monitors their operational integrity and provides for automatic control of equipment and transmission of information necessary to prepare the facility for fire, based on a predetermined sequence. An FACP may also supply electrical energy to operate any associated sensor, control, transmitter, or relay. There are currently four basic types of FACP panels on the market: coded panels, conventional panels, addressable panels, and multiplex systems.

Figure 9.35 A Siemens MXL fire-alarm control panel (top) and graphic annunciator (bottom) for Potomac Hall at James Madison University; photograph by Ben Schumin.
Source: Wikipedia.

With the introduction of increasingly sophisticated technology, today's alarm systems have the ability to provide more information to the fire department and first responders. In many cases, they can do more than just tell them that there has been an alarm in the building; they can be directed by the kind of alarm and where the alarm is. Moreover, many modern systems now include speakers that provide alerts in place of (or in addition to) traditional bell-type alarms. These speakers also can be used in emergencies other than fires to instruct and inform occupants of the situation.

These voice-actuated systems can include prerecorded or live messages that play in the event of fire or another emergency. Typical prerecorded messages tell occupants that an alarm has been sounded and that they should remain in their designated area for further instruction. Building management can then manually use the system to deliver additional information and prepare occupants for an evacuation if necessary. Alert systems can also close fire doors, recall elevators, and interface and monitor the installed suppression systems, such as sprinklers. It should be noted that when fire-alarm systems are properly installed and maintained, they perform very well. But when they are not, the public and fire service may be subject to unnecessary false alarms that puts everyone at risk.

Alarm systems can also connect with a building's ventilation, smoke-management, and stairwell-pressurization systems – all of which are critical to life safety. Again, these features are dependent on the building in which the system is installed. In addition, the marriage of mass-notification systems (MNS) and fire-alarm control systems is a growing trend that will hopefully continue and be applied in larger varieties of facilities and multibuilding properties, including schools, high-rise buildings, mass-transit hubs, and even public gathering places such as theaters, restaurants, and places of worship.

An annunciator panel is sometimes employed to monitor the status of the different areas in a designated fire zone for theft protection and control of a facility's alarm devices. There may be several fire zones in a building. Each fire zone is clearly marked on the panel. The annunciator panel identifies the different zones and their specific security status. Should a fire occur, an indicator light flashes on the panel and identifies the fire's location. For example, the light on the panel might indicate that a fire has occurred in fire zone 4. This information allows the Fire Department to quickly locate the fire.

9.6.4 Codes and Standards

Code compliance is the first objective in any design. There are a number of relevant national codes that relate to green-building fire-protection systems that are published by the National Fire Protection Association (NFPA). It should be noted that fire codes can vary substantially from one jurisdiction to another, and, while these codes are not mandatory in all jurisdictions, they should nevertheless be adhered to whenever possible because they provide maximum safety for property and personnel and can help guide system design and installation.

NFPA 72, National Fire Alarm Code, governs the design, installation, operation, and maintenance of fire-detection and -alarm systems. It includes requirements for detector spacing, occupant notification, and control-panel functionality.

NFPA 750, Standard on Water Mist Fire Protection Systems, governs water-mist-system classification and incorporates requirements for water-mist-system design, installation, operation, and maintenance.

NFPA 2001, Standard on Clean Agent Fire Extinguishing Systems, governs the design, installation, operation, and maintenance of clean-agent systems. It additionally includes requirements for assessing design concentrations, safe personnel-exposure levels, and system-discharge times. The standard also stipulates that an agent be included on the U.S. Environmental Protection Agency's Significant New Alternatives Policy list.

Finally, green buildings today have numerous fire-protection-system options that can be employed. Careful consideration of a building and its anticipated hazards will help determine which areas require protection. Due to the recent advances in technology, fire-detection and -suppression systems can now adequately support and sustain a modern green-building philosophy. The methodical selection of a clean-agent or water-mist system can also help contribute to LEED™ certification credits for building owners and developers.

10 Economics of Green Design

10.1 General Overview

The current financial crisis and global economic recession are putting unprecedented strain on the nation's construction industry. Peter Morris, principal of the construction consultancy Davis Langdon, believes that the drastic reduction in construction activity is encouraging increased competition among bidders and lower escalation pressure on projects to the extent that cost trends have become negative, leading to moderate construction price deflation. But one of the biggest causes of concern, according to Morris, is the issue of contractor financing and working capital. Many contractors are finding it increasingly difficult to maintaining adequate cash flow for operations, and none have the resources to manage significant expansion of working capital. This has obliged bidders to be very cautious and judicious in selection of projects, with a strong preference towards those that have good cash flows.

Looking at the bright side, there has emerged an increasingly broad awareness of the significant benefits that green buildings have to offer. For example, the U.S. military, including the Air Force and Navy, now requires that their new buildings be LEED™ green buildings. This may be in part because they recognize the linkage between wasteful energy consumption and the exposure of military forces to military confrontation related to oil resources. As Boston mayor Thomas M. Menino put it, "High-performance green building is good for your wallet. It is good for the environment. And it is good for people."

Benefits such as reduced energy bills and lower potable-water consumption are easily quantified, while other benefits including green design's impact on occupant health or security are usually much more difficult to quantify. Incisive Media's 2008 Green Survey: Existing Buildings found that nearly 70 percent of commercial building projects in the United States have already incorporated some kind of energy-monitoring system. The survey also found that energy conservation is the most widely implemented green program in commercial buildings, followed by recycling and water conservation. Moreover, approximately 65 percent of building owners who have implemented green-building features claim their investments have already resulted in a positive return.

Likewise, Turner Construction Company's 2008 Green Building Barometer states that roughly 84 percent of respondents maintain that their green buildings have resulted in lower energy costs, and 68 percent recorded lower overall operating costs. Additionally, "75 percent of executives said that recent developments in the credit markets would not make their companies less likely to construct green buildings." In fact, the survey contends that 83 percent would be "extremely" or "very" likely to seek LEED™ certification for buildings they are planning to build within the next three years. In the

DOI: 10.1016/B978-1-85617-691-0.00010-2

same survey executives reported that green buildings have better financial performance than nongreen buildings, especially in the following sectors:

- Higher building values (72 percent)
- Higher asking rents (65 percent)
- Greater return on investment (52 percent)
- Higher occupancy rates (49 percent)

A similar study maintains that 60 percent of commercial-building owners offer education programs to assist tenants in implementing green programs in their space, reflecting a growing understanding of the importance of environmental awareness among employees and customers in addition to the use of green materials and systems.

In a comparative study by Davis Langdon (2006) in which the construction costs of 221 buildings were analyzed, it was found that 83 buildings were constructed with the intent of achieving LEED™ certification and 138 did not have any sustainable-design intentions. The study found that a majority of the buildings analyzed were able to achieve LEED™ certification without increased funding. Another investigation conducted by Davis Langdon of a wide and diverse range of studies by other organizations found that the average construction cost premium required to achieve a moderate level of green features, equivalent to a Silver LEED™ certification was roughly between one and two percent. However, Davis Langdon also found that often half or more of the green projects in these studies revealed no increase in construction costs at all.

Yet, despite the effusive praise for and increased awareness of sustainable design, property owners and developers have sometimes been slow to embrace green-building practices. This appears to be the prevailing sentiment of CB Richard Ellis' Green Downtown Office Markets: A Future Reality," a report depicting the general progress of the green-building movement. The report looks at the obstacles preventing a broad-based acceptance of sustainable design in office construction. Perhaps the main obstacle to embracing design sustainability is the perception of initial outlay compared to long-term benefits; even though an increasing number of studies similar to the one conducted by Davis Langdon in 2006 clearly show that there is no significant difference in average costs for green buildings as compared to conventionally constructed buildings. The other hurdle that required addressing was the lack of data on development, construction costs, and time needed to recoup costs – making education the most crucial tool for sustainable design.

Nevertheless, this general slowdown in moving forward has not caused total cessation of green development. On the contrary, increased awareness and education have had a very positive effect. For example, Alex Wilson, president of BuildingGreen, LLC, notes: "Given that we manage 1.7 billion square feet globally, we have a real opportunity to help our clients understand how they can make changes in operations in their own facilities to reduce greenhouse-gas emissions and be more energy-efficient."

Green-building research currently constitutes an estimated $193 million per year, or roughly 0.2 percent of federally funded research. This represents only 0.02 percent of the estimated $1 trillion value of annual U.S. building construction, despite the fact that the building-construction industry represents approximately 9 percent of the

U.S. GDP. The construction industry is currently reinvesting only 0.6 percent of sales back into research, which is markedly less than the average for other U.S. industries and private-sector construction research investments in other countries.

The U.S. Green Building Council (USGBC) suggests that, unless we move decisively toward increasing and improving green-building practices, we are likely to soon be confronted with a dramatic backlash in adverse impact of the built environment on human and environmental health. Building operations today are estimated to account for 38 percent of U.S. carbon-dioxide emissions, 71 percent of electricity use, and 40 percent of total energy use. If the energy required to manufacture building materials and construct buildings is included, this number goes up to an estimated 48 percent. Buildings also consume roughly 12 percent of the country's water in addition to rapidly increasing amounts of land. Moreover, construction and remodeling of buildings account for 3 billion tons, or 40 percent, of raw material used globally each year, which in turn has a negative impact on human health; in fact, up to 30 percent of new and remodeled buildings may experience acute indoor-air quality problems.

10.2 Costs and Benefits of Green Design

Peter Morris opines: "clearly there can be no single, across the-board answer to the question 'what does green cost?' On the other hand, any astute design or construction professional recognizes that it is not difficult to estimate the costs to go green for a specific project. Furthermore, when green-building concepts and features are incorporated early in the design process, it greatly increases the ability to construct a certified green building at a cost comparable to a code-compliant one. This means that it is possible today to construct green buildings or buildings that meet the US Green Building Council's (USGBC) Leadership in Energy and Environmental Design (LEED™) third-party certification process with minimal increase in initial costs (Figure 10.1)."

Davis Langdon also suggests that, to be successful in building green and to keep the costs of sustainable design under control, three critical factors must be understood and implemented: "First, clear goals are critical for managing the cost. It is not enough to simply state, 'we want our project to be green'; the values should be determined and articulated as early in the design process as possible. Second, once the sustainability goals have been defined, it is essential to integrate them into the design and to integrate the design team so that the building elements can work together to achieve those goals. Buildings can no longer be broken down and designed as an assemblage of isolated components. This is the major difference between traditional building techniques and the new sustainable-design process. Third, integrating the construction team into the project team is critical. Many sustainable-design features can be defeated or diminished by poor construction practices. Such problems can be eliminated by engaging the construction team, including subcontractors and site operatives, in the design and procurement process."

Another noteworthy study, which represents the largest international research of its kind, Greening Buildings and Communities: Costs and Benefits, is based on extensive

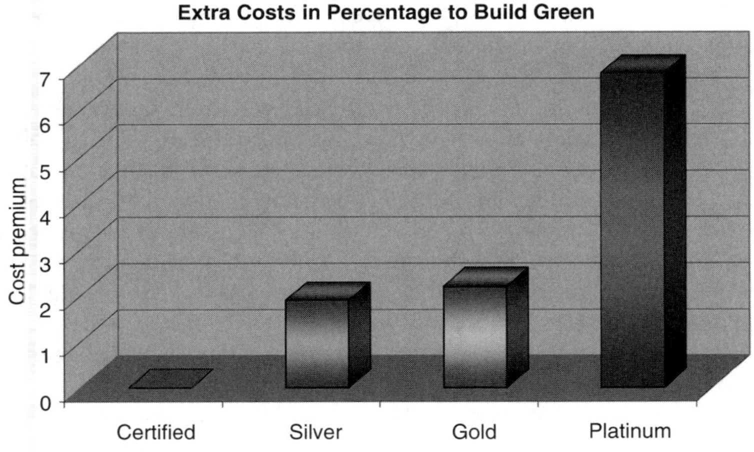

Figure 10.1 Diagram showing one study's estimated cost of building green for the various LEED™ Rating Systems.
Source: USGBC.

financial and technical analysis of 150 green buildings across the U.S. and in 10 other countries and provides the most detailed findings to date on the costs and financial benefits of building green. The study found that benefits of building green consistently outweigh any potential cost premium. Among the main conclusions arrived at in the study:

1. Most green buildings cost zero to four percent more than conventional buildings, with the largest concentration of reported "green premiums" between zero and one percent. Green premiums increase with the level of greenness, but most LEED™ buildings, up through Gold level, can be built for the same cost as conventional buildings. This stands in contrast to a common misperception that green buildings are much more expensive than conventional buildings.
2. Energy savings alone make green building cost effective. Energy savings outweigh the initial cost premium in most green buildings. The present value of 20 years of energy savings in a typical green office ranges from $7 per square foot (Certified) to $14 per square foot (Platinum), more than the average additional cost of $3 to $8 per square feet for building green.
3. Green-building design goals are associated with improved health and with enhanced student and worker performance. Health and productivity benefits remain a major motivating factor for green-building owners but are difficult to quantify. Occupant surveys generally demonstrate greater comfort and productivity in green buildings.
4. Green buildings create jobs by shifting spending from fossil-fuel-based energy to domestic energy efficiency, construction, renewable energy, and other green jobs. A typical green office creates roughly one-third of a permanent job per year, equal to $1 per square foot of value in increased employment compared to a similar nongreen building.
5. Green buildings are seeing increased market value (higher sales/rental rates, increased occupancy, and lower turnover) compared to comparable conventional buildings. CoStar,

for example, reports an average increased sales price from building green of more than $20 per square foot, providing a strong incentive to build green even for speculative builders.

6. Roughly 50 percent of green buildings in the study's data set see the initial "green premium" paid back by energy and water savings in five years or less. Significant health and productivity benefits mean that over 90 percent of green buildings pay back an initial investment in five years or less.
7. Green community design (e.g., LEED™ ND) provides a distinct set of benefits to owners, residents, and municipalities, including reduced infrastructure costs, transportation and health savings, and increased property value. Green communities and neighborhoods have a greater diversity of uses, housing types, job types, and transportation options and appear to better retain value in the market downturn than conventional sprawl.
8. Annual gas savings in walkable communities can be as much as $1000 per household. Annual health savings (from increased physical activity) can be more than $200 per household. CO_2 emissions can be reduced by 10 to 25 percent.
9. Upfront infrastructure development costs in conservation can be reduced by 25 percent, approximately $10,000 per home.
10. Religious and faith groups build green for ethical and moral reasons. Financial benefits are not the main motivating factor for many places of worship, religious educational institutions, and faith-based nonprofits. A survey of faith groups building green found that financial cost effectiveness of green building makes it a practical way to enact the ethical/moral imperative to care for the Earth and communities. Building green has also been found to energize and galvanize faith communities.

Even when a green building costs more up front due to inefficient planning and execution, the costs are quickly recouped through lower operating costs over the life of the building. The green-building approach is a method that applies to a project's life-cycle cost analytical tools for determining the appropriate up-front expenditure.

10.2.1 The Economic Benefits of Green Buildings

The numerous savings in costs can only be fully realized when they are incorporated at the project's conceptual design phase and with the assistance of an integrated team of professionals. Using an integrated systems approach ensures that the building is designed in a holistic manner as one system rather than a number of stand-alone systems.

Some of the benefits of building green are not easy to quantify; how do you measure improving occupant health, comfort, and productivity or pollution reduction? This is one of the main reasons why they are not adequately considered in cost analysis. It would be prudent therefore to consider setting aside a small portion of the building budget (e.g., as a contingency) to cover differential costs associated with less tangible green-building benefits or to cover the cost of researching and analyzing green-building options. Even when experiencing difficult times, many green-building measures can be incorporated into a project with minimal or zero increased up-front costs and enormous savings (Figure 10.2).

What does "green" cost is one of the first questions often asked by owners and developers regarding sustainable design; typical translation: does it cost more? This

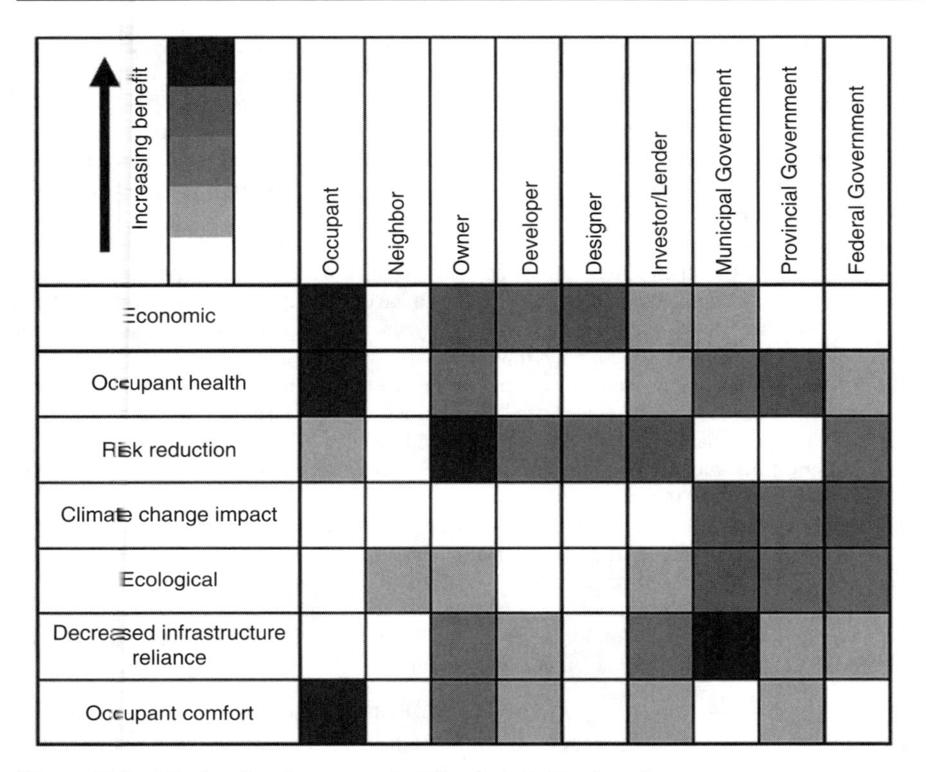

Figure 10.2 Matrix of various green-building stakeholder benefits.
Source: A Business Case for Green Buildings in Canada.

raises the question of more than what? For example, is the question asking more than the building would have cost without the sustainable-design features or more than available funds? The answers to these questions have thus far been largely elusive due to the lack of hard data. Nevertheless, due to considerable research undertaken in recent years, we now have substantial data that allow us to compare the cost of green buildings and nonsustainable buildings with comparable design programs.

When considering sustainable design and how it relates to construction costs, it is important to analyze the costs and benefits using a holistic approach. This means including evaluation of operations and maintenance costs, user productivity and health, and design and documentation fees, among other financial measurements. This is because empirical experience has repeatedly shown that it is the construction-cost implications that have the greatest impact to drive decisions about sustainable design. Assisting teams in comprehending the actual construction costs for real projects of achieving green and providing a methodology that will allow teams to viably manage these construction costs can go a long way to facilitate a team's ability to get past the question of whether to go green or not.

From such an analysis we can conclude that many projects are able to achieve sustainable design within the initial budget or with minimal supplemental funding. This suggests that developers continue to find ways to incorporate project goals and values, regardless of budget, by making choices. However, every building project is unique and should be considered as such when addressing the cost and feasibility of LEED™ certification. There is no one-size-fits-all answer, and benchmarking with other comparable projects can be valuable and informative but not predictive. Any estimate of cost relating to sustainable design for a specific building must be made with reference to that building's particular circumstances and goals.

The cost of green design has dropped significantly in the last few years as the number of green buildings has increased. The trend of declining costs associated with increased experience in green-building construction has manifested itself in a number of states throughout the country. Moreover, studies have shown that LEED™-certified buildings constructed at the Certified or Silver level have often been built without an increase in preliminary costs.

In fact, Davis Langdon explored this topic in a report issued in 2004, Costing Green: A Comprehensive Cost Database and Budgeting Methodology. The report concluded that many LEED™-rated projects cost the same as projects with no stated green goals. The study actually found that many projects with LEED™ level ratings often cost less than non-LEED™ projects. Davis Langdon also discovered that the opposite was often true, in that no direct correlation was found between higher costs for going green and lower costs for conventional building.

Some recent studies, however, suggest that Gold- and Platinum-level projects indicate an increase of one to five percent, with two percent being the average increased additional first cost. But even in cases that show a small increase in upfront costs, these costs are usually recouped within a short period – one to two years during the life cycle of the building – through tax incentives and operational savings resulting from decreased energy and water use.

A recent study by Greg Kats of Capital E Analysis provides a summary of financial benefits from going green, as shown in Figure 10.3. The report concluded that financial benefits for building green are estimated to be between $50 and $70 per square foot in a LEED™ building; this represents more than 10 times the additional cost associated with building green. These financial benefits come in the form of lower energy, waste, and water costs; lower environmental and emissions costs; lower operational and maintenance costs; lower absenteeism; increased productivity and health; higher retail sales; and easier reconfiguration of space, resulting in less downtime and lower costs. Cost estimates that are based on a sample of 33 office and school buildings suggested only 0.6 percent greater costs for LEED™ certification, 1.9 percent for Silver, 2.2 percent for Gold, and 6.8 percent for Platinum. Although these estimates are direct costs, they nevertheless closely reflect those provided by the USGBC.

The principal drivers of green building appear to be: increasing energy costs (75 percent). government regulations/tax incentives (40 percent). and global influences (26 percent).

Category	20-Year Net Present Value
Energy Savings	$5.80
Emissions Savings	$1.20
Water Savings	$0.50
Operations and Maintenance Savings	$8.50
Productivity and Health	$36.90 to $55.30
Subtotal	$52.90 to $71.30
Average Extra Cost of Building Green	(−3.00 to −$5.00)
Total 20-Year Net Benefit	**$50.00 to $65.00**

Figure 10.3 Financial benefits of green buildings: summary of findings (per square foot). The financial benefits of going green are related mainly to productivity.
Source: Capital E analysis.

Lacking clear data pertaining to development, construction costs, and time required to recoup costs is the most obvious obstacle in the industry's slow acceptance of green construction, which is why education has become the most important tool in promoting the green-construction initiative. The main obstacles measured in the study include:

- Too multidisciplinary (41 percent)
- Not convinced of increased ROI (37 percent)
- Failure to understand benefits (26 percent)
- Lack of service providers (20 percent)
- Too difficult (17 percent)
- Greenwashing (16 percent)
- Lack of shareholder support (10 percent)

Most building owners and developers build to sell in the sense that they construct or revamp an office building or lease its space with the hope of selling the asset within a three- to five-year time frame to repay debt and ensure a profit. The speed with which the process is completed impacts the amount of profit generated upon consummation of the sale of the building. Uninformed developers who are under the misperception that green construction costs more spurs fears, as they are already concerned about the cost of short-term debt and conventional building materials.

Typical benefits for building owners and tenants alike, as deduced by Davis Langdon, include:

- Reduced risk of building obsolescence
- Ability to command higher lease rates, therefore increasing profits
- Potentially higher occupancy rates
- Higher future capital value
- Less need for refurbishment in the future
- Higher demand from institutional investors
- Lower operating costs
- Mandatory for government tenants

- Lower tenant turnover
- Enhance occupant comfort and health
- Improve air/thermal/acoustic environment, thereby improving employee productivity and satisfaction

10.2.2 Cost Considerations

The economic considerations with respect to green buildings can be divided into direct capital Costs and direct operating costs.

Direct capital costs are costs associated with the original design and construction of the building and generally include interest during construction (IDC). The general perception of some building stakeholders is that the capital costs of constructing green buildings are significantly higher than those of conventional buildings, whereas many others within the green-building field believe that green buildings actually cost less or no more than conventional buildings. The green-building industry believes that savings are achieved by the downsizing of systems through better design and the elimination of unnecessary systems, which will offset any increased costs caused by implementing more advanced systems.

Capital and operational costs are normally relatively easy to measure, because the required data are readily quantifiable and available. Productivity effects, on the other hand, are difficult to quantify yet are nevertheless important to consider due to their potential impact. Other indirect and external effects can be wide-reaching and may also be difficult to quantify.

Direct operating costs consist of all necessary expenditures required to operate and maintain a building over its full life. They include the total costs of building operation, such as energy use, water use, insurance, maintenance, waste, and property taxes. The primary costs are those associated with heating and cooling and maintenance activities (painting, roof repairs and replacement, etc.). Included in this cost category are less obvious items such as churn (the costs of reconfiguring space and services to accommodate occupant moves). Excluded from this category are costs relating to major renovations, cyclical renewal, and residual value or demolition costs, which are considered to be direct capital investments.

10.2.2.1 Insurance Rates

Insurance is a direct operating cost, and green buildings have many tangible benefits that reduce or mitigate a variety of risk, and that should be reflected in the insurance rates for the building. The fact that green buildings generally provide a healthier environment for occupants should be reflected in health-insurance premiums. Indeed, the general attributes of green buildings (e.g., the incorporation of natural light, off-grid electricity, and commissioning) should reduce a broad range of liabilities, and the general site locations also potentially reduce risks of property loss due to natural disasters.

A fully integrated design of a building will typically reduce the risk of inappropriate systems or materials being employed, which could have a positive impact on other insurable risks. In some cases, insurance companies offer premium reductions for certain green features, such as commissioning or reduced reliance on fossil-fuel-based heating systems (resulting in reduced fire threat). The list of premium reductions will undoubtedly increase with further education and awareness as the broad range of benefits is more fully recognized.

10.2.2.2 Churn Rates

Churn rate reflects the frequency with which building occupants are moved, either internally or externally, including occupants who move but remain within a company and those who leave a company and are replaced. Green buildings typically have lower churn rates than conventional buildings due to increased occupant comfort and satisfaction.

10.2.3 Life-Cycle Costing

Life-cycle costing is basically a method of combining both capital and operating costs to determine the net economic effect of an investment and to evaluate the economic performance of additional investments that may be required for green buildings. It is based on discounting all future costs and benefits to dollars of a specific reference year, which are referred to as present-value (PV) dollars. This makes it feasible to intelligibly quantify costs and benefits and compare alternatives based on the same economic criterion or reference dollar.

Peter Morris, a principal with Davis Langdon, says: "Perhaps a measure of the success of the LEED™ system, which was developed to provide a common basis for measurement, is the recent proliferation of alternative systems, each seeking to address some perceived imbalance or inadequacy of the LEED™ system, such as the amount of paperwork, the lack of weighting of credits, or the lack of focus on specific issues. Among these alternative measures are broad-based approaches, such as Green Globes, and more narrowly focused measures, such as calculations of a building's carbon footprint or measurements of a building's energy efficiency (the ENERGY STAR® rating). All these systems are valid measures of sustainable design, but each reflects a different mix of environmental values, and each will have a different cost impact."

Effectively integrating sustainable-design elements into a project during its early development and design phases can reduce building costs. On the other hand, if green-design elements are not considered until late in the design process or during construction, and if parts of the project or the entire project has to be redesigned, then a significant increase in costs can be contemplated. However, according to a recent study by the World Business Council for Sustainable Development (WBCSD), key players in real estate and construction, unfortunately, often misjudge the costs and benefits of "green" buildings.

10.2.3.1 First Cost

In documented research the most often cited reason for not incorporating green elements into building designs is the assumed increase in first cost. Some aspects of design have little or no first cost, including site orientation and window and overhang placement. Other sustainable systems that incorporate additional costs in the design phase, such as an insulated shell, can be offset, for example, by the reduced cost of a smaller mechanical system. Material costs can be reduced during the construction phase of a project by the use of dimensional planning and other material efficiency strategies. Such strategies can reduce the amount of building materials needed and cut construction costs, but they require forethought on the part of designers to plan a building that creates less construction waste solely due to its dimensions and structural design. An example of dimensional planning is designing rooms in 4-foot multiples, since wallboard and plywood sheets come in 4- and 8-foot lengths. Furthermore, one dimension of a room can be designed in 6- or 12-foot multiples to correspond with the length of carpet and linoleum rolls, and, whenever possible, rooms should be designed with 2-foot incremental dimensions.

Nevertheless, construction projects typically have initial or up-front costs that may include capital investment costs related to land acquisition, construction, or renovation and to the equipment needed to operate a facility. Land acquisition costs are normally included in the initial cost estimate if they differ among design alternatives. A typical example of this would be when comparing the cost of renovating an existing facility with new construction on purchased land.

A 2007 study by Davis Langdon, updating an earlier study, states: "It is clear from the substantial weight of evidence in the marketplace that reasonable levels of sustainable design can be incorporated into most building types at little or no additional cost. In addition, sustainable materials and systems are becoming more affordable, sustainable design elements are becoming widely accepted in the mainstream of project design, and building owners and tenants are beginning to demand and value those features."

Ashley Katz, a communications coordinator for the U.S. Green Building Council, says: "Costs associated with building commissioning, energy modeling, and additional professional services typically turn out to be a risk \-mitigation strategy for owners. While these aspects might add on to the project budget, they will end up saving projects money in the long run and are also best practices for building design and construction."

Many experienced users of LEED™ have found it possible to build at Silver and Gold LEED™ levels for the same cost as conventional buildings, although some studies have concluded that higher levels of certification add roughly one to two percent of the overall budget to the cost of construction. LEED™ registration and certification fees are not substantial, averaging about $2000, depending on a project's size. In any case, registration is required for projects pursuing LEED™ certification, and it provides access to a variety of online resources including LEED™ Online, which is an online project-management tool that project teams use to prepare their documentation for LEED™ certification.

10.2.4 Life-Cycle Cost Method

Life-cycle cost analysis (LCCA) is a method used to estimate the total cost of facility ownership. It takes into consideration all costs relating to acquiring, owning, and disposing of a building or building system. Sieglinde Fuller of the National Institute of Standards and Technology (NIST) says: "LCCA is especially useful when project alternatives that fulfill the same performance requirements,but differ with respect to initial costs and operating costs have to be compared in order to select the one that maximizes net savings. For example, LCCA will help determine whether the incorporation of a high-performance HVAC or glazing system, which may increase initial cost but result in dramatically reduced operating and maintenance costs, is cost-effective or not." However, LCCA is not useful when it comes to budget allocation.

Despite the general consensus on the valid reliable basis for adopting a life-cycle approach, the majority of building stakeholders prefer to focus on minimizing direct costs or, at best, applying short time frame payback periods. Many developers, building owners, and other stakeholders hold the view that basing opinions on anything other than a reduced direct-cost approach is fiscally irresponsible, when in reality the opposite is often the case. This lack of adoption is largely due to the typical corporate structure that dissociates direct and operating costs and to the fact that most construction managers often lack the mandate to reduce operating costs, although they are mandated to reduce construction cost. This unfortunate reality is also evidenced by owner/developers, who oversee construction of buildings for their own use. Other causal factors identified with this issue include the following:

- Not comprehending the life-cycle concept
- Cash-flow constraints
- Difficulty in calculating performance as opposed to the relative ease in calculating direct cost
- Lack of adequate support from lending institutions

The main objective of an LCCA is to calculate the overall costs of project alternatives and to select the design that safeguards the ability of the facility to provide the lowest overall cost of ownership in line with its quality and function. The LCCA should be performed early in the design process to allow any needed design refinements or modifications to take place before finalization to optimize the life-cycle costs (LCC).

Perhaps the most important and challenging task of an LCCA (or any economic evaluation method, for that matter), is to evaluate and determine the economic effects of alternative designs of buildings and building systems and to be able to quantify these effects and depict them in dollar amounts.

Lowest life-cycle cost (LCC) provides a straightforward and easy-to-interpret measure of economic evaluation. Other commonly used methods include net savings (or net benefits), savings-to investment ratio (or savings benefit-to-cost ratio), internal rate of return, and payback period. Fuller sees them as being consistent with the lowest LCC measure of evaluation if they use the same parameters and length of study period. Almost identical approaches can be made to making cost-effective choices for building-related projects irrespective of whether it is called cost estimating, value engineering, or economic analysis.

Sustainable buildings can be assessed as cost-effective through the life-cycle cost method, a way of assessing total building cost over time. After identifying all costs by year and amount and discounting them to present value, they are added to arrive at total life-cycle costs for each alternative: The elements include:

- Initial design and construction costs
- Operating costs including energy, water/sewage, waste, recycling, and other utilities.
- Maintenance, repair, and replacement costs
- Other environmental or social costs/benefits including but not limited to impacts on transportation, solid waste, water, energy, infrastructure, worker productivity, and outdoor air emissions

Because sustainable buildings are also considered healthy buildings, they can have a positive impact on worker illness costs.

10.2.4.1 Supplementary Measures

Sieglinde Fuller says: "Supplementary measures of economic evaluation are net savings (NS), savings-to-investment ratio (SIR), adjusted internal rate of return (AIRR), and simple payback (SPB) or discounted payback (DPB). They are sometimes needed to meet specific regulatory requirements. For example, the FEMP LCC rules (10 CFR 436A) require the use of either the SIR or AIRR for ranking independent projects competing for limited funding. Some federal programs require a payback period to be computed as a screening measure in project evaluation. NS, SIR, and AIRR are consistent with the lowest LCC of an alternative if computed and applied correctly, with the same time-adjusted input values and assumptions. Payback measures, either SPB or DPB, are only consistent with LCCA if they are calculated over the entire study period, not only for the years of the payback period." Supplementary measures are considered to be relative measures – i.e., they are computed for an alternative relative to a base case.

The design of buildings requires the integration and incorporation of different kinds of data into a synthetic whole. This integrated process, or "whole building" design approach, requires the active and full participation of many players – users, code officials, building technologists, cost consultants, specifications specialists, architects, civil engineers, mechanical and electrical engineers, structural engineers, and consultants from many specialized fields. The best buildings are produced through active, deliberate, and full collaboration among all players.

Building-related investments typically involve a great deal of uncertainty relating to their costs and potential savings. Performing an LCCA greatly increases the ability and likelihood of deciding on a project that can save money in the long run. Yet this does not alleviate some of the potential uncertainty associated with the LCC results, mainly because LCCAs are typically conducted early in the design process when only estimates of costs and savings are available rather than specific dollar amounts. This uncertainty in input values means that actual results may differ from estimated outcomes.

Deterministic techniques, such as sensitivity analysis and breakeven analysis, are easily conducted without the need for additional resources or information. They will produce a single-point estimate highlighting how uncertain input data affects the analysis outcome. However, the basis of the probabilistic method is that of quantifying risk exposure by deriving probabilities of achieving different values of economic worth from probability distributions for input values that are uncertain. Probabilistic techniques possess greater informational and technical requirements than do deterministic ones. The final choice for using one or the other method depends on several factors, mainly the size of the project, its significance, and the available resources. Both sensitivity analysis and breakeven analysis are approaches that are simple to perform and should be integral components of every LCCA.

10.2.5 Increased Productivity

One of the pivotal benefits of green buildings is their positive effect on productivity. There are numerous studies that clearly illustrate that green buildings dramatically affect productivity. However, because these studies are often broad in nature and rarely focus on unique green-building attributes, they need to be supplemented by other thorough, accurate, and statistically sound research to fully comprehend the effects of green buildings on occupant productivity, building performance, and sales. It seems prudent, however, that any productivity gains attributable to a green building should be included in the life-cycle cost analysis, particularly for an owner-occupied building. The difficulty of properly attributing gains such as reduced absenteeism and staff turnover rates appears to have made this the exception rather than the rule.

Some of the key features of green buildings that relate to increased productivity include controllability of systems relating to ventilation, temperature, and lighting; daylighting and views; natural and mechanical ventilation; pollution-free environments; and vegetation. It is not always clear why these features produce improved productivity, although healthier employees typically mean happier employees, and studies show that a feeling of wellbeing produces increased worker satisfaction, improved morale, reduced absenteeism, and increased productivity.

One study by Lawrence Berkeley National Laboratory concluded that improvements to indoor environments such as are commonly found in green buildings could reduce healthcare costs and work losses:

- from communicable respiratory diseases by 9 to 20 percent
- from reduced allergies and asthma from 18 to 25 percent
- from nonspecific health and discomfort effects by 20 to 50 percent

Hannah Carmalt, a project analyst with Energy Market Innovations, says: "The most intuitive explanation is that productivity increases are due to better occupant health and therefore decreased absenteeism. When workers are less stressed, less congested, or do not have headaches, they are more likely to perform better." High-performance buildings have many potential benefits including increased market value, lower operating and maintenance costs, improved occupancy of commercial buildings, and increased employee satisfaction and productivity. Occupants of green buildings have consistently recorded increased productivity and healthier working and living environments.

In a William Fisk study, green buildings were found to add $20 to $160 billion in increased worker productivity annually. This is due to the fact that LEED™-certified buildings were found to yield significant productivity and health benefits, such as heightened employee productivity and satisfaction, fewer sick days, and fewer turnovers. Moreover, other independent studies have shown that better climate control and improved air quality can increase employee productivity by an average of 11 to 15 percent annually. However, it should be noted that in commercial and institutional buildings payroll costs generally significantly overshadow all other costs, including those involved in the design, construction, and operation of a building.

10.2.5.1 Productivity

While theoretical explanations may help explain why productivity increases have been documented to occur in green buildings, it is still necessary to define the particular elements of green buildings that are directly related to productivity. Sound control, for example, while recognized as increasing productivity, is often excluded from green-related studies, mainly because it is not necessarily a green-building feature. Likewise, the presence of biological pollutants such as molds are associated with decreased productivity but are also excluded because typical green buildings do not automatically eliminate the presence of such materials (although their presence is reduced by improved ventilation in green buildings).

As mentioned previously, among the most significant elements of green buildings that affect productivity and thermal comfort is occupant control over temperature, ventilation, and lighting. Green buildings usually try to incorporate this feature because it can noticeably decrease energy use by ensuring areas are not heated, cooled, or lit more than is required. These measures are decisive to maintaining energy efficiency and occupant satisfaction within a building. The Urban Land Institute and the Building Owners and Managers Association conducted a survey that found that occupants rated air temperature (95 percent) and air quality (94 percent) most crucial in terms of tenant comfort. The study also determined that 75 percent of buildings did not have the option or capability to adjust features and that many individuals were willing to pay higher rents in order to obtain such measures.

Various studies have associated increased productivity with increased emotional wellbeing. In a study conducted by the Heschong Mahone Group (HMG) it was found that higher test scores in daylit classrooms were achieved due to students being happier. HMG also found that when teachers were able to control the amount of daylighting in classrooms, students progressed 19 to 20 percent faster than students in classrooms that lacked controllability. Similar studies that were performed in office settings clearly showed that increases in individual control over temperature, ventilation, and lighting produced significant increases in productivity.

Office space with views to the outdoors is another common feature of green buildings that is associated with productivity. The 2003 office study conducted by the consultants HMG, mentioned previously, also found correlations between productivity and access to outdoor views: test scores were generally 10 to 15 percent higher, and calling performance increased by 7 to 12 percent. This reinforces HMG's earlier 1999 study findings of schools that children in classrooms of the Capistrano School District progressed

15 percent faster in math and 23 percent faster in reading when they were located in the classrooms with the largest windows.

Ventilation, too, is a feature of green buildings that is associated with productivity. The importance of ventilation is that it facilitates the introduction of fresh air to cycle through the building, removing stale or pollutant air from the interior. Germs, molds, and various VOCs, such as those emitted by paints, carpets, and adhesives, can often be found within buildings lacking adequate ventilation, causing sick-building syndrome (SBS). Typical symptoms include inflammation, asthma, and allergic reaction. Ventilation can play a critical role in worker productivity, as evidenced by the extensive research that has been conducted to address these issues. This is also why it is imperative to minimize the use of toxic materials inside the building. Many of the products used in conventional office buildings, such as carpets and copying machines, contain toxic materials, and minimizing them decreases the potential hazards associated with their use and disposal. Furthermore, because these materials are known to leak pollutants into the indoor air, they can have an adverse impact on worker productivity if the building is not properly ventilated.

Productivity can therefore be influenced by many factors. Even the simple inclusion of vegetation inside and around the building has been shown to have a positive impact on productivity. Plants inside buildings help control indoor-air pollution, whereas outdoor vegetation aids in outdoor-air pollution and also helps improve water quality, control erosion, and provide more natural habitats.

The fact that green buildings improve job satisfaction, resulting in increased productivity, is undeniable. Additionally, job satisfaction generates less staff turnover, thereby improving the overall productivity of a firm. Less time spent on job training allows more time to be spent on productive work. Staff retention is one of the decisive factors for many firms that make the decision to green their office buildings.

As for speculative or leased facilities, it is more problematic to commit a market value to occupant productivity gains and have them accurately reflected in the business case at the decision-making point. Nevertheless, there is now adequate data and evidence quantifying the effects to support taking them into account on some basis. And while an owner of a leased facility may not financially benefit directly from increased user productivity, there could be indirect benefits in the form of increased rental fees and occupancy rates. For the majority of commercial buildings, the use of a conservative estimate for the potential reduction in salary costs and productivity gains will loom large in any calculation, as indicated by various case studies.

10.2.6 Improved Tenant/Employee Health

LEED™ standards are intended to improve indoor-air quality and reduce potential health risks, especially allergies and other sensitivities. Moreover, LEED™-certified buildings use critical resources more efficiently when compared to conventional buildings, which are only built to meet code requirements. LEED™-certified buildings have healthier work and living environments, which contribute to higher productivity and improved employee health and comfort (Figure 10.4).

Figure 10.4 LEED™-certified buildings provide healthier work and living environments. Photo of interior space in Ohlone College, Newark Center for Health Sciences and Technology, Newark, CA, which received a LEED™ Platinum rating.
Source: Green Building Market Barometer, 2008.

Green buildings are typically expected to incorporate superior air quality, abundant natural light, access to views, and effective noise control. Each of these qualities is for the benefit of building occupants, making these building better places to work or live. Building occupants are increasingly seeking many green-building features, such as superior air quality and control of air temperatures and views. An extensive North American study focusing on office-building tenant satisfaction determined that tenants highly value comfort in office buildings. Respondents reportedly attributed the highest importance to comfortable air temperature (94 percent) and indoor-air quality (94 percent). These two features were the only ones that were considered "most important" and on the list of features with which tenants were currently least satisfied. This study also determined that the principal reasons why tenants move out are due to heating or cooling problems.

Improved indoor environmental quality is one of green buildings' core components. Building occupants understand that natural light, clean air, and thermal comfort are required elements to stay healthy and productive, in addition to providing enjoyable living and work spaces. The number of credible studies demonstrating an intrinsic connection between green-building strategies and occupant health and wellbeing is endless. In office settings and learning environments, this directly translates into increased productivity and reduced absenteeism. A 2000 study conducted by the Harvard School of Public Health and the Polaroid Corporation found that companies are losing billions of dollars annually because of employee absenteeism.

By building to LEED™ green standards as outlined above, tenants recognize the true value of a healthy indoor environment as it relates to enhanced employee productivity and decreased susceptibility to the VOCs, molds, and maladies that contribute to sick-building syndrome (SBS). This becomes a significant attribute when tenants take into account how much time their employees spend at work. The health benefits of leasing in a LEED™-certified building may even contribute to a firm's bottom line.

For most tenants, the enhanced productivity and health benefits provided by a LEED™-certified building add to cost savings generated by the operating efficiencies previously mentioned. This becomes viable as owners are able to pass utilities and maintenance savings on to leasing tenants – a feature that tenants of conventional, nongreen buildings don't enjoy.

William Fisk, in a 2002 study "How IEQ Affects Health, Productivity," estimates that 16 to 37 million cases of colds and flu could be avoided by improving indoor environmental quality. This translates into a $6 to $14 billion annual savings in the United States while at the same time reducing sick-building-syndrome (SBS) symptoms – a condition whereby occupants become temporarily ill – by 20 to 50 percent, resulting in $10 to $30 billion annual savings in the United States.

10.2.7 Enhancement of Property Value and Marketability

Enhancement of property value is a key factor for speculative developers who fail to directly achieve operating cost and productivity savings. To date, not many thorough and reliable studies have been conducted on the intrinsic relationship between property values and green buildings, even though this is an important aspect that should be quantified and included in the economic calculations. It is an element of particular relevance to speculative developers who intend to either sell or lease a new building, although it can also have a bearing on the decision process in general, including developers who intend to occupy a building while maintaining an eye on the market value of the asset.

There are many factors that will or could increase property values for green buildings. Much of the real-estate industry, unfortunately, still does not fully comprehend the benefits of green buildings and therefore is unable to correctly convey these benefits to prospective purchasers. Moreover, the benefits may not be properly reflected in the selling prices or lease rates.

Jerry Yudelson maintains that increased annual energy savings have been found to promote higher building values. As an example, a 75,000-square-foot building saves $37,500 per year in energy costs vs. a comparable building built to code. (This saving translates into a saving of 50 cents per square foot per year.) At capitalization rates of 6 percent, typical today in commercial real estate, green-building standards would add $625,000 ($8.33 per square foot) to the value of the building. This means that for a small upfront investment, an owner can reap benefits that typically offer a payback of three years or less and a rate of return exceeding 20 percent.

High-performance buildings can offer owners many important benefits ranging from higher market value to more satisfied and productive occupants. The benefits of green and high-performance buildings are not evident in higher base rents in some local markets. The primary reason for this is that majority of the benefits accrue to tenants, and tenants usually need proof before they are willing to participate in the cost of investments that they perceive will help them be more productive or save money. It is only very recently, due mainly to increased awareness, that tenants have begun to fully appreciate the benefits of cleaner air, more natural lighting, and easier-to-modify spaces.

10.2.7.1 Resale Values and Rental Rates

Several recent American studies have provided considerable evidence that green buildings, particularly green buildings with good-quality natural lighting, can have a dramatic effect on sales in commercial buildings. Furthermore, without exception, these studies report that there is a sound economic basis for green buildings but only when operational costs are included into the equation. More specifically, whole-building studies conclude that the net present values (NPV) for pursuing green buildings as opposed to conventional buildings range from \$50 to \$400 per square foot (540 to \$4300 per square meter). The NET depends on a building's length of time analyzed (e.g., 20 to 60 years) and the degree to which the building implements green strategies. One of the main conclusions from these studies is that, generally, the greener the building, the higher the NET.

With regard to rental sales, a CoStar study found that LEED™ buildings can command rent premiums of about \$11.24 per square foot versus their conventional peers in addition to a 3.8 percent increase in occupancy rate. The study also found that rental rates in ENERGY STAR® buildings can boast a \$2.38 per square foot premium versus comparable non-ENERGY STAR® buildings in addition to a 3.6 percent greater occupancy rate. However, what is perhaps more remarkable and what may prove to be a trend that could signal greater attention from institutional investors is that ENERGY STAR® buildings are commanding an average of \$61 per square foot more than their conventional counterparts, and LEED™ buildings are commanding a surprising \$171 more per square foot.

This is quite extraordinary since most leasing arrangements, particularly in the office/commercial sectors, provide little incentive to undertake changes that might be construed as being beneficial to the environment. For example, leases often have fixed rates with no regard to energy or water consumption, even though the lessees have control over most energy- and water-consuming devices. This can prove very frustrating for owners and lessees, since consumption patterns are often difficult to determine due to a lack of detailed metering by space.

10.2.8 Other Indirect Benefits

There are numerous other indirect benefits that include improved image, risk reduction, futureproofing. and self-reliance. These and other similar benefits of green buildings may be captured by investors and should not be discarded in economics considerations. Although they may not be easily quantified and in some cases may even be intangible, they should nevertheless be factored into the business case because they are intrinsically connected to sustainable design and they can significantly affect the value of a green building.

10.2.8.1 Improved Image

Even if we disregard the financial benefits, green buildings are generally perceived by the public as modern, dynamic, and altruistic. Another key message conveyed by sustainable buildings is concern for the environment. Green buildings can therefore send a strong symbolic message of an owner's commitment to sustainability. Some of the benefits that companies can enjoy from these perceptions include employee pride, satisfaction, and wellbeing, which translate into reduced turnover, recruitment advantages, and improved morale. These powerful images can be an important consideration in a company's decision to pursue occupancy in a green building.

10.2.8.2 Risk Reduction

Peter Morris, a principal with Davis Langdon, commenting on the recent downturn in the U.S. economy, tells us: "Risk remains a serious concern for construction projects. Delay and cancellation of projects, even projects under construction, is a growing trend." Morris also proposes a key theme for project owners in the current market turmoil, which is that "the successful adoption of a competitive procurement strategy, in order to secure lower costs, will depend on active steps by the owner and the project team to ensure that the contractor is in a position to provide a realistic and binding bid and that contractors' bidding costs are minimized."

However, with that in mind, many risks can be mitigated through the application of green-building principles. In this regard, the Environmental Protection Agency currently classifies indoor-air quality as one of the top five environmental health risks. Sick-building syndrome (SBS) and building-related illness (BRI), both of which are discussed in Chapter 7, are among the issues of major concern and often end up being resolved in the courts. Business owners and operators are increasingly facing legal action from tenants blaming the building for their health problems.

The base cause of SBS and BRI is poor building design and/or construction, particularly with respect to the building envelope and mechanical systems. Green buildings should accentuate and promote not only safe but also exceptional air quality, and no functioning green building should ever have to suffer from SBS or BRI. A similar argument may be made for mold-related issues, as they too are facing increased litigation today.

10.2.8.3 Futureproofing

According to Davis Langdon, "Going green is 'futureproofing' your asset." Green buildings are inherently efficient and safe and as such help ensure that they will not be at a competitive disadvantage in the future. In this respect, there are a number of potential future risks that are significantly mitigated in green buildings:

- Energy conservation protects against future energy price increases.
- Water conservation shields against water increases.
- Occupants of green buildings are generally more comfortable and contented, so it can be assumed that they will generally be less likely to be litigious.
- A documented effort to build or occupy a healthy green building demonstrates a level of due diligence that could stand as an important defense against future lawsuits or changes in legislation, even when faced with currently unknown problems.

10.2.8.4 Self-Reliance

The fact that green buildings often incorporate natural lighting and ventilation and internal energy and water generation makes them less likely to rely on external grids and less likely to be affected by grid-related problems or failures such as blackouts, water shortages, or contaminated water. This element is acquiring increasing importance globally because of the potential risk of terrorism.

10.2.9 External Economic Effects

External effects generally consist of costs or benefits of a project that accrue to society and are not readily captured by the private investor. Examples of this are reduced reliance on infrastructure (sewers, roads, etc.), reduced greenhouse gases, and reduced health costs. The extent to which these benefits can be factored into a business case relies on the extent to which they can be converted from the external to the internal side of the ledger. This constitutes a vital factor in any assessment of the costs and benefits of green buildings. Thus the costs of green vegetated roofs are borne by the developer or investor, while many of the benefits, such as reduced heat-island effects and reduced storm-water runoff, accrue at a broader societal level.

If the investor is a government agency or if a private developer is compensated for including features that produce benefits at a societal level, then the business case can encompass a much broader range of effects. For example, there are jurisdictions, such as in the state of Oregon, that offer tax incentives for green building, thereby providing a direct business-case payoff to the investor. Another example is Arlington, Virginia, which allows higher floor-space to land-coverage ratios for green buildings.

Infrastructure costs such as water use and disposal are typically provided by governments and are rarely cost-effective or even cost-neutral, and in many instances, governments are required to heavily subsidize water use and treatment.

External environmental costs consist mainly of pollutants in the form of emissions to air, water, and land and the general degradation of the ambient environment. Buildings can be singled out as having the largest indirect environmental impact on human health. Other critical impacts, such as damage to ecosystems, crops, and structures/monuments and resource depletion, should also be considered even though they do not have a large associated indirect cost relative to human health.

10.2.9.1 Job Creation

There are significant environmental impacts associated with the transportation of materials for the construction industry. Accordingly, LEED™ rating systems promote the use of local or regional materials, which in turn promote local or regional job creation. In addition to the above, green-building attributes are often labor intensive rather than material- or technology-intensive.

10.2.9.2 International Recognition and Export Opportunities

Green building can also have economic ramifications on a much broader scale as a result of increased international recognition and related export sales. The 2005 Environmental Sustainability Index, prepared by Yale and Columbia Universities, benchmarked the ability of nations to protect the environment by integrating data sets including natural-resource endowments, pollution levels, and environmental management efforts into a smaller set of indicators of environmental sustainability. This study ranked the United States 45th.

10.2.10 Increased Recruitment and Retention

Providing a healthy and pleasant work environment increases employee satisfaction, productivity, and retention. It also increases the ability to compete for the most qualified employees as well as for business. Statistical data and other evidence indicate that high-performance green buildings can increase a company's ability to recruit and retain employees because buildings with good air quality, abundant amounts of natural light, and better circulated heat and air conditioning are more pleasant, healthier and more productive places to work.

Willingness to join and remain with an organization are aspects often overlooked when considering how green buildings affect employees. It is estimated that the cost of losing a single good employee is roughly between \$50,000 and \$150,000, and many organizations experience a 10 to 20 percent annual turnover, some of it from persons whom they would have liked to retain. In some cases people decide to leave due to poor physical and working environments. In a workforce of, say, 100 people, turnover at this level implies 10 to 20 people leaving per year.

If green buildings are able to reduce turnover by only 10 percent for example – i.e., 1 to 2 people out of 10 to 20, that value would range from \$50,000 to possibly as much as \$300,000, which is more than enough to justify the costs of certifying a building project. Figure 10.5 compares the difference in occupancy rates between ENERGY STAR® and non-ENERGY STAR® buildings.

10.2.11 Tax Benefits

Many states now offer tax incentives for green buildings. States such as New York and Oregon offer state tax credits, while others, such as Nevada, offer property- and

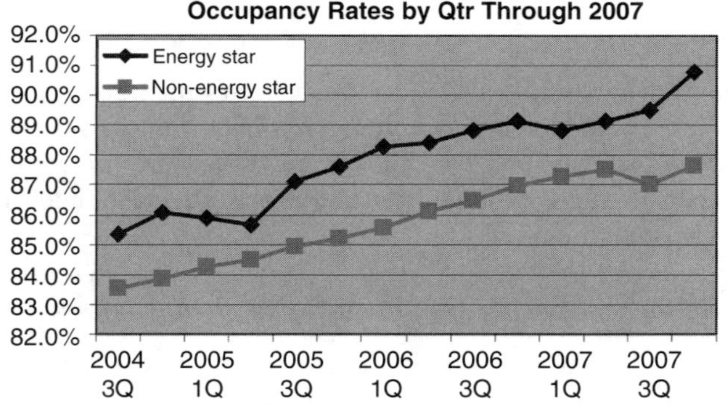

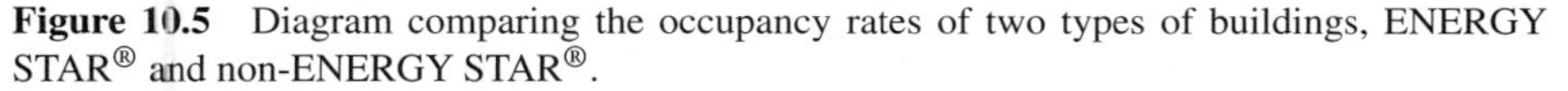
Figure 10.5 Diagram comparing the occupancy rates of two types of buildings, ENERGY STAR® and non-ENERGY STAR®.
Source: *Does Green Pay Off?* by Norm Miller, Jay Spivey, and Andy Florance.

sales-tax abatements. The federal government also offers tax credits. The state of Oregon credits vary and are based on building area and LEED™ certification level. At the Platinum level, for example, a 100,000-square-foot building in Oregon can expect to receive a net-present-value tax credit of up to $2 per square foot, which is transferrable from public or nonprofit entities to private companies (e.g., contractors or benefactors), making it even more attractive than a credit that applies only to private owners.

The state of New York has a tax credit that allows builders who meet energy goals and use environmentally preferable materials to apply up to $3.75 per square foot for interior work and $7.50 per square foot for exterior work against their state tax bill. In order to qualify for this credit, a building needs to be certified by a licensed architect or engineer in addition to meeting specific requirements for energy use, water use, indoor-air quality, waste disposal, and materials selection. This means that the energy used in new buildings must not exceed 65 percent of that allowed under the New York energy code; in rehabilitated buildings, energy use cannot exceed 75 percent.

The state of Nevada offers a property-tax abatement of up to 35 percent for up to 10 years to private development projects that achieve LEED™ Silver certification. This means that if the property tax represents one percent of value, it could be worth as much as five percent of the building cost, which translates to much more than the actual cost of achieving LEED™ Silver certification of a large project. This has encouraged a large number of Nevada projects to pursue LEED™ certification, including the $7 billion, 17-million-square-foot Project City Center in Las Vegas, which is one of the world's largest private development projects to date. The state of Nevada also provides for sales-tax abatement for green materials used in LEED™ Silver certified buildings.

In addition to the state tax incentives, there are federal tax incentives such as the 2005 federal Energy Policy Act, which offers two major tax incentives for differing aspects of green buildings: a tax credit of 30 percent for use of both solar thermal and electric systems and a tax deduction of up to $1.80 per square foot for projects that reduce energy use for lighting, HVAC, and water-heating systems by at least 50 percent compared with the 2001 baseline standard. These tax deductions may be taken by the design team leader (typically the architect) when applied to government projects.

10.3 Miscellaneous Green-Building Costs

10.3.1 Energy Costs

Operational expenses for energy, water, and other utilities depend to a large extent on consumption, current rates, and price projections. But since energy and to a lesser extent water consumption and building configuration and envelope are interdependent elements, energy and water costs are usually assessed for the building as a whole rather than for individual building systems or components.

To accurately predict energy costs during the preliminary design phase of a project is rarely simple. Assumptions must be made regarding use profiles, occupancy rates, and schedules, all of which can have a dramatic impact on energy consumption. There are numerous computer programs currently on the market such as Energy-10 and eQuest that can provide the required data assumptions on the amount of energy consumption for a building. Alternatively, the information and data can come from engineering analysis. Other software packages, such Energy Plus™ (DOE), DOE-2.1E, and BLAST, are excellent programs, but they require more detailed input not normally available until later in the design process.

Before selecting a program, it is important to determine whether you require annual, monthly, or hourly energy-consumption estimates and whether the program is capable of adequately tracking savings in energy consumption even when design changes take place or when different efficiency levels are simulated. Figure 10.6 is an example of the typical costs incurred by an HVAC system over 30 years (useful life).

Energy represents a substantial and widely recognized cost of building operations that can be reduced through energy efficiency and related measures that are part of green-building design. Although estimates vary slightly, the consensus is that green buildings, on average, use 30 percent less energy than conventional buildings. A detailed survey of 60 LEED™-rated buildings demonstrates that green buildings, when compared to conventional buildings are:

- On average more energy-efficient by approximately 25 to 30 percent
- Characterized by lower electricity peak consumption
- More likely to generate renewable energy on-site
- More likely to purchase grid power generated from renewable-energy sources (green power and/or tradable renewable certificates)

Green-building energy savings come mainly from reduced electricity purchases and secondarily from reduced peak energy demand. On average, green buildings are estimated to be 28 percent more efficient than conventional buildings and on average generate 2 percent of their power on-site from photovoltaics (PV). The financial benefits that accrue from a 30 percent reduced consumption at an electricity price of \$0.08/kWh come to about \$0.30/foot^2/year, with a 20-year NPV (net present value) of over \$5/foot^2, equal to or more than the average additional cost associated with building green.

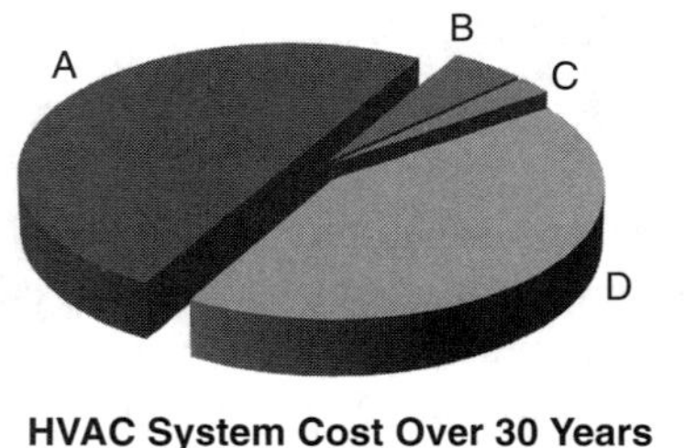

Figure 10.6 Pie chart illustrating typical costs (in percentage terms) incurred by an HVAC system over 30 years, its useful life.
Source: Washington State Department of General Administration.

	Certified	Silver	Gold	Average
Energy Efficiency (above standard code)	18%	30%	37%	28%
On-site Renewable Energy	0%	0%	4%	2%
Green Power	10%	0%	7%	6%
Total	28%	30%	48%	36%

Figure 10.7 Table showing reduced energy use in green buildings as compared with conventional buildings.
Source: USGBC, Capital E Analysis.

Jerry Yudelson, author of *The Green Building Revolution*, says: "Many green buildings are designed to use 25 to 40 percent less energy than current codes require; some buildings achieve even higher efficiency levels. Translated to an operating cost of $1.60 to $2.50 per square foot for electricity (the most common energy source for building), this energy savings could reduce utility operating costs by 40 cents to $1 per square foot per year. Often, these savings are achieved for an added investment of just $1 to $3 per square foot. With building costs reaching $150 to $300 per square foot, many developers and building owners are seeing that it is a wise business decision to invest 1 to 2 percent of capital cost to secure long-term savings, particularly with a payback of less than three years. In an 80,000-square-foot building, the owner's savings translates into $32,000 to $80,000 per year, year after year, at today's prices."

The environmental and health costs associated with air pollution caused by nonrenewable electric power generation and on-site fossil-fuel use are generally excluded when making investment decisions. Figure 10.7 highlights the reduced energy used in green buildings as compared with conventional buildings.

10.3.2 Operation, Maintenance, and Repair Costs

Scores of studies on sustainability have generally shown that LEED™-certified buildings typically both cost less and are easier to operate and maintain over the life of the structure than conventional buildings. This puts them in a position to command higher lease rates than conventional buildings in their markets.

However, operation (nonfuel), maintenance, and repair (OM&R) costs are often more difficult to estimate than other building expenditures. Operating schedules and maintenance standards will vary from one building to the next; the variation in these costs is significant even when the buildings are of the same type and age. It is therefore important to use common sense and good judgment in estimating these costs.

Supplier quotes and published estimating guides can sometimes provide relevant information on maintenance and repair costs. Some of the data-estimation guides, such as R. S. Means and BOMA, typically report, for example, average owning and operating costs per square foot by age of the building, geographic location, number of stories, and number of square feet.

Once the project is operational, buildings tend to recoup any added costs within the first one to two years of the life cycle of the building. LEED™-certified buildings use considerably less energy and water then a conventional building. Green buildings typically use 30 to 50 percent less energy and 40 percent less water than their conventional counterparts, yielding significant savings in operational costs. The New Buildings Institute (NBI) recently released a research study indicating that new buildings certified under the U.S. Green Building Council's (USGBC) LEED™ certification system are, on average, performing 25 to 30 percent better than non-LEED™-certified buildings in terms of energy use. The study also suggests that Gold and Platinum LEED™ buildings have average energy savings approaching 50 percent.

10.3.3 Replacement Costs

Sieglinde Fuller says that the number and timing of capital replacements of building systems depend largely on the estimated life of the system and the length of the study period. He further states that the same sources providing the cost estimates for the initial investments should be used to obtain estimates of replacement costs and expected useful lives. A good starting point for estimating future replacement costs is to use their cost as of the base date. The LCCA method is designed to escalate base-year amounts to their future time of occurrence.

10.3.3.1 Residual Values

The residual value of a system (or component) is basically the value it will have after it has been depreciated – i.e. its remaining value at the end of the study period or at the time it is replaced during the study period. According to Fuller, residual values can be based on value in place, resale value, salvage value, or scrap value, net of any selling, conversion, or disposal costs. For rule-of-thumb calculations, the residual value of a system with remaining useful life in place can be determined by linearly prorating its initial costs.

10.3.4 Other Costs

10.3.4.1 Finance Charges

Neither finance charges nor taxes usually apply to federal projects. However, finance charges and other payments do apply if a project is financed through an Energy Savings Performance Contract (ESPC) or Utility Energy Services Contract (UESC). These charges are normally included in the contract payments negotiated with the energy-service company (ESCO) or the utility.

10.3.4.2 Nonmonetary Benefits and Costs

Nonmonetary benefits or costs are project-related features for which there is no meaningful way of assigning a dollar value; despite efforts to develop quantitative measures of benefits, there are situations that simply do not lend themselves to such analysis. For example, projects may provide certain benefits such as improved quality of the working environment or preservation of cultural and historical resources. By their nature, these benefits are external to the LCCA and difficult to assess, but if considered significant they should be taken into account in the final investment decision, portrayed in the project documentation, and included in a life-cycle cost analysis.

The analytical hierarchy process (AHP) can be used to formalize the inclusion of nonmonetary costs or benefits in the decision-making process. It is one of a set of multi-attribute decision analysis (MADA) methods that are used when considering qualitative and quantitative non-monetary attributes in addition to common economic evaluation measures when evaluating project alternatives. The ASTM E 1765 Standard Practice for Applying Analytical Hierarchy Process (AHP) to Multi-attribute Decision Analysis of Investments Related to Buildings and Building Systems presents a general procedure for calculating and interpreting AHP scores of a project's total overall desirability when making building-related capital-investment decisions. The WBDG Productive Committee is an excellent source of information for estimating productivity costs.

10.4 Design and Analysis Tools and Methods

The federal government is the nation's largest owner and operator of built facilities; during the energy crises of the 1970s and 1980s, it was faced with increasing initial construction costs and ongoing operational and maintenance expenses. As a result, facility planners and designers decided to use economic analysis to evaluate alternative construction materials, assemblies, and building services with the objective of lowering costs. In today's difficult economic climate, building owners wishing to reduce expenses or increase profits are again employing economic analysis to improve their decision making during the course of planning, designing, and constructing a building. Moreover, federal, state, and municipal entities have all enacted legislative mandates (in varying degrees) requiring the use of building economic analysis to determine the most economically efficient or cost-effective choice among building alternatives. Figure 10.8 illustrates the general steps taken in an economic-analysis process.

10.4.1 Present-Value Analysis

Present-value analysis is a standard method for using the time value of money to appraise long-term projects. The basic concept is that the value of a dollar profit today is greater than the value of a dollar profit next year. How much greater is determined by what is called the discount rate, as in how much of a discount you would expect if you were buying a dollar's worth of next year's profit. Net present value (NPV) allows decision makers to compare various alternatives on a similar time scale by converting the various options to current dollar figures. A project is generally considered acceptable if the net present value is positive over the expected lifetime of the project.

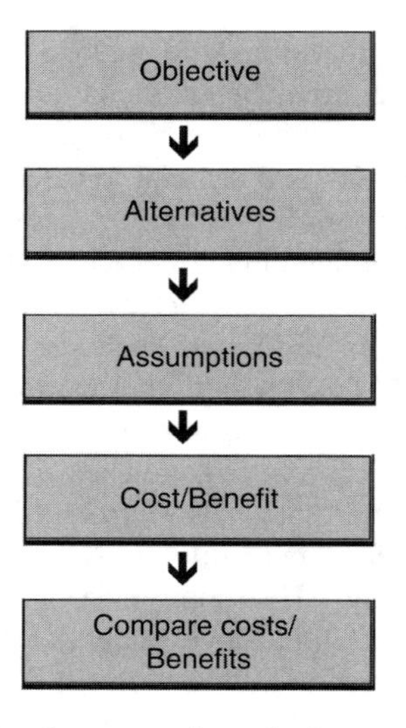

Figure 10.8 Diagram illustrating the economic-analysis process.
Source: Whole Building Design Guide.

As an example, let us take a building that is considering a lighting change from traditional incandescent bulbs to fluorescents. The initial investment to change the lights themselves is estimated to be $40,000. After the initial investment, it is estimated to cost $2,000 to operate the lighting system but will yield $15,000 in savings each year. This thus produces an annual cash flow of $13,000 every year after the initial investment. If, for simplicity, a discount rate of 10 percent is assumed and it is calculated that the lighting system will be utilized over a five-year time period, this scenario would yield the following NPV calculations:

$$t = 0\,\text{NPV} = (-40{,}000)/(1 + .10)0 = -40{,}000.00$$

$$t = 1\,\text{NPV} = (13{,}000)/(1.10)1 = 11{,}818.18$$

$$t = 2\,\text{NPV} = (13{,}000)/(1.10)2 = 10{,}743.80$$

$$t = 3\,\text{NPV} = (13{,}000)/(1.10)3 = 9{,}767.09$$

$$t = 4\,\text{NPV} = (13{,}000)/(1.10)4 = 8{,}879.17$$

$$t = 5\,\text{NPV} = (13{,}000)/(1.10)5 = 8{,}071.98$$

Based on the above information, the total NPV over the lifetime of the project would come to $9,280.22.

10.4.1.1 Discount Rate

Discounting adjusts costs and benefits to a common point in time. Thus, in order to add and compare cash flows that are incurred at different times during the life cycle of a building, they need to be made time-equivalent. To make cash flows time equivalent, the LCC method converts them to present values by discounting them to a common point in time, which is usually the base date. The interest rate used for discounting essentially represents the investor's minimum acceptable rate of return. To some extent, the selection of the discount rate is dependent on the use to which it will be put.

The discount rate for federal energy and water-conservation projects is determined annually by the DOE's Federal Energy Management Program (FEMP); for other federal projects, those not primarily concerned with energy or water conservation, the discount rate is determined by the Office of Management and Budget (OMB). These discount rates, however, represent real discount rates that do not include the general rate of inflation.

10.4.1.2 Length of Study Period

The study period begins with the base date, which is the date to which all cash flows are discounted. The study period includes any planning, construction, and implementation periods as well as the service or occupancy period. The study period remains the same for all of the alternatives considered.

10.4.1.3 Service Period

The service period essentially begins when the completed building is occupied or when a system is taken into service. This is the period over which operational costs and benefits are evaluated. In FEMP analyses, the service period cannot exceed 25 years.

10.4.1.4 Contract Period

The contract period in ESPC and UESC projects lies within the study period, starting when the project is formally accepted, energy savings begin to accrue, and contract payments begin to be due. The contract period generally ends when the loan is paid off.

10.4.1.5 Discounting Convention

In OMB and FEMP studies, annually recurring cash flows such as operational costs are normally discounted from the end of the year in which they are incurred. In MILCON studies they are typically discounted from the middle of the year. All single amounts such as replacement costs and residual values are discounted from their dates of occurrence.

10.4.1.6 Application

With regard to the application of LCCA, Sieglinde Fuller states that it can be applied to any capital-investment decision in which relatively higher initial costs are traded for reduced future-cost obligations. Fuller also maintains: "It is particularly suitable for the evaluation of building design alternatives that satisfy a required level of building performance but may have different initial investment costs; different operating, maintenance, and repair costs; and possibly different lives." LCCA is an approach that provides a much better assessment of the long-term cost-effectiveness of a project than alternative economic methods that mainly focus on first costs or on short-term operating-related costs.

According to Fuller, LCCA can be performed at various levels of complexity, and its scope can vary from a "back-of-the envelope" study to a detailed analysis with thoroughly researched input data, supplementary measures of economic evaluation, complex uncertainty assessment, and extensive documentation. The comprehensiveness of the exercise should be tailored to the requirements of the project.

10.4.1.7 Inflation

An LCCA can be performed in either constant dollars or current dollars. Both methods of calculation produce identical present-value life-cycle costs. However, a constant-dollar analysis does not include the general rate of inflation, which means that it has the advantage of not requiring an estimate of the rate of inflation for the years in the study period. A current-dollar analysis, on the other hand, does include the rate of general inflation in all dollar amounts, discount rates, and price-escalation rates.

Constant-dollar analysis is generally recommended for federal projects, except for projects financed by the private sector through the Energy Savings Performance Contract (ESPC) and the Utility Energy Services Contract (UESC). There are several alternative financing studies available that are usually performed in current dollars if the analyst wants to compare contract payments with actual operational or energy-cost savings from year to year.

10.4.2 Sensitivity Analysis

Sensitivity analysis is a technique recommended by FEMP for energy and water-conservation projects. Critical assumptions should be varied and net present value and other outcomes recomputed to determine how sensitive outcomes are to changes in the assumptions. The assumptions that deserve the greatest attention will rely on the dominant benefit and cost elements and the areas of greatest uncertainty of the program being analyzed. In general, a sensitivity analysis is used for estimates of benefits and costs, the discount rate, the general inflation rate, and distributional assumptions. Models used in the analysis should be well documented and, where possible, available to facilitate independent review. Sensitivity analysis is useful for:

- Identifying which of a number of uncertain input values has the greatest impact on a specific measure of economic evaluation
- Determining how variability in the input value affects the range of a measure of economic evaluation
- Testing different scenarios to answer "what if" questions

10.4.3 Breakeven Analysis

Breakeven analysis focuses on the relationship between fixed cost, variable cost, and profit. It is mostly used when decision-makers want to know the maximum cost of an input that will allow the project to still break even or, conversely, what minimum benefit a project can produce and still cover the cost of the investment. To perform a breakeven analysis, benefits and costs are set equal, all variables are specified, and the breakeven variable is solved mathematically.

10.4.4 Computer Programs

The use of computer programs can considerably reduce the time and effort spent on formulating the LCCA, performing the computations, and documenting the study. There are many LCCA-related software programs available, all of which can be found on the Internet. Below are some of the more popular packages:

- Building Life-Cycle Cost (BLCC) Program, version 5.3-08, is a program and economic-analysis tool developed by the National Institute of Standards and Technology (NIST) for the U.S. Department of Energy Federal Energy Management Program (FEMP). It is designed to provide computational support for the analysis of capital investments in buildings.
- ECONPACK (Economic Analysis Package) for Windows is a comprehensive economic-analysis computer package incorporating calculations, documentation, and reporting capabilities. It is structured to permit its use by noneconomists to prepare complete, properly documented economic analysis (EA) in support of DoD funding requests. The program was developed by the U.S. Army Corps of Engineers.
- ENERGY-10™ is a cost-estimating program tool that assists architects, builders, and engineers to rapidly (within 20 minutes) identify the most cost-effective, energy-saving measures to employ in designing a low-energy building. Using the software at the early phases of a design can reportedly result in energy savings of 40 percent to 70 percent with little or no increase in construction cost. The software is available through the Sustainable Buildings Industry Council (SBIC).
- Life-Cycle Cost in Design WinLCCID Program was originally developed for MILCON by the Construction Engineering Research Laboratory of the U.S. Army Corps of Engineers. The program is a life-cycle-costing tool that is used to evaluate and rank design alternatives for new and existing buildings and carry out "what if" analyses based on variables such as present and future costs and/or maintenance and repair costs.
- Success Estimator Estimating and Cost Management System is a cost-estimating tool available from U.S. Cost that gives estimators, project managers and owners real-time, simultaneous access to their cost data and estimating projects from any Internet-connected computer.

10.4.5 Relevant Codes and Standards

There are many standards that are relevant to green building. These include:

- 10 CFR 436 Subpart A: Federal Energy Management and Planning Programs, Methodology, and Procedures for Life-Cycle Cost Analyses
- ASTM E2432: Standard Guide for the General Principles of Sustainability Relative to Building
- Circular No. A-94 Revised: Guidelines and Discount Rates for Benefit-Cost Analysis of Federal Programs
- Energy Policy Act of 2005

- Executive Order 13123: Greening the Government through Efficient Energy Management, DOE Guidance on Life-Cycle Cost Analysis Required by Executive Order 13123
- Facilities Standard for the Public Buildings Service, P100 (GSA): Chapter 1.8: Life-Cycle Costing
- NAVFAC P-442 Economic Analysis Handbook
- Standards on Building Economics, 6th ed., ASTM, 2007.
- Sustainable Building Technical Manual (DOE/EPA)

10.5 Liability Issues

10.5.1 General Overview

Liability can be an extremely complex matter, and this section cannot address the many concerns and legal matters that may arise with regard to liability issues. Builders, manufacturers and designers are strongly advised to consult their attorneys and professional liability insurance carriers for advice on these matters. While building owners and managers are not expected to guarantee the safety or wellbeing of their tenants, visitors, and guests, they are required to exercise reasonable care to protect them from foreseeable events. The number of liability lawsuits filed against American companies has increased dramatically over the last decade.

Thus, while the green-building movement is having a very significant and positive impact on the construction industry, there are aspects such as the risk of liability that are causing considerable concern and what need to be addressed. For example, green buildings are generally more efficient users of energy and materials, resulting in reduced safety factors for the different systems. But green buildings are sometimes prone to using nonstandard materials and systems, which may result in an increased risk of failure. To minimize these risks, qualified designers should be employed to ensure that the design process is correctly implemented. In this respect, Ward Hubbell, president of Green Building, says: "One of our most pressing issues is the fact that some buildings designed to be green fail to live up to expectations. And in business, as we all know, where there are failed expectations, there are lawsuits. All practices and/or products that could possibly result in a firm's exposure and liability should be clearly identified. The good news is that this period of increased legal action or the threat thereof will in fact motivate the kind of clarity and measurement that both reduces liability risks and results in better buildings."

Nevertheless, acronyms and phrases such as IAQ (indoor-air quality), IEQ (indoor environmental quality), SBS (sick-building syndrome), and BRI (building-related illness) are tossed around to such a degree that building owners and managers just shrug them off. This is unfortunate, since recent studies indicate that the incidence of commercial buildings with poor IAQ and the frequency of litigation over the effects of poor IAQ are increasing substantially. These increases will have obvious ramifications for insurance carriers, which pay for many of the costs of healthcare and general commercial liability.

Some of this increased concern about liability issues may be due to the enormous surge in interest in green buildings, which has resulted in many misconceptions and exaggerations put forth by owners, designers, manufactures, and distributors. "Greenwash" is the general term used within the industry for this form of misconception. It can apply to building materials, systems, buildings, or companies, among other things. Greenwash can ultimately discredit the entire green-building industry in addition to being the source of numerous lawsuits because the ultimate goals of green buildings are not achieved through their use. Attorneys often categorize these claims into two basic groups, materials and performance.

10.5.1.1 Material Claims

Because of the lack of precision in what constitutes a green building, material, or system providers will frequently find a material property with limited green characteristics and market this property and the material as being "green". As an example, a material that uses high recycled content might also contain unacceptable amounts of urea-formaldehyde in its production; this material may be erroneously marketed as green even though its overall impact on the environment is negative. Such false claims have also occurred when material or system providers base their claims on unreliable and inaccurate information. But as the green-building field matures and as processes such as life-cycle analysis become more mainstream, these risks should decline. The employment of a reliable material-rating system would also reduce the plethora of false material or system claims.

10.5.1.2 Performance Claims

A not uncommon phenomenon within the green-building industry is the misrepresentation of a person's or company's knowledge and expertise regarding green building. When building owners and other stakeholders rely on this expertise, the result can be a dismal failure of the green building to achieve its ultimate goal. Misconceptions of this type permeate many building stakeholder groups.

Problems often arise when building owners, designers, and builders differ in their interpretation of what constitutes a successful green building and particularly when building owners fail to explicitly communicate their thoughts at the commencement of the project. Issues of this kind are compounded when the parties are relatively new to the concepts of the green-building process. The two main areas that typically need to be addressed are a building's failure to achieve a promised level of green-building certification and a building's operational performance.

The first area may be problematic in that it could impact the building owner's ability to qualify for a grant or tax incentive on which the owner may have relied to assist in offsetting the project's initial costs. For public buildings, certification may indeed be required by law.

In considering a building's operational performance, there is some expectation that green buildings will, in addition to reducing environmental impacts, reduce energy and water costs, require less maintenance, and provide other long-term benefits to the building owner. The point is, however, that while the design may incorporate a wide range of green features, there are numerous variables among a building's design and occupancy that can invariably impact its operational performance. These potential areas of misunderstanding can be minimized by following good business practices that facilitate clear communication and common expectations between building owners, designers, and rating organizations.

An action may be brought against the building owner, the builder, the architect or engineer, or the product manufacturer. In such cases, experts will likely be required to give opinions as to whether there has been negligence in the design, execution, or performance of duty. But often the investigation involves much more than expert opinion; for example, laboratory and other tests may be recommended to help determine the cause of failed performance. The role that experts are required to play will therefore vary depending on the case. Sometimes an expert will serve solely as a consultant to the lawyer and remain in the background without his or her name ever being known to the other side. At other times an expert may be used in the pretrial stages, perhaps to give an affidavit supporting one or more issues of the case. In other cases the expert may serve solely as an expert witness at trial. Sometimes an expert will play a combination of these roles.

Herbert Leon MacDonell, author of *The Evidence Never Lies,* rightly said, "You can lead a jury to the truth, but you can't make them believe it," which is why good field notes and photographs are necessary, as they are the basis of solid documentation. Field notes should be written in a clear, legible, and articulate manner, as they provide firsthand recorded observations and are thus irreplaceable. Moreover, they should be self-explanatory in addition to being accurate. Photographs provide a visual record and are cardinal to forensic investigations, particularly with issues such as mold, filters, etc. Photographs should be of the highest quality and taken from different positions to give a comprehensive overview of the scene in question. Photographs should be well annotated and filed appropriately. Digital photographs may be stored on the computer or on CD. The forensic expert may also decide to supplement documentation of the project or scene with video photography. This will depend largely on the circumstances prevailing at the time.

Assigning culpability for green disputes often boils down to a matter of negligence, ignorance, or incompetence. American courts often require that qualified experts testify to the standard of care that is applicable to the case in dispute and to the professional's performance as measured by that applicable standard of care. The principle of standard of care may not apply to building contractors, since they are not deemed to be "professionals" in the sense of making independent evaluations and judgments based on learning and skill. However, builders are held to a "duty to perform" – that is, to strictly follow the plans, specifications, and provisions of their contracts.

10.5.2 Traditional Litigation: Pretrial and Trial Procedures

Traditional civil litigation is expensive and continues to escalate; it is complex and can drag on for years. The traditional litigation process requires the observance of specific protocols regarding rules of evidence and procedures for reports, pretrial discovery techniques, motion practices, interrogatories, depositions, hearsay rules, direct and cross-examination, and redirect and re-cross-examination. The consultant, building owner, or product manufacturer need not be fully conversant with the procedural details of civil lawsuits, although a basic understanding of the different stages is desirable. Litigation procedures are typically governed by statute in each jurisdiction. If we take New York as an example, procedure is governed by the Civil Practice Law and Rules; in federal courts it is governed by the Federal Rules of Civil Procedure. Lawsuits are typically divided into two basic phases, pretrial and trial.

10.5.2.1 Pretrial

In pretrial procedures, the parties to a lawsuit are identified and the issues in dispute are clarified. In addition, each party is offered an opportunity to learn about the other party's witnesses and potential evidence.

10.5.2.2 Trial

Before a trial can actually begin, a jury has to be selected. The method of jury selection depends on the jurisdiction of the trial. It can either be accomplished by the attorneys themselves or by the judge, depending on the rules of the court where the case is being heard. Upon jury selection, the trial can start with the attorneys of the plaintiff and defendant giving opening statements. These statements typically outline the strategy to be followed by the respective attorneys to prove their case in the trial.

10.5.3 Alternative Dispute Resolution

General concern and dissatisfaction with the present state of traditional civil litigation have provided impetus to the development of various alternative dispute-resolution (ADR) techniques for the construction industry, including sustainable construction and manufacturing industries. ADR techniques such as arbitration, mediation, negotiation, and other out-of-court settlement procedures are now employed in the vast majority of such disputes. They are essentially based on the premise that disputes can best be resolved through negotiation or mediation immediately after a conflict comes to light rather than through the tedious, costly, and time-consuming route of traditional civil litigation. When one party to the dispute insists on litigation, it may be because legal precedents have been shown to be favorable to that party. In the United States construction disputes are basically resolved in one of several ways (Figure 10.9):

- Settlement discussion among the parties: this often requires some compromise by all parties to the dispute.
- Mediation: this is usually nonbinding; an impartial mediator who is mutually agreed upon by the disputing parties is employed in the hope of bringing about an acceptable resolution through diplomacy and conciliation. This can be successful and satisfying in its outcome, providing the parties act in good faith.
- Arbitration can be binding or nonbinding; one or more arbitrators are agreed upon to hear the disputing parties' arguments, take and record testimony and evidence, ask probing questions, and finally arrive at a decision on a resolution through a quasijudicial process.

Resolution Process	Advantages	Disadvantages
Negotiation/assisted Negotiation/unassisted	• Parties have control • Confidential	• No structure • Entrenched bargaining positions likely
Mediation	• Structured • Skilled mediator helps avoid entrenched positions • Control and resolution lies with parties • Helps maintain future commercial relationship for parties • Costs less than litigation • Quick result • Confidential	• No decision if parties do not agree • A resolution may not be reached
Arbitration	• Structured • Can be quick, timetable controlled by parties • Costs may be less than litigation • Confidential	• Parties do not have control • Imposed decision • May jeopardize future relationship of parties
Litigation (Court Action)	• Structured	• Timetable controlled by Court • Costs may be significant • Parties do not have control • Imposed decision • May jeopardize future relationship of parties • Long waiting times • Goes on public record (no confidentiality)

Figure 10.9 Table showing the main advantages and disadvantages of different forms of resolution.

- Trial in court before a judge and usually a six-member jury that results in a binding verdict. Only a small percentage of cases of failure incidents end up in court in the United States.
- Settlement discussions are the best, least bruising, most private, and least expensive ways of resolving disputes.

Arbitration and mediation are viable, cost-effective alternatives to litigation. According to the American Arbitration Association (AAA): "Arbitration is the submission of a dispute to one or more impartial persons for a final and binding decision, known as an 'award.' Awards are made in writing and are generally final and binding on the parties in the case. Mediation, on the other hand, is a process in which an impartial third party facilitates communication and negotiation and promotes voluntary decision making by the parties to the dispute. This process can be effective for resolving disputes prior to arbitration or litigation."

A number of jurisdictions have now made it mandatory to use ADR methods prior to accepting a case and then only if ADR methods have failed to resolve the dispute. Most attorneys, however, advise caution in choosing ADR methods over traditional litigation; when ADR is chosen, preference is usually given to methods that are voluntary and non-binding.

A1 Acronyms and Abbreviations

A1.1 Organizations and Agencies

ACEEE	American Council for an Energy Efficient Economy
AFA	American Forestry Association
AIA	American Institute of Architects
ANSI	American National Standards Institute
APCA	Air Pollution Control Association
ASAE	American Society of Architectural Engineers
ASCE	American Society of Civil Engineers
ASHRAE	American Society of Heating, Refrigeration, and Air-conditioning Engineers
ASID	American Society of Interior Designers
ASME	American Society of Mechanical Engineers
ASNT	American Society for Non-Destructive Testing
ASPE	American Society of Plumbing Engineers
ASTM	American Society for Testing Materials
AWEA	American Wind Energy Association
AWMA	Air and Waste Management Association
BBRS	Board for Building Regulations and Standards
BCDC	Bay Conservation and Development Commission
BIFMA	Business and Institutional Furniture Manufacturer's Association
BLM	Bureau of Land Management
BOCA	Building Officials and Code Administrators
BREEM	Building Research Establishment Environmental Assessment Method
CEC	California Energy Commission
CFR	Code Federal Regulation
CIBSE	Chartered Institution of Building Services Engineers
CIWMB	California Integrated Waste Management Board

DOI: 10.1016/B978-1-85617-691-0.00011-4

CRI Carpet and Rug Institute
CRS Center for Resource Solutions
CSI Construction Specifications Institute
CUWCC California Urban Water Conservation Council

DoD Department of Defense
DGS Department of General Services (CA)
DOE U.S. Department of Energy
DOF Department of Finance (CA)
DPW Directorate of Public Works
DSA Division of the State Architect (CA)
DWR Department of Water Resources (CA)

EIA Energy Information Administration
EPA U.S. Environmental Protection Agency
ERDC USACE Engineer Research and Development Center
ESI European Standards Institute

FEMA U.S. Federal Emergency Management Agency
FSC Forest Stewardship Council

GBCI Green Building Certification Institute

IEA International Energy Agency
IEEE Institute of Electrical and Electronics Engineers, Inc.
IESNA Illuminating Engineering Society of North America
IMEX Industrial Material Exchange
IPMVP International Performance Measurement and Verification Protocol
ISO International Organization for Standardization

NAE National Academy of Engineering
NAS National Academy of Sciences
NBI New Building Institute
NCARB National Council of Architectural Registration Boards
NFRC National Fenestration Rating Council
NIST National Institute of Standards and Technology

OECD	Organization of Economic Cooperation and Development
OEE	Office of Energy Efficiency
OSHA	Occupational Safety and Health Administration (or Act)
OSWER	U.S. EPA Office of Solid Waste and Emergency Response
SBIC	Sustainable Building Industry Council
SCAQMD	South Coast Air Quality Management District
SEC	Securities and Exchange Commission
SMACNA	Sheet Metal and Air-Conditioning Contractors' National Association
UL	Underwriters Laboratories
USACE	U.S. Army Corps of Engineers
USDA	United States Department of Agriculture
USEPA	U.S. Environmental Protection Agency
USFS	U.S. Forest Service
USGBC	United States Green Building Council

A1.2 Referenced Standards and Legislation

ADA	Americans with Disabilities Act
ASHRAE 90.1	Building energy standard covering design, construction, operation, and maintenance.
ASHRAE 52.2	Standardized method of testing building ventilation filters for removal efficiency by particle size.
ASHRAE 55	Standard describing thermal and humidity conditions for human occupancy of buildings
ASHRAE 62	Standard that defines minimum levels of ventilation performance for acceptable indoor-air quality
ASHRAE 192	Standard for measuring air-change effectiveness
ASTM E408	Standard of inspection-meter test methods for normal emittance of surfaces
ASTM E903	Standard of integrated-sphere test method for solar absorptance, reflectance, and transmittance
CAA	Clean Air Act; Compliance Assurance Agreement
CASBEE	Comprehensive Assessment for Building Environmental Efficiency

CERCLA	Comprehensive Environmental Response, Compensation, and Liability Act
CERL	Construction Engineering Research Lab; part of USACE
CWA	Clean Water Act (aka FWPCA)
EISA	Energy Independence and Security Act of 2007
EPAct	Energy Policy Act of 2005
EPAct	U.S. Energy Policy Act of 1992
FFHA	Federal Fair Housing Act
FOIA	Freedom of Information Act; also similar state statutes
GS	Green Seal
MERV	Minimum Efficiency Reporting Value
NFPA	National Fire Protection Association
PURPA	Federal Public Utilities Regulatory Policy Act of 1978
RCRA	Resource Conservation and Recovery Act
TSCA	Toxic Substances Control Act
UBC	Uniform Building Code; the International Conference of Building Officials model building code

A1.3 Abbreviated General Terminology

A or AMP	Ampere
AAQS	Ambient Air Quality Standards
AC	Alternating Current
A/C	Air-Conditioning Unit
ACH	Air Change per Hour
ACM	Asbestos Containing Material
ACT AGE	Actual Age
ADAAG	ADA Architectural Guidelines
AE	Architect-Engineer firm (typically contracted for design services)
AEI	Advanced Energy Initiative
AEO	Annual Energy Outlook; DOE/EIA publication
AF	Acre-Foot (of water)
AFC	Application for Certification

AFV	Alternative-Fueled Vehicle
AFY	Acre-Feet per Year
AGMBC	LEED™ NC Application Guide for Multiple Buildings and On-Campus Building Projects (USGBC document)
AHM	Acutely Hazardous Materials
AHU	Air Handling Unit
AIB	Air Infiltration Barrier
AIRR	Adjusted Internal Rate of Return
AL	Aluminum
APPA	America Public Power Association
AQMD-	Air Quality Management District
AQMP	Air Quality Management Plan
ARB	Air Resources Board (CA)
ATC	Acoustical Tile Ceiling
A/V	Audiovisual
BAS	Building Automation System
BAAQMD	Bay Area Air Quality Management District
BC	Bill Calculator
BCF	Billion Cubic Feet
Bcfd	Billion Cubic Feet per Day
BEA	U.S. Bureau of Economic Affairs
BEEP	BOMA Energy Efficiency Program
BEES	Building for Environmental and Economic Sustainability Support
BG	Biomass Gasification
BIM	Building Information Model
BIPV	Building Integrated Photovoltaics
BMP	Best Management Practices
BOD	Basis of Design
BOD	Beneficial Occupancy Date
BRAC	Base Realignment and Closure
BRI	Building-Related Illness
BT	Building Technologies
BTU	British Thermal Unit
BTUH	British Thermal Unit per Hour
BUR	Built-Up Roofing

C&I	Commercial and Industrial
CAA	U.S. Clean Air Act
CAAQS	California Ambient Air Quality Standards
CalEPA	California Environmental Protection Agency
CAPM	Capital Asset Pricing Model
CARB	California Air Resources Board
CBC	California Building Code
CBECS	Commercial Building Energy Consumption Survey
CD	Construction Division
CDVR	Corrected Design Ventilation Rate
CEERT	Coalition for Energy Efficiency and Renewable Technologies
CEP	Central Energy Plant
CEU	Continuing Education Unit
CFCs	Chlorofluorocarbons
CFM	Cubic Feet per Minute
CFR	Code of Federal Regulations
CFS	Cubic Feet per Second
CHPS	Collaborative for High-Performance Schools
CIR	Credit Interpretation Ruling/Request
CO_2	Carbon Dioxide
CO	Carbon Monoxide
COC	Chain-of-Custody; proper accounting of material flows, as used by the FSC
COS	Center of Standardization
CMBS	Commercial Mortgage-Backed Securities
CMU	Concrete Masonry Unit
CPG	Comprehensive Procurement Guidelines
CSA	Canadian Standards Association
CWA	Clean Water Act
CxA	Commissioning Authority
DASA (I&H)	Deputy Assistant Secretary of the Army, Installations and Housing
dB	Decibel
DB	Design-Build; single contract
DBB	Design-Bid-Build
DC	Direct Current
DEC	Design Energy Cost

DHWH	Domestic Hot Water Heater/Water Heater
DOC	Determination of Compliance
DOR	Designer of Record
DPB	Discounted Payback
DS	Daylight Sensing Control
EA	Each
E&C	Engineering & Construction
ECB	Engineering and Construction Bulletin
ECB	Energy Cost Budget; a method of demonstrating compliance with ASHRAE 90.1
ECBEMS	Energy Management System
ECMs	Energy Conservation Measures
EER	Energy-Efficiency Ratio
EFF AGE	Effective Age
EIA	Energy Information Administration
EIFS	Exterior Insulation and Finish System
EIR	Environmental-Impact Report
EIS	Environmental-Impact Statement
EMCS	Energy Monitoring and Control System
EMF	Electromagnetic Fields
EMP	LEED™ Energy Modeling Protocol
EO	Executive Order
EPCA	Energy Policy and Conservation Act
EPDM	Etheylic-Propylene Dian-Monomer
ESA	Endangered Species Act (federal)
ESA	Environmental Site Assessment
ESC	Erosion and Sedimentation Control Plan
ESP	Energy Service Providers
ETS	Environmental Tobacco Smoke
EUL	Expected Useful Life
FCAA	Federal Clean Air Act
FEMA	Federal Emergency Management Agency
FEMP	Federal Energy Management Program
FERC	Federal Energy Regulatory Commission
FF&E	Finishes, Furniture (Fixtures), and Equipment

FIO	For Information Only
FOIL	Freedom of Information Letter
FSC	Forest Stewardship Council
FTC	Federal Trade Commission
FTE	Full-Time Equivalent
FTE	Full-Time Employee
FY	Fiscal Year
GBI	Green Building Initiative
GC	General Contractor
GD	Geographic District
GDP	Gross Domestic Product
GEP	Good Engineering Practice
GF	Glazing Factor
GHG	Greenhouse Gases
GIS	Geographic Information System
GPD	Gallons per Day
GPF	Gallons per Flush
GPM	Gallons per Minute
GW	Gigawatt
GW(h)	Gigawatt (hour); 1 billion watts
GWP	Global Warming Potential
H_2S	Hydrogen Sulfide
HAZMAT	Hazardous Materials
HCFCs	Hydrochlorofluorocarbons
HFCs	Hydrofluorocarbons
HP	Horsepower
HRA	Health Risk Assessment
HV	High Voltage
HVAC	Heating, Ventilating, and Air Conditioning
HVACR	Heating, Ventilation, Air Conditioning, and Refrigerants
Hz	Hertz
IAQ	Indoor-Air Quality
IDG	Installation Design Guide
IDIQ	Indefinite Delivery Indefinite Quantity

IEPR	Integrated Energy Policy Report
IEQ	Indoor Environmental Quality; one of the six LEED™ credit categories
IFMA	International Facilities Management Association
IPCC	Intergovernmental Panel on Climate Change
IPLV	Integrated Part Load Value
IRR	Internal Rate of Return
IS	Initial Study
ISO	Independent System Operator
Km	Kilometer
kW(h)	Kilowatt (hour); 1000 watts
kV	Kilovolt
kVA	Kilovolt-Ampere; transformer size rating
kVAR	Kilovolt-Ampere Reactive
kW	Kilowatt
LADWP	Los Angeles Department of Water and Power
LAN	Local Area Network
LAV	Lavatory
LBNL	Lawrence Berkeley National Laboratory
Lbs	Pounds
Lbs/Hr	Pounds per Hour
LCA	Life-Cycle Assessment
LCC	Life-Cycle Cost
LCCA	Life-Cycle Cost Analysis
LCGWP	Life-Cycle Global Warming Potential
LCODP	Life-Cycle Ozone Depletion Potential
LEED™	Leadership in Energy and Environmental Design; USGBC program
LEED™ AP	LEED™ Accredited Professional
LEED™ EB	LEED™ tool for Existing Buildings
LEED™ Homes	LEED™ tool for Homes
LEED™ NC	LEED™ tool for New Construction and Major Renovations
LEED™ ND	LEED™ tool for Neighborhood Development
LE/FE	Low-Emission/Fuel-Efficient Vehicle
LID	Low-Impact Development
LPDLr	Lighting Power Density Refrigerant Leakage Rate

LPG	Liquified Petroleum Gas; propane and butane
LQHC	Low-Quality Hydrocarbons; i.e., tar sands and oil shale
LTV	Loan-to-Value
LV	Low Voltage
LZ	Lighting Zone
M	Meter, Million, Mega, Milli, or Thousand
MDF	Medium-Density Fiberboard
MEP	Mechanical, Electrical, and Plumbing
MERV	Minimum Efficiency Reporting Value
M/F	Male/Female
MMT	Million Metric Tons
MOU	Memorandum of Understanding
M/S	Meters per Second
MSDS	Material Safety Data Sheet
MV	Megavolt
MVA	Megavolt-amperes
MW	Megawatt; 1 million watts
MWD	Metropolitan Water District
MWh	Megawatt (hour);1 million watts
NAAQS	National Ambient Air-Quality Standards
NAICS	North American Industry Classification System
NC	New Construction
NCPA	Northern California Power Agency
NEMA	National Electrical Manufacturers Association
NEPA	National Environmental Policy Act; federal "equivalent" of CEQA of 1969
NES	National Energy Savings
NG	Natural Gas
NO_2	Nitrogen Dioxide
NO	Nitrogen Oxide
NOx	Nitrogen Oxides
NPDES	National Pollutant Discharge Elimination System
NPV	Net Present Value
NREL	National Renewable Energy Labs
NS	Net Savings

O_3	Ozone
OACSIM	Office of the Assistant Chief of Staff for Installation
O&M	Operation and Maintenance
OCONUS	Outside the Continental United States
ODP	Ozone-Depleting Potential
OPR	Owner's Project Requirements Document
OSA	Outside Air
OSB	Oversight Board
PBP	Payback Period
PCA	Property Condition Assessment
PDT	Project Development Team
PM	Project Manager
PM	Particulate Matter
PML	Probable Maximum Loss
PMO	Project Management Oversight
PMV	Predicted Mean Vote
POC	Point of Contact
PPM	Parts per Million
PPT	Parts per Thousand
PRM	Performance Rating Method
PSI	Pounds per Square Inch
PTO	Permit to Operate
PU	Per Unit
PUC	Public Utilities Commission
PV	Photovoltaics
PV	Present Value
QA/QC	Quality Assurance/Quality Control
RA	Return Air
Rc	Refrigerant Charge
RD&D	Research, Development, and Demonstration
REC	Renewable-Energy Certificate
RFP	Request for Proposals
RFQ	Request for Qualifications
RH	Relative Humidity

RIT	Regional Integration Team
RTU	Rooftop Package Unit
RUL	Remaining Useful Life
SBS	Sick-Building Syndrome
SBTF	Sustainable Building Task Force (CA)
SCE	Southern California Edison
SDG&E	San Diego Gas and Electric Company
SEER	Seasonal Energy-Efficiency Ratio
SHGC	Solar Heat-Gain Coefficient
SIR	Savings-to-Investment Ratio
SMUD	Sacramento Municipal Utility District
SOG	Slab-on-Grade
SOx	Sulfur Oxides
SA	Supply Air
SDD	Sustainable Design and Development
SF	Square Feet
SHGC	Solar Heat-Gain Coefficient
SO_2	Sulfur Dioxide
SOx	Sulfur Oxides
SOW	Scope of Work
SPB	Simple Payback
SPIRIT	Army-developed point/credit-based system for measuring sustainability of buildings/development; modified version of LEED™ version 2.0
SRI	Solar Reflectance Index
T	Ton
TAC	Toxic Air Contaminant
TL	Total Losses
TOG	Total Organic Gases
TP	Total Phosphorous
TPD	Tons per Day
TPY	Tons per Year
TSP	Total Suspended Particulate Matter
TSS	Total Suspended Solids
TVOC	Total Volatile Organic Compounds

UFGS Unified Facilities Guide Specifications
UMCS Utility Monitoring and Control System
USGS United States Geological Survey
UST Underground Storage Tank
UV Ultraviolet Radiation

V Volts
VAV Variable Air Volume
VCAPCD Ventura County Air Pollution Control District
VMT Vehicle Miles Traveled
VOC Volatile Organic Compound

W Watt
WB Wet Bulb
WBDG Whole Building Design Guide
WC Water Closet

ZEV Zero Emissions Vehicle; minimum ENERGY STAR® rating of 40

A2 Glossary

Absorption The process by which incident light energy is converted to another form of energy, usually heat.

Accessible Describes a site, building, facility, or portion thereof that complies with accessibility guidelines.

Accessible Route A continuous, unobstructed path connecting all accessible elements and spaces of a building or facility. Interior accessible routes may include corridors, floors, ramps, elevators, lifts, and clear floor space at fixtures. Exterior accessible routes may include parking access aisles, curb ramps, crosswalks at vehicular ways, walks, ramps, and lifts.

Acid Aerosol Acidic liquid or solid particles that are small enough to become airborne. High concentrations of acid aerosols can be irritating to the lungs and have been associated with some respiratory diseases, such as asthma.

Adapted Plants Plants that reliably grow well in a given habitat with minimal attention from humans in the form of winter protection, pest protection, water irrigation, or fertilization once root systems are established in the soil. Adapted plants are considered to be low-maintenance but not invasive.

Addendum A written or graphic instruction issued by the architect prior to the execution of the contract that modifies or interprets the bidding documents by additions, deletions, clarifications, or corrections. An addendum becomes part of the contract documents when the contract is executed.

Adhesive A material used to bond two materials together.

Adobe A heavy clay soil used in many southwestern states to make sun-dried bricks.

Aggregate Fine, lightweight, coarse, or heavyweight grades of sand, vermiculite, perlite, or gravel added to cement for concrete or plaster.

Air Conditioning A process that simultaneously controls the temperature, moisture content, distribution, and quality of air.

Air Filter A device designed to remove contaminants and pollutants from air passing through the device.

Air-Handling Unit A mechanical unit used for air conditioning or movement of air, as in direct supply or exhaust of air within a structure.

Aligned Section A section view in which some internal features are revolved into or out of the plane of the view.

Allergen A substance capable of causing an allergic reaction because of an individual's sensitivity to it.

Alligatoring A pattern of rough cracking on a coated surface, similar in appearance to alligator skin.

DOI: 10.1016/B978-1-85617-691-0.00012-6

Alternating Current (AC) Electrical current that continually reverses direction of flow. The frequency at which it reverses is measured in cycles per second, or Hertz (Hz). The magnitude of the current itself is measured in amps (A).

Alternator A device for producing alternating-current (AC) electricity, usually driven by a motor but sometimes by other means, including water and wind power.

Ambient Lighting Lighting from any source that produces general illumination, as opposed to task lighting.

Ambient Temperature The temperature of the surroundings.

American Bond Brickwork pattern consisting of five courses of stretchers followed by one bonding course of headers.

Ammeter A device used for measuring current flow at any point in an electrical circuit.

Ampere The unit for the electric current; the flow of electrons. One amp is one coulomb passing in one second. One amp is produced by an electric force of one volt acting across a resistance of one ohm.

Analog The processing of data by continuously variable values.

Anemometer: A device used to measure wind speed.

Angle of Incidence Angle between a surface and the direction of incident radiation; applies to the aperture plane of a solar panel. There are only minor reductions in power output within plus/minus 15 degrees.

Animal Dander Tiny scales of animal skin.

ANSI The American National Standards Institute. ANSI is an umbrella organization that administers and coordinates the national voluntary consensus standards system.

Appeal A formal written request to review the content of an exam question for accuracy, validity, or errors in content and grammar. Appeals must be specific to an exam question and must be submitted by the candidate to GBCI's Accreditation Department within 10 days of the exam appointment. The appeal must describe the content of the exam question and, if possible, the nature of error. Exam scores are not modified under any conditions.

Arc A portion of the circumference of a circle.

Architect's Scale The scale used when dimensions or measurements are to be expressed in feet and inches.

Array A number of solar modules connected together in a single structure.

Asphalt Shingles Shingles made of asphalt or tar-impregnated paper with a mineral material embedded; very fire-resistant.

Assumed Liability Liability that arises from an agreement between people, as opposed to liability that arises from common or statutory law.

ASTM International Formerly the American Society for Testing Materials. An organization that develops and publishes testing standards for materials and specifications used by industry.

Authority Having Jurisdiction (AHJ) The governmental body responsible for the enforcement of any part of the standard codes, the official or agency designated to exercise such a function, and/or the Architect.

Axial Load A weight that is distributed symmetrically to a supporting member, such as a column.

Axonometric Projection A set of three or more views in which the object appears to be rotated at an angle so that more than one side is seen.

Backfill Any deleterious material (sand, gravel, etc.) used to fill an excavation.

Baffle A single opaque or translucent element used to diffuse or shield a surface from direct or unwanted light.

Ballast Electrical "starter" required by certain lamp types, especially fluorescents.

Balloon Framing A system in wood framing in which the studs are continuous without an intermediate plate for the support of second-floor joists.

Baluster A vertical member that supports handrails or guardrails.

Balustrade A horizontal rail held up by a series of balusters.

Banister The part of the staircase that fits on top of the balusters.

Bar Chart A calendar that graphically illustrates a projected time allotment to achieve a specific function.

Base A trim or molding piece found at the interior intersection of the floor and the wall.

Beam A weight-supporting horizontal member.

Base Building The core (common areas) and shell of the building and its systems, which typically are not subject to improvements to suit tenant requirements.

Base Flashing Flashing that covers the edges of a membrane.

Batten A narrow strip of wood used to cover a joint.

Batt Insulation An insulating material formed into sheets or rolls with a foil or paper backing, to be installed between framing members.

Bearing Wall A wall that supports any vertical loads in addition to its own weight.

Benchmark A point of known elevation from which the surveyor can establish all the grades.

Bend Allowance An additional amount of metal used in a bend in metal fabrication.

Bill of Material A list of standard parts or raw materials needed to fabricate an item.

Biodiversity The tendency in ecosystems, when undisturbed, to have a great variety of species forming a complex web of interactions. Human population pressure and resource consumption tend to reduce biodiversity dangerously; diverse communities are less subject to catastrophic disruption.

Black Water Waste water generated from toilet flushing. Black water has a higher nitrogen and fecal coliform level than gray water. Some jurisdictions include water from kitchen sinks or laundry facilities.

Blistering The condition that paint presents when air or moisture is trapped underneath and makes bubbles that break into flaky particles and ragged edges.

Blocking The use of internal members to provide rigidity in floor and wall systems. Also used for fire draft stops.

Blueprints Documents containing all the instructions necessary to manufacture a part. The key sections of a blueprint are the drawings, dimensions, and notes. Although blueprints used to be blue, modem reproduction techniques now permit printing of black on white as well as colors.

Body Plan An end view of a ship's hull, composed of superimposed frame lines.

Board Foot A unit of lumber of measure equaling 144 cubic inches; the base unit is 1 inch thick and 12 inches square, or $1 \times 12 \times 12 = 144$ cubic inches.

Boiler Equipment designed to heat water or generate steam.

Bond In masonry, the interlocking system of brick or block to be installed.

Boundary Survey A mathematically closed diagram of the complete peripheral boundary of a site, reflecting dimensions, compass bearings, and angles. It should bear a licensed land surveyor's signed certification and may include a metes and bounds or other written description.

Breezeway A covered walkway with open sides between two different parts of a structure.

Brick Pavers A term used to describe special brick to be used on the floor surface.

British Thermal Unit (BTU) The amount of heat energy required to raise one pound of water from a temperature of 60 degrees F to 61 degrees F at one atmosphere pressure. One watt-hour equals 3413 BTU.

Building Codes Rules and regulations adopted by the governmental authority having jurisdiction over commercial real estate, which govern ots design, construction, alteration, and repair. In some jurisdictions trade or industry standards may have been incorporated into such building codes by the governmental authority. Building codes are interpreted to include structural, HVAC, plumbing, electrical, life safety, and vertical transportation.

Building Envelope The enclosure of the building that protects the building's interior from outside elements, namely the exterior walls, roof, and soffit areas.

Building Inspector A representative of a governmental authority employed to inspect construction for compliance codes, regulations, and ordinances.

Building Line An imaginary line determined by zoning departments to specify on which area of a lot a structure may be built (also known as a setback).

Building Permit A permit issued by an appropriate governmental authority allowing construction of a project in accordance with approved drawing and specifications.

Building-Related Illness A discrete, identifiable disease or illness that can be traced to a specific pollutant or source within a building. Also see **Sick-Building Syndrome**.

Building Systems Interacting or independent components or assemblies forming single integrated units and comprising a building and its sitework, such as pavement and flatwork, structural frame, roofing, exterior walls, plumbing, HVAC, electrical, etc.

Build-Out The interior construction and customization of a space (including services and layout) to meet the tenant's requirements in either new construction or renovation (also referred to as fit-out or fit-up).

Caisson A below-grade concrete column for the support of beams or columns.

Candela A common unit of light output from a source.

Cantilever A horizontal structural condition in which a member extends beyond a support, such as a roof overhang.

Capacitor An electronic component used for the temporary storage of electricity as well as for removing unwanted noise in circuits. A capacitor will block direct current but will pass alternating current.

Capillary The action by which the surface of a liquid, where it is in contact with a solid, is elevated or depressed.

Carbon Footprint A measure of an individual's, family's, community's, company's, industry's, product's, or service's overall contribution of carbon dioxide and other greenhouse gases to the atmosphere. It takes into account energy use, transportation methods, and other means of emitting carbon. A number of carbon calculators have been created to estimate carbon footprints, including one from the U.S. Environmental Protection Agency.

Casement A type of window hinged to swing outward.

Casting A metal object made by pouring melted metal into a mold.

Catch Basin A complete drain box made in various depths and sizes; water drains into a pit, then through a pipe connected to the box.

Caulk Any type of material used to seal walls, windows, and doors to keep out the weather.

Cavity Wall A masonry wall formed with an air space between each exterior face.

Cement Plaster A plaster that is comprised of cement rather than gypsum.

Central HVAC System A system that produces a heating or cooling effect in a central location for subsequent distribution to satellite spaces that require conditioning.

Centrifugal Chiller A vapor-compression chiller that utilizes a centrifugal compressor; most commonly used in systems with cooling capacities from 80 to 10,000 tons.

Centrifuge A particular type of fluid moving device that imparts energy to the fluid by high-velocity rotary motion through a channel; fluids enter the device along one axis and exit along another axis.

Certificate for Payment A statement from the architect to the owner confirming the amount of money due the contractor for work accomplished, materials and equipment suitably stored, or both.

Certificate of Insurance A document issued by an authorized representative of an insurance company stating the types, amounts, and effective dates of insurance in force for a designated insured.

Certificate of Occupancy Document issued by a governmental authority certifying that all or a designated portion of a building complies with the provisions of applicable statutes and regulations and permitting occupancy for its designated use.

Certificate of Substantial Completion A certificate prepared by the architect on the basis of an inspection stating that the building or a designated portion thereof is substantially complete. It establishes the date of substantial completion; states the responsibilities of the owner and the contractor for security, maintenance, heat, utilities, damage, and insurance; and taxes the time within which the contractor shall complete the items listed therein.

Cesspool An underground catch basin for the collection and dispersal of sewage.

Change Order A written and signed document between the owner and the contractor authorizing a change in the work or an adjustment in the contract sum or time. The contract sum and time may be changed only by change order. A change order may be in the form of additional compensation or time or less compensation or time, referred to as a deduction.

Checklist List of items used to check drawings.

Chiller Equipment designed to produce chilled water; see also **Vapor Compression Chiller**.

Circuit A continuous system of conductors providing a path for electricity.

Circuit Breaker An automatic switch that can shut power off when it senses too much current.

Circumference The length of a line that forms a circle.

Clear Floor Space The minimum unobstructed floor or ground space required to accommodate a single, stationary wheelchair and occupant.

Clerestory A window or group of windows placed above the normal window height, often between two roof levels.

Coefficient of Utilization (CU) The ratio of light energy (lumens) from a source, calculated as lumens received on the work plane relative to the light energy emitted by the source alone.

Column A vertical weight-supporting member.

Column Pad An area of concrete in the foundation to support weight distributed into a column.

Combustion An oxidation process that releases heat; on-site combustion is a common heat source for buildings.

Commissioning A systematic process to verify that building components and systems function as intended and required; systems may need to be recommissioned at intervals during a building's life cycle.

Common Use Refers to those interior and exterior rooms, spaces, or elements that are made available for the use of a restricted group of people (for example, occupants of a homeless shelter or office building or the guests of such occupants).

Component A fully functional portion of a building system, piece of equipment, or building element.

Composite Wood A product consisting of wood or plant particles or fibers bonded together by a synthetic resin or binder. Examples include plywood, particle board, OSB, MDF, and composite door cores.

Compressor A device designed to increase the density of a compressible fluid, refrigerant, or air.

Computer-Aided Drafting (CAD) A method by which engineering or architectural drawings may be developed on a computer.

Computer-Aided Manufacturing (CAM) A method by which a computer uses a design to guide a machine that produces parts.

Concrete Block A rectangular concrete form with cells in it.

Condensation The process by which moisture in the air becomes water or ice on a surface (such as a window) whose temperature is colder than the air's temperature.

Condenser A device designed to condense a refrigerant; an air-to-refrigerant or water-to-refrigerant heat exchanger; part of a vapor compression or absorption refrigeration cycle.

Conductor A material used to transfer electricity, often in the form of wires.

Conduit A pipe or elongated box used to house and protect electrical cables.

Construction Documents All drawings, specifications, addenda, and other pertinent information associated with the construction of a specific project.

Contingency Allowance A sum included in the project budget designated to cover unpredictable or unforeseen items of work or changes in the work subsequently required by the owner.

Contour Line A line that represents the change in level from a given datum point.

Contract A legally enforceable promise or agreement among two or more persons.

Convection Transfer of heat through the movement of a liquid or gas.

Cooling Tower Equipment designed to reject heat from a refrigeration cycle to the outside environment through an open-cycle evaporative process; an exterior heat-rejection unit in a water-cooled refrigeration system.

Cornice The projecting or overhanging structural section of a roof.

Cost Appraisal Evaluation or estimate (preferably by a qualified professional appraiser) of the market or other value, cost, utility, or other attribute of land or a facility.

Cost Estimate A preliminary statement of approximate cost, determined by one of the following methods: area and volume method relates to the cost per square foot or cubic foot of the building; unit-cost method relates to the cost of one unit multiplied by the number of units in the project–for example, in a hospital, the cost of one patient unit multiplied by the number of patient units in the project; in-place unit method refers to the cost in-place of a unit, such as a door, cubic yards of concrete, or squares of roofing.

Coving The curving of the floor material against the wall to eliminate the open seam between floor and wall.

Cradle-to-Grave Analysis Analysis of the impact of a product from the sourcing of its materials through the end of its useful life to disposal of all waste products. Cradle-to-cradle is a related term signifying the recycling or reuse of materials at the end of their first useful life.

Crawl Space The area under a floor that is only excavated sufficiently to allow a person to crawl under it to reach electrical or plumbing devices.

Cross-Section A slice through a portion of a building or member that depicts its various internal conditions.

CSI Construction Specifications Institute; a membership organization of design professionals, construction professionals, product manufacturers, and building owners that develops and promotes industry communication standards and certification programs.

Current The flow of electric charge in a conductor between two points that differ in electrical potential (voltage); it is measured in amperes.

Curtain Wall An exterior wall that provides no structural support.

Cut-Off Voltage The voltage level at which the charge controller (regulator) disconnects the PV array or the load from the battery.

Damper A device designed to regulate the flow of air in a distribution system.

Dangerous or Adverse Conditions Conditions that may pose a threat or possible injury to the field observer and may require the use of special protective clothing, safety equipment, access equipment, or any other precautionary measures.

Date of Agreement A date stated in an agreement. If no date is stated, it may refer to the date on which the agreement is actually signed or the date established by an award.

Date of Commencement of the Work The date established in a notice to the contractor to proceed; in the absence of such notice, the date of the owner-contractor agreement or such other date that may be established.

Date of Substantial Completion The date certified by the architect when the work or a designated portion of it is sufficiently complete, in accordance with the contract documents, that the owner can occupy the premise for the use for which it is intended.

Datum Point Reference point.

Daylight Factor (DF) The ratio of the illuminance at a particular point within an enclosure to the simultaneous unobstructed outdoor illuminance under the same sky conditions, expressed as a percentage. Direct sunlight is excluded for both values of illumination. The daylight factor is the sum of the sky component, the external reflected component, and the internal reflected component. The interior plane is usually a horizontal work plane. If the sky condition is the CIE standard overcast condition, then the DF will remain constant regardless of absolute exterior illuminance.

Daylighting The controlled admission of natural light into a space through glazing with the intent of reducing or eliminating electric lighting. By utilizing solar light, daylighting creates a stimulating and productive environment for building occupants.

Dead Load The weight of a structure and all its fixed components.

Decibel (dB) Unit of sound level or sound-pressure level. It is ten times the logarithm of the square of the sound pressure divided by the square of reference pressure, 20 micropascals.

Defective Work Work not conforming to the contract requirements.

Deferred Maintenance Physical deficiencies that cannot be remedied with routine maintenance, normal operating maintenance, etc., excluding minimal conditions that generally do not present a material physical deficiency to the subject property.

Design-Build Construction An agreement in which an owner contracts with a prime or main contractor to provide both design and construction services for the entire construction project. Use of the design-build project-delivery system grew from

5 percent of U.S. construction in 1985 to 33 percent in 1999 and is projected to surpass low-bid construction in 2005. If a design-build contract is extended further to include the selection, procurement, and installation of all furnishings, furniture, and equipment, it is called a turnkey contract.

Detail An enlarged drawing showing a structural aspect, an aesthetic consideration, a solution to an environmental condition, or the relationship among materials or building components.

Diffuser A device designed to supply air to a space while providing good mixing of supply and room air and avoiding drafts; normally ceiling-installed.

Digital The processing of data by numerical or discrete units.

Dimension Line A thin, unbroken line (except in the case of structural drafting) with each end terminating in an arrowhead; used to define the dimensions of an object. Dimensions are placed above the line except in structural drawing, where the line is broken and the dimension placed in the break.

Direct Costs (Hard Costs) The aggregate costs of all labor, materials, equipment, and fixtures necessary for the completion of construction.

Direct-Costs Loan; Indirect-Costs Loan The portion of the loan amount applicable and equal to the sum of the loan budget amounts for direct costs and indirect costs, respectively, shown on the borrower's project cost statement.

Direct Current (DC) Electrical current that flows only in one direction, although it may vary in magnitude; contrasts with alternating current.

Discount Factor A factor that translates expected benefits or costs in any given future year into present-value terms. The discount factor is equal to $1/(1+i)t$ where i is the interest rate and t is the number of years from the date of initiation for the program or policy until the given future year.

Discount Rate The interest rate used in calculating the present value of expected yearly benefits and costs.

Dormer A structure that projects from a sloping roof to form another roofed area. This new area is typically used to provide a surface to install a window.

Downspouts Pipes connected to the gutter to conduct rain water to the ground or sewer.

Drip Irrigation System An irrigation system that slowly applies water to the root system of plants to maximize transpiration while minimizing wasted water and top-soil runoff. Drip irrigation usually involves a network of pipes and valves that rests on the soil or underground at the root zone.

Drywall An interior wall covering installed in large sheets made from gypsum board.

Duct Sheet-metal forms used for the distribution of cool or warm air throughout a structure.

Due Diligence The process of conducting a walk-through survey and making appropriate inquiries into the physical condition of a project's improvements, usually in connection with a commercial real-estate transaction. The degree and type of such survey or other inquiry may vary for different properties and purposes.

Dwelling Unit A single unit that provides a kitchen or food-preparation area in addition to rooms and spaces for living, bathing, sleeping, and the like. Dwelling units include single-family homes; townhouses used as transient group homes; apartment buildings used as shelters; guestrooms in a hotel that provide sleeping accommodations and food-preparation areas; and other similar facilities used on a transient basis.

Easement The right to access to or through another piece of property, such as a utility easement.

Eave That portion of the roof that extends beyond the outside wall.

Egress A continuous and unobstructed way of exit travel from any point in a building or facility to a public way. A means of egress can refer to both vertical and horizontal travel and may include intervening room spaces, doorways, hallways, corridors, passageways, balconies, ramps, stairs, enclosures, lobbies, horizontal exits, courts, and yards. An accessible means of egress is one that complies with ADA guidelines and does not include stairs, steps, or escalators. Areas of rescue assistance or evacuation elevators may be included as part of accessible means of egress.

Elastomeric A material having characteristics that allow it to expand or contract its shape and still return to its original dimensions without losing stability.

Electric Current The flow of electrons measured in amperes.

Electrical Grid A network for electricity distribution across a large area.

Electricity The movement of electrons produced by a voltage through a conductor.

Electrode An electrically conductive material, forming part of an electrical device, often used to lead current into or out of a liquid or gas. In a battery the electrodes are also known as plates.

Element An architectural or mechanical component of a building, facility, space, or site–e.g., telephone, curb ramp, door, drinking fountain, seating, or water closet.

Elementary Wiring Diagram A wiring diagram showing how each individual conductor is connected within the various connection boxes of an electrical circuit system; a simplified schematic diagram.

Embodied Energy The total energy that a product may be said to "contain," including all energy used in growing, extracting, manufacturing, and transportation to the point of use. The embodied energy of a structure or system includes the embodied energy of its components plus the energy used in construction.

Energy Power consumed multiplied by the duration of use; for example, 1000 watts used for four hours equal 4000 watt-hours.

Engineer's Scale A scale with dimensions in feet and decimal parts of a foot or a scale ratio that is a multiple of 10.

Environmental Tobacco Smoke (ETS) Mixture of smoke from the burning end of a cigarette, pipe, or cigar or exhaled by a smoker (secondhand smoke or passive smoke).

Epicenter The point of the earth's surface directly above the focus or hypocenter of an earthquake.

Expansion Joint A joint often installed in concrete construction to reduce cracking and to provide workable areas.

Expected Useful Life (EUL) The average amount of time in years that an item, component, or system is estimated to function when installed new and assuming routine maintenance is practiced.

Exploded View A pictorial view of a device in a state of disassembly, showing the appearance and interrelationship of parts.

Extension Line A line used to visually connect the ends of a dimension line to the relevant feature on the part. Extension lines are solid and are drawn perpendicular to the dimension line.

Façade The exterior covering of a structure.

Face of Stud (FOS) Outside surface of a stud, a term used most often in dimensioning or as a point of reference.

Fascia A horizontal member located at the edge of a roof overhang.

Facility All or any portion of buildings, structures, site improvements, complexes, equipment, roads, walks, passageways, parking lots, or other real or personal property located on a site.

Falsework Temporary supports of timber or steel sometimes required in the erection of difficult or important structures.

Felt A tar-impregnated paper used for water protection under roofing and siding materials, sometimes used under concrete slabs for moisture resistance.

Fiberoptics Optical, clear strands that transmit light without electrical current, sometimes used for outdoor lighting.

Fillet A concave internal corner in a metal component, usually a casting.

Filter A device designed to remove impurities from a fluid passing through it; see also **Air Filter**.

Final Completion Term denoting that the work has been completed in accordance with the terms and conditions of the contract documents.

Final Inspection Final review of the project by the architect to determine final completion, prior to issuance of the final certificate for payment.

Final Payment The payment of the entire unpaid balance of the contract sum as adjusted by change orders made by the owner to the contractor upon issuance by the architect of the final certificate for payment.

Finish Grade The soil elevation in its final state upon completion of construction.

Finish Marks Marks used to indicate the degree of smoothness of finish to be achieved on surfaces to be machined.

Fire Barrier A continuous membrane such as a wall, ceiling, or floor assembly that is designed and constructed to a specified fire-resistant rating to hinder the spread of fire and smoke. This rating is based on a time factor. Only fire-rated doors may be used in these barriers.

Fire Compartment or Fire Zone An enclosed space in a building that is separated from all other parts of the building by the construction of fire separations having fire-resistance ratings.

Fire Department Records Records maintained by or in the possession of the local fire department in the area in which the subject property is located.

Fire Door A door used between different types of construction that has been rated as being able to withstand fire for a certain amount of time.

Fire Draft Stop A member placed in walls, floors, and ceilings to prevent a rapid spread of fire.

Fire-Resistance Rating Sometimes called fire rating, fire-resistance classification, or hourly rating; a term defined in building codes for various types of construction and occupancies, usually given in half-hour increments.

Fire-Stop Blocking placed between studs or other structural members to resist the spread of fire.

Firewall A type of fire separation of noncombustible construction that subdivides a building or separates adjoining buildings to resist the spread of fire and that has a fire-resistance rating and structural ability to remain intact under fire conditions for the required fire-rated time.

Flashing A thin, impervious sheet of material placed in construction to prevent water penetration or to direct the flow of water. Flashing is used especially at roof hips and valleys, roof penetrations, joints between a roof and a vertical wall, and in masonry walls.

Floor Joist Structural member for the support of floor loads.

Floor Plan A horizontal section taken at approximately eye level.

Flue Liner A terracotta pipe used to provide a smooth flue surface so that unburned materials will not cling to the flue.

Flush Even, level, or aligned.

Fly Ash The fine ash waste collected from the flue gases of coal combustion, smelting, or waste incineration.

Footcandle A common unit of illumination used in the U.S. The metric unit is the lux.

Footings Weight-bearing concrete construction elements poured in place in the earth to support a structure.

Footlambert The U.S. unit for luminance. The metric unit is the nit.

Format The general makeup or style of a drawing.

Foundation Plan A drawing that graphically illustrates the location of various foundation members and conditions that are required for the support of a specific structure.

Framing Connectors Metal devices, varying in size and shape, for the purpose of joining wood framing members together.

Frieze A decoration or ornament shaped to form a band around a structure.

Frost Line The depth at which frost penetrates the soil.

Full Section A sectional view that passes entirely through the object.

Fungi Any of a group of parasitic lower plants that lack chlorophyll, including molds and mildews.

Furred Ceiling A separate surface beyond the main ceiling or wall that provides a desired air space and modifies interior dimensional conditions.

Furring Wood strips attached to structural members that are used to provide a level surface for finishing materials when different-sized structural members are used.

Fuse A device used to protect electrical equipment from short circuits. Fuses are made with metals that are designed to melt when the current passing through them is high enough. When the fuse melts, the electrical connection is broken, interrupting power to the circuit or device.

Gable A type of roof with two sloping surfaces that intersect at the ridge of the structure.

Galvanized Steel products that have had zinc applied to the exterior surface to provide protection from rusting.

Gambrel Roof A barn-shaped roof.

Gauge The thickness of metal or glass sheet material.

General Conditions Construction project activities and associated costs that are not usually assignable to a specific material installation or subcontract, such as temporary electrical power; the contract document that spells out the relationships among the parties to the contract.

General Contract Any contract (together with all riders, addenda, and other instruments referred to therein) that requires the general contractor or such other person to provide, supervise, or manage the procurement of all labor and material needed for completion of a project.

Generator A mechanical device used to produce DC electricity. Power is produced by coils of wire passing through magnetic fields. Most alternating-current generating devices are also referred to as generators.

Gigawatt (GW) A measurement of power equal to a thousand million watts.

Gigawatt-Hour (GWh) A measurement of energy; one gigawatt-hour is equal to one gigawatt being used for a period of one hour or one megawatt being used for 1000 hours.

Girder A horizontal structural beam for the support of secondary members such as floor joists.

Glare The effect produced by luminance within one's field of vision that is sufficiently greater than the luminance to which one's eyes are adapted; it can cause annoyance, discomfort, or loss in visual performance and visibility.

Glue-Laminated Beam A beam comprised of a series of wood members glued together.

Grade The designation of the quality of a manufactured piece of wood.

Grading The moving of soil to effect the elevation of land at a construction site.

Gray Water Waste water that does not contain toilet wastes and can be reused for irrigation after simple filtration. Waste water from kitchen sinks and dishwashers may not be considered gray water in all cases.

Greenhouse Gas A gas in the atmosphere that traps some of the sun's heat, preventing it from escaping into space. Greenhouse gases are vital for making the Earth habitable, but increasing rates contribute to climate change. Greenhouse gases include water vapor, carbon dioxide, methane, nitrous oxide, and ozone.

Grid An electrical-utility distribution network.

Grid-Connected An energy-producing system connected to the utility transmission grid, also called grid-tied.

Ground Floor Any occupiable floor less than one story above or below grade with direct access to grade. A building or facility always has at least one ground floor and may have more than one if a split-level entrance has been provided or if a building is built into a hillside.

Grout A mixture of cement, sand, and water used to fill joints in masonry and tile construction.

Guardrail A horizontal protective railing used around stairwells, balconies, and changes of floor elevation greater than 30 inches.

Gusset A plate added to the side of intersecting structural members to help form a secure connection and to reduce stress.

Gutter A metal or wooden trough set below the eaves to catch and conduct water from rain and melting snow to a downspout.

Half Section A combination of an orthographic projection and a section view to show two halves of a symmetrical object.

Halogen Lamp A special type of incandescent globe made of quartz glass and a tungsten filament, enabling it to run at a much higher temperature than a conventional incandescent globe. Efficiency is better than a normal incandescent but not as good as a fluorescent light.

Harmonic Content Frequencies in the output waveform in addition to the primary frequency (usually 50 or 60 Hz). Energy in these harmonics is lost and can cause undue heating of the load.

Head The top of a window or door frame.

Header A horizontal structural member spanning over openings, such a doors and windows, to support weight above them.

Header Course In masonry, a horizontal masonry course of brick laid perpendicular to the wall face; used to tie a double-wythe brick wall together.

Heat Exchanger A device designed to efficiently transfer heat from one medium to another (for example, water-to-air, refrigerant-to-air, refrigerant-to-water, or steam-to-water).

Heat Island Effect The incidence of higher air and surface temperatures caused by solar absorption and reemission from roads, buildings, and other structures.

Heat Pump A device that uses a reversible-cycle vapor-compression refrigeration circuit to provide cooling and heating from the same unit at different times.

Heat Recovery A process whereby heat is extracted from exhaust air before the air is dumped to the outside environment; the recovered heat is normally used to preheat incoming outside air; this may be accomplished by heat-recovery wheels or heat-exchanger loops.

Hertz (Hz) Unit of measurement for frequency, normally 50 Hz in Europe and 60 Hz in the U.S. The magnitude of the current is measured in amperes.

Hip Roof A roof shape with four sloping sides.

Hose Bibb A faucet used to connect a hose.

Hydronic System A heating or cooling system that relies on the circulation of water as the heat-transfer medium. A typical example is a boiler with hot water circulated through radiators.

Illuminance The density of the luminous flux incident on a surface, expressed in footcandles or lux. This term should not be confused with illumination–i.e., the act of illuminating or state of being illuminated.

Incandescent Light An electric lamp that is evacuated or filled with an inert gas and contains a filament (commonly tungsten). The filament emits visible light when heated to extreme temperatures by passage of electric current through it.

Incident Light Light that shines on to the surface of a PV cell or module.

Indemnification A contractual obligation by which one person or entity agrees to secure another against loss or damage from specified liabilities.

Indirect Costs (Soft Costs) Certain costs (other than direct costs) of completion of the project, including but not limited to architects', engineers', and lender's attorneys' fees, ground rents, interest and title charges in respect of building loan mortgages, real-estate taxes, water and sewer rents, survey costs, loan-commitment fees, insurance and bond premiums, and such other nonconstruction costs as are part of the cost of improvements.

Indirect-Cost Statement A statement by the borrower in a form approved by the lender of indirect costs incurred and to be incurred.

Indirect Lighting Illumination achieved by reflection, usually from wall and/or ceiling surfaces.

Indoor-Air Quality (IAQ) A term that takes into account the introduction and distribution of adequate ventilation air; control of airborne contaminants; and maintenance of acceptable temperature and relative humidity.

Infill Site A site that is largely located within an existing community. For the purposes of LEED™ for Homes credits, an infill site is defined as having at least 75 percent of its perimeter bordering land that has been previously developed.

Inscribed Figure A figure that is completely enclosed by another figure.

Insolation The amount of sunlight reaching an area, usually expressed in watt-hours per square meter per day.

Inspection Examination of work completed or in progress to determine its conformance with the requirements of the contract documents. The architect ordinarily makes only two inspections, one to determine substantial completion and the other to determine final completion. These inspections should be distinguished from the more general observations made by the architect on visits to the site during the progress of the work. The term is also used to mean examination of the work by a public official, owner's representative, or others.

Insulation Any material capable of resisting thermal, sound, or electrical transmission.

Interconnection Diagram A diagram showing the cabling between electronic units as well as how the terminals are connected.

Internal Rate of Return The discount rate that sets the net present value of the stream of net benefits equal to zero. The internal rate of return may have multiple values when the stream of net benefits alternates from negative to positive more than once.

Inverter An device that converts DC power from the PV array/battery to AC power, used for both stand-alone and grid-connected systems.

Irradiance The solar power incident on a surface, usually expressed in kilowatts per square meter. Irradiance multiplied by time gives insolation.

Isometric Drawing A form of pictorial drawing in which the main lines are equal in dimension, normally drawn using a 30- or 90-degree angle.

Isometric Wiring Diagram A diagram showing the outline of a structure and the location of equipment such as panels, connection boxes, and cable runs.

Jamb The side portion of a door, window, or other opening.

Joist A horizontal beam used to support a ceiling.

Joule (J) The energy conveyed by one watt of power for one second, equal to 1/3600 kilowatt-hour.

Junction box A protective enclosure on a PV module where PV strings are electrically connected and electrical protection devices such as diodes can be fitted.

Key Plan A plan, reduced in scale, used for orientation purposes.

Kilowatt (kW) A unit of electrical power equal to 1000 watts.

Kilowatt-hour (kWh) The amount of energy that derives from a power of 1000 watts acting over a period of 1 hour.

Knee Wall A wall of less than full height.

Lattice A grille made by criss-crossing strips of material.

Ledger A structural framing member used to support ceiling and roof joists at the perimeter walls.

LEED™ Leadership in Energy and Environmental Design. A sustainable-design building certification system promulgated by the United States Green Building Council. Also an accrediting program for professionals who have mastered the certification system.

LEED™ Accredited Professional The credential earned by candidates who passed the exam between 2001 and June 2009.

Legend A description of any special or unusual marks, symbols, or line connections used in a drawing.

Lien A monetary claim on a property.

Life-Cycle Cost A sum of all costs of creation and operation of a facility over a period of time.

Life-Cycle Cost Analysis A technique used to evaluate the economic consequences over a period of time of mutually exclusive project alternatives.

Light Shelf A horizontal element positioned above eye level to reflect daylight onto the ceiling.

Limit of Liability The maximum amount that an insurance company agrees to pay in case of loss.

Lintel A load-bearing structural member supported at its ends, usually located over a door or window.

Live Load A temporary and changing load superimposed on structural components by the use and occupancy of a building, not including wind, earthquake, or dead loads.

Load The electrical power being consumed at any given moment or averaged over a specified period. The load that an electric generating system supplies varies greatly with time of day and to some extent season of year. In an electrical circuit, the load is any device or appliance that is using power.

Load-Bearing Wall A support wall that holds floor or roof loads in addition to its own weight.

Load Circuit The wiring including switches and fuses that connects the load to the power source.

Lumen (lm) The luminous flux emitted by a point source with a uniform luminous intensity of one candela.

Luminaire A complete electric-lighting unit, including housing, lamp, and focusing and/or diffusing elements, informally referred to as a fixture.

Lux The International System (SI) unit of illumination. It is the illumination on a surface one square meter in area on which there is a uniformly distributed flux of 1 lumen.

Major Diameter The largest diameter of an internal or external thread.

Manifold A fitting that has several inlets or outlets to carry liquids or gases.

Mansard A four-sided, steep-sloped roof.

Masonry Opening The actual distance between masonry units where an opening occurs. It does not include the wood or steel framing around the opening.

Masonry Veneer A layer of masonry units that is bonded to a frame or masonry wall.

Master Specification A resource specification section containing options for selection, usually created by a design professional firm, which, once edited for a specific project, becomes a contract specification.

MasterFormat Industry standard for organizing specifications and other construction information, published by CSI and Construction Specifications Canada, consisting of a six- or eight-digit numbering system with 49 divisions.

Masterspec® Subscription specification library published by ARCOM and owned by the American Institute of Architects.

Mastic An adhesive used to hold tiles in place; an adhesive used to glue many types of materials in the building process.

Mechanical Drawing A scale drawing of a mechanical object.

Mechanics' Lien A lien on real property created by statute in all states in favor of a person supplying labor or materials for a building or structure for the value of labor or materials supplied by the person. In some jurisdictions a mechanic's lien also exists for the value of professional services. Clear title to the property cannot be obtained until the claim for the labor, materials, or professional services is settled.

Megawatt (MW) A measurement of power equal to one million watts.

Megawatt-Hour (MWh) A measurement of power with respect to time (i.e., energy). One megawatt-hour is equal to one megawatt used for a period of one hour or one kilowatt used for 1000 hours.

Mesh A metal reinforcing material placed in concrete slabs and masonry walls to help resist cracking.

Mezzanine or Mezzanine Floor That portion of a story that is an intermediate floor level placed within the story and having occupiable space above and below its floor.

Module A system based on a single unit of measure.

Modulus of Elasticity (E) The degree of stiffness of a beam.

Moisture Barrier Typically a plastic material used to restrict moisture vapor from penetrating into a structure.

Molding Trim mounted around windows, floors, doors, and closets.

Monolithic One-Pour System A method of poured-in-place concrete construction that has no joints.

Mortar The mixture of cement, sand, lime, and water that provides a bond for the joining of masonry units.

Multifamily Dwelling Any building consisting of more than two dwelling units.

Multizone HVAC System A central-air HVAC system that utilizes an individual supply air stream for each zone; warm and cool air are mixed at the air-handling unit to provide supply air appropriate to the needs of each zone. A multizone system requires the use of several separate supply air ducts.

Native Vegetation Plants whose presence and survival in a specific region are not due to human intervention. Certain experts argue that plants imported to a region by prehistoric peoples should be considered native. The term for plants that are imported and then adapt to survive without human cultivation is naturalized vegetation.

NEC National Electrical Code, which contains guidelines for all types of electrical installations.

Negligence Failure to exercise due care under the circumstances. Legal liability for the consequences of an act or omission frequently depends upon whether or not there has been negligence.

Net Metering The practice of exporting surplus solar power (to actual power needs) during the day to the electricity grid, which either causes the electric meter to (physically) go backwards and/or creates a financial credit. (At night, the system draws from the electricity grid in the normal way.)

Net Size The actual size of an object.

Noise Reduction Coefficient (NRC) Average of the sound-absorption coefficients of the four octave band's 250, 500, 1,000, and 2,000 Hertz, rounded to the nearest 0.05.

Nominal Discount Rate A discount rate that includes the rate of inflation.

Nominal Size The call-out size; may not be the actual size of the item.

Nonbearing Wall A wall that supports no loads other than its own weight. Some building codes consider walls that support only ceiling loads as nonbearing.

Nonconforming Work Work that does not fulfill the requirements of the contract documents.

Nonferrous Metal A metal such as copper or brass that contains no iron.

Oblique Drawing A type of pictorial drawing in which one view is an orthographic projection and the views of the sides have receding lines at an angle.

Occupiable A room or enclosed space designed for human occupancy in which individuals congregate for amusement, educational, or similar purposes or in which occupants are engaged at labor and that is equipped with means of egress, light, and ventilation.

Offset Section A section view of two or more planes in an object to show features that do not lie in the same plane.

Ohm The resistance between two points of a conductor when a constant potential difference of one volt applied between these points produces in the conductor a current of one ampere.

Ohm's Law A mathematical formula that allows either voltage, current, or resistance to be calculated when the other two values are known. The formula is: $V = I \times R$, where V is the voltage, I is the current, and R is the resistance.

Opinion of Probable Costs Determination of probable costs, a preliminary budget, for a suggested remedy.

Operating Cost Any cost of the daily function of a facility.

Organic Compounds Chemicals that contain carbon. Volatile organic compounds vaporize at room temperature and pressure. They are found in many indoor sources, including many common household products and building materials.

Orientation: Position with respect to the cardinal directions.

Orthographic Projection A view produced when projectors are perpendicular to the plane of the object. It gives the effect of looking straight at one side.

Outlet An electrical receptacle that allows for current to be drawn from the system.

Packaged Air Conditioner A self-contained unit designed to provide control of air temperature, humidity, distribution, and quality.

Parapet A portion of a wall extending above the roof level.

Parging A thin coat of plaster used to smooth a masonry surface.

Partial Occupancy The occupancy by the owner of a portion of a project prior to final completion.

Partial Section A sectional view consisting of less than a half section, used to show the internal structure of a small portion of an object; also known as a broken section.

Partition An interior wall.

Party Wall A wall dividing two adjoining spaces such as apartments or offices.

Passive Solar Home A house that utilizes part of the building as a solar collector, as opposed to active solar, such as PV.

Patent Defect A defect in materials or equipment in completed work that reasonably careful observation could have discovered; distinguished from a latent defect, which could not be discovered by reasonable observation.

Peak Load The maximum usage of electrical power occurring in a given period of time, typically a day. The electrical supply must be able to meet the peak load if it is to be reliable.

Performance Specifications The written material containing the minimum acceptable standards and actions as may be necessary to complete a project.

Phase An impulse of alternating current. The number of phases depends on the generator windings. Most large generators produce a three-phase current that must be carried on at least three wires.

Photometer An instrument for measuring light.

Photovoltaic (PV) Any device that produces free electrons when exposed to light.

Photovoltaic (PV) Array A number of PV modules connected together in a single structure.

Photovoltaic (PV) Panel A term often used interchangeably with PV module (especially in single module systems).

Photovoltaic System All the parts connected together that are required to produce solar electricity.

Physical Deficiency Conspicuous defect or significant deferred maintenance of a subject property's material systems, components, or equipment as observed during the field observer's walkthrough survey. Included within this definition are material life-safety/building-code violations and material systems, components, or equipment that are approaching, have reached, or have exceeded their typical EUL or whose RUL should not be relied upon in view of actual or effective age, abuse, excessive wear and tear, exposure to the elements, lack of proper or routine maintenance, etc. This definition specifically excludes deficiencies that may be remedied with routine maintenance, miscellaneous minor repairs, normal operating maintenance, etc., and excludes minimal conditions that generally do not constitute a material physical deficiency of the subject property.

Pictorial Wiring Diagram A diagram showing actual pictorial sketches of the various parts of a piece of equipment and the electrical connections between the parts.

Pier A vertical support for a building or structure, usually designed to hold substantial loads.

Pile A steel or wooden pole driven into the ground sufficiently to support the weight of a wall and building.

Pillar A pole or reinforced wall section used to support the floor and consequently the building.

Planking A term for wood members having a minimum rectangular Section of 1.5 to 3.5 inches in thickness, used for floor and roof systems.

Plans All final drawings and specifications prepared by the owner, architect, the general contractor, or major subcontractors, and approved by the lender and the construction consultant; they describe and show the labor, materials, equipment, fixtures, and furnishings necessary for construction, including all amendments and modifications thereof made by approved change orders (and also showing minimum grade of finishes and furnishings for all areas to be leased or sold in ready-for-occupancy conditions).

Plat A map or plan view of a lot showing principal features, boundaries, and location of structures.

Plenum An air space above the ceiling for transporting air from the HVAC system.

Polarity The direction of flow of current.

Polyvinyl Chloride (PVC) A plastic material commonly used for pipe and plumbing fixtures and as an insulator on electrical cables. A toxic material, which is being replaced with alternatives made from more benign chemicals.

Portico A roof supported by columns instead of walls.

Post A vertical wood structural member generally 4 × 4 inches or larger.

Post-and-Beam Construction A type of wood-frame construction using timber for the structural support.

Potable Water Water suitable for drinking, generally supplied by the municipal water system.

Power Basic unit of electricity equal to the product of current and voltage (in DC circuits); the rate of doing work, expressed in watts (W). For example, a generator rated at 800 watts can provide that amount of power continuously.

Precast A concrete component that has been cast in a location other than the one in which it will be used.

Present Value The current value of a past or future sum of money as a function of an investor's time value of money.

Pressed Wood Products A group of materials used in building and furniture construction that are made from wood veneers, particles, or fibers bonded together with an adhesive under heat and pressure.

Primer The first coat of paint or glue when more than one coat will be applied.

Progress Payment Partial payment made during progress of the work on account of work completed and or materials suitably stored.

Progress Schedule A diagram, graph, or other pictorial or written schedule showing proposed and actual dates of starting and completion of the various elements of the work.

Project Cost Total cost of the project including construction cost, professional compensation, land costs, furnishings and equipment, financing, and other charges.

Projection A technique for showing one or more sides of an object to give the impression of a solid object.

Project Manual The volume(s) prepared by the architect for a project, which may include the bidding requirements, sample forms, conditions of the contract, and specifications.

Purlin A horizontal roof member that is laid perpendicular to rafters to help limit deflections.

Quarry Tile An unglazed, machine-made tile.

Quick Set A fast-curing cement plaster.

Rafter A sloping or horizontal beam used to support a roof.

Radius A straight line from the center of a circle or sphere to its circumference or surface.

Radon (Rn) and Radon-Decay Products Radon is a radioactive gas formed in the decay of uranium. Radon-decay products (also called radon daughters or progeny) can be breathed into the lungs, where they continue to release radiation as they further decay.

Rainscreen A method of constructing walls in which the cladding is separated from a membrane by an air space that allows pressure equalization to prevent rain from being forced in, often used for high-rise buildings or for buildings in windy locations.

Rake Joint A recessed mortar joint.

Ramp A walking surface that has a running slope greater than 1:20.

Remote Area Power Supply (RAPS) A power-generation system used to provide electricity to remote and rural homes, usually incorporating power generated from renewable sources such as solar panels and wind generators as well as nonrenewable sources such as petrol-powered generators.

Readily Accessible Describes areas of the subject property that are promptly made available for observation by the field observer at the time of the walkthrough survey, do not require the removal of materials or personal property such as furniture, and are safely accessible in the opinion of the field observer.

Record Drawings Construction drawings revised to show significant changes made during the construction process, usually based on marked-up prints, drawings, and other data furnished by the contractor to the architect; sometimes referred to as as-built drawings.

Rectifier A device that converts AC to DC current, as in a battery charger or converter.

Recycled Material Material that would otherwise be destined for disposal but is diverted or separated from the waste stream, reintroduced as material stock, and processed into another end product.

Reference Designation A combination of letters and numbers to identify parts on electrical and electronic drawings. The letters designate the type of part, and the numbers designate the specific part. For example, reference designator R-12 indicates the twelfth resistor in a circuit.

Reference Numbers Numbers used on a drawing to refer the reader to another drawing for more detail or other information.

Reference Plane The normal plane to which all information is referenced.

Reflectance The ratio of energy (light) bouncing away from a surface to the amount striking it, expressed as a percentage.

Refrigerant A heat-transfer fluid employed by a refrigerating process, selected for its beneficial properties (stability, low viscosity, high thermal capacity, appropriate state-change points).

Register An opening in a duct for the supply of heated or cooled air.

Regulator A device used to limit the current and voltage in a circuit, normally to allow the correct charging of batteries from power sources such as solar panels and wind generators.

Relative Humidity The amount of water vapor in the atmosphere compared to the maximum possible amount at the same temperature.

Release of Lien Instrument executed by a person or entity supplying labor, materials, or professional services on a project that releases that person's or entity's mechanic's lien against the project property.

Remaining Useful Life (RUL) A subjective estimate based upon observations; average estimates of similar items, components, or systems; or a combination thereof of the number of remaining years that an item, component, or system is estimated to be able to function in accordance with its intended purpose before warranting replacement. Such period of time is affected by the initial quality of an item, component, or system, the quality of the initial installation, the quality and amount of preventive maintenance exercised, climatic conditions, extent of use, etc.

Removed Section A drawing of an object's internal cross section, located near the basic drawing of the object.

Renewable Energy Energy that is produced from a renewable source.

Requisition A statement prepared by the borrower in a form approved by the lender setting forth the amount of the loan advance requested in each instance.

Resistance (R) The property of a material that resists the flow of electric current when a potential difference is applied across it, measured in ohms.

Resistor An electronic component used to restrict the flow of current in a circuit; sometimes used specifically to produce heat, such as in a water-heater element.

Retainage A sum withheld from progress payments to the contractor in accordance with the terms of the owner-contractor agreement.

Retaining Wall A masonry wall supported at the top and bottom, designed to resist soil loads.

R-Factor A unit of thermal resistance applied to the insulating value of a specific building material.

Return Air Air that has circulated through a building as supply air and has been returned to the HVAC system for additional conditioning or release from the building.

Revision Block A block is located in the upper right corner of a print to provide a space to record any changes made to the original print.

Revolved Section A drawing of an object's internal cross section superimposed on the basic drawing of the object.

Rheostat An electrical control device used to regulate the current reaching a light fixture; a dimmer switch.

Ridge The highest point on a sloped roof.

Residual Value The value of a building or building system at the end of the study period.

Roof Drain A receptacle for removal of roof water.

Roof Pitch The ratio of total span to total rise, expressed as a fraction.

Rotation A view in which the object is apparently rotated or turned to reveal a different plane or aspect, all shown within the view.

Rough In To prepare a room for plumbing or electrical additions by running wires or piping for a future fixture.

Rough Opening A large opening made in a wall or roof frame to allow the insertion of a door or window.

R-Value The unit that measures thermal resistance (the effectiveness of insulation); the higher the number, the better the insulation qualities.

Sanitary Sewer A conduit or pipe carrying sanitary sewage.

Scale The relation between the measurement used on a drawing and the measurement of the object it represents; a measuring device, such as a ruler, having special graduations.

Schedule of Values A statement furnished by the contractor to the architect reflecting the portions of the contract sum allocated to the various stages of the work and used as the basis for reviewing the contractor's applications for payment.

Schematic Diagram A diagram using graphic symbols to show how a circuit functions electrically.

Scratch Coat The first coat of stucco, which is scratched to provide a good bond surface for the second coat.

Scupper An opening in a parapet wall attached to a downspout for water drainage from the roof.

Scuttle Attic or roof access with a cover or door.

Section A view showing internal features as if the viewed object has been cut or sectioned.

Section Lines Thin, diagonal lines used to indicate the surface of an imaginary cut in an object.

Seismicity The worldwide or local distribution of earthquakes in space and time; a general term for the number of earthquakes in a unit of time or for relative earthquake activity.

Septic Tank A tank in which sewage is decomposed by bacteria and dispersed by drain tiles.

Sheet Steel Flat steel weighing less than five pounds per square foot.

Shear Distribution The distribution of lateral forces along the height or width of a building.

Shear Wall A wall construction designed to withstand shear pressure caused by wind or earthquake.

Shoring Temporary support made of metal or wood used to support other components.

Short-Term Costs Opinions of probable costs to remedy physical deficiencies, such as deferred maintenance, that may not warrant immediate attention but require repairs or replacements that should be undertaken on a priority basis in addition to routine preventive maintenance. Such opinions of probable costs may include costs for testing, exploratory probing, and further analysis should this be deemed warranted by the consultant. Generally, the time frame for such repairs is within one to two years.

Sick-Building Syndrome Term that refers to a set of symptoms that affect some number of building occupants during the time they spend in the building and diminish or go away during periods when they leave the building; the symptoms cannot be traced to specific pollutants or sources within the building. See also **Building-Related Illness**.

Sill A horizontal structural member supported by its ends.

Single-Line Diagram A diagram using single lines and graphic symbols to simplify a complex circuit or system.

Single Prime Contract The most common form of construction contracting. In this process, the bidding documents are prepared by the architect/engineer for the owner and made available to a number of qualified bidders. The winning contractor then enters into a series of subcontract agreements to complete the work.

Site A parcel of land bounded by a property line or a designated portion of a public right-of-way.

Site Improvement Landscaping, paving for pedestrian and vehicular ways, outdoor lighting, recreational facilities, and the like added to a site.

Skylight A relatively horizontal, glazed roof aperture for the admission of daylight.

Slab-on-Grade The foundation construction for a structure with no basement or crawl space.

Solar Energy Energy from the sun.

Solar Heat-Gain Coefficient Solar heat gain through the total window system relative to the incident solar radiation.

Solar Module A device used to convert light from the sun directly into DC electricity by using the photovoltaic effect. Usually consists of multiple solar cells bonded between glass and a backing material. A typical solar module would have 100 watts of power output (but module powers can range from 1 to 300 watts) and measure two by four feet.

Solar Panel A device that collects energy from the sun and converts it into electricity or heat.

Solar Power Electricity generated by conversion of sunlight either directly through the use of photovoltaic panels or indirectly through solar-thermal processes.

Solar Thermal A form of power generation using concentrated sunlight to heat water or other fluid that may then used to drive a motor or turbine.

Sole Plate A horizontal structural member used as a base for studs or columns.

Spandrel Beam The beam in an exterior wall of a structure.

Special Conditions A section of the conditions of the contract, other than general and supplementary conditions, which may be prepared to describe conditions unique to a particular project.

Specifications A part of the contract documents contained in the project manual consisting of written requirements for material, equipment, construction systems, standards, and workmanship.

Specific Gravity The ratio of the weight of a solution to the weight of an equal volume of water at a specified temperature; used with reference to the sulfuric acid-electrolyte solution in a lead-acid battery as an indicator of battery state of charge; more recently called relative density.

Split-Level A house that has two levels, one about a half level above or below the other.

Stack A vertical plumbing pipe.

Standpipe A vertical pipe generally used for the storage and distribution of water for fire-extinguishing purposes.

Statute of Limitations A stature specifying the period of time with which legal action must be brought for alleged damage, injury, or other legal relief. The lengths of the periods vary from state to state and depend upon the types of legal action. Ordinarily, the period commences with the occurrence of the damage or injury or discovery of the act resulting in the alleged damage or injury. In construction-industry cases, many jurisdictions define the period as commencing with completion of work or services performed in connection therewith.

Steel Plate Flat steel weighing more than five pounds per square foot.

Storm Sewer A sewer used for conveying rain water, surface-water condensate, cooling water, or similar liquid wastes exclusive of sewage.

Stretch-Out Line The base or reference line used in marking a development.

Stringer The inclined support member of a stair that supports the risers and treads.

Stucco A type of plaster made from Portland cement, sand, water, and a coloring agent, applied to exterior walls.

Structural Frame The component or building system that supports the building's non-variable forces or weights (dead loads) and variable forces or weights (live loads).

Stud A light vertical structure member, usually of wood or light structural steel, used as part of a wall and for supporting moderate loads.

Subcontract Agreement between a prime contractor and a subcontractor for performing a portion of the work at the site.

Subcontractor A person or entity who has a direct or indirect contract with a contractor to perform any of the work at the site.

Substitution A material, product, or item of equipment offered in lieu of that specified.

Superintendent Contractor's representative at the site, responsible for continuous field supervision, coordination, and completion of the work; unless another person is designated in writing by the contractor to the owner and the architect, also responsible for the prevention of accidents.

Supervision Direction of the work by contractor's personnel.

Surety Bond A legal instrument under which one party agrees to answer to another party for the debt, default, or failure to perform of a third party.

Surge An excessive amount of power drawn by an appliance when it is first switched on. An unexpected flow of excessive current, usually caused by excessive voltage, can damage appliances and other electrical equipment.

Survey Observations made by the field observer during a walkthrough to obtain information concerning the subject property's readily accessible and easily visible components or systems.

Sustainable The condition of being able to meet the needs of present generations without compromising needs of future generations; achieving a balance among extraction, renewal, and environmental inputs and outputs so as to cause no overall net environmental burden or deficit.

Symbol Stylized graphical representation of commonly used component parts shown in a drawing.

Synergy Action of two or more substances to achieve an effect of which each is individually incapable. As applied to toxicology, two exposures together (for example, asbestos and smoking) are far more risky than the combined individual risks.

System A combination of interacting or interdependent components assembled to carry out one or more functions.

Tangent A straight line from one point to another that passes over the edge of a curve.

Task Lighting Light provided for a specific task, as compared to general or ambient lighting.

T-Beam A beam with a T-shaped cross section.

Tee A fitting, either cast or wrought, that has one side outlet at right angles to the run.

Temper To harden steel by heating and sudden cooling by immersion in oil, water, or other coolant.

Template A piece of thin material used as a true-scale guide or as a model for reproducing various shapes.

Tensile Strength The maximum stretching of a piece of metal (rebar, etc.) before breaking, calculated in kps.

Termite Shield Sheet metal placed in or on a foundation wall to prevent termite intrusion.

Terrazzo A mixture of concrete, crushed stone, calcium shells, and/or glass, polished to a tilelike finish.

Thermal Comfort Individually, an expression of satisfaction with the thermal environment; statistically, such expression of satisfaction from at least 80 percent of the occupants in a space.

Thermal Resistance (R) A unit used to measure a material's resistance to heat transfer. The formula for thermal resistance is: R = thickness (in inches)/k.

Thermostat An automatic device controlling the operation of HVAC equipment.

Three-Phase Power A combination of three alternating currents in a circuit with their voltages displaced at 120 degrees, or one-third of a cycle.

Threshold The beveled member directly under a door.

Tie A soft metal wire that is twisted around a rebar or rod and chair to hold in place until concrete is poured.

Timely Access Entry provided to the consultant at the time of the site visit.

Timely Completion Completion of the work or designated portion thereof on or before the date required.

Title Block A blocked area in the lower right corner of the print, providing information to identify the drawing, its subject matter, origins, scale, and other data. Title blocks are unique to each manufacturer.

Title Insurer The issuer(s), approved by interim and permanent lenders, of the title insurance policy or policies insuring the mortgage.

Tolerance The amount that a manufactured part may vary from its specified size.

Top Chord The topmost member of a truss.

Topographic Survey The configuration of a surface, including its relief and the locations of its natural and man-made features, usually recorded on a drawing showing surface variations by means of contour lines indicating height above or below a fixed datum.

Top Plate A horizontal member at the top of an outer building wall, used to support a rafter.

Toxicity A reflection of a material's ability to release poisonous particulate.

Transformer A device that changes voltage from one level to another; a device used to transform voltage levels to facilitate the transfer of power from the generating plant to the customer.

Transient Lodging A building, facility, or portion thereof, excluding inpatient medical-care facilities and residential facilities, that contains sleeping accommodations. Transient lodging may include but is not limited to resorts, group homes, hotels, motels, and dormitories.

Transistor A semiconductor device used to switch or otherwise control the flow of electricity.

Trap A fitting designed to provide a liquid seal that will prevent the back passage of air without significantly affecting the flow of waste water through it.

Triangulation A technique for making complex sheet metal forms using geometrical constructions to translate dimensions from the drawing to the pattern.

Trimmer A piece of lumber, usually a 2 × 4, that is shorter than the stud or rafter but used to fill in where the longer piece would have been normally spaced to allow for a window, door, or some other opening in the roof or floor or wall.

Truss A prefabricated sloped roof system incorporating a top chord, bottom chord, and bracing.

Turbulence Any deviation from parallel flow in a pipe due to rough inner wall surfaces, obstructions, etc.

UL Underwriters Laboratories, Inc., a private testing and labeling organization that develops test standards for product compliance. UL standards appear throughout specifications, often in roofing requirements, and always in equipment utilizing or delivering electrical power.

Unfaced Insulation Insulation that does not have a facing or plastic membrane over one side of it.

Union Ell An ell with a male or female union at one end.

Union Joint A pipe coupling, usually threaded, which permits disconnection without disturbing other sections.

Union Tee A tee with a male or female union at one end of the run.

Utility Plan A floor plan of a structure showing locations of heating, electrical, plumbing, and other service-system components.

Vacuum Any pressure less than that exerted by the atmosphere.

Valley The area of a roof where two sections come together and form a depression.

Valve A device designed to control water flow in a distribution system; common valve types include globe, gate, butterfly, and check.

Vapor Barrier A moisture barrier.

Vapor Compression Chiller Refrigeration equipment that generates chilled water via a mechanically driven process using a specialized heat-transfer fluid as refrigerant; comprised of four major components: a compressor, condenser, expansion valve, and evaporator; operating energy is input as mechanical motion.

Variable Air Volume (VAV) HVAC System A central-air HVAC system that utilizes a single supply air stream and a terminal device at each zone to provide appropriate thermal conditions through control of the quantity of air supplied to the zone.

Vegetated Roof A roof partially or fully covered by vegetation. By creating roofs with a vegetated layer, the roof can counteract the heat-island effect as well as providing additional insulation and cooling during the summer.

Vehicular Way A route intended for vehicular traffic, such as a street, driveway, or parking lot.

Veneer A thin layer or sheet of wood.

Veneered Wall A single-thickness (one-wythe) masonry-unit wall with a backup wall of frame or other masonry, tied but not bonded to the backup wall.

Vent A hole in the eaves or soffit to allow the circulation of air over an insulated ceiling; usually covered with a piece of metal or screen.

Ventilation The exchange of air or the movement of air through a building, either naturally through doors and windows or mechanically by motor-driven fans.

Ventilation Rate The rate at which indoor air enters and leaves a building, expressed in the number of changes of outdoor air per unit of time (air changes per hour, or ach) or in the rate at which a volume of outdoor air enters per unit of time (cubic feet per minute, or cfm).

Vent Stack A system of pipes used for air circulation and to prevent water from being suctioned from the traps in the waste-disposal system.

Vertical Pipe Any pipe or fitting installed in a vertical position or that makes an angle of not more than 45 degrees with the vertical.

View A drawing of a side or plane of an object as seen from one point.

Volt (E) or (V) The unit of electric potential and potential difference; the amount of work done per unit charge in moving a charge from one place to another; the potential difference across a resistance of one ohm when a current of one amp is flowing.

Voltage Drop The voltage lost along a length of wire, conductor, or resistor due to the resistance of that conductor. The voltage drop is calculated by using Ohm's Law.

Voltage Protection A sensing circuit on an inverter that will disconnect the unit from the battery if input voltage limits are exceeded.

Voltage Regulator A device that controls the operating voltage of a photovoltaic array.

Voltmeter An electrical or electronic device used to measure voltage.

Waiver of Lien An instrument by which a person or organization that has or may have a right of mechanic's lien against the property of another relinquishes such right.

Walk An exterior pathway with a prepared surface intended for pedestrian use, including general pedestrian areas such as plazas and courts.

Warranty Legally enforceable assurance of quality or performance of a product or work or of the duration of satisfactory performance.

Waste Pipe Discharge pipe from any fixture, appliance, or appurtenance in connection with a plumbing system that does not contain fecal matter.

Water-Cement Ratio The ratio between the weight of water to cement.

Water Hammer The noise and vibration that develop in a piping system when a column of noncompressible liquid flowing through a line at a given pressure and velocity is abruptly stopped.

Water Main The water-supply pipe for public or community use.

Waterproofing Materials used to protect below- and on-grade construction from moisture penetration.

Watt (W) The unit of electrical power commonly used to define the electricity consumption of an appliance; the power developed when a current of one ampere flows through a potential difference of one volt.

Watt Hour (Wh) A unit of energy equal to one watt of power being used for one hour.

Weep Holes Small holes in a wall to permit water to exit from behind.

Wetland In storm-water management, a shallow, vegetated, ponded area that serves to improve water quality and provide wildlife habitat.

Wind Lift (Wind Load) The force exerted by the wind against a structure.

Wiring (Connection) Diagram A diagram showing the individual connections within a unit and the physical arrangement of the components.

Working Drawings A set of drawings that provides the necessary details and dimensions to construct the object; may include specifications.

Wye (Y) A fitting that has one side outlet at any angle other than 90 degrees.

Wythe A continuous masonry-wall width.

X Brace Cross-brace for joist construction.

Xeriscape™ A trademarked term referring to water-efficient choices in planting and irrigation design. It refers to seven basic principles for conserving water and protecting the environment: planning and design, use of well-adapted plants, soil analysis, practical turf areas, use of mulches, appropriate maintenance, and efficient irrigation.

Zenith Angle The angle between directly overhead and a line through the sun. The elevation angle of the sun above the horizon is 90 degrees minus the zenith angle.

Zinc Noncorrosive metal used for galvanizing other metals.

Zone Numbers Numbers and letters on the border of a drawing to provide reference points to aid in indicating or locating specific points on the drawing.

Zoning The legal restriction that deems that parts of cities are restricted to particular uses, such as residential, commercial, industrial, and so forth.

Zoning Permit A permit issued by an appropriate governmental authority allowing land to be used for a specific purpose.

A3 Sample Exam Questions

According to the GBCI: "The LEED™ Green Associate exam is designed to measure your skills and knowledge against criteria developed by subject-matter experts and to assess your knowledge and skill to understand and support green design, construction, and operations. The LEED™ Green Associate exam is comprised of 100 randomly delivered multiple-choice questions and must be completed in two hours; total seat time for the LEED™ Green Associate exam will be two hours and 20 minutes including a tutorial and short satisfaction survey." In order to pass any given LEED™ professional credentialing exam, you are required to achieve a minimum competency score of 170 out of 200. It is always important to check the GBCI website (www.gbci.org) for the latest updates.

The questions below are provided for your convenience to assist you in familiarizing yourself with the format and general content of the exam. However, while the content of this sample exam is believed to be representative of the type of questions on the LEED™ Green Associate exam, it does not necessarily reflect the content that will appear on the actual exam. Moreover, your ability to correctly answer these sample questions does not necessarily guarantee your ability to successfully answer questions on the actual LEED™ Green Associate exam. It is primarily intended to provide an indication of areas requiring more attention prior to taking the LEED™ AP Exam.

A3.1 Sustainable Sites

1. What are the requirements of SS Credit 8, Light Pollution Reduction? Choose two.
   A. Interior lighting automatically controlled to shut off during nonbusiness hours
   B. Full-cutoff luminaires and shades used for all exterior lamps
   C. Conformance with zoned requirements as given in IESNA RP-33
   D. Exterior lamps meet a maximum 75 percent of the lighting power densities for exterior areas and building facades as defined by IESNA

2. A project that earns SS2 Development Density certification and looks to utilize a vegetated roof can count towards which of the following credits? Choose two.
   A. SS5.1 Protect and restore habitat
   B. SS7.1 Heat-island effect
   C. MR5.1 Regional materials
   D. WE1.1 Water-efficient landscaping

3. A large food manufacturer in White Plains, N.Y., has the following staff: 2 full-time managers, 58 full-time laborers, 16 part-time student workers who average 20 hours per week, and a cleaning staff of 3 who work an 8-hour shift at night. The LEED™ team is pursuing

DOI: 10.1016/B978-1-85617-691-0.00013-8

SS4.2 Alternate transportation–Bikes. How many bike-rack locations will need to be available and how many showers will the food manufacturer need to add to its site?

A. 4 bike storage locations and 3 showers
B. 2 bike storage locations and 2 showers
C. 2 bike storage locations and 1 shower
D. 3 bike storage locations and 3 showers

4. An open-grid pavement system with an SRI of 39 could count toward which of the following credits?

A. SS5.2 Reduced site disturbance–Maximize open space
B. SS7.1 Heat-island effect non-roof
C. SS6.2 Storm-water design quality control
D. MR6 Rapidly renewable materials

5. A 20,000-square-foot structure is being built on an open 50,000-square-foot site. The applicable zoning ordinances have no open-space requirements. The project should provide how many square feet of open space at a minimum to meet SS5.2 Site development–Maximize open space?

A. 0
B. 5000
C. 10,000
D. 20,000
E. 30,000

6. Which is not a requirement of SS Prerequisite 1, Construction activity pollution prevention? Choose one

A. Create and implement an erosion and sedimentation control (ESC) plan for all construction activities associated with the project
B. Conform to the 2003 EPA construction general permit or local erosion and sedimentation control standards, whichever is more stringent
C. Prevent sedimentation of storm sewers or receiving streams
D. Prevent polluting the air with dust and particulate matter
E. Protect receiving stream channels from excessive erosion by implementing a stream-channel protection strategy and quantity-control strategies

7. What is the intent of SS Credit 5.2 Site development–Maximize open space?

A. Conserve existing natural areas and restore damaged areas to provide habitat and promote biodiversity
B. Maintain open areas amid development to protect ecosystems and provide recreational space
C. Provide a high ratio of open space to development footprint to promote biodiversity
D. Channel development to urban areas with existing infrastructure, protect greenfields, and preserve habitat and natural resources

8. The requirements for SS Credit 2.1 Development density are based on which factors? Choose two.

A. A typical two-story downtown development
B. A typical urban mixed-use development with retail, high-rise office, residential buildings, city streets. and parks
C. A 60,000-square-foot-per-acre net density
D. A 45,000-square-foot-per-acre net density
E. Density required for provision of basic community services

9. A large downtown industrial-building renovation project has met the requirements for SS Credit 2 Development density. The owner wants to achieve additional credits and is asking for your advice. Installation of a vegetated roof with native and adapted plants would help the project achieve which credits? Chose three.

A. SS Credit 3 Brownfield redevelopment
B. SS Credits 5.1 and 5.2 Site development
C. SS Credit 6.1 and 6.2 Stormwater design
D. SS Credit 7.1 Heat-island effect
E. SS Credit 7.2 Heat-island effect

10. Your team is designing an office building with 200 full-time staff, 200 half-time staff, and 20 peak-use visitors. There will be a parking lot with 100 spaces. You determine that you could meet the requirements for SS Credit 4.3, Alternative transportation by doing the following. Choose three.

A. Providing ZEV-classified vehicles and designated spaces close to the main entrance for nine occupants
B. Providing five designated parking spaces for low-emitting or fuel-efficient vehicles
C. Providing three plug-in stations for electric vehicles anywhere in the parking lot
D. Ensuring that the company's fleet of vehicles achieves a score of 40 or more in the ACEEE guide
E. Providing ZEV-classified vehicles and designated spaces close to the main entrance for five occupants
F. Provide 21 designated parking spaces for low-emitting or fuel-efficient vehicles

11. What is a potential advantage of purchasing an office building located on a brownfield site?

A. Lower property costs
B. Treating contaminants on-site
C. Limited liability
D. Using solar detoxification technologies
E. No previous development

12. Which of the following can help a project achieve SS Credit 7.2 Heat-island reduction–Roof?

A. 100 percent ENERGY STAR® roof
B. Low-sloped roof covered in SRI 29 material for 95 percent of the roof surface
C. Low-sloped roof covered in SRI 29 material for 70 percent of the roof surface
D. 100 percent green roof

13. A commercial office project requires 500 parking spaces. There are 1000 available spaces for the project. What site feature would most qualify for SS Credit 7.1 Heat-island effect–Non-Roof?

A. Above-ground parking that uses natural asphalt
B. Reduce the parking to 250 spaces
C. An underground parking deck for 250 parking spaces
D. Half of the parking spaces are shaded

14. What is the maximum distance a building's entrance needs to be from a rail or bus line to qualify for SS Credit 4.1 Public transportation access?

A. $1/4$-mile walking distance from one bus line
B. $1/4$-mile walking distance from two or more separate bus lines
C. $1/2$-mile walking distance from one rail line

D. ½-mile walking distance from two or more separate rail lines
E. ½-mile walking distance from two or more separate bus lines

15. Which of the following best defines light pollution? Choose one.
A. Glare from sunlight reflecting off building windows in urban areas
B. Pollution of an ecosystem with light, limiting visibility of night-sky views
C. Pollution caused by chemicals and gases given off by artificial lighting
D. Pollution by lightweight materials carried by wind

A3.2 Water Efficiency

1. When targeting WE2 Innovative waste-water technologies, which of the following strategies can be used to treat 50 percent of waste water on site to tertiary standards?
A. Packaged biological nutrient systems
B. Constructed wetlands
C. Biological treatment
D. High-efficiency filtration systems
2. According to LEED™, potable water is defined as:
A. Water from kitchen sinks
B. Water from irrigation systems
C. Untreated household waste water that has not come into contact with toilet waste
D. Water suitable for dinking and supplied from well or municipal water systems
3. What standard mandated the use of water-conserving plumbing fixtures to reduce water usage?
A. ASHRAE/IESNA 90.1
B. Montreal Protocol
C. Energy Policy Act of 1992
D. 2003 EPA Construction General Permit
4. You are required to provide which data for compliance with WE Credit 3 Water-use reduction? Choose three.
A. Dishwasher flow rates
B. Number of fixtures
C. Lavatory flow rates
D. Fixture model and manufacturer
E. Default baseline reduction of 20 percent
F. Male-to-female ratio in special-occupancy situations
5. A residential high-rise building uses all of the following strategies or fixtures to reduce potable water use. Which do not contribute to compliance with WE Credit 3.1 Water-use reduction? Choose three.
A. Efficient horizontal-axis clothes washers
B. Waterless urinals
C. Efficient janitor sink
D. Efficient hand-wash fountains in a gym
E. Reduction of water used during cooling-tower blowdown
F. Use of rain water for toilet flushing
G. Efficient lavatory sinks
H. Efficient dishwasher in a commercial kitchen

6. The project team on a renovation project wants to reuse some of the water fixtures. To earn WE Credit 3 Water-use reduction, the existing faucets would need a maximum gpm rate of ____ , and the water closets would need a maximum gpf rate of _____.
   A. 1.6, 1.0
   B. 2.2, 1.6
   C. 2.5, 1.0
   D. 1.0, 1.6
   E. 2.2, 2.2
7. Gray water is most often used for:
   A. Dishwashers
   B. Drinking
   C. Irrigation
   D. Swimming pools
8. By U.S. Green Building Council estimates, how many more billions of gallons of water do Americans extract than they return to the natural water system each year?
   A. 2100
   B. 4600
   C. 3700
   D. 3300
9. The highest form of waste-water treatment, which removes organics, solids, and nutrients in addition to providing biological or chemical polishing, is which of the following?
   A. Underground effluence treatment
   B. Potable system filtering
   C. Biological treatment
   D. Tertiary treatment

A3.3 Materials and Resources

1. Select four of the following materials that contribute to the rapidly renewable content calculation under MR Credit 6 Rapidly renewable content
   A. Straw-plastic composite decking
   B. Cotton-insulated ductwork
   C. Landscaping plants with a less than 10-year life cycle
   D. Cork flooring
   E. Linoleum
   F. Composite-wood panels
   G. Bamboo cabinets
   H. Oak plywood flooring
2. Construction waste is estimated to make up ____percent of the total solid-waste stream in the United States.
   A. 50
   B. 40
   C. 25
   D. 20
3. A project team for a 18,000-square-foot commercial building in a suburban office park wants to earn as many anticipated credits as possible during the LEED™ design submittal.

At the same time, it also wants to coordinate the entire construction process as much as possible in advance to help ensure compliance with credits that can only be submitted for the construction submittal. In which area of LEED™ does the latter concern predominate?
A. Sustainable sites
B. Water efficiency
C. Energy and atmosphere
D. Materials and resources
E. Environmental quality

4. When pursuing MR1.1 Building reuse, 55 percent (based on surface area) of the building structure must be maintained. Which three items can be included?
A. Exterior skin and framing systems
B. Roof decking
C. Window assemblies
D. Structural floor

5. Requirements of MR Prerequisite 1, Storage and collection of recyclables, include which of the following? Choose two.
A. Collection of organic waste
B. Specific sizing of recycling collection areas
C. Recycling areas within 200 yard of entrances
D. Collection of corrugated cardboard and paper
E. Collection of metal

6. You are renovating and adding to a 15,000-square-foot historic building and seeking LEED™ certification for the resultant 50,000-square-foot building. You will reuse 95 percent of the existing building structure. Many of the floors, trim, and doors in the existing building are not suitable for its new use but will be used in the addition, defraying 6 percent of the total materials cost. Which one of the following credits would apply?
A. MR Credits 1.1, 1.2, and 1.3 Building reuse
B. MR Credits 1.2 and 1.3 Building reuse
C. MR Credit 1.1 Building reuse
D. MR Credit 1.3 Building reuse
E. None of the above

7. You are working on a project and the owner decides to pursue LEED™ certification after construction has already started. You are trying to document MR Credit 4 Recycled content and have asked a representative of your supplier about the recycled content in the steel framing members. She tells you that all of her firm's suppliers attempt to provide steel with 80 percent postconsumer recycled content or better. Using this information, you may count which percentage of recycled content in the steel toward the credit?
A. 0
B. 5
C. 25
D. 40
E. 80

8. A renovation project is attempting MR Credit 3 Materials reuse. The materials used to achieve this credit may also be applied to which other credit?
A. MR Credit 5 Regional materials
B. MR Credit 2 Construction waste management

- **C.** MR Credit 4 Recycled content
- **D.** MR Credit 3 Materials reuse
- **E.** MR Credit 1 Building reuse

9. In order to receive a credit point for MR credit 3.2 Material reuse, which one of the following must be achieved?

- **A.** Use salvaged, refurbished, or reused materials for an additional 10 percent beyond MR Credit 3.1 (15 percent total, based on cost).
- **B.** Use salvaged, refurbished, or reused materials for an additional 3 percent beyond MR Credit 3.1 (10 percent total, based on cost).
- **C.** Use salvaged, refurbished, or reused materials for an additional 5 percent beyond MR Credit 3.1 (10 percent total, based on cost).
- **D.** Use salvaged, refurbished, or reused materials for an additional 8 percent beyond MR Credit 3.1 (10 percent total, based on cost).

10. MR Credit 1.1 Building reuse stipulates that a minimum surface area (walls, roofs, and floors) must be maintained. This total percentage is:

- **A.** 25
- **B.** 50
- **C.** 75
- **D.** 100

11. A project team has included regionally purchased new plumbing products in the plan to achieve MR Credit 5.1 Regional materials. The plumbing products will need to be included in which other credit?

- **A.** MR Credit 4 Recycled content
- **B.** MR Credit 1.2 Building reuse
- **C.** MR Credit 2.1 Construction waste management
- **D.** MR Credit 3.1 Resource reuse

A3.4 Energy and Atmosphere

1. Which four commissioning activities should, at a minimum, be completed in order to fulfill EA 1 Fundamental commissioning of building energy systems?

- **A.** Renewable energy systems
- **B.** CO_2 monitoring controls
- **C.** Lighting and daylighting controls
- **D.** Domestic hot-water systems
- **E.** Building envelope
- **F.** HVACR systems

2. How many points would a building that engages in renewable-energy contracts for 80 percent of its electricity receive based on EA6 Green power?

- **A.** 1
- **B.** 2
- **C.** 3
- **D.** None

3. In EA1 Optimize energy performance, process energy is considered to include but is not limited to which of the following items? Choose three.

A. Office equipment
B. Parking-garage ventilation
C. Laundry washing
D. Elevators
E. Lights for the building grounds

4. Which of the following are requirements for EA Prerequisite 1, Fundamental commissioning of the building energy systems? Choose three.

A. Review contractor submittals applicable to systems being commissioned
B. Develop basis of design
C. Incorporate commissioning requirements into the construction documents
D. Review building operation within 10 months after substantial commissioning
E. Verify the installation and performance of commissioned systems

5. Using rooftop photovoltaic panels supplying 17.5 percent of a building's energy needs would help a project comply with EA Credit 2,On-site renewable energy and which other credit?

A. EA Credit 6 Green power
B. EA Credit 1 Optimize energy performance
C. ID Credit 1 Innovation in design for exemplary performance under EA Credit 2
D. SS Credit 7 Heat-island effect

6. A manufacturing facility in Colorado is seeking LEED™ certification and installs backup generators powered by corn-based biodiesel, with a capacity to provide 50 percent of the facility's electricity needs. According to EPA's Emissions and Generation Resource Integrated Database, an average of 60 percent of the facility's utility electricity comes from hydroelectric generation. How many points in EA Credit 2 On-site renewable energy is the facility eligible for?

A. 0
B. 1
C. 2
D. 3

7. An owner of a 20,000-square-foot office building wants to earn as many points as possible under EA Credit 1 Optimize energy performance but cannot afford an energy-modeling process. A compliance path using which standard would be most appropriate for this situation?

A. ASHRAE 90.1-2007
B. Advanced Buildings Core Performance Guide
C. Building Performance Rating Method
D. Advanced Energy Design Guide

8. To achieve EA Credit 4 Enhanced refrigerant management requires that the project team calculate which factors? Choose two.

A. Ozone-depletion potential of CFC refrigerants used
B. LCODP and LCGWP
C. Ozone-depletion potential of halons used
D. Pounds of refrigerant and tons of cooling required by the project
E. Refrigerant atmospheric impact

9. Which item that must be performed by an independent commissioning agent to receive EA Credit 3 Enhanced commissioning?

A. Perform one commissioning design review of the project requirements, basis of design, and design documents prior to 50 percent construction-document phase

B. Review of contractor submittals applicable to the systems being commissioned

C. Develop a systems manual that provides the necessary information to operate the commissioned systems

D. Verify that the requirements for training operating personnel and building occupants are completed

E. Review building operation within 10 months of substantial completion and provide a plan for resolving and commissioning related issues

10. Which of the following systems are not eligible for EA Credit 2 On-site renewable energy? Choose two.

A. Geo-exchange system

B. Geothermal energy systems

C. Architectural features such as passive-solar and daylighting strategies

D. Solar thermal systems

11. To earn points for EA Credit 5 Measurement and verification, the measurement and verification period must cover a length of time not less than:

A. Six months following construction completion

B. One year of postconstruction occupancy

C. 18 months following construction completion

D. Two years of postconstruction occupancy

12. EA Credit 4 Enhanced refrigerant management sets a maximum threshold for which environmental impact of the built environment?

A. Ozone depletion

B. The combination of chlorofluorocarbons and ozone depletion

C. The combination of ozone depletion and global-warming potential

D. Global-warming potential

13. For the purposes of EA Credit 2 On-site renewable energy, which of the following are considered on-site renewable-energy systems? Choose three.

A. Renewable-energy credits

B. Wave and tidal power systems

C. Photovoltaic systems

D. Geothermal-heating systems

E. Ground-source heat pumps

F. Passive-solar features

14. Buildings in the United States consume roughly what percentage of energy annually?

A. 52

B. 60

C. 37

D. 25

15. Which of the following refrigerants has the lowest ozone-depletion potential (ODP)

A. HCFC

B. CFC

C. HFC
D. Halon

16. Which strategy can earn EA Credit 1.1 Optimize energy performance–lighting power?
A. Comply with ASHRAE 90.1-2007
B. Reduce lighting-power density to 16 percent below the standard
C. Provide lighting controls to at least 90 percent of occupants
D. For at least 90 percent of regularly occupied areas achieve a minimum daylight factor of 2 percent and provide glare control

A3.5 Indoor Environmental Quality

1. To meet the requirements of IEQ4.3 Low-emitting materials –carpet systems, the carpet will need to meet the testing and product requirements of:
A. Carbon Trust Good Practice Guide
B. Green Label Plus Program
C. South Coast Air Quality Management District Rule 1168
D. Green Seal Standard GS11

2. Which of the following items would not be required for a project to achieve IEQ3.1 Construction IAQ management plan?
A. An HVAC subcontractor on a high-rise building ensures that all return ducts are covered with plastic so not to allow contaminants to enter the system
B. If HVAC equipment is used prior to startup, the same subcontractor ensures that MERV8 filters are installed on all return ducts
C. Construction processes that create large amounts of dust including sanding dry-wall or sawing wood must occur only on days when the ventilation system is not running
D. During construction, the subcontractor ensures that the project meets or exceeds the control measures of SMACNA IAQ Guidelines for Occupied Building under Construction

3. Which two actions taken during construction of a three-story office building would allow the project to achieve IEQ Prerequisite 2, Environmental tobacco-smoke control?
A. Interior smoking areas exhausted directly out the building
B. Smoking not allowed except in designated smoking areas
C. Posting a "No Smoking" sign on the exterior of the building
D. Exterior smoking areas located 10 feet from the building

4. To meet IEQ Credit 8.1 Daylight and views, the calculation option requires that you achieve a minimum glazing factor of ___percent in a minimum of ___percent of regularly occupied spaces. The measurement option requires demonstrating illumination of ___ footcandles throughout ___percent of regularly occupied spaces:
A. 5, 75, 20, 75
B. 5, 90, 20, 90
C. 2, 75% 25, 75
D. 2, 75, 25, 90
E. 2% 75, 20, 80

5. Products certified under which two programs can help a project earn IEQ Credit 4.3 Low-emitting materials?
- **A.** GreenGuard
- **B.** GreenSpec
- **C.** Green Label Plus
- **D.** Green Seal
- **E.** FloorScore

6. Which one of the following credits requires submission of photos demonstrating implementation of requirements?
- **A.** IEQ Credit 8.2 Daylight and views
- **B.** SS Prerequisite 1, Construction-activity pollution prevention
- **C.** SS Credit 5.1 Site development
- **D.** IEQ Credit 3.1 Construction IAQ management plan
- **E.** SS Credit 1 Site selection

7. Which standard does IEQ Credit 1 Outdoor-air delivery monitoring reference?
- **A.** ASHRAE 90.1-2007
- **B.** ASHRAE 55-2004
- **C.** ASHRAE 62.1-2004
- **D.** CIBSE Applications Manual 10: 2005

8. You are an advisor on a LEED™ application for a multistory office building in Virginia. Internally generated cooling loads dominate design considerations. Which credit should the project focus on to reduce the cooling load?
- **A.** IEQ Credit 8.1 Daylighting and views
- **B.** WE Credit 1 Water-efficient landscaping
- **C.** EA Credit 4 Enhanced refrigerant management
- **D.** IEQ Credit 1 Outdoor-air delivery monitoring

9. On a project seeking LEED™ certification, a subcontractor insists on assembling the doors using adhesives that do not comply with SCAQMD Rule #1168. What is the best approach relative to IEQ Credit 4.1 Low-emitting materials?
- **A.** Assemble the doors in a ventilated area of the building using MERV 8 filters
- **B.** Comply instead with GS-11
- **C.** Assemble the doors off-site
- **D.** Do not attempt compliance with the credit

10. When calculating the glazing factor of a design to verify compliance with IEQ Credit 8.1 Daylight and views, which of the following factors should be taken into consideration in your calculation? Choose four.
- **A.** Window areas
- **B.** Effect of light shelves
- **C.** Square footage of all regularly occupied spaces
- **D.** Visible transmittance of windows
- **E.** Effect of glare control
- **F.** Geometry factor by glazing type

11. According to IEQ 7.2 Thermal comfort, corrective action must be taken if the percentage of occupants who are dissatisfied with the thermal comfort level exceeds:
- **A.** 10
- **B.** 15

C. 20
D. 25
E. 30

12. In using the flush-out method (Option 1) to earn IEQ 3.2 Construction IAQ management plan (before occupancy), what is the total volume of outdoor air that must be supplied to the building before occupancy?
A. 7000 cubic feet per square foot of floor area
B. 10,000 cubic feet per cubic foot of interior space
C. 14,000 cubic feet per cubic foot of interior space
D. 14,000 cubic feet per square foot of floor area

13. What is the minimum distance a designated smoking area must be away from any outdoor-air intakes to achieve IEQ Prerequisite 2, ETS control?
A. 15 feet
B. 20 feet
C. 25 feet
D. 30 feet
E. 50 feet

14. Which standard sets VOC limits for commercial interior nonflat paints?
A. ASHRAE Standard 62.1-2004
B. SCAQMD Rule 1113
C. GC-03
D. GC-11
E. GS-11
F. GS-03

15. A tenant wants to minimize energy use on a project. What is the best strategy to achieve this?
A. Increase ventilation
B. Signing a two-year contract to purchase green power
C. Task lighting with occupancy sensors
D. Purchasing tradable renewable certificates
E. Daylighting 90 percent of spaces

A3.6 Innovative Design and Miscellaneous

1. Name one aspect of the LEED™ Accredited Professional's role on a LEED™ project.
A. To ensure the integrity of the LEED™ process
B. To efficiently provide interpretations of LEED™ requirements when needed
C. To support an integrated design process
D. To advocate for the best environmental choices on a project

2. Using public transportation and promised access to a municipal parking garage, a multiuse project is able to reduce its parking capacity by 55 percent compared with the original design while still offering 10 percent more parking spaces than required by city ordinances. Through the implementation of multiple transport and parking options, the project is able to demonstrate a quantitative reduction in automobile use. Which credit does the project comply with?

A. SS Credit 4.4 Alternative transportation
B. ID Credit 1 Innovation in design
C. SS Credit 7.1 Heat-island effect: nonroof
D. SS Credit 7.2 Heat-island effect: roof
E. SS Credit 2 Development density

3. Which of the following are valid criteria for achieving an innovation credit in a category not otherwise specifically addressed by LEED™ ? Choose three.
A. The innovation includes an educational aspect
B. A point was awarded for another project for the same innovation
C. The project demonstrates quantitative performance improvements for environmental benefit
D. The process or specification is comprehensive for the project being certified
E. The formula developed for the credit is applicable to other projects

4. Which of the following are determining factors in a project's cost for LEED™ certification? Choose two.
A. USGBC membership, project size, and project type (office, mixed use, education, government)
B. LEED™ rating system and project type
C. Project size and LEED™ rating system
D. USGBC membership and certification level being sought
E. Certification level earned

5. For which credits do you need to calculate FTE building occupants? Choose three.
A. IEQ Credit 8 Daylight and views
B. WE Credit 3 Water-use reduction
C. IEQ Credit 2 Increased ventilation
D. SS Credit 4 Alternative transportation
E. EA Credit 1 Optimize energy performance
F. WE Credit 2 Innovative waste-water technologies

6. Which are conditions for submitting an appeal to LEED™ ? Choose three.
A. After a credit has been denied in the preliminary LEED™ review
B. After a credit has been denied in the final LEED™ review
C. Upon paying $500 to USGBC per credit appealed
D. Within 60 business days of denial
E. Upon submitting a project narrative and three project highlights

7. When can a project be referred to as LEED™-certified?
A. When a project has been registered with LEED™
B. When the final LEED™ review is delivered
C. When the project team accepts the final LEED™ review
D. When the LEED™ certificate and plaque are delivered

8. Which of the following should be included with a project's LEED™ application? Choose three.
A. Building measurements with metric conversions
B. A check or credit-card payment to the U.S. Green Building Council
C. A list of all credits approved through CIRs

D. Project narrative including three highlights
E. Photos or renderings of the project

9. Before submitting a CIR, a project team should first do all of the following except:
A. Consult the LEED™ Reference Guide
B. Self-evaluate whether the project meets the credit intent
C. Determine if there is an existing CIR similar to your situation
D. Prepare a full project narrative for submittal with the CIR

10. An existing three-story building undergoing major renovations is most likely a candidate for certification under which LEED™ rating system?
A. LEED™ for Existing Buildings
B. LEED™ for New Construction
C. LEED™ for Core and Shell
D. LEED™ for Commercial Interiors

11. Which of the following are requirements for submittal documentation for ID Credit 2 LEED™ Accredited Professional? Choose two.
A. List number of LEED™ APs on project
B. Provide a copy of the LEED™ AP certificate
C. Provide list of project experience during last two years, including LEED™ certification levels achieved or pending
D. Provide documentation with the design submittal
E. Give the name of the LEED™ AP's company

12. *Environmental Building News* is best described as a:
A. Monthly periodical on new EPA regulations
B. Monthly periodical on sustainable design and construction
C. Monthly periodical produced by Building Green
D. Monthly periodical produced by USGBC and GBCI

13. Designers of a new manufacturing facility seek to earn an Innovation in Design credit in the LEED™ NC Rating system by complying with building-industry acoustical standards. Which of the following organizations provide these standards?
A. American Society of Heating, Refrigeration, and Air-Conditioning Engineers
B. Architectural National Standards Institute
C. American National Standards Institute
D. Environmental Protection Agency

14. The typical costs for registering and certifying a building for LEED™ are based on:
A. Location and lot size
B. Building square footage
C. Project type
D. Type of construction

15. A high-rise building project has a $10 million materials budget. What estimated value of materials is needed to earn an Innovation in Design point for regional extracted materials?
A. $3.6 million
B. $1 million
C. $1.8 million
D. $2 million
E. $3 million
F. $4 million

16. Which factors impact the cost of building certification?
A. Number of credits applied for
B. Square footage
C. Number of credits earned
D. Having more than one LEED™ AP on the project
E. Number of CIRs submitted

A3.7 Answer Key

Sustainable Sites	Water Efficiency	Materials and Resources	Energy and Atmosphere	Indoor Environmental Quality	Innovative Design and Miscellaneous
1. A, C	1. A, B, D	1. A, D, E, G	1. A, C, D, F	1. B	1. C
2. A, B	2. D	2. B	2. B	2. C	2. B
3. A	3. C	3. D	3. A, C, D	3. A, B	3. C, D, E
4. B, C	4. C, D, F	4. A, B, D	4. B, C, E	4. C	4. C, E
5. C	5. A, E, H	5. D, E	5. B	5. C, E	5. B, D, F
6. E	6. B	6. E	6. A	6. D	6. B, C, E
7. C	7. C	7. C	7. B	7. C	7. C
8. A, C	8. C	8. A	8. B, D	8. A	8. B, D, E
9. B, C, E	9. D	9. C	9. B	9. C	9. D
10. A, B, C		10. C	10. A, C	10. A, C, D, F	10. B
11. A		11. A	11. B	11. C	11. B, E
12. D			12. C	12. D	12. A
13. C			13. B, C, D	13. C	13. C
14. B, C			14. C	14. E	14. B
15. B			15. C	15. C	15. D
			16. B		16. B, C, E

Bibliography

Allen, Edward and Iano, Joseph, *Fundamentals of Building Construction: Materials and Methods*, Fourth Edition, John Wiley, New York, 2003.

Apgar, M., "The Alternative Workplace: Changing Where and How People Work," *Harvard Business Review*, pp. 121–135, May–June 1998.

Armer, G.S.T., *Monitoring and Assessment of Structures*, Spon, New York, 2001.

Arnold, C., "Green Movement Sweeps U.S. Construction Industry," NPR, July 2, 2006.

ASHRAE, Proposed Standard 189P: Standard for the Design of High-Performance Green Buildings, Second public review, Atlanta, July 2008.

ASTM, "Standard Guide for Property Condition Assessments: Baseline Property Condition Assessment, Designation: E2018-01."

Bady, S., "Green Building Programs More About Bias than Science, Expert Argues," *Professional Builder*, 2008.

Bezdek, R.H., In: "Green Building: Balancing Fact and Fiction," Cannon, S.E. and Vyas, U.K., moderators. *Real Estate Issues*, Vol. 33, No. 2, pp. 2–5, 2008.

Bonda, P. and Sosnowchik, K., *Sustainable Commercial Interiors*, John Wiley & Sons, 2007.

Build it Green, "New Home Construction: Green Building Guidelines," 2007 edition.

Burr, A.C., "CoStar Green Report: The Big Skodowski," CoStar Group, January 30, 2008.

Butters, F., In: "Green Building: Balancing Fact and Fiction," Cannon, S.E. and Vyas, U.K., moderators. *Real Estate Issues*, Vol. 33, No. 2, pp. 10–11, 2008.

Calow, P., *Handbook of Environmental Risk Assessment and Management*, Blackwell Science Ltd, Oxford, 1998.

Coggan, D.A., "Intelligent Building Systems (Intelligent Buildings Simply Explained)," website: http://www.coggan.com/intelligent-building-systems.html, viewed June 1, 2009.

Craiger, Philip and Shenoi, Sujeet (Editors), *Advances in Digital Forensics*, Springer, 2007.

Davis Langdon, *Costing Green: A Comprehensive Cost Database and Budgeting Methodology*, 2004.

Deasy, C.M., *Designing Places for People: A Handbook for Architects, Designers, and Facility Managers*, Whitney Library of Design, New York, 1985.

De Chiara, Joseph, and Panero, Julius, *Time-Saver Standards for Interior Design and Space Planning*, McGraw-Hill, New York, 2001.

Del Percio, S., "What's Wrong with LEED?", *American City*, Issue 14, Spring 2007, website: http://www.americancity.org/, accessed April 2007.

DiLouie, C., "Why Do Daylight Harvesting Projects Succeed or Fail?" Lighting Controls Association, March 2006.

Dorgan, C., Cox, R. and Dorgan, C., "The Value of the Commissioning Process: Costs and Benefits," Farnsworth Group, Madison, WI, paper presented at the 2002 US Green Building Council Conference, Austin, TX.

Dworkin, Joseph F., "Waterproofing Below Grade," *The Construction Specifier*, March 1990.

Edwards, Sandra, *Office Systems: Designs for the Contemporary Workspace*, PBC International Inc., New York, 1986.

EIA, Annual Energy Outlook, Environmental Information Administration, 2008; Assumptions to the Annual Energy Outlook, Energy Information Administration, 2008; cited in "Green Building Facts," U.S. Green Building Council, November 2008.

Federal Emergency Management Agency, 426, "Reference Manual to Mitigate Potential Terrorist Attacks in High Occupancy Buildings," 2004.

Fisk, W.J. and Rosenfeld, A.H., "The Indoor Environment–Productivity and Health–and $$$," *Strategic Planning for Energy and the Environment*, 17(4), 53–57, 1998.

Global Green Building Trends, "SmartMarket Report," *McGraw Hill Construction*, 2008.

Goldsmith, Selwyn, *Designing for the Disabled: The New Paradigm*, Architectural Press, Oxford, 1999.

Gore, Al. "The Future is Green," *Vanity Fair,* May 2006.

Gottfried, D., "Sustainable Building Technical Manual," U.S. Green Building Council, U.S. Department of Energy and U.S. Environmental Protection Agency, 2000.

Green Building Certification Institute, LEED™ Green Associate Candidate Handbook, May 2009.

Green Building Certification Institute, *LEED™ Professional Accreditation Handbook*, Revised Edition, May 2009.

GSA, "LEED Cost Study: Final Report," submitted to the U.S. General Services Administration, Steven Winter Associates, Inc., October 2004.

Gunn, R.A. and Burroughs, M.S., "Work spaces that Work: Designing High-Performance Offices," *The Futurist*, pp. 19–24, March–April 1996.

Haasl, Theresia Powell and Claridge, D., "The Cost Effectiveness of Commissioning," *HPAC Engineering*, pp. 20–24, October 2003.

Harmon, Sharon Koomen, and Kennon, Katherine E., *The Codes Guidebook for Interiors,* Second Edition, John Wiley & Sons, New York, 2001.

Hellier, Charles J., *Handbook of Nondestructive Evaluation*, McGraw-Hill, New York, 2001.

Hess-Kosa, K., *Environmental Site Assessment Phase I*, Second Edition, CRC Press, 1997.

Horman, M.J., Riley, D.R., Pulaski, M.H., Magent, C., Dahl, P., Lapinski, A.R., Korkmaz, S., Luo, L. and Harding, N.G., "Delivering Green Buildings: Process Improvements for Sustainable Construction," *Journal of Green Building*, 1(1), 123–140, 2006.

Houghton-Evans, Robert William, "Well Built?: A Forensic Approach to the Prevention, Diagnosis and Cure of Building Defects," RIBA Enterprises, 2005.

Kats, Greg, "The Costs and Benefits of Green," Capital E Analytics, 2007.

Kennett, S., "Making BREEAM Robust," Building, website: http//www.building.co.uk/story, accessed March 2008.

Kibert, C.J. *Sustainable Construction: Green Building Design and Delivery*, John Wiley, Hoboken, NJ, 2005.

Koomen-Harmon, S. and Kennon, K.E., *The Codes Guidebook for Interiors*, Second Edition, John Wiley, New York, 2001.

Kubal, Michael T., *Construction Waterproofing Handbook*, McGraw Hill, New York, 2000.

Kubba, Sam A.A., *Space Planning for Commercial and Residential Interiors,* McGraw-Hill, New York, 2003.

Kubba, Sam A.A., *Property Condition Assessments*, McGraw-Hill, New York, 2007.

Kubba, Sam A.A., *Architectural Forensics*, McGraw-Hill, New York, 2008.

LEED™, LEED™ NC Green Building Rating System for New Construction and Major Renovations, Version 2.2, USGBC, website: www.usgbc.org, accessed August 2008.

Levy, Matthys, Salvadori, Mario G., and Woest, Kevin (Illustrator), *Why Buildings Fall Down: How Structures Fail*, W.W. Norton, 1994.

Lewis, B.T. and Payant, R. *Facility Inspection Field Manual: A Complete Condition Assessment Guide*, McGraw-Hill, New York, 2001.

Lippiatt, B.C. and Norris, G.A., "Selecting Environmentally and Economically Balanced Building Materials," sponsored by National Institute of Standards and Technology, 1998.

Loveland, J., "Daylight by Design," *LD + A*, October 2003.

Luhmann, Thomas, *Close Range Photogrammetry: Principles, Techniques and Applications*, John Wiley, Hoboken, NJ, 2007.

Lupton, M. and Croly, C., "Designing for Daylight," *Building Sustainable Design*, February 2004.

Macaluso, J., "An Overview of The LEED™ Rating System," Empire State Development, Green Construction Data, May 2009.

Macdonald, Susan (Editor), *Concrete Building Pathology*, Blackwell, 2002.

Madsen, J.J., "The Realization of Intelligent Buildings," *Buildings*, March 2008.

Mago, S. and Syal, M., "Impact of LEED® – NC Projects on Construction Management Practices," M.S. Thesis, Construction Management Program, Michigan State University, MI, 2007.

Markovitz, M., "The Differences Between Green Globes and LEED™," *PROSALES*, August 27, 2008.

Martín, C. and Foss, A., "All That Glitters Isn't Green and Other Thoughts on Sustainable Design for the PATH Partners," National Building Museum, Fall 2006.

May, S., "Do Green Design Strategies Really Cost More?", DCD Construction, 2009.

Mendler, Sandra & Odell, William, *The HOK Guidebook to Sustainable Design*, John Wiley, New York, 2000.

Morris, P., 2009 Market Update: A Guide to Working in a Recession, Davis Langdon, January 2009.

Morris, P. and Matthiessen, L.F., Cost of Green Revisited, Davis Langdon, July 2007.

Muldavin, S., "A Strategic Response to Sustainable Property Investing," *PREA Quarterly*, pp. 33–37. Summer 2007, website: www.muldavin.com.

Murphy, Brian L. and Morrison, Robert D., *Introduction to Environmental Forensics*, Second Edition, Academic Press, 2007.

Nadel, Barbara A., *Building Security Handbook for Architectural Planning and Design*, McGraw-Hill, New York, 2004.

National Institute of Building Sciences, The Whole Building Design Guide; website: www.wbdg.org/sustainable.php.

Needy, K.L., Ries, R., Gokhan, N.M., Bilec, M. and Rettura. B., "Creating a Framework to Examine Benefits of Green Building Construction," Proceedings from the American Society for Engineering Management Conference, October 20–23, Alexandria, Virginia, pp. 719–724, 2004.

Piper, J.E., *Handbook of Facility Assessment*, Fairmont Press, 2004.

Poynter, Dan, *Expert Witness Handbook: Tips and Techniques for the Litigation Consultant*, Second Edition, Para Publishing, 1997.

Propst, Robert, *The office:- A facility based on change*, Business Press, Elmhurst, IL, 1968.

Prowler, D., Donald Prowler & Associates, Revised and updated by Vierra, S., Steven Winter Associates, Inc., Whole Building Design, last updated: 08-07-2008.

Pulaski, M.H., Horman, M.J. and Riley, D.R., "Constructability Practices to Manage Sustainable Building Knowledge," *Journal of Architectural Engineering*, 12(2), 2006.

Ratay, R.T., *Structural Condition Assessment*, John Wiley, New York, 2005.

Reznikoff, S.C., *Specifications for Commercial Interiors*, Whitney Library of Design, New York, 1989.

Rogers, E. and Kostigen, T.M., *The Green Book*, Three Rivers Press, New York, 2007.

Samaras, C., Sustainable Development and the Construction Industry: Status and Implications, Carnegie Mellon University, website: http://www.andrew.cmu.edu/user/csamaras/, 2004.

Sampson, Carol A., *Estimating for Interior Designers*, Watson-Guptill Publications, New York, 2001.

Sara, M.N., *Site Assessment and Remediation Handbook,* Second Edition, CRC Press, Boca Raton, FL 2003.

Scheer, R. and Woods, R., "Is There Green in Going Green?", *SBM*, April 2007.

Smith, W.D. and Smith, L.H., *McGraw-Hill On-Site Guide to Building Codes 2000: Commercial and Interiors*, McGraw-Hill, New York, 2001.

Starr, J. and Nicolow, J., How Water Works for LEED™, BNET, CBS Interactive Inc., October 2007.

Steven Winter Associates, *Accessible Housing by Design*, McGraw-Hill, New York, 1997.

Stodghill, A., LEED™ vs. Green Globes, It's the Environment Stupid, August 27, 2008.

Sullivan, Patrick J., Agardy Franklin J. and Traub, Richard K., *Practical Environmental Forensics: Process and Case Histories*, John Wiley, New York, 2000.

Syal, M., Impact of LEED™ Projects on Constructors, AGC Klinger Award proposal, Michigan State University, Construction Management Program, School of Planning Design and Construction, E. Lansing, MI, 2005.

Tilton, Rita, Jackson, Howard J. and Rigby, Sue Chappell., *The Electronic Office: Procedures and Administration*, Eleventh Edition, South-Western Publishing Co., Cincinnati, OH, 1996.

Turner, C. and Frankel, M., Energy Performance of LEED™ for New Construction Buildings, New Buildings Institute, March 4, 2008.

United Nations, Report of the World Commission on Environment and Development, General Assembly Resolution 42/187, December 11, 1987.

USACE Army LEED™ Implementation Guide, Headquarters, U.S. Army Corps of Engineers, January 15, 2008.

USGBC, *United States Green Building Council*, website: www.usgbc.org, accessed in 2008–2009.

U.S. Green Building Council, LEED™ Reference Guide for Green Building Design and Construction, for the Design, Construction and Major Renovations of Commercial and Institutional Buildings, Including Core and Shell, and K-12 School Projects, 2009 Edition.

U.S. Green Building Council, LEED™ 2009 for New Construction and Major Renovations, November 2008.

Vermont Green Building Network, Green building rating systems, website: www.vgbn.org/gbrs.php, accessed June 2009.

Woods, J.E., In: "Green Building: Balancing Fact and Fiction," Cannon, S.E. and Vyas, U.K., moderators. *Real Estate Issues*, Vol. 33, No. 2, p. 10, 2008.

Woods, J.E., Expanding the principles of performance to sustainable buildings, Focus on Green Building, Entrepreneur Media, Inc., Fall 2008.

Yoders, J., "Integrated Project Delivery Using BIM," *Building Design and Construction*, April 1, 2008.

Yudelson, J., The Green Building Revolution, US Green Building Council, 2008.

Yudelson, J., The Business Case for Green, Yudelson Associates, March 2008.

Zelinsky, Marilyn, *New Workplaces for New Workstyles*, McGraw-Hill, New York, 1998.

Index